High Temperature Solid Oxide Fuel Cells: Fundamentals, Design and Applications

High Temperature Solid Oxide Fuel Cells: Fundamentals, Design and Applications

Edited by:
Subhash C Singhal and Kevin Kendall

ELSEVIER

Elsevier
Linacre House, Jordan Hill, Oxford, OX2 8DP, UK
Radarweg 29, PO Box 211, 1000 AE Amsterdam, The Netherlands

First edition 2003
Reprinted 2004, 2006, 2007

Copyright © 2003 Elsevier Ltd. All rights reserved

No part of this publication may be reproduced, stored in a retrieval system
or transmitted in any form or by any means electronic, mechanical, photocopying,
recording or otherwise without the prior written permission of the publisher

Permissions may be sought directly from Elsevier's Science & Technology Rights
Department in Oxford, UK: phone (+44) (0) 1865 843830; fax (+44) (0) 1865 853333;
email: permissions@elsevier.com. Alternatively you can submit your request online by
visiting the Elsevier web site at http://elsevier.com/locate/permissions, and selecting
Obtaining permission to use Elsevier material

Notice
No responsibility is assumed by the publisher for any injury and/or damage to persons
or property as a matter of products liability, negligence or otherwise, or from any use
or operation of any methods, products, instructions or ideas contained in the material
herein. Because of rapid advances in the medical sciences, in particular, independent
verification of diagnoses and drug dosages should be made

British Library Cataloguing in Publication Data
High temperature solid oxide fuel cells: fundamentals,
 design and applications
 1. Solid oxide fuel cells
 I. Singhal, Subhash C. II. Kendall, Kevin, 1943–
 621.3'12429

Library of Congress Cataloging in Publication Data
High temperature solid oxide fuel cells: fundamentals, design and
applications / edited by Subhash C. Singhal and Kevin Kendall.
 p.cm.
 Includes bibliographical references and index.
 ISBN 1-85617-387-9 (hardcover)
 1. Solid oxide fuel cells. I. Singhal, Subhash C. II. Kendall, Kevin, 1943–
TK2931 .H54 2002
621.31'2429-dc21   2002040761

ISBN: 978-1-8561-7387-2

For information on all Elsevier publications
visit our website at books.elsevier.com

Printed and bound in *Great Britain*

07 08 09 10 10 9 8 7 6 5 4

Working together to grow
libraries in developing countries

www.elsevier.com | www.bookaid.org | www.sabre.org

ELSEVIER   BOOK AID International   Sabre Foundation

Contents

List of Contributors		xi
Preface		xv

Chapter 1 Introduction to SOFCs
1.1	Background	1
1.2	Historical Summary	2
1.3	Zirconia Sensors for Oxygen Measurement	4
1.4	Zirconia Availability and Production	5
1.5	High-Quality Electrolyte Fabrication Processes	7
1.6	Electrode Materials and Reactions	8
1.7	Interconnection for Electrically Connecting the Cells	11
1.8	Cell and Stack Designs	12
1.9	SOFC Power Generation Systems	14
1.10	Fuel Considerations	15
1.11	Competition and Combination with Heat Engines	17
1.12	Application Areas and Relation to Polymer Electrolyte Fuel Cells	18
1.13	SOFC-Related Publications	19
	References	19

Chapter 2 History
2.1	The Path to the First Solid Electrolyte Gas Cells	23
2.2	From Solid Electrolyte Gas Cells to Solid Oxide Fuel Cells	26
2.3	First Detailed Investigations of Solid Oxide Fuel Cells	29
2.4	Progress in the 1960s	32
2.5	On the Path to Practical Solid Oxide Fuel Cells	40
	References	44

Chapter 3 Thermodynamics
3.1	Introduction	53
3.2	The Ideal Reversible SOFC	56
3.3.	Voltage Losses by Ohmic Resistance and by Mixing Effects by Fuel Utilisation	62

	3.4	Thermodynamic Definition of a Fuel Cell Producing Electricity and Heat	66
	3.5	Thermodynamic Theory of SOFC Hybrid Systems	69
	3.6	Design Principles of SOFC Hybrid Systems	77
	3.7	Summary	80
		References	81

Chapter 4 Electrolytes

4.1	Introduction	83
4.2	Fluorite-Structured Electrolytes	83
4.3	Zirconia-Based Oxide Ion Conductors	89
4.4	Ceria-Based Oxide Ion Conductors	92
4.5	Fabrication of ZrO_2 and CeO_2-Based Electrolyte Films	94
4.6	Perovskite-Structured Electrolytes	96
	4.6.1 $LaAlO_3$	97
	4.6.2 $LaGaO_3$ Doped with Ca, Sr and Mg	99
	4.6.3 $LaGaO_3$ Doped with Transition Elements	104
4.7	Oxides with Other Structures	106
	4.7.1 Brownmillerites (e.g. $Ba_2In_2O_6$)	106
	4.7.2 Non-cubic Oxides	108
4.8	Proton-Conducting Oxides	110
4.9	Summary	112
	References	112

Chapter 5 Cathodes

5.1	Introduction	119
5.2	Physical and Physicochemical Properties of Perovskite Cathode Materials	120
	5.2.1 Lattice Structure, Oxygen Nonstoichiometry, and Valence Stability	120
	5.2.2 Electrical Conductivity	123
	5.2.3 Thermal Expansion	125
	5.2.4 Surface Reaction Rate and Oxide Ion Conductivity	126
5.3	Reactivity of Perovskite Cathodes with ZrO_2	130
	5.3.1 Thermodynamic Considerations	130
	5.3.1.1 Reaction of Perovskites with the Zirconia Component in YSZ	130
	5.3.1.2 Reaction of perovskite with the yttria (dopant) component in YSZ	130
	5.3.1.3 Interdiffusion between Perovskite and Fluorite Oxides	131
	5.3.2 Experimental Efforts	132
	5.3.3 Cathode/Electrolyte Reactions and Cell Performance	134
	5.3.4 Cathodes for Intermediate Temperature SOFCs	136

	5.4	Compatibility of Perovskite Cathodes with Interconnects	138
		5.4.1 Compatibility of Cathodes with Oxide Interconnects	138
		5.4.2 Compatibility of Cathodes with Metallic Interconnects	139
	5.5	Fabrication of Cathodes	142
	5.6	Summary	143
		References	143
Chapter 6	**Anodes**		
	6.1	Introduction	149
	6.2	Requirements for an Anode	150
	6.3	Choice of Cermet Anode Components	151
	6.4	Cermet Fabrication	153
	6.5	Anode Behaviour Under Steady-State Conditions	156
	6.6	Anode Behaviour Under Transients Near Equilibrium	158
	6.7	Behaviour of Anodes Under Current Loading	160
	6.8	Operation of Anodes with Fuels other than Hydrogen	164
	6.9	Anodes for Direct Oxidation of Hydrocarbons	165
	6.10	Summary	168
		References	169
Chapter 7	**Interconnects**		
	7.1	Introduction	173
	7.2	Ceramic Interconnects (Lanthanum and Yttrium Chromites)	174
		7.2.1 Electrical Conductivity	174
		7.2.2 Thermal Expansion	177
		7.2.3 Thermal Conductivity	178
		7.2.4 Mechanical Strength	178
		7.2.5 Processing	179
	7.3	Metallic Interconnects	181
		7.3.1 Chromium-Based Alloys	181
		7.3.2 Ferritic Steels	182
		7.3.3 Other Metallic Materials	186
	7.4	Protective Coatings and Contact Materials for Metallic Interconnects	187
	7.5	Summary	189
		References	190
Chapter 8	**Cell and Stack Designs**		
	8.1	Introduction	197
	8.2	Planar SOFC Design	197
		8.2.1 Cell Fabrication	205

		8.2.1.1	Cell Fabrication Based on Particulate Approach	205
		8.2.1.2	Cell Fabrication Based on Deposition Approach	206
	8.2.2	Cell and Stack Performance		208
8.3	Tubular SOFC Design			210
	8.3.1	Cell Operation and Performance		213
	8.3.2	Tubular Cell Stack		214
	8.3.3	Alternative Tubular Cell Designs		216
8.4	Microtubular SOFC Design			219
	8.4.1	Microtubular SOFC Stacks		222
8.5	Summary			224
	References			225

Chapter 9. Electrode Polarisations

9.1	Introduction	230
9.2	Ohmic Polarisation	232
9.3	Concentration Polarisation	233
9.4	Activation Polarisation	237
	9.4.1 Cathodic Activation Polarisation	243
	9.4.2 Anodic Activation Polarisation	249
9.5	Measurement of Polarisation (By Electrochemical Impedance Spectroscopy)	251
9.6	Summary	257
	References	257

Chapter 10 Testing of Electrodes, Cells and Short Stacks

10.1	Introduction	261
10.2	Testing Electrodes	262
10.3	Testing Cells and 'Short' Stacks	267
10.4	Area-Specific Resistance (ASR)	272
10.5	Comparison of Test Results on Electrodes and on Cells	277
	10.5.1 Non-activated Contributions to the Total Loss	280
	10.5.2 Inaccurate Temperature Measurements	280
	10.5.3 Cathode Performance	281
	10.5.4 Impedance Analysis of Cells	282
10.6	The Problem of Gas Leakage in Cell Testing	283
	10.6.1 Assessment of the Size of the Gas Leak	284
10.7	Summary	286
	References	286

Chapter 11 Cell, Stack and System Modelling

11.1	Introduction	293
11.2	Flow and Thermal Models	294

	11.2.1	Mass Balance		295
	11.2.2	Conservation of Momentum		295
	11.2.3	Energy Balance		296
11.3	Continuum-Level Electrochemistry Model			299
11.4	Chemical Reactions and Rate Equations			303
11.5	Cell- and Stack-Level Modelling			308
11.6	System-Level Modelling			314
11.7	Thermomechanical Model			315
11.8	Electrochemical Models at the Electrode Level			318
	11.8.1	Fundamentals and Strategy of Electrode-Level Models		319
	11.8.2	Electrode Models Based on a Mass Transfer Analysis		321
	11.8.3	One-Dimensional Porous Electrode Models Based on Complete Concentration, Potential, and Current Distributions		322
	11.8.4	Monte Carlo or Stochastic Electrode Structure Model		324
		11.8.4.1	Electrode or Cell Models Applied to Ohmic Resistance-Dominated Cells	324
		11.8.4.2	Diagnostic Modelling of Electrodes to Elucidate Reaction Mechanisms	324
		11.8.4.3	Models of Mixed Ionic and Electronic Conducting (MIEC) Electrodes	325
11.9	Molecular-Level Models			325
11.10	Summary			326
	References			327

Chapter 12 Fuels and Fuel Processing

12.1	Introduction	333
12.2	Range of Fuels	335
12.3	Direct and Indirect Internal Reforming	338
	12.3.1 Direct Internal Reforming	340
	12.3.2 Indirect Internal Reforming	341
12.4	Reformation of Hydrocarbons by Steam, CO_2 and Partial Oxidation	342
12.5	Direct Electrocatalytic Oxidation of Hydrocarbons	346
12.6	Carbon Deposition	347
12.7	Sulphur Tolerance and Removal	351
12.8	Anode Materials in the Context of Fuel Processing	352
12.9	Using Renewable Fuels in SOFCs	354
12.10	Summary	355
	References	356

x High Temperature Solid Oxide Fuel Cells: Fundamentals, Design and Applications

Chapter 13 Systems and Applications
- 13.1 Introduction — 363
- 13.2 Trends in the Energy Markets and SOFC Applicability — 365
- 13.3 Competing Power Generation Systems and SOFC Applications — 367
- 13.4 SOFC System Designs and Performance — 370
  - 13.4.1 Atmospheric SOFC Systems for Distributed Power Generation — 370
  - 13.4.2 Residential, Auxiliary Power and Other Atmospheric SOFC Systems — 373
  - 13.4.3 Pressurised SOFC/Turbine Hybrid Systems — 374
  - 13.4.4 System Control and Dynamics — 376
  - 13.4.5 SOFC System Costs — 378
  - 13.4.6 Example of a Specific SOFC System Application — 379
- 13.5 SOFC System Demonstrations — 380
  - 13.5.1 Siemens Westinghouse Systems — 380
    - 13.5.1.1 100 kW Atmospheric SOFC System — 380
    - 13.5.1.2 220 kW Pressurised SOFC/GT Hybrid System — 383
    - 13.5.1.3 Other Systems — 384
  - 13.5.2 Sulzer Hexis Systems — 385
  - 13.5.3 SOFC Systems of Other Companies — 386
- 13.6 Summary — 388
- References — 389

Index — 393

List of Contributors

Harlan U Anderson
Electronic Materials Applied Research Center, 104 Straumanis Hall, University of Missouri-Rolla, Rolla, MO 65410-1240, USA
harlanua@umr.edu

Rob J F van Gerwen,
KEMA Power Generation & Sustainables, KEMA Nederland BV, PB 9035, 6800 ET Arnhem, The Netherlands
rob.vangerwen@kema.nl

Peter Vang Hendriksen
Materials Research Department, Risø National Laboratory, DK-4000 Roskilde, Denmark
peter.hendriksen@risoe.dk

Teruhisa Horita
National Institute of Advanced Industrial Science and Technology, AIST Tsukuba Central No. 5, Higashi 1-1-1, Tsukuba, Ibaraki 305-8565, Japan
t.horita@aist.go.jp

Tatsumi Ishihara
Department of Applied Chemistry, Faculty of Engineering, Kyushu University, Hakozaki 6-10-1, Higashi-ku, Fukuoka 812-8581, Japan
ishihara@cstf.kyushu-u.ac

Ellen Ivers-Tiffée
Institut für Werkstoffe der Elektrotechnik IWE, Universität Karlsruhe (TH), Adenauerring 20, 76131 Karlsruhe, Germany
ellen.ivers@iwe.uni-karlsruhe.de

Kevin Kendall
School of Chemical Engineering, The University of Birmingham, Edgbaston, Birmingham B15 2TT, UK
k.kendall@bham.ac.uk

Mohammad A Khaleel
Pacific Northwest National Laboratory, PO Box 999, Richland, WA 99352, USA
moe.khaleel@pnl.gov

Augustin J McEvoy
Institute of Molecular and Biological Chemistry, Faculty of Basic Sciences, Ecole Polytechnique Fédérale de Lausanne, CH-1015 Lausanne, Switzerland
augustin.mcevoy@epfl.ch

Nguyen Q. Minh
General Electric Power Systems, Hybrid Power Generation Systems, 19310 Pacific Gateway Drive, Torrance, CA 90502-1031, USA
nguyen.minh@ps.ge.com

Hans-Heinrich Möbius
Ernst-Moritz-Arndt-Universität Greifswald (Emeritus)
Rudolf-Breitscheid Strasse 25, D 17489 Greifswald, Germany
vmoebius@uni-griefswald.de

Mogens Mogensen
Materials Research Department, Risø National Laboratory, DK-4000 Roskilde, Denmark
mogens.mogensen@risoe.dk

R Mark Ormerod
Birchall Centre for Inorganic Chemistry and Materials Science, School of Chemistry and Physics, Keele University, Staffordshire ST5 5BG, UK
r.m.ormerod@keele.ac.uk

Nigel M Sammes
Connecticut Global Fuel Cell Center, University of Connecticut, 44 Weaver Road, Unit-5233, Storrs, CT 06269-5233, USA
sammes@engr.uconn.edu

J Robert Selman
Center for Electrochemical Science and Engineering, Illinois Institute of Technology, Chicago, IL 60616, USA
selman@iit.edu

Subhash C Singhal
Pacific Northwest National Laboratory, PO Box 999, Richland, WA 99352, USA
singhal@pnl.gov

Frank Tietz
Forschungszentrum Jülich GmbH, Institut fur Werkstoffe und Verfahren der Energietechnik (IWV-1), D-52425 Jülich, Germany
f.tietz@fz-juelich.de

Anil V Virkar
Department of Materials Science and Engineering, University of Utah, Salt Lake City, Utah 84112, USA
anil.virkar@m.cc.utah.edu

Wolfgang Winkler
Fuel Cells Laboratory, Hamburg University of Applied Sciences, Faculty of Mechanical Engineering and Production, Berliner Tor 21, 20099 Hamburg, Germany
winkler@rzbt.haw-hamburg.de

Osamu Yamamoto
Aichi Institute of Technology, 13-1, Kamo-gome, Chaiki-cho, Ichinomiya, Aichi 491-0801, Japan
osyamamo@alles.or.jp

Harumi Yokokawa
National Institute of Advanced Industrial Science and Technology, AIST Tsukuba Central No 5, Higashi 1-1-1, Tsukuba, Ibaraki 305-8565, Japan
h-yokokawa@aist.go.jp

PREFACE

High temperature solid oxide fuel cells (SOFCs) are the most efficient devices for the electrochemical conversion of chemical energy of hydrocarbon fuels into electricity, and have been gaining increasing attention in recent years for clean and efficient distributed power generation. The technical feasibility and reliability of these cells, in tubular configuration, has been demonstrated by the very successful operation of a 100 kW combined heat and power system without any performance degradation for over two years. The primary goal now is the reduction of the capital cost of the SOFC-based power systems to effectively compete with other power generation technologies. Toward this end, several different cell designs are being investigated and many new collaborative programs are being initiated in the United States, Europe, and Japan; noteworthy among these are the Solid State Energy Conversion Alliance (SECA) program in the United States, the Framework 6 programs in the European Union, and the New Energy and Industrial Technology Development Organization (NEDO) programs in Japan. The funding for SOFC development worldwide has risen dramatically and this trend is expected to continue for at least the next decade. In addition to cost reduction, these development programs are also investigating wider applications of SOFCs in residential, transportation and military sectors, made possible primarily because of the fuel flexibility of these cells. Their application in auxiliary power units utilizing gasoline or diesel as fuel promises to bring SOFCs into the 'consumer product' automotive and recreational vehicle market.

This book provides comprehensive, up-to-date information on operating principle, cell component materials, cell and stack designs and fabrication processes, cell and stack performance, and applications of SOFCs. Individual chapters are written by internationally renowned authors in their respective fields, and the text is supplemented by a large number of references for further information. The book is primarily intended for use by researchers, engineers, and other technical people working in the field of SOFCs. Even though the technology is advancing at a very rapid pace, the information contained in most of the chapters is fundamental enough for the book to be useful even as a text for SOFC technology at the graduate level.

As in any book written by multiple authors, there may be some duplication of information or even minor contradiction in interpretation of various electrochemical phenomena and results from chapter to chapter. However, this has been kept to a minimum by the editors. Also, in the interest of making the book available in a reasonable time, it has not been possible to provide uniformity in the nomenclature and symbols from chapter to chapter; we apologise for that.

Many of our colleagues in the SOFC community provided useful comments and reviews on some of the chapters and we are thankful to them. The encouragement and financial support of the United States Department of Energy-Fossil Energy (through Dr. Mark Williams, National Energy Technology Laboratory) to one of the editors (SCS) is deeply appreciated. We are also grateful to Ms. Jane Carlson, Pacific Northwest National Laboratory, for her administrative support during the editing of the chapters.

Subhash C. Singhal
Richland, Washington, USA

Kevin Kendall
Birmingham, UK

September 2003

Chapter 1

Introduction to SOFCs

Subhash C. Singhal and Kevin Kendall

1.1 Background

Solid oxide fuel cells (SOFCs) are the most efficient devices yet invented for conversion of chemical fuels directly into electrical power. Originally the basic ideas and materials were proposed by Nernst and his colleagues [1–3] in Gottingen at the end of the nineteenth century, as described in Chapter 2, but considerable advances in theory and experiment are still being made over 100 years later.

Figure 1.1 shows an SOFC scheme. It contains a solid oxide electrolyte made from a ceramic such as yttria-stabilised zirconia (YSZ) which acts as a conductor of oxide ions at temperatures from 600 to 1000°C. This ceramic material allows oxygen atoms to be reduced on its porous cathode surface by electrons, thus being converted into oxide ions, which are then transported through the ceramic body to a fuel-rich porous anode zone where the oxide ions can react, say with hydrogen, giving up electrons to an external circuit as shown in Figure 1.1. Only five components are needed to put such a cell together: electrolyte, anode, cathode and two interconnect wires.

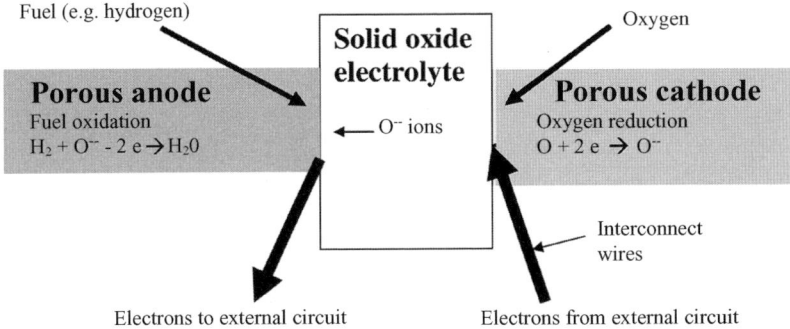

Figure 1.1   Schematic of solid oxide fuel cell (SOFC).

This is almost magical in its elegance and simplicity, and it is astonishing that this process has not yet been commercialised to supplant the inefficient and polluting combustion heat engines which currently dominate our civilization. Largely, this failure has stemmed from a lack of materials knowledge and the absence of chemical engineering skills necessary to develop electrochemical technology. Our belief is that this knowledge and expertise is now emerging rapidly. The purpose of this book is to present this up-to-date knowledge in order to facilitate the inventions, designs and developments necessary for commercial applications of solid oxide fuel cells.

An essential aspect of SOFC design and application is the heat produced by the electrochemical reaction, not shown in Fig. 1.1. As Chapter 3 shows, heat is inevitably generated in the SOFC by ohmic losses, electrode overpotentials etc. These losses are present in all designs and cannot be eliminated but must be integrated into a heat management system. Indeed, the heat is necessary to maintain the operating temperature of the cells. The benefit of the SOFC over competing fuel cells is the higher temperature of the exhaust heat which makes its control and utilization simple and economic.

Because both electricity and heat are desirable and useful products of SOFC operation, the best applications are those which use both, for example residential combined heat and power, auxiliary power supplies on vehicles, and stationary power generation from coal which needs heat for gasification. A residential SOFC system can use this heat to produce hot water, as currently achieved with simple heat exchangers. In a vehicle the heat can be used to keep the driver warm. A stationary power system can use the hot gas output from the SOFC to gasify coal, or to drive a heat engine such as a Stirling engine or a gas turbine motor.

These ideas, from fundamentals of SOFCs through to applications, are expanded in the sections below to outline this book's contents.

1.2 Historical Summary

The development of the ideas mentioned above has taken place over more than a century. In 1890, it was not yet clear what electrical conduction was. The electron had not quite been defined. Metals were known to conduct electricity in accord with Ohm's law, and aqueous ionic solutions were known to conduct larger entities called ions. Nernst then made the breakthrough of observing various types of conduction in stabilised zirconia, that is zirconium oxide doped with several mole per cent of calcia, magnesia, yttria, etc. Nernst found that stabilised zirconia was an insulator at room temperature, conducted ions in red hot conditions, from 600 to 1000°C and then became an electronic and ionic conductor at white heat, around 1500°C. He patented an incandescent electric light made from a zirconia filament and sold this invention which he had been using to illuminate his home [1–3]. He praised the simultaneous invention of the telephone because it enabled him to call his wife to switch on the light device while he travelled back from the university. The heat-up time was a problem even then [4].

The zirconia lighting filament was not successful in competing with tungsten lamps and Nernst's invention languished until the late 1930s when a fuel cell concept based on zirconium oxide was demonstrated at the laboratory scale by Baur and Preis [5]. They used a tubular crucible made from zirconia stabilised with 15 wt% yttria as the electrolyte. Iron or carbon was used as the anode and magnetite (Fe_3O_4) as the cathode. Hydrogen or carbon monoxide was the fuel on the inside of the tube and air was the oxidant on the outside. Eight cells were connected in series to make the first SOFC stack. They obtained power from the device and speculated that this solid oxide fuel cell could compete with batteries. But several improvements were necessary before this would be possible. For example, the electrolyte manufacturing process was too crude and needed optimising, especially to make the electrolyte thinner to reduce the cell resistance from around 2 Ω. In addition, the electrodes were inadequate, especially the cathode Fe_3O_4 which readily oxidised. Also, the power density was small with the stacking arrangement used, the connections between many cells had to be developed, and the understanding of fuel reactions and system operation needed much attention.

It was not until the 1950s that experiments began on pressed or tape-cast discs of stabilised zirconia when a straightforward design of test system was developed which is still in use today. The essentials of the apparatus are shown in Figure 1.2a. A flat disc of stabilised zirconia, with anode and cathode on its two sides, was sealed to a ceramic tube and inserted in a furnace held at red heat [6]. A smaller diameter tube was inserted into the ceramic tube to bring fuel to the anode, and another tube brought oxidant gas to the cathode side. Current collector wires and voltage measurement probes were brought out from the electrode surfaces. Once a flat plate of electrolyte had been used, it was easy to see how the flat plate voltaic stack could be built up with interconnecting separator plates to build a realistic electrochemical reactor, as shown in Figure 1.2b. The interconnect plate is essentially made from the anode current collector and the

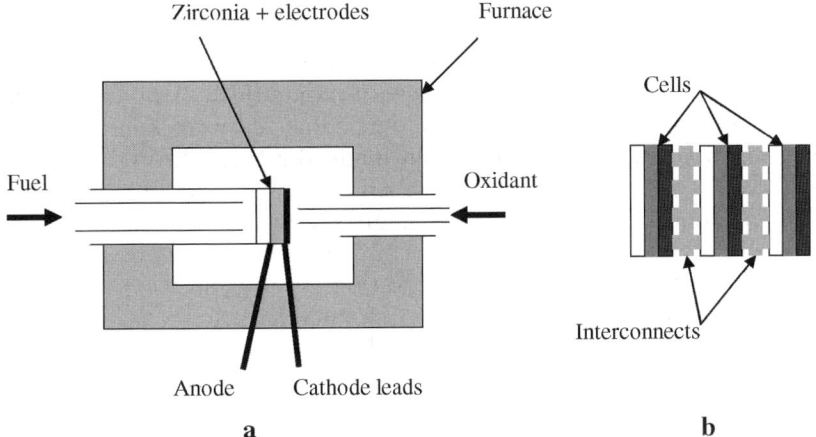

Figure 1.2 (a) Flat plate test cell; (b) planar stack of cells and interconnects.

cathode current collector joined together into one sheet, thus combining the two components. Additionally, the interconnector can contain gas channels which supply fuel to the anode and oxidant to the cathode as well as electrically connecting the anode of one cell to the cathode of the next.

It turned out that there were several problems with flat plate stacks as they were made larger to generate increased power, including sealing around the edges and thermal expansion mismatch which caused cracking. Consequently, tubular designs have had greater success in recent years. However, the configuration of Figure 1.2 has been prominent in zirconia sensors, discussed in the next section, which are now manufactured in large numbers.

1.3 Zirconia Sensors for Oxygen Measurement

An SOFC in reality already exists on every automobile: it is the oxygen sensor device which sits in the exhaust manifold in order to control the oxygen content of the effluent mixture entering the exhaust catalyst. The composition of the effluent mixture must be controlled to near stoichiometric if the catalyst is to operate at its optimum performance. Yttria-stabilised zirconia (YSZ) is generally used as the electrolyte because this uniquely detects oxygen, and platinum is normally painted on its surface using proprietary inks to provide the electrodes.

In the original design, which was used from the 1970s, the configuration was similar to that of Baur and Preis [5]. A thimble of YSZ, containing typically 8 wt% yttria, was pressed from powder and fired to 1500°C to densify it. Platinum electrodes were applied and the unit then fixed in a steel plug which could be screwed into the car exhaust manifold, so that the YSZ +anode was protruding into the hot gases. Air was used as the oxygen reference on the cathode side. A wire connection supplied the voltage from the inner electrode to the engine management system, while the other electrode was grounded to the chassis.

Once the exhaust warmed up, above 600°C, the voltage from the sensor reflected the oxygen concentration in the exhaust gas stream. This voltage varied with the logarithm of oxygen level, giving the characteristic λ-shaped curve of voltage versus oxygen concentration, hence the name 'lambda sensor'. The control system then used the oxygen sensor signal to manage the engine so that the exhaust composition was optimised for the catalyst. Various improvements have been made to this basic system over the years; for example, a heating element can be built within the thimble, in order to obtain a rapid heating sensor.

The major improvement introduced by Robert Bosch GmbH in 1997 was to redesign the zirconia sensor and to manufacture it by a different method. Instead of pressing a thimble from dry powder, a wet mix of zirconia powder with polymer additives was coated and dried like a paint film on a moving belt in a tape-casting machine. The film dried to a thickness of around 100 μm and could be screen printed with the platinum metallisation before pressing three or four sheets together to form a planar sensor array which was fired and then sectioned to size before inserting in the metal boss which screwed into the engine manifold.

This process was rather like the ceramic capacitor process developed for the electronic ceramics industry.

There were several benefits of this new device:

- There was much less zirconia in it, about 2.8 g;
- The thinner ceramic electrolyte gave much faster response;
- Heaters and other circuits could readily be printed onto the flat sheets.

An immediate bonus of this technology was the possibility of producing linear response sensors as opposed to the logarithmic response of the thimble type, so as to match the electronic control system more easily. This was achieved by setting the oxygen reference by using one of the sheets as an oxygen pump which could then leak from the cathode compartment through a standard orifice.

Oxygen sensors are now widely used in food storage, in metal processing and in flame controllers, but the main market is automobiles. Zirconia technology for sensors has been very successful in the marketplace, and it has pushed forward the development of solid oxide fuel cell materials. The main difference is that the power output of sensors is low so that partially stabilised zirconia can be used. At higher power, fully stabilised zirconia must be used if the electrolyte is to remain stable for long periods. The supply of this electrolyte material is discussed next.

1.4 Zirconia Availability and Production

The main electrolyte material used in SOFCs at present is YSZ, as described more fully in Chapter 4. Although many other oxide materials conduct oxide ions, some rather better than zirconia, this material has a number of significant attributes which make it ideal for this application, including abundance, chemical stability, non-toxicity and economics. Against these one can mention several drawbacks, including the high thermal expansion coefficient, and the problems of joining and sealing the material.

Low-grade stabilised zirconia already commands a large market, especially in refractories, pigment coatings and colours for pottery, but it is only recently that technical-grade zirconias have been produced for applications such as thermal barrier coatings on gas turbine components, hip joint implants and cutting tools. Much of this technology has stemmed from the study of pure zirconia and the effects of small amounts of dopants on the crystal structure and properties. Large effects were seen in the early 1970s, pointing the way to substantial applications of this material [7].

Figure 1.3 shows the trend in worldwide production levels of ionic conductor-grade yttria-stabilised zirconia over time. It is evident that in 1970 there was very small production at a rather high price. However, the introduction of the zirconia lambda sensor to control the emissions of automobiles in the 1970s had a large effect on the production rate, and price has dropped steadily since that time. The price in 2000 was about $50 per kg in 50 kg lots but this is expected to

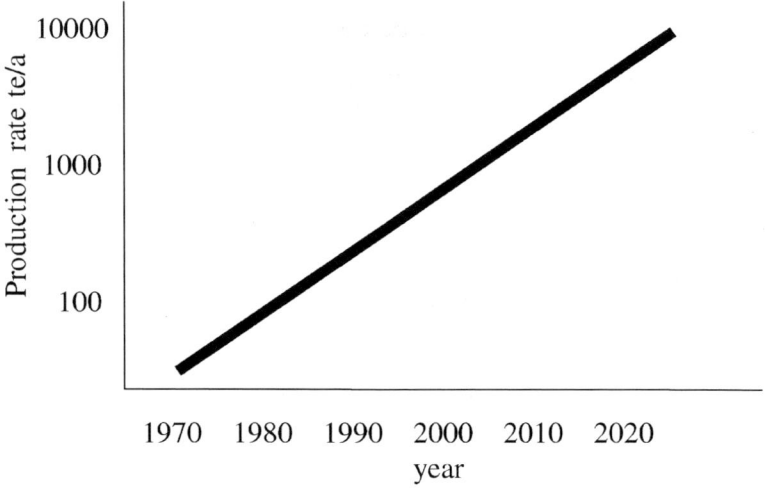

Figure 1.3   Trend in the production of ionic conducting yttria-stabilised zirconia powder.

fall steadily with time towards $13 per kg in 2020 as production rises to many thousands of tons per year. In 2000, the sensor application of YSZ was dominant with an estimated world production of 500 metric tons, but it is expected that fuel cell power systems will rapidly rise to overtake sensors in demanding YSZ by about 2010.

There is little doubt that large quantities of zirconia will be needed for SOFC applications in the years to come, with annual requirements rising to more than 1 Mte per year, rather as titania expanded in the last century for pigment applications. Fortunately, zirconia is one of the most common materials in the earth's crust, being much more available than copper or zinc, for example. Large deposits exist in Australia, Africa, Asia and America, usually as the silicate, zircon ($ZrSiO_4$). In terms of cost, the greatest difficulty is purifying this raw material, especially to remove SiO_2 which tends to block the ionic and electron paths in fuel cell systems. A typical zirconia powder for electrolyte application should contain less than 0.1% by weight of silica, and the highest quality YSZ electrolytes contain only 0.005% by weight. Other impurities, like alumina and titania, can be useful in gettering the damaging silica, so that levels of 0.1% by weight are normal. The main impurity, hafnia, is usually present at several wt% but causes no problem because it is an ionic conductor itself. Often, zirconia contains small amounts of radioactive α emitter impurities, and this could pose a potential health problem during processing, but otherwise there are no significant toxic hazards known.

Yttria is the principal stabiliser used at present, though both the more expensive scandia and ytterbia give better ionic conductivity. Typically, yttria is added at 13–16% by weight (8–10.5 mol%) to give a fully stabilised cubic material. Details of these materials are given in Chapter 4. Supply of scarce dopants such as scandia could be a problem in future. However, a more significant issue is the processing of the electrolyte material into a functional device.

1.5. High-Quality Electrolyte Fabrication Processes

One of the main issues in slowing down the advances in SOFCs has been the difficulty of making good cells. The electrolyte has to meet several criteria for success:

- It must be dense and leak-tight.
- It has to have the correct composition to give good ionic conduction at the operating temperatures.
- It must be thin to reduce the ionic resistance.
- It must be extended in area to maximise the current capacity.
- It should resist thermal shocks.
- It must be economically processable.

These requirements are not easily reconciled. Industrial ceramic processing has traditionally focused on the pressing of dry powders in metal dies or in rubber moulds to make spark plugs, for example. Although zirconia sensors have been made by this technique, and although much academic research has used this method, it is difficult to make thin-walled parts of large area in this way. A stacked tubular design made by powder pressing had been demonstrated in the 1960s but this proved to be expensive because of diamond grinding and of high resistance due to the 500 μm thick electrolyte [8]. It was far better to move towards the advanced ceramic processes such as chemical vapour deposition, tape casting and extrusion (see Figure 1.4) to make the required thin films of electrolyte.

In the late 1970s, electrochemical vapour deposition began to be used to make tubular cells at Westinghouse [9,10]. A porous tubular substrate, around 15–20 mm in diameter, made originally from calcia-stabilised zirconia but later from the cathode material, doped lanthanum manganite, was placed in a low-pressure furnace chamber, and zirconium chloride plus yttrium chloride

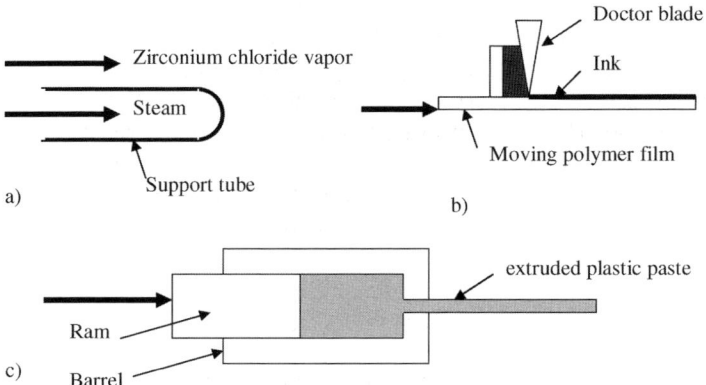

Figure 1.4 Schematic of three electrolyte fabrication processes: (a) electrochemical vapour deposition; (b) tape casting; (c) extrusion.

vapour was passed along the outside of the tube, while water vapour passed down the inside. This deposited a layer of yttria-doped zirconia which first blocked the pores at the surface of the substrate tube and then subsequently grew to about 40 µm in thickness to form the electrolyte layer [10]. The interconnect strip could also be formed from magnesia-doped lanthanum chromite by the same principle [11]. Although tubular SOFCs give good electrochemical performance, the process is lengthy and expensive when compared with tape casting. Also, the heavy tubes cannot be heated rapidly and require a 4–6 hour start-up time.

Tape casting was originally used to make thin tape materials for electronic applications [12] especially using organic solvents. A slurry of the YSZ powder, with solvent and dispersing agent, for example methyl ethyl ketone/ethanol mixture with KD1 (Uniqema), was ball milled for 24 hours to finely grind the particles and remove agglomerates [13]. Then a polymer and plasticiser mixture was prepared by milling polyethylene oxide and dibutylphthalate with the solvent, mixed with the particle dispersion, and followed by further ball milling. After filtering and vacuum deairing, the slurry was tape cast on a polymer film and dried for 3 hours before firing at 1300°C.

Water-based tape casting is much more desirable than the organic solvent system for environmental reasons, and this has been developed by Viking Chemicals who prepared their own pure zirconia by solvent extraction techniques [14]. The calcined zirconia powder was bead milled in water with ammonium polyacrylate solution (Darvan 821A, Vanderbilt) to give a very stable dispersion. To this suspension, a solution of purified ethyl cellulose was added, followed by filtering and deairing. This was tape cast onto polymer film, then dried and fired at 1450°C. Similar dispersions have been screen printed onto tape cast anode tapes made by a similar casting procedure to give co-fired supported electrolyte films of reduced thickness which gave enhanced current capacity [15]. Such results were originally reported by Minh and Horne [16] who used the tape calendering method which is similar to tape casting but with a plastic composition [17]. They also corrugated the plates and made monolithic designs by sticking corrugated pieces together in a stacked structure. Of course, the problem with flat plate designs is the thermal shock which prevents rapid heating or cooling. This was a particular problem for monolithic structures which cracked very easily when made more than a few centimetres in length.

To prevent the thermal shock problem, smaller diameter tubes have been produced by extrusion as described in Chapter 8 [18]. Again, these compositions were prepared by mixing zirconia powder with water and polymer, for example polyvinyl alcohol. Extrusion through a die gave tubes which could be as little as 2 mm in diameter and 100–200 µm in wall thickness, sinterable at 1450°C.

1.6 Electrode Materials and Reactions

Having produced the YSZ electrolyte membrane, it is then necessary to apply electrodes to the fuel contact surface (anode) and the oxidant side (cathode).

These electrodes are usually made from particulate materials which are partially sintered to form porous conducting layers. Often, several layers are laid down because this allows a gradient of properties ranging from nearly pure YSZ at the electrolyte surface to almost pure electrode composition at the interconnect contact, as illustrated in Figure 1.5 for a typical anode structure. In addition the expansion coefficients can then be better matched across the layers.

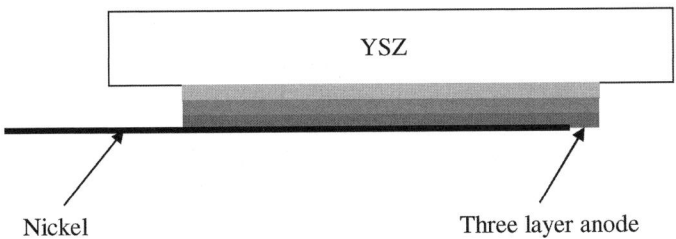

Figure 1.5 Three-layer anode made by printing three inks of different composition onto YSZ [12].

Nickel is the main anode material used in SOFC anodes since 1964, largely because of its known performance and economics. Unfortunately, nickel metal does not adhere strongly to YSZ and flakes off unless it is mixed with zirconia. This flaking is driven by the large difference in expansion coefficient between metal and ceramic; YSZ expands at around 11×10^{-6}/K whereas nickel expands much more at 13.3×10^{-6}/K. By powder mixing 30 vol% nickel oxide with YSZ, followed by firing at 1300°C to give a porous anode layer by reduction in hydrogen, this mismatch can be reduced. The expansion coefficient of this 'nickel cermet' anode is about 12.5×10^{-6}/K, allowing much better adhesion to the electrolyte. Sandwiching this anode cermet between two slightly different compositions, one nearest the zirconia with less nickel, the other near the gas stream with more nickel, can give excellent anode properties, both from the catalytic and the electronic conduction points of view. The two main requirements of the anode are to allow rapid, clean reactions with the fuel and to provide good conduction to the interconnect.

The main problem with the nickel-based anodes is their propensity to coke, that is to become coated with a carbon layer on reacting with hydrocarbon fuel. This carbon layer has two deleterious effects: it can disrupt the anode by pushing the nickel particles apart; and it can form a barrier at the nickel surface, preventing gas reactions. Typically, if a hydrocarbon such as methane is fed directly into an SOFC anode, then it may not remain functional after as little as 30 minutes as the coking proceeds. Additives to the Ni+YSZ cermet such as 5% ceria or 1% molybdena can inhibit this process [19]. Alternatively, metals other than nickel can be employed [20].

Cathodes present the main electrode issues in designing and operating SOFCs, as described in Chapter 5. Since these operate in a highly oxidising environment, it is not possible to use base metals and the use of noble metals is cost prohibitive. Consequently, semiconducting oxides have been the most prominent candidates since 1966 when doped lanthanum cobaltites began to be used, followed in

1973 by lanthanum manganites. Typically, $La_{0.8}Sr_{0.2}MnO_3$ (LSM) gives a good combination of electronic conductivity and expansion coefficient matching, and is now available commercially for SOFC applications. Higher conductivity can be obtained at higher dopant levels, but the expansion coefficient then becomes too high. Lanthanum cobaltite is a much better material from the catalytic and conduction standpoints but is too reactive with zirconia and also expands too much. Even the manganite reacts with zirconia above 1400°C and produces an insulating layer of lanthanum zirconate which increases the resistance enormously. Therefore, firing of the cathode materials on YSZ tends to be kept below 1300°C, and a minor excess of manganese is used to inhibit the reaction. The manganese can be seen diffusing into the YSZ at high temperatures, a blackened region gradually penetrating the normally white electrolyte.

In order to minimise the resistance at the LSM cathode, especially as the operating temperature of the SOFC is reduced below 1000°C, it has become normal practice to mix the LSM powder with YSZ powder, roughly in 50/50 proportion, to form the first layer of cathode material at the electrolyte surface. This allows a larger 'three-phase boundary' (the line where the gas phase meets both electrolyte and electrode phases) to exist between the oxygen molecules in the gas phase, the LSM particle and the YSZ electrolyte as shown in Figure 1.6. By this means, the cathode contribution to cell resistance can be brought down to about 0.1 Ω for 1 cm^2 of electrode [21]. Alternatively, various doping layers such as ceria can be applied to the YSZ electrolyte before printing on the electrode composition.

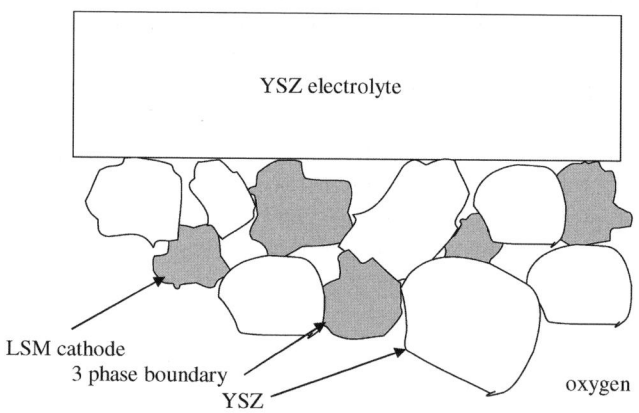

Figure 1.6 Concept of extended three-phase boundary at cathode/electrolyte interface.

The electrode layers have been applied using numerous methods, ranging from vapour deposition and solution coating to plasma spraying and colloidal ink methods such as screen printing and paint spraying, which is perhaps the most economic method. This process is widely used in the traditional ceramic industry to lay down glaze layers from particulate inks to give electrode thicknesses of 50–100 μm. It is advantageous to reduce the number of fabrication steps by adopting composite processes whereby several layers are produced at one time. One such technique was developed by Minh to tape calender the two electrodes and the electrolyte together as one sheet [22]. Another method, that of co-extrusion, was developed to make the anode and electrolyte in one step [23].

1.7 Interconnection for Electrically Connecting the Cells

The interconnection requires two interconnect wires but these are often combined into a single material which makes contact with the anode on one side and the cathode on the other.

Ideally an inert and impervious conducting material is needed to withstand both the oxidising potential on the air electrode and the reducing condition at the fuel side as described in Chapter 7. In the SOFC, since 1974, lanthanum chromite has been used to carry out this function for the systems operating near 1000°C. This material has almost exactly the same thermal expansion coefficient as YSZ, depending on doping. Typically strontium dopant has been used at 20 mol% to give an expansion coefficient of about 11×10^{-6}/K. For systems operating at lower temperatures, 700–850°C, it is conceivable that metallic alloys like ferritic stainless steel could be used. Other chromium-based alloys have also been tested [24].

Magnesium-doped lanthanum chromite has been the material most used by Westinghouse (now Siemens Westinghouse) to produce single cells and stacks of their tubular design [11]. The material was initially deposited by an electrochemical vapour deposition process to form a strip along the lanthanum manganite tube and is now deposited by plasma spraying. This made contact with the anode of the neighbouring cell to give series connection along a stack as shown in Chapter 8. This material has worked very well and has provided single cell lifetimes of up to 70,000 h in hydrogen.

The problem is that the lanthanum chromite is not quite inert. It expands in hydrogen as shown by the results of Figure 1.7 [25]. In particular, strontium doped material can expand by 0.3% in length, sufficient to cause large distortion

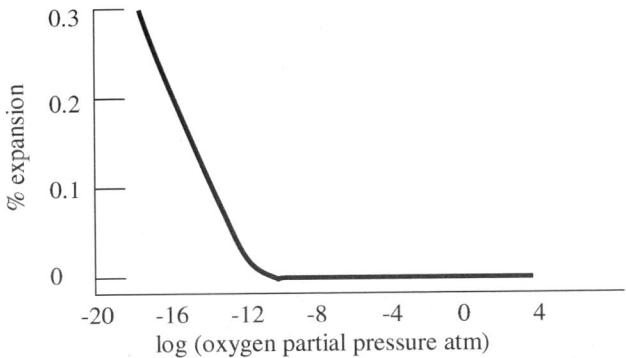

Figure 1.7 Expansion of $La_{0.79}Sr_{0.20}CrO_3$ as the oxygen is extracted by hydrogen [25].

and cracking in 100 mm × 100 mm plates. This has caused difficulties as larger planar stacks have been constructed with thick interconnect plates.

Such large lanthanum chromite interconnect plates have generally been made by powder processing methods. A fine powder of the desired composition is prepared by mixing lanthanum, strontium and chromium nitrates, then reacted with glycine at a high temperature [26]. The reaction mixture fluffs up into a fine powder which can be readily compacted to form interconnect plates, or extruded to make tube structures [27]. For example, calcia-doped lanthanum chromite was co-extruded with YSZ to make an electrolyte tube containing an interconnect strip along its length. This was co-fired to give a dense composite.

A major difficulty with such interconnects is the difficulty of sintering to full density. Lanthanum chromite powders do not sinter easily, especially in oxidising atmospheres. Strontium-doped materials require a low partial pressure of oxygen during sintering to become leak-tight, e.g. 10^{-12} bar at a temperature of 1720°C. Calcia-doped materials are better and can sinter in air at 1600°C. In this case, especially with chromium deficiency, liquid phases appear during the process and these help to pull the particles together. The downside is that these liquids can soak away into surrounding porous materials, as Minh found when co-firing his monolithic tape calendered composites [28].

To avoid these problems, Sulzer has used metal interconnects in their small-scale residential SOFC heat and power unit [29]. The alloy is largely chromium, with 5 wt% iron and 1 wt% yttria to give dispersion strengthening, made by Plansee AG in Austria. This alloy has almost the same expansion coefficient as YSZ and has the benefit of improved strength and toughness when compared with lanthanum chromite. But it requires coating to prevent chromium migration and is also expensive at the present time.

Another approach is to adopt a design similar to the lead acid battery and to use wires brought out from the electrodes and connected externally. This is the approach adopted by Adelan in their microtubular design. Clearly, the design of the cells and how they fit into the overall stack is vitally important in deciding such issues.

1.8 Cell and Stack Designs

A solid oxide fuel cell is a straightforward five-component entity as described in Figure 1.1. The main problem, which has been exercising engineers for the past 30 years, is that of designing cells which can be stacked to produce significant power output. This power output is directly proportional to the cell area, so the maximum area of YSZ membrane must be packed into the SOFC stack. This is similar to a heat exchanger design exercise. Two plausible solutions are obvious: a stack of flat plates or an array of parallel tubes. Typical heat exchanger problems of joining, cracking and leakage are evident in the SOFC stacks because of the complex materials and the high expansion coefficient. Of course the difficulties are greater because of the temperature of operation. Additional difficulties arise because of the low toughness of the ceramic components and the necessity of making electrical connections between all the cells.

Planar cells have the advantage that they can be readily electroded by screen printing, they can be stacked together with narrow channels to achieve high power densities and they can provide short current pathways through the interconnect. If p is the power in Watts per square cm of membrane and g is the gap in cm between planar electrolyte sheets, then the stack volumetric power density is p/g kW/litre, typically 1 kW/litre for a Sulzer planar stack where p is 0.5 W/cm^2 [29]. In a tubular stack packed in a square array, as in the Westinghouse design, the power density depends on the diameter D of cells and the gap g between them according to $\pi Dp/(D + g)^2$ which gives 0.6 kW/litre for Westinghouse tubes 2 cm in diameter with a 0.2 cm gap. This is lower than the planar stack because of the relatively large diameter of the tubes. Obviously, high power density depends on having small diameters and less gaps. The micro-tubular design gives 6 times better power density at 0.15 cm diameter of electrolyte tube with 0.1 cm spacing as shown in Figure 1.8. All these figures exclude the volume of thermal insulation and other ancillary parts.

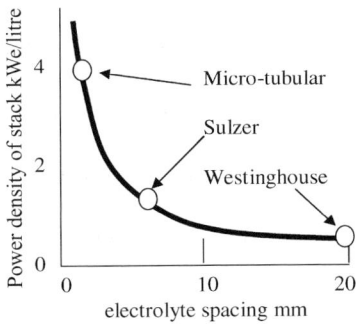

Figure 1.8   Power density of three different stacking geometries.

Many companies, including General Electric Power Systems (formerly Honeywell), McDermott Technologies/SOFCo, Ceramic Fuel Cells Ltd, Delphi/Battelle and Sulzer are currently developing planar SOFCs because of the known merits of that design, as explained in Chapter 8. However, two problems are still significant: one of heat-up and the other of sealing. The slow heat-up of existing planar designs is a consequence of the high thermal expansion coefficient and brittleness of YSZ. If the planar stack is heated to 800°C too rapidly, then it may crack, causing catastrophic failure. Any large YSZ structure will suffer the same problem as thermal gradients are set up through the ceramic. The large Westinghouse tubular cells require up to several hours to heat up safely. Thus it is important to use smaller plates or tubes to resist thermal shock. The downside of this is the greater assembly problem for large numbers of small cells.

Large planar cells display two other problems which cause concern. The first is the difficulty of making and handling large areas of delicate sheets; the maximum size that has been successfully fabricated is about 30 by 30 cm, far smaller than that possible with polymer membranes. The second problem is the gas sealing around the edges of the planar cells; this can be achieved with metal or glass seals but the required tolerance of around 10 µm in the membrane dimensions causes high cost. Sulzer avoided this issue by having discs without seals at the outer circumference. Of course, any rigid bonding together of a large ceramic structure also exacerbates the thermal shock issue. Monolithic designs have not been successful for that reason.

The Westinghouse tubular design is ingenious because 1.5–2 m long cells could be manufactured and handled as a result of the inherent strength of the tube structure. A 100 kWe generator could then be built from 1152 such cells. Moreover, the sealing problem was eliminated by inserting an air feeder tube down the cell tube. Although the Westinghouse tubular design is large and expensive, it did demonstrate several important features which have lent credence to the SOFC technology:

- The cells can run for long periods without much deterioration
- The efficiency can be impressive, around 50%
- Methane can be used as fuel after desulphurising and pre-reforming
- The SOFC exhaust can drive a gas turbine
- Emissions are low

In order to understand and predict the performance of such complex stack structures, various mathematical models have been developed, as described in Chapter 11. The most fundamental model starts from the reaction diffusion equations, assuming constant temperature conditions, and calculates the gradient of reactants, products and potentials along a tube or plate of electrolyte [30]. This gives a very sharp reaction front under normal operating conditions if the tube or plate is open ended. The chemical gradient along the SOFC can also be predicted as oxide ions permeate through the electrolyte [31]. Another important model sets out to calculate temperature and current distributions in a stack of cells [32]. Many such models for different geometries including planar and tubular have been published.

1.9 SOFC Power Generation Systems

Typically 25% of the volume of a fuel cell system is made up of the cell stack. The rest of the reactor is the balance of plant (BOP) which includes thermal insulation, pipework, pumps, heat exchangers, heat utilisation plant, fuel processors, control system, start-up heater and power conditioning, as described in Chapter 13. Arguably, this BOP is the dominant part of the system and should be treated with some concern. One of the major problems of the original Westinghouse design for a 100 kWe cogenerator was its large 16 m^2 footprint and huge weight of 9.3 te [33]. This was not competitive with a standard diesel engine combined heat and power unit.

Originally, SOFCs were designed to compete with large power generation units like central power stations, ships and locomotives, especially to run on coal gas or heavy fuels. During the last 10 years, the realisation has steadily dawned that SOFCs can work well in small, portable, residential and auxiliary power systems, particularly running on natural gas, propane or biogas [34]. Typical examples of such developments are those of Sulzer [29], Adelan, Delphi, General Electric and Siemens Westinghouse.

A typical schematic for a small SOFC system is given in Figure 1.9. The electrical power output for a mobile power application could be 100 We for communications up to 5 kWe to power a house or to supply air conditioning and auxiliary power in a vehicle. The heat output is less important for such devices because electrical efficiency is not the main performance criterion.

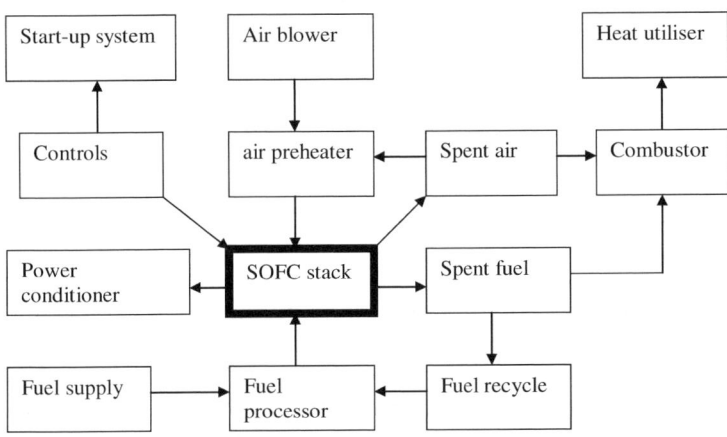

Figure 1.9   Flow sheet showing the BOP surrounding the SOFC stack.

The main moving part in this plant is the air blower, together with a fuel pump if pressurised fuel is not supplied. All the other parts except valves are solid state and should give the system low maintenance cost and high reliability over a life of many thousands of hours. In a small system, the reliability is the key competitive feature which gives advantage over internal combustion engines. Such heat engines are dominated by moving parts which require oil changes, new spark plugs, rebores, etc. Below 50 kWe, combustion engines are not usually economic because of maintenance costs, so SOFCs have a ready market.

1.10 Fuel Considerations

One of the great benefits of the SOFC is that it can utilise a wide range of fuels, as described in Chapter 12. The fastest reaction at the nickel anode is that of hydrogen. But other fuels can also react directly on the anode, depending on catalyst composition. For example, carbon monoxide can react on Ni/YSZ, but has a higher overpotential than hydrogen [35]. Also, methane can react on the anode but requires ceria or other catalysts to provide suitable sites for direct oxidation [36].

Fuel reforming can also take place on nickel at the anode. This occurs when steam is added to the hydrocarbon fuel, typically at a ratio of 3 parts steam to 1 part of fuel. The reaction of methane is then given by

$$CH_4 + H_2O \rightarrow CO + 3H_2$$

The hydrogen and carbon monoxide released by this reaction can then react individually with oxide ions emerging from the electrolyte. Usually the CO conversion is sluggish so the shift reaction also occurs on the anode to produce more hydrogen:

$$CO + H_2O \rightarrow CO_2 + H_2$$

It was demonstrated in the 1960s that hydrocarbons could be injected directly into SOFCs if steam was supplied [37]. The steam can beneficially be obtained from the spent fuel stream. The main problem with direct use of hydrocarbons is that coke can form to block up and contaminate the anode. There are two damaging reactions which can occur on the nickel:

$$2CO \rightarrow CO_2 + C$$

$$CH_4 \rightarrow 2H_2 + C$$

When carbon formation was investigated in detail, by temperature-programmed reaction, three different types of material were discovered on the nickel, as indicated by the temperature required for oxidation [38]. The most stable carbon could not be removed below 1100 K and tended to form when current was flowing through the cell.

The other damaging mechanism of SOFC failure stems from fuel impurities. Sulphur is the most prevalent impurity and can be present up to 1% level in marine diesel fuel. SOFCs cannot operate with this amount of sulphur. More typically, natural gas often has 'odorant' sulphur compounds added to make leaks more easily detectable. Even the lower levels of such additions, about 10 ppm, are damaging for SOFC nickel anodes, and the upper limits around 100 ppm could cause failure in about 1 h of operation. There are two approaches to solving this problem: adding a sulphur absorber to the fuel processing unit; and using anode metals which are less affected by sulphur. Fortunately, the levels of sulphur in gasoline and diesel fuel are now being reduced for environmental reasons, with the best formulations containing less than 10 ppm.

The second difficulty is the number of additives in conventional fuels which have been formulated for other technologies. For example, regular gasoline contains more than 100 different molecules, some added as lubricants or surfactants. Moreover, the mixture can change with time and place because the standard is dictated by octane number and not composition. Consequently, it is unlikely that SOFCs will be able to run directly on gasoline, although this has been attempted. An objective of current research is to formulate a fuel which can operate in an SOFC and a vehicle engine simultaneously [39,40].

1.11 Competition and Combination with Heat Engines

If the SOFC is to be successful commercially, then it must compete with existing heat engines that are currently used to produce electricity from hydrocarbon combustion. Such engines operate by burning fuel to heat a volume of gas, followed by expansion of the hot gas in a piston or turbine device driving a dynamo. These are inefficient and polluting when compared with fuel cells but can be surprisingly economic as a result of a century's development, optimisation and mass production. Ostwald got it famously wrong in 1892 when he said that 'the next century will be one of electrochemical combustion'. Fuel cells are still significantly more costly than conventional engines which can be manufactured for less than $50 per kWe. The SOFC advantages of efficiency, modularity, siting and low emissions count for little if they cost $10,000 per kWe. These arguments are considered more fully in Chapter 13.

In the 1980s, it was envisaged that SOFCs could compete commercially with other power generation systems, including large centralised power stations and smaller cogeneration units [41]. This has not yet happened because costs have remained high despite large injections of government funding for SOFCs development in the USA, Japan and Europe. It has been estimated that costs of $400 per kWe could be achieved with mass production using powder methods [42]. Such costs would be competitive with present large power station costs.

One of the most promising applications of SOFCs for the future is in combination with a gas turbine as indicated in Chapter 3. The flow scheme is shown in Figure 1.10. The SOFC stack forms the combustor unit in a gas turbine system. Compressed air is fed into the SOFC stack where fuel is injected and electrical power drawn off. Operating near 50% conversion of fuel to electrical power, this SOFC then provides pressurised hot gas to a turbine operating at 35% efficiency. The overall electrical conversion efficiency of this system can approach 75%, and this could be further improved by adding a steam turbine [43].

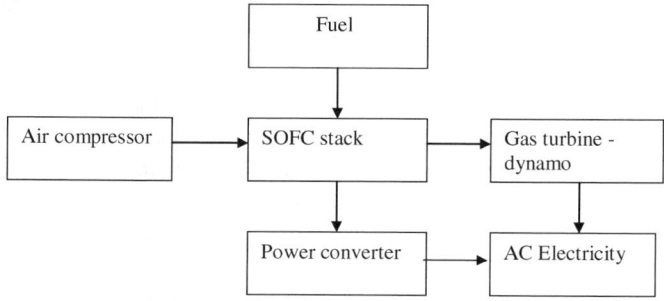

Figure 1.10   Combination of SOFC with a gas turbine generator.

Westinghouse carried out a paper study in 1995 and tested this concept in a 200 kWe class hybrid system in 2001–2002. The SOFC stack was operated in a pressure chamber at 3.5 bar abs. A 50 kW microturbine was used to utilise the hot gas exiting the SOFC stack. The overall efficiency of this first-of-a-kind, proof-of-concept prototype was measured at 57% [44]. This is the highest conversion efficiency device for fuel to power yet devised. Efficiencies up to 75% are expected in larger hybrid systems when fully developed.

1.12 Application Areas and Relation to Polymer Electrolyte Fuel Cells

There are many possible applications for SOFCs as described in Chapter 13. The stationary power market has been most investigated up to this point, since there is a great need for clean and quiet distributed power generation units, e.g. in hospitals, hotels and sports facilities located in cities. Traditionally, such demands have been served by 200 kWe diesel engine or gas turbine packages, with the heat supplied to the building for hot water or steam. Other fuel cell systems based on phosphoric acid, molten carbonate and solid polymer electrolytes have also been developed to fill this niche. Heat engine generators are cheaper at the moment and fuel cell devices will generally not be preferred unless regulations for low emissions are imposed.

The other application which has been much studied is that of integration of SOFCs with coal gasification. An SOFC is eminently suited to integration with a coal gasifier plant in large power stations and should result in highest overall conversion compared to other fuel cell types. Unfortunately, the investment required to build such plants is large.

Therefore other smaller applications have emerged in recent years, especially to compete with polymer electrolyte fuel cells (PEFCs) which have been rapidly evolving to satisfy the zero emission electric vehicle market. A typical PEFC for a vehicle is 30 kWe, runs on pure hydrogen, requires significant quantities of platinum, and is still significantly more costly than an internal combustion engine. It does have rapid heat-up and can deliver significant power on a cold-start but this advantage is destroyed if pure hydrogen has to be obtained through hot reforming of methanol or gasoline, which introduces delay and sluggishness into the system. The SOFC can be useful in vehicles, not to replace the engine, but to supply auxiliary power to supplement or replace the existing battery system. Typically, 1–5 kWe is required, mainly to drive air conditioning. The benefits of SOFCs for this application are:

- it can run on the same hydrocarbon fuel as the internal combustion engine (e.g. gasoline, diesel);
- it can provide useful heat;
- it can run when the engine is switched off;
- it is much more efficient than the existing electrical system; and
- it has low emissions.

Similar arguments apply to truck cabs which have to be heated.

Smaller, portable SOFC power units which can replace batteries are also being considered. These can deliver power in the range from 20 We to 10 kWe and can run directly on a wide range of fuels from natural gas, to propane, methanol and isooctane. If the cells are small to avoid thermal shock, then the start-up can be quick [45]. The other application for such units is in residential cogeneration using pipeline gas. Installing SOFCs in every home will cut residential carbon dioxide emissions by up to 50%.

1.13 SOFC-Related Publications

SOFC technology has been advancing at a rapid rate over the past years since Minh and Takahashi published their excellent monograph [46]. Large numbers of new developments have occurred and many more scientists and engineers are working in this field. Bringing all the research and development information together in this one volume should help to unify the subject and produce further breakthroughs.

Several conference proceedings are published each year containing material on SOFCs but these tend to be collections of individual research papers at a particular time rather than a complete compendium of the technology. These include The Electrochemical Society Proceedings series on SOFCs which has been edited by Singhal et al [47–54], and the Proceedings of the European SOFC Forums [55–59].

It is believed that the publication of this volume will provide detailed up-to-date information for the researchers who are about to make SOFCs commercial in the near future.

References

[1] W. Nernst, Z. Elektrochem., **6** (1899) 41.
[2] W. Nernst, DRP 104872, 1897.
[3] W. Nernst, US Patent 685 730, 1899.
[4] K. Mendelssohn, *The World of Walther Nernst, the Rise and Fall of German Science, 1864–1941*, Macmillan, London, 1973.
[5] E. Baur and H. Preis, Z Elektrochem., **43** (1937) 727.
[6] H. Peters and H. H. Mobius, Z Physik. Chem., **209** (1958) 298.
[7] A. H. Heuer and L W. Hobbs (eds), *Science and Technology of Zirconia*, American Ceramic Society, Columbus, Ohio, 1981; see also *Science and Technology of Zirconia*, vols II–V, American Ceramic Society, Columbus, Ohio, 1984–1993.
[8] D. H. Archer, L. Elikan and R. L. Zahradnik, in *Hydrocarbon Fuel Cell Technologies*, ed. B. S. Baker, Academic Press, New York, 1965. p. 51.
[9] A. O. Isenberg, Solid State Ionics, **3** (1981) 431.
[10] U. B. Pal and S. C. Singhal, J. Electrochem. Soc., **137** (1990) 2937–2941.

[11] U. B. Pal and S. C. Singhal, *High Temperature Science*, **27** (1990) 251.
[12] C. Bagger, Improved production methods for YSZ electrolyte and Ni-YSZ Anode for SOFC, in *1992 Fuel Cell Seminar Abstracts*, Tucson, AZ, 1992.
[13] K. Kendall, *Powder Technology*, **58** (1989) 151–161.
[14] M. Prica, K. Kendall and S. A. Markland, *J. Am. Ceram. Soc.*, **81** (1998) 541–548.
[15] M. Cassidy, K. Kendall and G. Lindsay, in *Proceedings of the 2nd European SOFC Forum*, ed. B. Thorstensen, Switzerland, 1996, pp. 667–676.
[16] N. Q. Minh and C. R. Horne, in *Proceedings of the 14th Risø International Symposium on Materials Science, High Temperature Electrochemical Behaviour of Fast Iron and Mixed Conductors*, eds. F. W. Poulsen, J. J. Benzen, T. Jacobsen, E. Skou and M. J. L. Ostergard, Risø National Labortory, Roskilde, Denmark, 1993, p. 337.
[17] K. Kendall, Revolution in ceramic processing, in *3rd European Symposium on Engineering Ceramics*, ed. F. L. Riley, Elsevier Applied Science, London, 1989, pp. 97–107.
[18] K. Kendall, in *International Forum on Fine Ceramics*, Japan Fine Ceramics Center, Nagoya, 1992, pp.143–148.
[19] C. Finnerty, T. Alston, R. M. Ormerod and K. Kendall, in *Portable Fuel Cells*, ed. F. N. Buchi, Oberrohrdorf, Switzerland, 1999, pp. 27–34.
[20] S. Park, J. M. Vohs and R. T. Gorte, *Nature*, **404** (2000) 265–267.
[21] M. Cassidy, C. Bagger, N. Brandon and M. Day, in *Proceedings of the 4th European SOFC Forum*, ed. A. J. McEvoy, Switzerland, 2000, pp. 637–646.
[22] N. Q. Minh, in *Proceedings of the 37th Sagamore Army Materials Research Conference*, ed. D. J. Viechnicki, Army Materials Technology Lab, Watertown, MA, 1990, p. 213.
[23] K. Kendall, E. Wright and A. Golds, in *Solid Oxide Fuel Cells IV*, eds. M. Dokiya, O. Yamamoto, H. Tagawa and S. C. Singhal, The Electrochemical Society Proceedings, Pennington, NJ, PV95-1, 1995, pp. 229–233.
[24] W. J. Quadakkers, H. Greiner and W. Kock, in *Proceedings of the 1st European SOFC Forum*, ed. U. Bossel, Switzerland, 1994, p. 525.
[25] S. Srilomsak, D. P. Schilling and H. U. Anderson, in *Solid Oxide Fuel Cells I*, ed. S. C. Singhal, The Electrochemical Society Proceedings, Pennington, NJ, PV89-11, 1989, p. 129.
[26] L. A. Chick, J. L. Bates, L. R. Pederson and H. E. Kissinger, in *Solid Oxide Fuel Cells I*, ed. S. C. Singhal, The Electrochemical Society Proceedings, Pennington, NJ, PV89-11, 1989, pp. 170–179.
[27] K. Kendall and M. Prica, in *Proceedings of the 1st European SOFC Forum*, ed. U. Bossel, Switzerland, 1994, pp. 163–170
[28] N. Q. Minh, T. R. Armstrong, J. R. Esopa, J. V. Guiheen, C. R. Horne and J. J. van Ackeren, in *Solid Oxide Fuel Cells III*, eds. S. C. Singhal and H. Iwahara, The Electrochemical Society Proceedings, Pennington, NJ, PV93-4, 1993, pp. 801–808.
[29] W. Glatz, M. Janousek, E. Batawi and K. Honegger, in *Proceedings of the 4th European SOFC Forum*, ed A. J. McEvoy, Switzerland, 2000, pp. 855–864.

[30] A. C. King, R. C. Copcutt and K. Kendall, *Proc. R. Soc. Lond.*, **A452** (1996) 2639–2653.

[31] J. Billingham, A. C. King, R. C. Copcutt and K. Kendall, *SIAM J. Appl. Math.*, **60** (2000) 574–601.

[32] S. Ahmed, C. C. McPheeters and R. Kumar, *J. Electrochem. Soc.*, **138** (1991) 2712.

[33] W. L. Lundberg, *SOFC cogeneration system conceptual design*, Report No GRI 89/0162, Gas Research Institute, Chicago, IL, 1989.

[34] G. A. Tompsett, C. M. Finnerty, K. Kendall, T. Alston and N. M. Sammes, *J. Power Sources*, **86** (2000) 376–382.

[35] R. L. Zahradnik, *J. Electrochem. Soc.*, **117** (1970) 1443.

[36] S. J. Livermore, J. W. Cotton and R. M. Ormerod, *J. Power Sources*, **86** (2000) 411–416.

[37] J. Weissbart and R. Ruka, *J. Electrochem. Soc.*, **109** (1962) 723.

[38] R. M. Ormerod, *Stud. Surf. Sci. Catal.*, **122** (1999) 35.

[39] G. J. Saunders and K. Kendall, in *Solid Oxide Fuel Cells VIII*, eds. S. C. Singhal and M. Dokiya, The Electrochemical Society Proceedings, Pennington, NJ, PV2003-07, 2003, pp. 1305–1314.

[40] G. J. Saunders and K. Kendall, *J. Power Sources*, **106** (2002) 258–263.

[41] P. D. Lilley, E. Erdle and F. Gross, *Market Potential of SOFC*, Report No EUR 12249 EN, CEC, Luxembourg, 1989.

[42] J. Cotton, *DTI cost study of SOFCs*, ETSU Report, 1997.

[43] A. L. Dicks, R. J. Carpenter, E. Erdle, D. F. Lander, P. D. Lilley, A. G. Melman and N. Woudstra, *SOFC systems study*, Vol. I, eds A. G. Melman and N. Woudstra, Report No EUR 13103 EN, CEC, Luxembourg, 1991.

[44] S. E. Veyo and W. L. Lundberg, in *Proceedings of the 2nd European SOFC Forum*, ed. B. Thorstensen, Switzerland, 1996, pp. 69–78.

[45] M. Prica, T. Alston and K. Kendall, in *Solid Oxide Fuel Cells V*, eds. U. Stimming, S. C. Singhal, H. Tagawa and W. Lehnert, The Electrochemical Society Proceedings, Pennington, NJ, PV97-40, 1997, pp. 619–625.

[46] N. Q. Minh and T. Takahashi, *Science and Technology of Ceramic Fuel Cells*, Elsevier, Amsterdam, 1995.

[47] S. C. Singhal (ed.), *Solid Oxide Fuel Cells I*, The Electrochemical Society Proceedings, Pennington, NJ, PV89-11, , 1989.

[48] F. Grosz, P. Zegers, S. C. Singhal and O. Yamamoto (eds.), *Solid Oxide Fuel Cells II*, Commission of the European Communities, Luxembourg, 1991.

[49] S. C. Singhal and H. Iwahara (eds.), *Solid Oxide Fuel Cells III*, The Electrochemical Society Proceedings, Pennington, NJ, PV93-4, 1993.

[50] M. Dokiya, O. Yamamoto, H. Tagawa and S. C. Singhal (eds.), *Solid Oxide Fuel Cells IV*, The Electrochemical Society Proceedings, Pennington, NJ, PV95-1, 1995.

[51] U. Stimming, S. C. Singhal, H. Tagawa and W. Lehnert (eds.), *Solid Oxide Fuel Cells V*, The Electrochemical Society Proceedings, Pennington, NJ, PV97-40, 1997.

[52] S. C. Singhal and M. Dokiya (eds.), *Solid Oxide Fuel Cells VI*, The Electrochemical Society Proceedings, Pennington, NJ, PV99-19, 1999.

[53] H. Yokokawa and S. C. Singhal (eds.), *Solid Oxide Fuel Cells VII*, The Electrochemical Society Proceedings, Pennington, NJ, PV2001-16, 2001.
[54] S. C. Singhal and M Dokiya (eds.), *Solid Oxide Fuel Cells VIII*, The Electrochemical Society Proceedings, Pennington, NJ, PV2003-07, 2003.
[55] U. Bossel (ed.), *Proceedings of the First European SOFC Forum*, Oberrohrdorf, Switzerland, 1994.
[56] B. Thorstensen (ed.), *Proceedings of the Second European SOFC Forum*, Oberrohrdorf, Switzerland, 1996.
[57] P. Stevens (ed.), *Proceedings of the Third European SOFC Forum*, Oberrohrdorf, Switzerland, 1998.
[58] A. J. McEvoy (ed.), *Proceedings of the Fourth European SOFC Forum*, Oberrohrdorf, Switzerland, 2000.
[59] J. P. P. Huijmans (ed.), *Proceedings of the Fifth European SOFC Forum*, Oberrohrdorf, Switzerland, 2002.

Chapter 2

History

Hans-Heinrich Möbius

2.1 The Path to the First Solid Electrolyte Gas Cells

Starting in 1800, Davy carried out many investigations into the electrolysis of water and aqueous solutions. Experiments using more and more concentrated solutions of alkali hydroxides led to melting flux electrolysis and in 1807 to the discovery of alkali metals [1]. Davy observed that dry solid alkali compounds were non-conductors but became electrically conducting through just a little moisture. For Faraday it seemed important that many electrically conducting liquids lost their conductivity during solidification [2]. In his continuing investigations, Faraday introduced the basic terminology of electrochemistry, and with the aid of many results concerning the concept 'electrolyte' in 1834 he classified substances into first and second types of conductors, metallic and electrolytic [3]; the first type now recognised as electronic and the second type as ionic conductors.

Faraday encountered problems with the classification of silver sulphide, which exhibited conductivities comparable to metals in the high-temperature range, but, in contrast to metals, lost its conductivity upon cooling [2]. Hittorf (1851) devoted himself to this special problem and proved that Ag_2S is electrolytically decomposable [4]. The generation of a counter voltage (polarisation by chemical precipitation) during the passage of a current was recognised as a characteristic feature of electrolytic conductivity of solids [4,5], and this led to the discovery of an increasing number of solid conductors of the second type (ionic).

As early as 1774, Cavendish [6] had observed an increase in the conductivity of glass on heating. The electrolytic nature of this conduction was discovered by Beetz [7] and Buff [8] in 1854. Using mercury, zinc amalgam, various solid metals, carbon, and pyrolusite (MnO_2) as electrodes, Buff demonstrated galvanic cells and batteries free of water 'in which glass takes over the role of the moist conductor', and he investigated the associated voltage and polarisability.

A short period before, Gaugain [9] and Bequerel [10] had published experiments on the thermoelectricity between metal contacts on glass and porcelain. Buff [8] reproduced the results, which turned out very differently depending on the position of the contacts in oxidising or reducing regions of flames, and he interpreted the voltages as a mixture of thermoelectric forces and voltages which he had previously observed between bare platinum wires in flames [11]. However, Gaugain investigated his experiment, which at first was constructed of two tubes of glass, platinum wires, air and alcohol vapour, in more detail [12]. He observed the delivery of current, the polarity of the electrodes and their behaviour when the electrode metals or the gas supply were changed. He also noted the large voltage alteration when different gases were mixed with oxygen beyond a certain proportion (known today as the jump at the stoichiometric point), and phenomena associated with an iron/air cell which convinced him of the decisive role of oxygen in the electrode reaction. Although restricted by the lack of sensitivity of the available measuring device (a leaf electroscope) so that small voltage differences could not be detected, Gaugain nevertheless found that 'the new source of electricity possesses all the characteristic features of an aqueous-electric cell', and thus he discovered in 1853 galvanic solid electrolyte gas cells.

Towards the end of the nineteenth century the term 'solid electrolyte' was in use, and many facts were known about the behaviour of these materials. *The Science of Electricity* by Wiedemann (1893–98) includes the chapters 'Conductivity of Solid Salts' and 'Determination of the Electromotive Force – Two Metals and Solid Electrolytes' and 'Electrolysis of Solid Electrolytes' [13]. However, in Ostwald's textbook on general chemistry, solid electrolytes are not mentioned [14].

Technological interest in solid ion conductors first arose in connection with the development of electric lighting devices. Early carbon filament lamps manufactured since about 1880 could not compete with the existing gas incandescent light. In 1897, Nernst suggested in a patent [15] that a solid electrolyte in the form of a thin rod could be made electrically conducting by means of an auxiliary heating appliance and then kept glowing by the passage of an electric current. At first Nernst mentioned only 'lime, magnesia, and those sorts of substances' as appropriate conductors. Later investigations stimulated by experiences with gas mantles led to his observation 'that the conductivity of pure oxides rises very slowly with temperature and remains relatively low, whereas mixtures possess an enormously much greater conductivity, a result in complete agreement with the known behaviour of liquid electrolytes' [16]. He pointed out that, for example, the conductivity of pure water and pure common salt is low but that of an aqueous salt solution is high. In a short time many of the mixed oxides which exhibit high conductivity at elevated temperatures, including the particularly favourable composition 85% zirconia and 15% yttria [17], the so-called Nernst mass [18,19], were identified. The thesis of Reynolds [20], inspired by Nernst and presented in 1902, expanded this field by measuring the conductivity in the range 800–1400°C of numerous binary and ternary systems, among others, formed by ZrO_2 with the oxides of the elements La, Ce, Nd, Sm, Ho, Er, Yb, Y, Sc, Mg, Ca, Th and U, including investigations on the role of composition, concentration, direction of temperature alteration (hysteresis) and other phenomena.

Figure 2.1 shows one of the many designs of the Nernst lamp [21]. When the lamp was switched on, the voltage was applied to the Nernst rod, h, and to the parallel heating resistor, i. Both these components were incorporated in a glass envelope containing air. After sufficient preheating, the current started flowing through the Nernst rod h and through the winding k of an electromagnet b. At a specified electric current the magnet switched off the heater by opening the contacts between m and l and then the Nernst rod carried all the current and emitted light due to resistive heat generation.

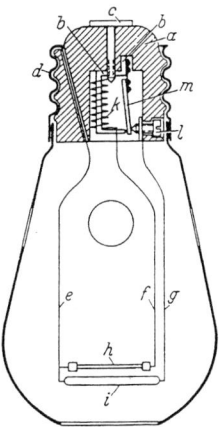

Figure 2.1 *Arrangement of a Nernst glower in a lamp (DRP 114 241, filed 9 April 1899).*

The light efficiency of the Nernst lamp exceeded that of the carbon filament lamp by nearly 80%. However, there were problems. It was difficult to fabricate reliable contacts to the glower, and the platinum leads and heater made the lamp expensive. The glowers had to be prevented from melting with the aid of special series resistors. It was necessary to wait in darkness for half a minute after switching on the lamp until the light appeared. In view of these and other disadvantages, interest in the Nernst lamp, although considerable for a few years, soon disappeared with the introduction of the first tungsten filament lamps, which were much simpler and permitted a substantial increase in the light efficiency by raising the filament temperature.

The Nernst zirconia rods were similar to metallic conductors in that decomposition did not occur with the passage of direct current. Nevertheless, Nernst was convinced that his filaments were ionic conductors, and he assumed that, e.g. in yttria-stabilised zirconia (YSZ), the yttria provided the necessary charge carriers [16]. He observed evidence of oxygen transport, but believed that metal cations were also deposited by the direct current, later oxidising and diffusing back into the filament.

It was not until 1943 that Wagner [22] (in memory of Walther Nernst who died on 18 November 1941) recognised the existence of vacancies in the anion sublattice of mixed oxide solid solutions and thus explained the conduction mechanism of the Nernst glowers. We now know that Nernst lamp filaments were oxide ion conductors and the platinum contacts behaved as air electrodes. It follows that Nernst lamps were the first commercially produced solid electrolyte gas cells.

2.2 From Solid Electrolyte Gas Cells to Solid Oxide Fuel Cells

Electrochemistry was given an important impetus when its connection with thermodynamics was explained by Helmholtz in 1882 [23]. Then, in 1894 Ostwald demonstrated that energy from coal could be produced much more efficiently with a galvanic cell than with a steam engine [24].

The agreement between the voltages measured with galvanic solid electrolyte gas cells and calculated thermodynamically was verified by Haber and co-workers in 1905. From 330 to 570°C they used glass and from 800 to 1100°C porcelain as the electrolyte, and partly platinum, partly gold as the material for the electrodes in cells, first with C, CO, CO_2 and O_2 [25], then in oxyhydrogen cells, and in hydrogen and oxygen concentration cells [26,27]. Typical phenomena such as the dependence of the voltage on the gas flux, deviations from zero ('asymmetry voltages'), and sluggishness in the establishment of constant voltages at low temperatures were observed. Parallel to the publication of the results, Haber filed the first patent on fuel cells with a solid electrolyte [28] (Figure 2.2). To compensate for alterations in the composition of the glass electrolyte by the migration of ions caused by current, he proposed to exchange the gases in the electrode chambers as soon as disturbing alterations were noticed.

The decomposition equilibria of metal oxides were investigated in 1916 by Treadwell in the region of 1000°C with quartz and porcelain as solid electrolytes and with a silver/oxygen electrode as the reference system [29]. After these investigations, Baur and Treadwell filed a patent on fuel cells with metal oxide electrodes and a molten salt, held in a porous ceramic, electrolyte [30]. Only after many fruitless experiments with liquid electrolytes of different types, Baur in 1937 came to the conclusion that fuel cells have to be made completely solid [31]. But the extensive empirical search by Baur [18,32,33] and other authors

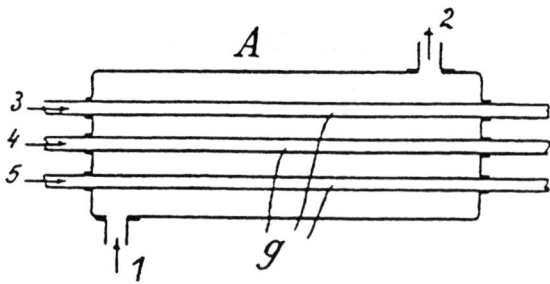

Figure 2.2 First diagram (Haber, 1905) of a fuel cell with solid electrolytes. Generator gas passed from 1 to 2 through chamber A (440°C) with parallel glass tubes g covered on both sides with thin layers of noble metal and swept inside by air.

up to the 1950s for suitable solid electrolytes, covering glasses, porcelains, clays and a great variety of oxide mixtures, was unsuccessful.

The empirical phase of the development of solid electrolyte fuel cells was overcome only after many general advances in research on solids. These included development of X-ray structure analysis, new knowledge on the ion conduction of solids from the measurements of transport numbers by Tubandt (first detection of unipolar conduction by anions), the establishment of the theory of disorder in solids by Frenkel, Schottky, Wagner and Jost, and the development of isotope methods for the investigation of diffusion processes in solids.

Starting from the observation of effects caused by small excesses of components in salts and oxides, Schottky investigated problems of fuel cells with solid electrolytes in 1935 [34] and suggested that a comprehensive patent should be applied for by Siemens and Halske [35] (Figure 2.3). He pointed out the advantages of solid over liquid electrolytes such as the feasibility of small layer thicknesses, less disturbance by ambipolar and neutral diffusion processes, and small absolute concentrations for the realisation of chemical potential differences in solids. He considered, among other things, porous metallic electrodes and electronic semiconductors forming intermediate or main electrodes (with the requirement that no continuous rows of mixed crystals with the electrolyte material should be formed), and he discussed cyclic processes for the continuous supply of the electrode chambers, self-regulation of the temperature, and repeated temporary chemical alternation of the polarity of

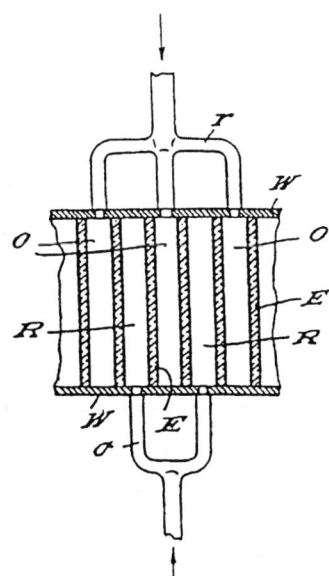

Figure 2.3 Solid electrolyte fuel cell specified in a patent by Siemens and Halske (inventor Schottky, 1935). W walls of a closed electrically isolating casing, E solid electrolyte discs with a thickness in the order of mm, O chambers with O_2 or air, R chambers with gas mixtures containing CO and CO_2, o and r gas lines to the chambers O and R.

the cells by changing the gas supply. Concerning the electrolytes, for which Schottky required a conductivity near 0.3 S/cm, halides, sulphates, carbonates and phosphates were considered but no oxides. An electrochemical exploitation of the combustion of coal seemed to be less feasible than that of the formation of hydrogen chloride.

Zirconia ceramics were first used in fuel cells in 1937 by Baur and Preis [18]. They wrote on the Degussa tube crucibles used (16 mm × 12 mm × 190 mm): 'Unsurpassed is the Nernst mass. But even this mixture is not satisfactory because the current enhances resistance considerably by electrolytic shift (migration away of the cations).' The problems were possibly caused by the material used for the cathode (Fe_3O_4, Figure 2.4), which oxidises at 1000°C in air to form poorly conducting Fe_2O_3 [29]. So the wrong conclusion was drawn: 'One has to look for an improvement of the Nernst mass or to put alongside it solid conductors of higher value.'

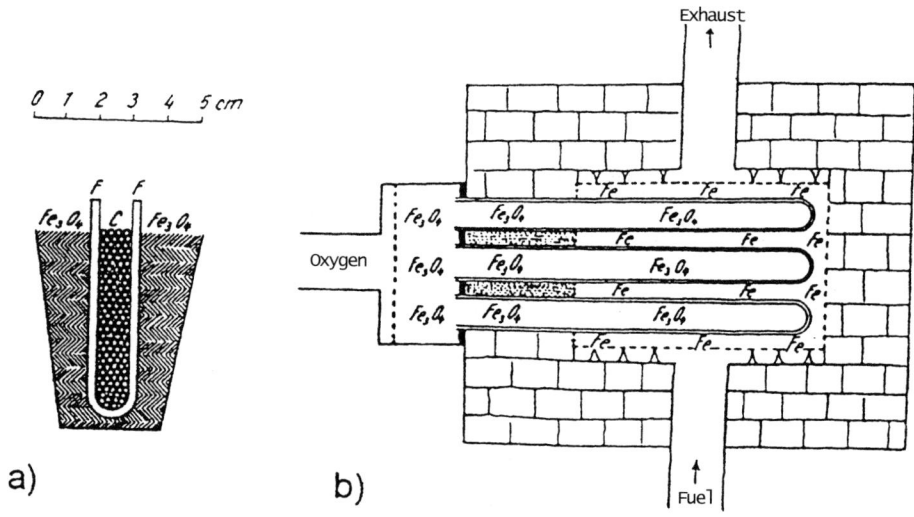

Figure 2.4 Fuel cell arrangement of Baur and Preis (1937): (a) Investigated cells with Nernst mass as solid electrolyte (F); (b) Proposal for the realisation of stacks of such fuel cells.

Baur preferred other ceramics containing tungsten oxide and cerium dioxide over the Nernst mass [18,32,33], and these were also used repeatedly by other investigators. However, as shown for tungstates of Ce, Ca and Zn [36], in oxidising and reducing conditions, electronic transport through such ceramics is so high that their application in fuel cells cannot be useful.

When Wagner had recognised the mechanism of conduction in the Nernst glower, he pointed out in 1943: 'For fuel cells with solid electrolytes anion conductors are to be considered exclusively. From this point of view a systematic investigation of the mixed crystal systems of the type of the Nernst mass with roentgenographic and electrical methods seems to be desirable' [22]. This was the start of concentrated work on solid oxide fuel cells (SOFCs).

2.3 First Detailed Investigations of Solid Oxide Fuel Cells

In 1951 Hund confirmed, with X-ray crystallographic and pycnometric investigations, the existence of oxide ion vacancies in the Nernst mass [37]. Hauffe, having worked from 1936 to 1940 in Darmstadt with Wagner [38], followed his suggestions and left to Peters (in 1951 in Greifswald) the investigation of the lattice structure and the electrical conductivity of some mixed oxides of the type of the Nernst mass (ZrO_2–Y_2O_3, ThO_2 and CeO_2 with Y_2O_3 and La_2O_3). After Hauffe had left for Berlin, Peters finished his thesis in Rostock in 1953 [39] and in 1954 gave Möbius the task of furthering the subject by investigation of galvanic cells using mixed-oxide phases.

The investigations [40], using model fuel cells (Figure 2.5a), were started with iron oxides, magnesium ferrite (following Biefeld [41]) and composites of iron and alumina as electrodes and were continued mainly with thin porous layers of platinum, nickel and iron. Very soon it was seen that completely gastight solid electrolyte discs of highly pure substances had to be produced if the

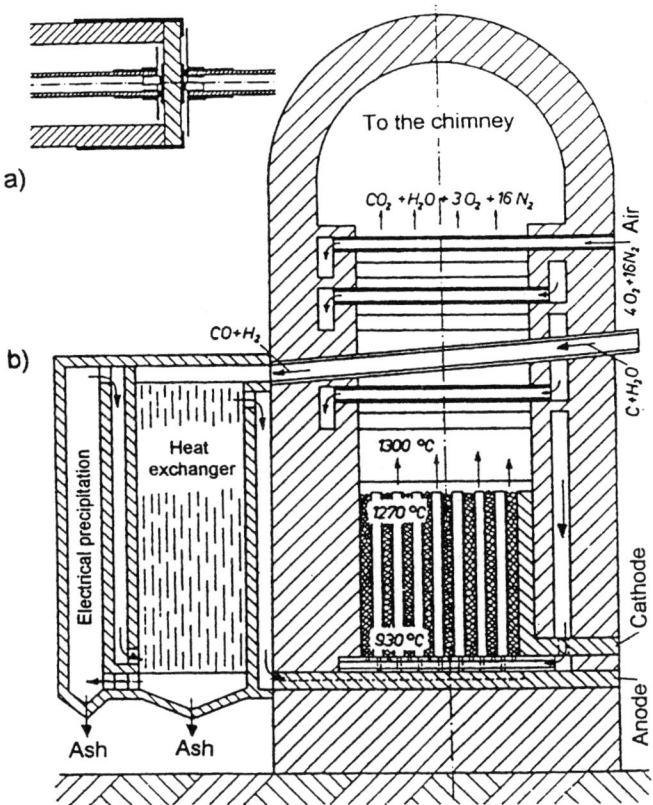

Figure 2.5 (a) Cross-section of galvanic cells with oxides as solid electrolyte investigated by Peters and Möbius. (b) Design of an SOFC plant as a basis for calculations (1958).

measurements were to lead to useful results. Compounds of thorium and cerium were effectively purified as ammonium double nitrates by crystallisation from hot concentrated nitric acid. Pure lanthanum oxide was prepared by fractional precipitation of hydroxides. Bearing in mind experience of oxide ceramics [42], powders of mixed oxides were pressed to produce gastight discs, 25 mm in diameter and 1–2 mm thick, which were sintered at temperatures up to 1920°C in a stream of oxygen on a support stack of $Al_2O_3/MgAl_2O_4/ThO_2$ in alumina tubes using a Tammann carbon tube furnace. ZrO_2 for this investigation was available at that time only in the form of a mixture with Y_2O_3 as a residue from the investigations of Peters because an embargo and the development of the nuclear industry made it difficult to obtain.

There were additional reasons for concentrating in Rostock on solid electrolytes based on ThO_2. In 1948, Ryschkewitsch [42] pointed out that a large-scale technical application of ThO_2 was still lacking. During the 1950s, it seemed that more zirconium than thorium was needed for the development of nuclear energy. Furthermore the mixed oxides with ThO_2 are crystallographically simpler than those with ZrO_2. Some stocks of ThO_2 existed for the fabrication of mantles for gaslight.

In the investigations, carried out from 1955 to 1957, for cells with different composition of the solid electrolyte, the electrode voltages were measured in the temperature range between 300 and 1350°C, and compared with thermodynamically calculated values.

Schottky had shown that the efficiency of solid electrolyte fuel cells with increasing load resistance decreases to zero if a noticeable part of the conductivity of the electrolyte is of electronic nature [34]. Therefore the efforts for purification and especially for separation of the polyvalent praseodymium cations from the solid electrolyte material were made. In the case of Th–La mixed oxides, with only 1 mol% $LaO_{1.5}$ the ion transport number 1 was reached, admittedly only with reducing gas on both electrodes (in CO,CO_2 concentration cells); in the oxygen/air cell even at 10 mol% $LaO_{1.5}$ this number was only near 0.8. A perfect disc of $Ce_{0.9}La_{0.1}O_{1.95}$ broke into pieces in $CO,CO_2/O_2$ between 700 and 840°C, reaching the mean ionic transport number 0.8. For the available ZrO_2 solid electrolyte (with 50 mol% $YO_{1.5}$) in the oxygen/air cell, the ion transport number was above 0.93.

On the basis of these results, the Boudouard equilibrium was investigated with $Th_{0.9}La_{0.1}O_{1.95}$ as solid electrolyte in the cell $CO,C,Fe/FeO,CO,CO_2$, using only the reactive carbon precipitated out of CO; iron in metallic or oxide form in the electrodes supported the establishment of the electrode potential catalytically. And with the ZrO_2 solid electrolyte in a $CO,CO_2,Fe_3O_4/Pt,O_2$ cell, the CO_2 dissociation equilibrium was investigated [43].

The good agreement between measured and thermodynamically calculated data in these cases led to the most important by-product of SOFC development: if solid electrolyte cells, charged with gases of known concentrations, deliver the theoretically expected cell voltages, it also must be possible to calculate unknown gas concentrations backwards from the cell voltages, measured between the cell terminals in gas phases, which can be oxidising or reducing.

Accordingly the potentiometric determination of gas concentrations with solid electrolyte cells and the first designs of probes for the *in situ* analysis of hot gases (reference electrode e.g. Ni,NiO or a gas with known oxygen partial pressure) were patented in 1958 [44]. The first calculations of oxygen partial pressures in purified nitrogen, using the measured cell voltages of solid oxide cells, were performed in 1955.

In 1957 Kiukkola and Wagner reported thermodynamic investigations on metal/metal oxide systems, for the first time using CaO-stabilised ZrO_2 (especially $Zr_{0.85}Ca_{0.15}O_{1.85}$) as solid electrolyte [45]. But they could not realise the intended measurements by using a gas reference electrode because their solid electrolytes (sintered at 1400–1450°C) were porous [46]. The investigations of Peters and Mann on metal/ metal oxide systems with gastight $Th_{0.9}La_{0.1}O_{1.95}$ solid electrolyte were performed using reference electrodes with CO,CO_2 mixtures [47].

The electronic part of conduction of ThO_2 electrolytes could be observed increasing with the oxygen partial pressure (oxidation semiconduction) even with pure white mixed oxides (purified from polyvalent cations). During the establishment of the electrode potentials there were signs of solubility of oxygen in the lattice. These facts led to the conclusion that the electronic conduction arose in the anion sublattice and that generally, in mixed oxides with oxide ion vacancies, holes can exist in the form of monovalent negative oxide ions.

For understanding the oxide ion conduction in mixed oxides, there was the problem, already seen by Wagner [22], that the radius of the oxide ions is larger than that of all cations in the crystals. Along with the concentration, it is always the mobility of the charge carriers which determines the conductivity of homogeneous bodies. By space-geometrical considerations it could be shown [40,48] that the fluorite lattice in particular offers better possibilities for the motion of the larger anions than it does for the smaller cations (Figure 2.6). Furthermore, from geometrical calculations it was clear that with decreasing radius of the cations down to a lower limit the possibilities for the motion of cations decrease and those of the anions increase.

The fact that the cations are firmly held in their places in the oxide ion conductors has much importance for the long-term stability of fuel cells. The comprehension of the low cation mobility supported the suggestion of incorporating polyvalent cations in mixed oxides with fluorite structure for obtaining electronic conducting layers and producing stable electrodes at the oxide ion conductors by sintering [49] (aiming at a continuous row of mixed crystals with the electrolyte material, contrary to the recommendation of Schottky [34]). The layers of mixed conductors should ensure ideal conditions for the conduction of oxide ions and electrons and also for the transfer reactions in the electrodes. After these ideas had been presented in the Class of Chemistry of the Academy of Sciences in Berlin in 1958 [50], there were substantial doubts in the discussion; a statistical mixing of all the different cations in a homogeneous solid phase at high temperatures was considered to be very probable. In this and other cases, important questions remained. For example, the cause of the relatively stable cell voltages, which were repeatedly observed

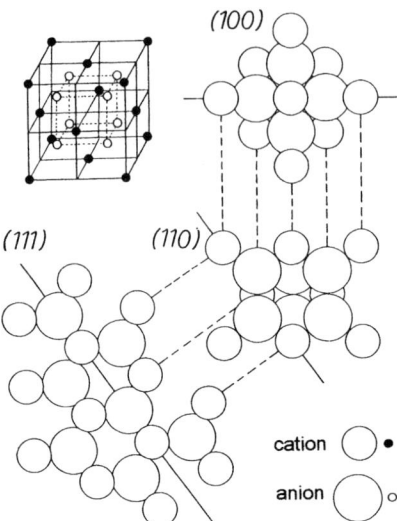

Figure 2.6 Cross-section through the ion ball model of the fluorite lattice (at the lower existence limit) which clearly demonstrate that the anions can, very much more easily than the cations, leave their places (1958).

below 600°C at the oxygen/air cell [40] and which sometimes far exceeded the expected thermodynamic values, remained totally unresolved.

Nevertheless, following promising results on the model fuel cells, the possibilities for SOFC applications in producing power from coal or fuel gases, for the electrolysis of water vapour and carbon dioxide, and for the separation of oxygen from air were considered in 1958 [40,51]. It became apparent that the attainment of high SOFC performance in these applications, discussed already by Schottky [34] and Baur [18], required high gas flows, which resulted in a large temperature difference, around 340 K, between the inlet and exit gases (Figure 2.5b) [51]. Considering the experience with ceramic bodies, such requirements could obviously only be met with tubular cells. In the proposed design of such cells, no series connection was intended. Current was to be taken from tubular cells, 1 m long with powdery electronic conducting material poured into and between the cells. This concept, corresponding to that of Baur (Figure 2.4), was not satisfactory because the conductivity of available electronic conductors was much too low, and the pressure drop across the conducting powder too high.

2.4 Progress in the 1960s

The paper of Kiukkola and Wagner [45] stimulated many activities in various parts of the world in the field of solid-state electrochemistry. In this development, zirconia-based solid electrolytes dominated immediately; e.g. $Zr_{0.85}Ca_{0.15}O_{1.85}$ was used by Weissbart and Ruka in the first device for the measurement of oxygen concentration in a gas phase using a high-temperature galvanic cell [52]. After 1960, a rapidly increasing number of applications for patents and of papers concerning SOFCs appeared in several countries. Some early contributions are described below.

In the USA, during a short period in 1961–62, four companies applied for patents on solid oxide fuel cells, partly with series connection [53–57]. The first publication in English about this subject, 'A Solid Electrolyte Fuel Cell' by Weissbart and Ruka came out in 1962 [58]. From February 1962 to April 1963 in Westinghouse Electric Corporation, a team of 16 people under the leadership of Archer developed solid oxide fuel cell stacks; parallel flat-plate and tubular cells with platinum electrodes were produced and connected in series, using a gold/nickel solder [59]. Flat discs did not lead to satisfactory success; they resulted in bulky stacks and had difficulties with sealing, in contrast to the tubular cells using bell-and-spigot joints (Figure 2.7). The main problem was at

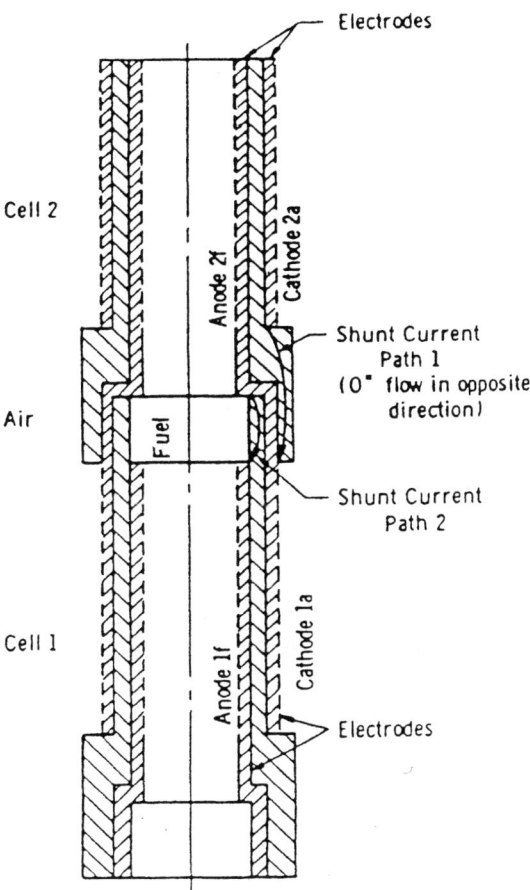

Figure 2.7 Cross-section of two solid electrolyte fuel cells of the bell-and-spigot type with shunt current paths in the seal region developed at Westinghouse (1963)

the platinum anodes, which did not withstand long-term current loading; they tended to peel off from the electrolyte, probably due to the water vapour formed between the electrolyte and the anode layer.

In Europe, in 1958 Palguyev and Volchenkova published conductivity measurements on $3ZrO_2 2CeO_2$ + 10 wt% CaO and other systems [60]. From 1960 onwards, results of a broadly based research programme on cells with solid oxide electrolytes appeared from the Ural branch of the Academy of Sciences of the USSR [61] under the leadership of Karpachov. Tannenberger et al., starting in 1959 at the Battelle Institute in Geneva, presented a thin film fuel cell concept in a 1962 patent, where a porous ceramic support tube was used as a structural member [62]. From the Battelle Institute in Frankfurt, Sandstede gave in September 1962 the first report on the use of hydrocarbons as a fuel in solid oxide cells, applying a converter containing Ni gauze as catalyst upstream of the cells (discs of $Zr_{0.85}Ca_{0.15}O_{1.85}$, diameter 22 mm, with porous Pt layers), and compared measurements with theoretical calculations [63]. At about the same time, fuel cell work was started in France by Kleitz [64], and in Britain, a patent was filed in August 1963 [65] to form fuel cells by depositing layers on a porous metallic carrier.

In Japan, Takahashi, after investigations with alkali carbonate electrolytes, published in 1964 his first results obtained on fuel cells with solid oxide electrolytes [66].

Surveys of these activities were presented at the international fuel cell meetings in 1965, 1967 and 1969 in Brussels. In 1965, results on solid oxide fuel cells were published by General Electric [67], by the Battelle Institute in Geneva [68,69] and by the universities of Grenoble [70], Nagoya [71] and Greifswald [51]. Most developments began with conductivity measurements for optimising the solid electrolytes. Even very expensive rare earths, such as ytterbium oxide, were used [72] to achieve highest conductivities, and ternary systems were investigated to reduce costs (ZrO_2–Y_2O_3–Yb_2O_3 [73], ZrO_2–Y_2O_3–MgO [74]). As a rule, Al_2O_3 was added to achieve gastight, dense sintering products [72–75]. This provoked investigations of the effect of grain boundary conductivity in electrolyte materials [76].

The mobility of the oxide ions in $Zr_{0.85}Ca_{0.15}O_{1.85}$ was determined using the $^{18}O/^{16}O$ isotope exchange between solid and gas phase by Kingery et al. in 1959 [77] and more precisely by Simpson and Carter in 1965 [78]. In 1962, Schmalzried showed by X-ray intensity measurements that the Zr and Ca cations occupy random sites in the cation sublattice of $Zr_{0.85}Ca_{0.15}O_{1.85}$ [79]. In 1963, decrease in conductivity with time was seen as a sign of aging of the oxide ion conductors, caused by disorder–order transitions, in which the random distribution of the cations and oxide ions in the lattice changed to an ordered state [80,81]. Alterations of the composition influenced the effect substantially [82].

Several measurements confirmed the influence of the cation size on the conductivity of mixed oxides with fluorite structure [68,83–85]. These results and also the determination of the ion mobilities in Na_2S, which possesses antifluorite structure and reaches the highest known sodium ion conductivity [86], supported the space-geometrical considerations [48] corresponding to the 'excluded volume model' [87]. Doubts concerning the rapid intermixing of the different cations of oxide electrolytes and oxide electrode layers were resolved by investigations with radionuclides [88–90], which confirmed the low mobility of the cations in mixed oxides with fluorite structure. (In CaO-stabilised zirconia the ratio of the self-diffusion coefficients of the anions to that of the cations is larger than 10^6 even at $1700°C$).

At high oxygen pressures, oxide phases show defect electron (hole) conduction (oxidation semiconduction) and at low oxygen pressures excess electron conduction (reduction semiconduction). The transport number of excess electrons in $Zr_{0.85}Ca_{0.15}O_{1.85}$ as a function of the oxygen partial pressure could be determined by measurements with a Ca,CaO/air cell [79]. The hole conduction of zirconia-based solid electrolytes was noticed for the first time when cells with Ni,NiO reference electrodes for gas potentiometry [44,91] were tested in air. The harmful oxygen permeability was measured potentiometrically in 1965 [92].

Also in 1965, the fundamentals of gas potentiometry were presented, including the range of free oxygen and of oxygen in equilibria, and the 'neutral' transition range [93]. Calculations and measurements in the case of potentiometric titrations of different gases were in good agreement in all three ranges [94]. (The sudden change of the cell voltage of a hydrogen/air cell at the equivalence point when oxygen was fed to the hydrogen had already been shown graphically by Archer et al. [59].) Thus the investigations started for SOFCs led to the development of oxygen sensors (lambda probes) now widely used in automobiles. (A zirconia cell working potentiometrically was first proposed by Loos in 1969 as a sensor for O_2 and CO for the regulation of the air/fuel ratio in cars [95].)

Another less well-known by-product of SOFC development was the electrochemical thermometry; i.e. the determination of elevated temperatures on the thermodynamic scale with CO, CO_2, H_2, H_2O [96] or O_2 concentration cells [97].

The first investigations of polarisation phenomena in solid oxide fuel cells were conducted by the research groups in Sverdlovsk [61,98], Frankfurt [63], Geneva [69], Grenoble [70] and Nagoya [71]. In the detailed investigations of fuel cells with cerium–lanthanum mixed oxides by Takahashi et al. [71] the polarisations observed were much smaller at the anode than at the cathode because by partial reduction of the solid electrolyte, a mixed conductor (solid solution of Ce_2O_3 in CeO_2) was formed at the anode giving a depolarising interlayer. Detailed investigations of the polarisation of solid electrolyte cells by determining the complex admittance were first conducted by Bauerle in 1969 [99].

The high conductivity of cerium-lanthanum mixed oxides and the favourable polarisability of electrodes on such solid electrolytes was already stimulating application ideas in the 1960s. But electronic conductivity of these electrolytes above $600°C$ was seen as a weighty problem [71]. The influence of electronic conductivity on the cell performance was investigated first by means of an equivalent circuit [40,100]. The results, shown in Figure 2.8, led to the conclusion that the ion transport number has to be greater than 0.9 if a solid electrolyte was to be successful in a SOFC [100].

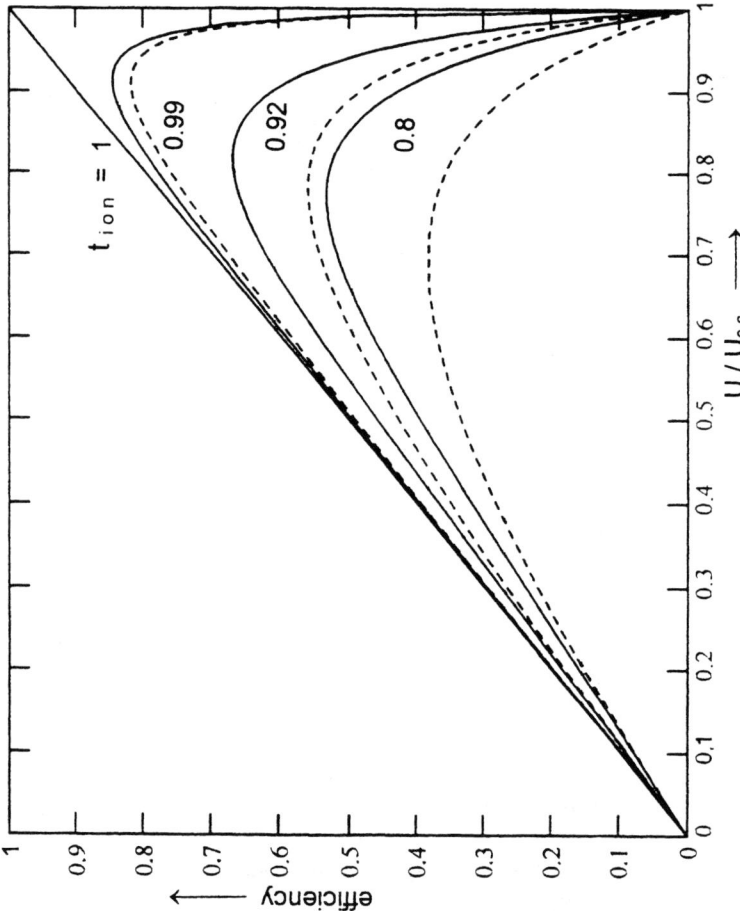

Figure 2.8 Efficiency of fuel cells (fraction of the change in the oxygen Gibbs enthalpy which is turned into useful energy) versus the quotient load voltage/open-circuit voltage: – – –, calculated using an equivalent circuit (1958–1967); ———, calculated using transport equations (1978–1981).

The focal point of work on solid oxide fuel cells during this period was the development of electrode materials. An early problem was poor adhesion of the anode layers, which became obvious in 1963 [59]. Spacil as early as 1964 found the now well-known solution of using layers of nickel closely mixed with solid electrolyte material [101].

It was considerably more difficult to find a suitable cathode material. It turned out to be a misconception to think that small concentrations of polyvalent cations in mixed oxides with fluorite structure produce a high electronic conductivity. Layers with a high concentration of praseodymium at the cathode and high concentration of cerium at the anode had to be realised in order to achieve anything near the desired conductivities [102]. Only mixed oxides with uranium proved to be a good material for stable layers with electronic and ionic conductivity, spreading the electrode reactions across the three phase boundaries of electrode/electrolyte/gas. This idea was confirmed by the result that thin mixed conducting interlayers between the pure electrolyte and the metallic conductor considerably reduced polarisation phenomena [103] and led to high current densities [104]. An optimised material with polyvalent uranium ions in mixed oxides with fluorite structure suitable for sintering on solid oxide electrolytes without phase boundaries was developed by Tannenberger in 1967 [105, 106]; it proved to be a favourable interlayer in cathodes and anodes [107, 108].

Indium oxide with different additives was proposed as a cathode material in 1966 [109] and frequently used (e.g. [110, 107, 108]). However, electronically conducting perovskites soon began to dominate the developments for both cathode and interconnect. The use of $La_{1-x}Sr_xCoO_3$ for the air electrode of solid oxide fuel cells marked the beginning [111], followed in 1967 by recommendations of $PrCoO_3$ [112] and of mixtures of the oxides of Pr, Cr, Ni and Co [113]. Strontium-doped lanthanum chromite, even now the most important ceramic interconnection material, was proposed by Meadowcroft in 1969 [114]. For cathodes, the situation in 1969 was summarised [115] as: 'It is apparent that a fully satisfactory air electrode for high temperature zirconia electrolyte fuel cells is still lacking.'

The SOFC stacks developed in 1963 were not safe enough for power generation aboard spacecrafts. SOFCs for electrolysis of the atmosphere in manned spacecrafts (recovery of oxygen from CO_2 and H_2O in stacks of bell-and-spigot cells, carbon deposition, and hydrogen separation [116]) were also investigated. For this application, electrolyte discs (6.3 cm diameter, 1.4–1.6 mm thick) and for the first time mixed oxides of Zr and Sc [117] were used. Flat-plate designs of SOFCs had been proposed earlier (Figure 2.9 [118, 119]).

Terrestrial applications aimed at an economic SOFC system for the production of electrical power from coal and air at an overall efficiency of 60% or greater. With a conceptual 100 kW coal-burning fuel cell power system (Figure 2.10 [120]), coal could be gasified using the heat and combustion products emerging from the fuel cell stacks.

In 1968 General Electric took up the idea of electrochemical dissociation of water vapour in solid oxide cells [121–123]. They hoped to produce cheap,

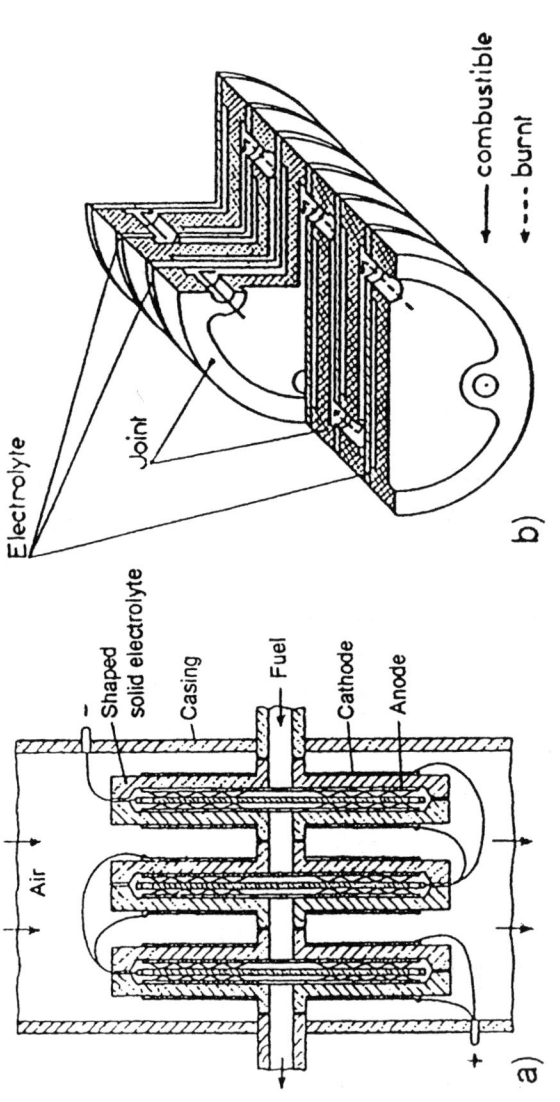

Figure 2.9 Solid oxide fuel cell concepts with flat-plate designs. (a) Shaped discs in a stack proposal of Brown, Boveri by Baukal (1965). (b) Flat discs between shaped junctions in a stack proposal from the University of Grenoble by Deportes et al. (1967).

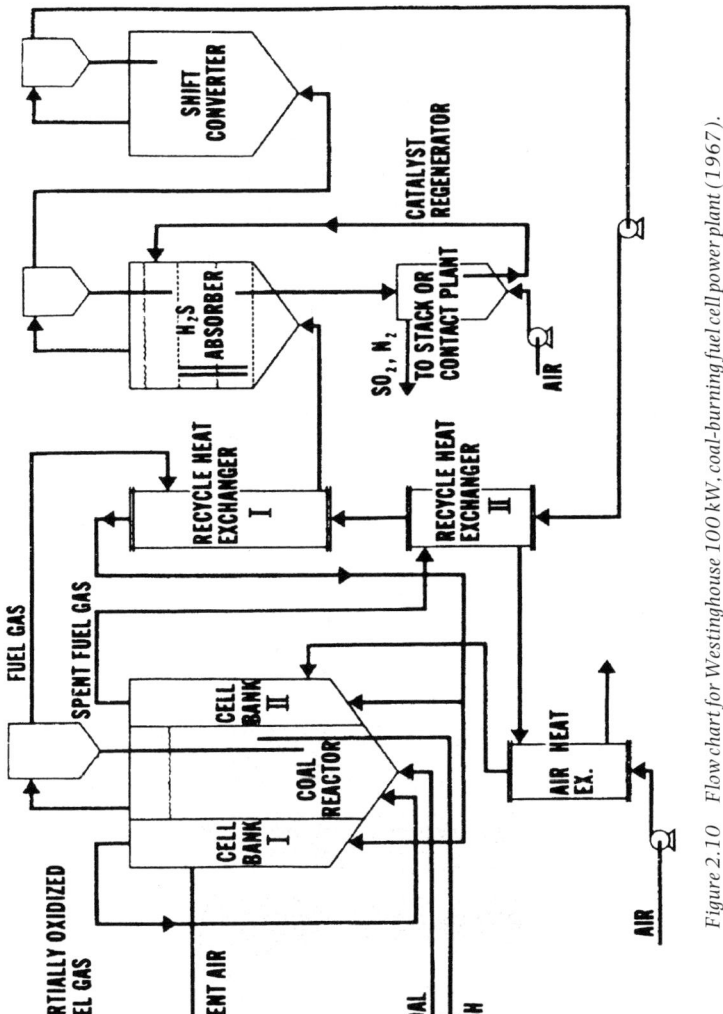

Figure 2.10 Flow chart for Westinghouse 100 kW, coal-burning fuel cell power plant (1967).

pure hydrogen by using anodes, depolarised by another reducing cheap gas and described the idea of gas production by internally short-circuited solid oxide cells [124].

Activities in associated SOFC technologies helped in testing new materials, designs and fabrication methods. An example was plasma spraying of complete cells on an aluminium mandrel which was removed by leaching in a KOH solution before high-temperature sintering [123]. For applications in fuel cells, electrolysers, gas separators and chemical reactors, many proposals for series connection of tubular solid oxide cells were also made at this time (Figure 2.11 [125]).

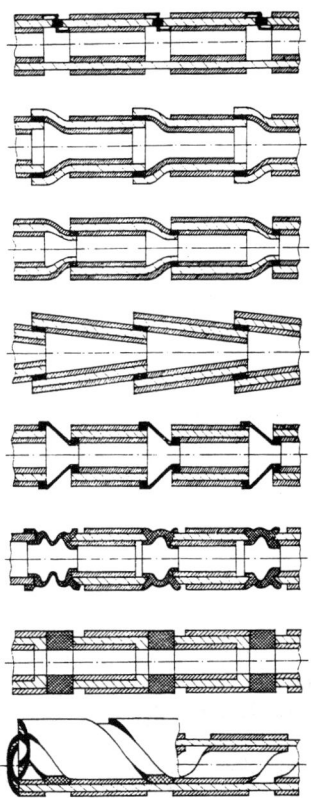

Figure 2.11   Proposals for connecting self-supporting tubular solid electrolyte cells in series (1974).

2.5 On the Path to Practical Solid Oxide Fuel Cells

By 1970 the results of investigations on properties and applications of solid oxide electrolytes were so numerous as to make them very difficult to survey. A comprehensive review by Etsell and Flengas included 674 references [126]. But commercial activity was confined to potentiometric oxygen sensors . In the USA, the programme for development of SOFCs ran out in 1970. The greatest obstacles to construction of a working SOFC were the cathode material and the electronically conducting interconnection material, together with the problems of suitable fabrication techniques for producing gastight thin films of electrolyte and interconnection, especially in their overlap regions [127].

These issues were intensively studied in the research laboratories of Brown Boveri, where under the leadership of Rohr from 1964, solid electrolyte fuel cells and oxygen sensors were investigated. Between 1969 and 1973 more than 100 oxide substances were synthesised and tested as electrode materials for SOFCs (Figure 2.12 [128, 129]). $LaNiO_3$ doped with Bi_2O_3 and $LaMnO_3$ doped with SrO proved to be particularly suitable, and from 1973 onwards, $La_{0.84}Sr_{0.16}MnO_3$ was used exclusively. With that, today's most commonly used cathode material was found. Special investigations were also devoted to the interconnection material [130]. Despite good results (successful tests of single cells at 1000°C over a period of more than 3 years; construction and tests of modules with 25 series-connected cells), the development was not continued after 1975 because the cost for the necessary manufacturing processes did not seem to be sufficiently economic [131].

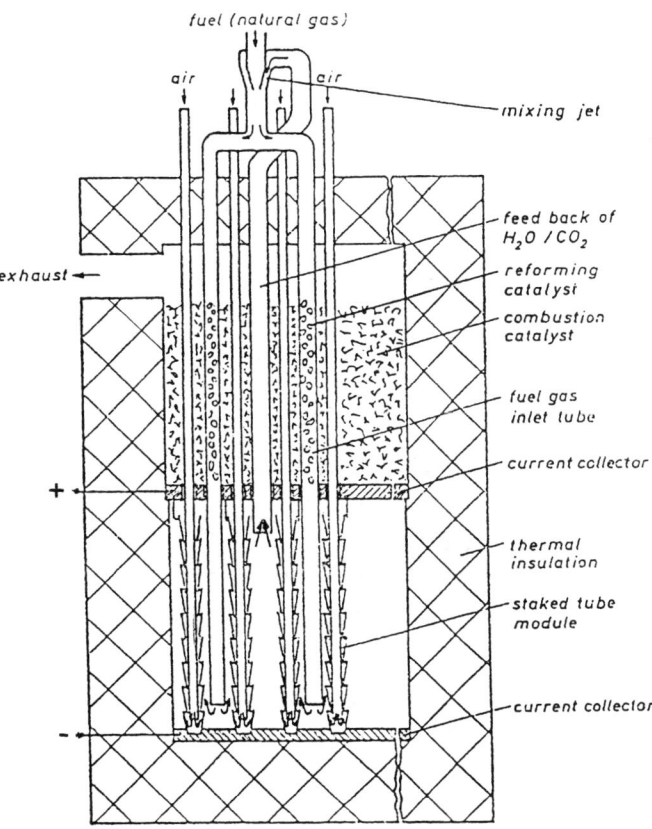

Figure 2.12 High-temperature fuel cell system for the conversion of methane by Brown, Boveri (1972).

A significant step forward was made by Isenberg in 1970 with the development of the electrochemical vapour deposition (EVD) method [127]. Then in 1978, Westinghouse started a new programme, in which EVD method was used for the perfect closing of the pores in the electrolyte and the interconnecting layers. A new cell design was instrumental in the breakthrough; long tubular cells (inside air, outside fuel), electrically interconnected by oxide materials and ductile metallic conductors, were combined together in tube bundles (Figure 2.13 [132]). This design led to the first 5 kW SOFC generator containing 324 cells (in 1986) [133] and to the 1152-cell 100 kW SOFC power system which began operation near Arnhem in the Netherlands in January 1998 [134].

Tubular cells are more stable against mechanical and thermal stresses than planar cells. But modern technologies (tape casting, screen printing, vapour deposition, plasma spraying, wet spraying and others) promise lower cost for the fabrication of planar cells. Therefore in the 1980s and 1990s, an increasing number of SOFC developments focused on planar designs [135]. In 1983 co-fired monolithic stacks of flat cells were fabricated and investigated at the Argonne National Laboratory [136]. Soon many possibilities were seen for the fabrication and arrangement of planar cells (Figure 2.14). In 2000, a 25 kW system with 3840 planar electrolyte-supported cells (11×9 cm^2) and with internal reforming anodes was fabricated for operation on natural gas by Ceramic Fuel Cells Ltd in Australia [137]. Many current developments are concentrated on anode-supported planar SOFCs.

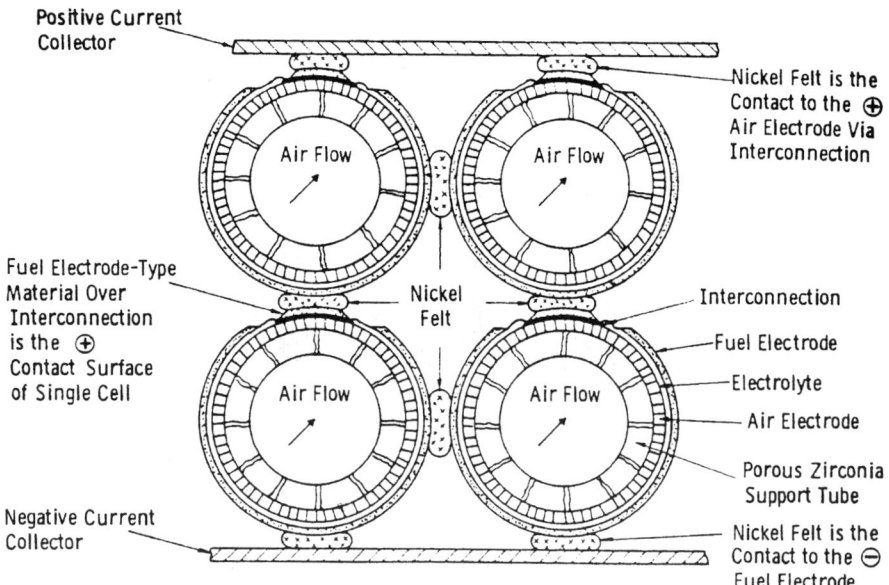

Figure 2.13 Cross-sections of Westinghouse multi-cell module concept, showing the components of the cells as well as series and parallel connection (1982).

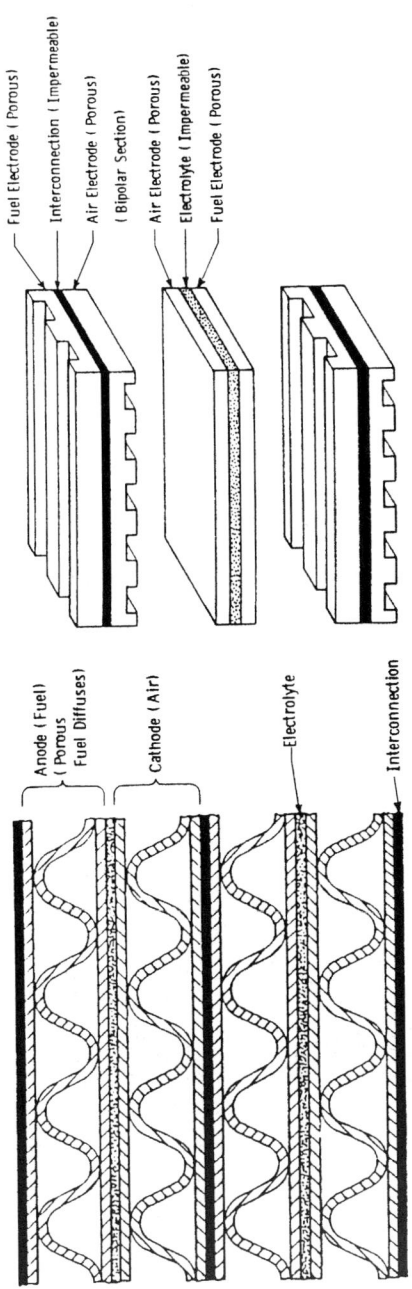

Figure 2.14 Different cross-flow designs for SOFC planar configuration containing alternate fuel and air layers (1986 and 1989)

All SOFC concepts would benefit if the cell stacks could be operated at intermediate temperatures of around 700–850°C, especially by using better oxide conductors. At first the highly conductive mixed oxides with cerium were totally excluded because of their electronic conductivity and mechanical instability between oxidising and reducing gases [40]. After the results of Takahashi in the 1960s, progress was made by theoretical treatment of the transport processes of ions and electrons in mixed conductors with the result that the electronic current decreases faster with decreasing load voltage than predicted by the equivalent circuit, and doped ceria remains a viable solid electrolyte, particularly for lower operation temperatures [138, 139]. Coatings of doped zirconia on the surface of doped ceria and dispersions of ceria and zirconia electrolyte particles were investigated. Recently, 500°C has been proposed as the optimum operation temperature of cells with $Ce_{0.9}Gd_{0.1}O_{1.95}$ (25 μm) electrolyte [140, 141].

In 1994, new solid electrolytes with high conductivity at low temperatures were found in the form of doped lanthanum gallates [142]. Since the self-diffusion coefficients of cations are apparently fundamentally larger in the perovskite-type oxides than in fluorite oxides, lanthanum gallate electrolyte and associated electrode materials tend to react too easily at the temperatures of fabrication and operation of cells.

At low operation temperatures, polarisation losses and the importance of catalysis of the electrode reactions increase. At the cathode, mixed potentials can arise when traces of combustible substances determine the electrode potential in competition with oxygen, an effect, mentioned near the end of Section 2.3, whose cause was recognised by Hartung in 1981 [143], today the basis of the development of hydrocarbon sensors. For the anodes growing interest is directed to materials which accelerate the electrochemical oxidation of CO and hydrocarbons and is stable against fuel impurities.

Many new cell and system ideas are currently being pursued. These include internal or *in situ* reforming of natural gas at the SOFC anode (1991); the HEXIS concept with stacks of circular cells arranged between plates of chromium alloys and without tight seals; a combination of electricity production, heat exchange and afterburning (1991); microtubular cells with high thermal shock resistance suitable for rapid start-up (1994); and high-efficiency hybrid SOFC/gas turbine power plants (1995) with SOFC operating under pressure. Many of these ideas are discussed elsewhere in this book.

References

[1] H. Davy, Elektrochemische Untersuchungen, Vorlesung am 19.11.1807, Verfahrensarten, um die feuerbeständigen Alkalien zu zersetzen, in *Ostwalds Klassiker der exakten Naturwissenschaften*, Nr. 45 (ed. W. Ostwald), Leipzig, 1893, pp. 52–55.

[2] M. Faraday, Experimental-Untersuchungen über Elektrizität (1833), IV. Reihe. Über ein neues Gesetz der Elektrizitätsleitung, in *Ostwalds Klassiker der exakten Naturwissenschaften*, Nr. 86 (ed. A. J. von Oettingen), Leipzig, 1897, pp. 39–55 and Note 9 of the Editor.

[3] M. Faraday, Experimental-Untersuchungen über Elektrizität (1834), VII. Reihe. Von der elektrochemischen Zersetzung, in *Ostwalds Klassiker der exakten Naturwissenschaften*, Nr. 87 (ed. A. J. von Oettingen), Leipzig, 1897, pp. 39–106.

[4] W. Hittorf, *Ann. Physik u. Chem.*, **84** (1851) 1–28.

[5] H. Buff, *Ann. Chem. Pharm.*, **96** (1855) 257–286.

[6] H. Cavendish, *Franklin Experiments and Observations on Electricity*, 5th edn., F. Newbery, London, 1774, p. 411; cited from [13].

[7] W. Beetz, *Ann. Physik u. Chem.*, **92** (1854) 452–466.

[8] H. Buff, *Ann. Chem. Pharm.*, **90** (1854) 257–283.

[9] J.-M. Gaugain, *C. R. Séances Acad. Sci.*, **37** (1853) 82–84.

[10] A. C. Becquerel, *C. R. Séances Acad. Sci.*, **38** (1854) 905–910.

[11] H. Buff, *Ann. Chem. Pharm.*, **80** (1851) 1–16.

[12] J.-M. Gaugain, *C. R. Séances Acad. Sci.*, **37** (1853) 584–588. Translation in English see H.-H. Möbius, On the history of solid electrolyte fuel cells. *J. Solid State Electrochem.*, **1** (1997) Appendix 12–13.

[13] G. Wiedemann, *Die Lehre von der Elektrizität*. F. Vieweg, Braunschweig, 2nd edn. 1893–98, vol. I, pp. 553–561, 815–819; vol. II, p. 491–493.

[14] W. Ostwald, *Lehrbuch der Allgemeinen Chemie*, Engelmann, Leipzig, 1st edn., 1885–87; 5th edn., 1917.

[15] W. Nernst, Verfahren zur Erzeugung von elektrischem Glühlicht. DRP 104 872, filed July 7, 1897.

[16] W. Nernst, *Z. Elektrochem.*, **6** (1899) 41–43.

[17] W. Nernst, Material for electric-lamp glowers. US Patent 685 730, filed August 24, 1899.

[18] E. Baur and H. Preis, *Z. Elektrochem.*, **43** (1937) 727–732.

[19] H.-H. Möbius, *Naturwissenschaften*, **52** (1965) 529–536.

[20] H. Reynolds, *Über die Leitfähigkeit fester Mischungen bei hohen Temperaturen. Ein spezieller Fall der festen Lösungen*. Thesis, Univ. Göttingen, 1902.

[21] H.-G. Bartel, G. Scholz and F. Scholz, *Z. Chem. (Leipzig)*, **23** (1983) 277–287.

[22] C. Wagner, *Naturwissenschaften*, **31** (1943) 265–268.

[23] H. von Helmholtz, Die Thermodynamik chemischer Vorgänge I, II, III. Sitzungsber. Akad. Wiss. Berlin 2.2.1882, 27.7.1882, 31.5.1883, in *Wiss. Abhandlungen*, Ambrosius Barth, Leipzig, vol. 2 (1883), pp. 958–978, 979–992; vol. 3 (1895), pp. 92–114.

[24] W. Ostwald, *Z. Elektrochem.*, **1** (1894) 81–84, 122–125; *Z. Physik. Chem.*, **15** (1894) 409–421.

[25] F. Haber and A. Moser, *Z. Elektrochem.*, **11** (1905) 593–609.

[26] F. Haber and F. Fleischmann, *Mitt. Z. Anorg. Chem.*, **51** (1906) 245–288.
[27] F. Haber and G. W. A. Foster, *Mitt. Z. Anorg. Chem.*, **51** (1906) 289–314.
[28] F. Haber, Verfahren zur Erzeugung von elektrischer Energie aus Kohle und gasförmigen Brennstoffen, Austrian Patent 27 743, filed August 5, 1905.
[29] W. D. Treadwell, *Z. Elektrochem.*, **22** (1916) 414–421.
[30] E. Baur and W. D. Treadwell, Brennstoffelement, DRP 325 783, filed September 20, 1916.
[31] E. Baur and R. Brunner, *Z. Elektrochem.*, **43** (1937) 725–727
[32] E. Baur and H. Preis, *Z. Elektrochem.*, **44** (1938) 695–698.
[33] E. Baur, *Brennstoff-Chemie*, **20** (1939) 385–387; *Ber. Ges. Kohlentechn.*, **5** (1940) 99–110.
[34] W. Schottky, *Wiss. Veröff. Siemens-Werke* **14** (1935) no. 2, 1–19 and Foreword.
[35] Siemens and Halske AG, W. Schottky named as inventor, Zur Stromlieferung geeignetes galvanisches Element bzw. Elementkette, DRP 650 224, filed April 6, 1935.
[36] H.-H. Möbius, H. Witzmann and D. Harzer, *Z. Chem. (Leipzig)*, **3** (1963) 157–158.
[37] F. Hund, *Z. Elektrochem. Angew. Physik. Chem.*, **55** (1951) 363–366.
[38] H.-J. Engell, Karl Hauffe zum 65. Geburtstag. *Ber. Bunsenges.Phys. Chem.*, **82** (1978) 351–352.
[39] H. Peters, *Untersuchungen der Gitterstruktur und der elektrischen Leitfähigkeit an Mischoxyden vom Fluorittyp*. Thesis, Univ. Rostock, 1953.
[40] H.-H. Möbius, *Theoretische und experimentelle Untersuchungen an Brennstoffelementen mit festen Elektrolyten*. Thesis, Univ. Rostock, 1958, *Mitteilungsbl. Chem. Ges. DDR* **6** (1959) 45–48.
[41] K. Biefeld, Brennstoffbatterie. DE-P 912 106, filed February 27, 1944.
[42] E. Ryschkewitsch, *Oxydkeramik der Einstoffsysteme vom Standpunkt der physikalischen Chemie*, Julius Springer, Berlin, 1948.
[43] H. Peters and H.-H. Möbius, *Z. Physik. Chem. (Leipzig)*, **209** (1958) 298–309.
[44] H. Peters and H.-H. Möbius, Verfahren zur Gasanalyse bei erhöhten Temperaturen mit Hilfe galvanischer Festelektrolytelemente, DDR-P 21673, filed May 20, 1958.
[45] K. Kiukkola and C. Wagner, *J. Electrochem. Soc.*, **104** (1957) 379–387.
[46] C. Wagner, private communication to H. Peters (1958).
[47] H. Peters and G. Mann, *Z. Elektrochem., Ber. Bunsenges. Physik. Chem.*, **63** (1959) 244–248.
[48] H.-H. Möbius, *Z. Chem. (Leipzig)*, **2** (1962) 100–106.
[49] H.-H. Möbius, Hochtemperaturelektroden für galvanische Festelektrolytzellen, insbesondere für Brennstoffelemente, DDR-P 22030, filed May 20, 1958.
[50] H.-H. Möbius, *Monatsber. dt. Akad. Wiss. Berlin*, **1** (1959) 34–36.

[51] H.-H. Möbius and B. Rohland, *Rev. Energ. Primaire. Journées Int. d'Etude Piles à Combustible, Bruxelles*, 1965, vol. 3, pp. 27–34.

[52] J. Weissbart and R. Ruka, *Rev. Sci. Instr.*, **32** (1961) 593–595.

[53] W. E. Tragert, Fuel cell. US Patent 3,138,487, filed February 2, 1961; US Patent 3,138,488, filed March 8, 1961; US Patent 3,296,030, filed November 1, 1962.

[54] W. E. Tragert, R. L. Fullman and R. E. Carter, Fuel cell, US Patent 3,138,490, filed February 28, 1961.

[55] W. Oser, Fuel cell, US Patent 3,281,273, filed May 5, 1961.

[56] R. J. Ruka and J. Weissbart, Electrochemical method for separating O_2 from a gas; generating electricity; measuring O_2 partial pressure; and fuel cell, US Patent 3,400,054, filed July 24, 1961.

[57] D. T. Bray, L. D. Lagrange, U. Merten and C. D. Park, Fuel cell having zirconia-containing electrolyte and ceramic electrodes. US Patent 3,300,344, filed June 27, 1962.

[58] J. Weissbart and R. J. Ruka, *J. Electrochem. Soc.*, **109** (1962) 723–726.

[59] D. H. Archer, E. F. Sverdrup, W. A. English and W. G. Carlson, An investigation of solid-electrolyte fuel cells. *Tech. Doc. Rept. ASD-TDR-63-448*, July 1963.

[60] S. F. Palguyev and Z. S. Volchenkova, *Tr. Inst. Khim. Akad. Nauk SSSR, Ural Filial*, **2** (1958) 183–200; C. A. 54 (1960) 9542 i.

[61] S. V. Karpachov, S. F. Palguyev, W. N. Chebotin, A. D. Neuimin, A. T. Filyayev, M. V. Perfilyev *et al. Tr. Inst. Elektrokhim. Akad. Nauk SSSR, Ural Filial*, **1** (1960 foll.).

[62] H. Tannenberger, H. Schachner and W. Simm, Festelektrolyt-brennstoffelement, DE-P 1 471 768, filed May 22, 1963; Swiss priority May 23, 1962.

[63] H. Binder, A. Köhling, H. Krupp, K. Richter and G. Sandstede, *Electrochim. Acta*, **8** (1963) 781–793.

[64] J. Besson, C. Deportes and M. Kleitz, 1er Colloque sur les piles à combustible. Bellevue, 6 déc 1962. Contrat de recherches 61 FR 136. 'Étude des électrolytes solides pour piles à combustible à haute température' and 'Utilisation des électrolytes dans les piles à combustible à haute température', in *Les Piles à Combustible*, Éditions Technip, Paris, 1965, pp. 87–102, 303–323.

[65] K. R. Williams and J. G. Smith, Fuel cell, GB Patent 1 049 428, filed August 15, 1963.

[66] T. Takahashi, *J. Electrochem. Soc.*, **34** (1966) 60–69.

[67] D. W. White, A zirconia electrolyte fuel cell. *Rev. Energ. Primaire. Journées Int. d'Etude Piles à Combustible, Bruxelles*, 1965, vol. 3, pp. 10–18.

[68] H. Tannenberger, H. Schachner and P. Kovacs, *Rev. Energ. Primaire. Journées Int. d'Etude Piles à Combustible, Bruxelles*, 1965, vol. 3, pp. 19–26.

[69] H. Schachner and H. Tannenberger, *Rev. Energ. Primaire. Journées Int. d'Etude Piles à Combustible, Bruxelles*, 1965, vol. 3, pp. 49–55.

[70] M. Kleitz, J. Besson and C. Deportes, *Rev. Energ. Primaire. Journées Int. d'Etude Piles à Combustible, Bruxelles*, 1965, vol. 3, pp. 35–41.

[71] T. Takahashi, K. Ito and M. Iwahara, *Rev. Energ. Primaire. Journées Int. d'Etude Piles à Combustible, Bruxelles*, 1965, vol. 3, pp. 42–48.

[72] H. Tannenberger, Electrolyte solide pour piles à combustible. Swiss Patent 400 264, filed November 23, 1962.

[73] F. J. Rohr, Festelektrolyt für Brennstoffzellen. DE-P 1 671 704, filed March 18, 1967.

[74] B. Rohland and H.-H. Möbius, *Abh. Sächs. Akad. Wiss. (Leipzig) Math.-nat. Kl.*, **49** (1988) 355–366.

[75] U. Neumeier and H. Tannenberger, Fester Elektrolyt für Brennstoffelemente. Swiss Patent 400 263, filed November 23, 1962. DE-P 1 471 770, filed November 22, 1963.

[76] T. Y. Tien, *J. Appl. Phys.*, **35** (1964) 122–124.

[77] W. D. Kingery, J. Pappis, M. E. Doty and D. C. Hill, *J. Amer. Ceram. Soc.*, **42** (1959) 393–398.

[78] L. A. Simpson and R. E. Carter, *J. Amer. Ceram. Soc.*, **49** (1966) 139–144.

[79] H. Schmalzried, *Z. Elektrochem. Ber. Bunsenges. Physik. Chem.*, **66** (1962) 572–576.

[80] T. Y. Tien and E. C. Subbarao, *J. Chem. Phys.*, **39** (1963) 1041–1047.

[81] E. C. Subbarao and P. H. Sutter, *J. Phys. Chem. Solids*, **25** (1964) 148–150.

[82] T. Takahashi and Y. Suzuki, *Rev. Energ. Primaire. Journées Int. d'Etude Piles à Combustible, Bruxelles*, 1967, p. 378.

[83] J. M. Dixon, L. D. LaGrange, U. Merten, C. F. Miller and J. T. Porter, *J. Electrochem. Soc.*, **110** (1963) 276–280.

[84] S. F. Palguev, A. D. Neuimin and V. N. Strekalovskij, *Tr. Inst. Elektrokhim. Akad. Nauk. SSSR, Ural Filial*, **9** (1966) 149–157.

[85] V. N. Chebotin and M. V. Perfiliev, *Elektrokhimiya tverdych elektrolytov*. Moskva izdatelstvo 'Chimiya', 1978, Chapter 4.1, pp. 125–129.

[86] H.-H. Möbius, H. Witzmann and R. Hartung, *Z. Physik. Chem. (Leipzig)*, **227** (1964) 40–55.

[87] J. B. Boyce and B. A. Huberman, *Physics Reports*, **51** (1979) 189–265 (p. 201 foll. and 211 foll.).

[88] H.-H. Möbius and J. Müller, *Z. Chem. (Leipzig)*, **1** (1961) 377.

[89] H.-H. Möbius, H. Witzmann and D. Gerlach, *Z. Chem. (Leipzig)*, **4** (1964) 154–155.

[90] W. H. Rhodes and R. E. Carter, *J. Amer. Ceram. Soc.*, **49** (1966) 244–249.

[91] H.-H. Möbius, S. Lang and K. Wilms, Verfahren zur Herstellung galvanischer Festelektrolytzellen für gaspotentiometrische Zwecke. DDR-P 43242, filed May 22, 1964.

[92] H.-H. Möbius and R. Hartung, *Silikattechnik (Berlin)*, **16** (1965) 276–280.

[93] H.-H. Möbius, *Z. Physik. Chem. (Leipzig)*, **230** (1965) 396–412.

[94] H.-H. Möbius, Z. Physik. Chem. (Leipzig), **231** (1966) 209–214.
[95] C. H. Loos, Vorrichtung zur Regelung des Luft-Brennstoff-Verhältnisses in einem Verbrennungsmotor. DE-P 2 010 793, filed March 6, 1970, priority NL March 22, 1969.
[96] H.-H. Möbius, Z. Chem. (Leipzig), **1** (1961) 63 and [19].
[97] W. T. Lindsay and R. J. Ruka, Electrochim. Acta, **13** (1968) 1867–1874.
[98] S. V. Karpachov, A. T. Filyayev and S. F. Palguyev, Electrochim. Acta, **9** (1964) 1681–85.
[99] J. E. Bauerle, J. Phys. Chem. Solids, **30** (1969) 2657–2670.
[100] T. Takahashi, K. Ito and H. Iwahara, Electrochim. Acta, **12** (1967) 21–30.
[101] H. S. Spacil, Electrical device including nickel-containing stabilized zirconia electrode, US Patent 3,503,809, filed October 30, 1964.
[102] H.-H. Möbius and B. Rohland, Method of producing fuel cells with solid electrolytes and ceramic oxide electrode layers. US Patent 3,377,203, filed November 18, 1964, DDR-P 46 300, filed June 27, 963.
[103] H.-H. Möbius and B. Rohland, Z. Chem. (Leipzig), **6** (1966) 158–159.
[104] B. Rohland and H.-H. Möbius, Naturwissenschaften, **55** (1968) 227–228.
[105] H. Tannenberger, Electrode pour pile à électrolyte solide. Swiss Patent, filed July 19, 1967; F-P 1.572.073, filed July 18, 1968.
[106] P. Shuk, P. Schmidt, R. Ruhle, S. Jakobs and H.-H. Möbius, Z. Chem., **24** (1984) 271–272.
[107] P. van den Berghe and H. Tannenberger, Festelektrolyt mit Elektrode. Swiss Patent, filed June 18, 1971; DE-P 22 28 770, filed June 13, 1972.
[108] H. Tannenberger and P. van den Berghe, Festelektrolyt mit Elektrode. Swiss Patent, filed November 19, 1974; DE-P 25 51 936, filed November 19, 1975.
[109] A. Isenberg, W. Pabst and G. Sandstede, Oxydisches Kathodenmaterial für galvanische Brennstoffzellen für hohe Temperaturen. DE-P 1 571 991, filed October 22, 1966.
[110] E. F. Sverdrup, A. D. Glasser and D. H. Archer, Fuel cell comprising a stabilized zirconium oxide electrolyte and a doped indium or tin oxide cathode. US Patent 3,558,360, filed January 8, 1968.
[111] D. D. Button and D. H. Archer, Development of $La_{1-x}Sr_xCoO_3$ air electrodes for solid electrolyte fuel cells. Amer. Ceram. Soc. Washington, Meeting, May 1966.
[112] S. P. Mitoff, High temperature electrical conductor comprising praseodymium cobaltate. US Patent 3,533,849, filed June 12, 1967.
[113] W. Pabst, G. Sandstede and G. Walter, Sauerstoffelektrode für galvanische Zellen, insbesondere Kathoden in Brennstoffzellen. DE-P 1 671 721, filed December 23, 1967.
[114] D. B. Meadowcroft, Brit. J. Appl. Phys. 2 (J. Phys. D), **2** (1969) 1225–1233.
[115] C. S. Tedmon Jr., H. S. Spacil and S. P. Mitoff, J. Electrochem. Soc., **116** (1969) 1170–1175.

[116] L. Elikan, J. P. Morris, C. K. Wu and C. G. Saunders, 180-day life test of solid electrolyte system for oxygen regeneration. ASME paper 71-Av-32 (1971) 1–12.

[117] J. Weissbart, W. H. Smart and T. Wydeven, Design and performance of a solid electrolyte oxygen generator test module. ASME paper 71-Av-8 (1971) 1–7.

[118] W. Baukal, Brennstoffbatterie, Swiss Patent 444 243, filed April 8, 1965.

[119] C. Deportes, J. Besson and G. Vitter, *Rev. Energ. Primaire. Journées Int. d'Etude des Piles à Combustible, Bruxelles*, 1967.

[120] D. H. Archer and R. L. Zahradnik, The design of a 100 kilowatt, coal burning, fuel cell power system. *Chem. Eng. Progr. Symp. Series* 63, No. 75 (1967) 55–62.

[121] General Electric, *Chem. Ing. News*, **47** (November 4,1968) 48–49.

[122] H. S. Spacil, Solid oxygen-ion electrolyte cell for the dissociation of steam. US Patent 3,635,812, filed July 5, 1968.

[123] H. S. Spacil and C. S. Tedmon, *J. Electrochem. Soc.*, **116** (1969) 1618–1633.

[124] H. S. Spacil and D. W. White, Internally short-circuited solid oxygen-ion electrolyte cell, US Patent 3,630,979, filed January 2, 1969.

[125] H.-H. Möbius, *Chem. Ges. DDR*, **21** (1974) 177–182.

[126] T. H. Etsell and S. N. Flengas, *Chem. Rev.*, **70** (1970) 339–376.

[127] W. Feduska and A. O. Isenberg, *J. Power Sources*, **10** (1983) 89–102.

[128] W. Fischer, H. Kleinschmager, F. J. Rohr, R. Steiner and H. H. Eysel, *Chem. Ing. Tech.*, **44** (1972) 726–732.

[129] F. J. Rohr, Entwicklung des Prototyps einer Hochtemperatur-Brennstoffzellen-Batterie, Forschungsabschlußbericht 1975, BMFT-FB-T 77-17 (1977), p. 265.

[130] W. Baukal, W. Kuhn, H. Kleinschmager and F. J. Rohr, *J. Power Sources*, **1** (1976) 91–97, 203–213.

[131] F. J. Rohr, in *Solid Electrolytes* (eds. P. Hagenmuller and W. van Gool), Academic Press, New York, 1978, pp. 431–450.

[132] A. O. Isenberg, Recent advancements in solid electrolyte fuel cell technology. in *1982 National Fuel Cell Seminar Abstracts*, Courtesy Associates, Washington, DC, (1982) 154–156.

[133] P. Reichner and J. M. Makiel, Development status of multi-cell solid oxide fuel cell generators, in *1986 National Fuel Cell Seminar Abstracts*, Courtesy Associates, Washington, DC, (1986) 32–35.

[134] J. Sukkel, in *Proceedings of the 4th European SOFC Forum*, ed. A. J. McEvoy, Switzerland, 2000, pp. 159–166.

[135] J. T. Brown, Solid oxide fuel cells in: *High Conductivity Solid Ionic Conductors*, (ed. T. Takahashi), World Scientific, Singapore, 1989, pp. 630–663.

[136] D. C. Fee *et al.*, Monolithic fuel cell development, in *1986 Fuel Cell Seminar Abstracts*, Courtesy Associates, Washington, DC, (1986) 40–47, 52–75.

[137] K. Foger and B. Godfrey, in *Proceedings of the 4th European SOFC Forum*, ed. A. J. McEvoy, Switzerland, 2000, pp. 167–173.
[138] D. S. Tannhauser, *J. Electrochem. Soc.*, **125** (1978) 1277–1282.
[139] I. Riess, *J. Electrochem. Soc.*, **128** (1981) 2077–2081.
[140] B. C. H. Steele and J. M. Floyd, *Proc. Brit. Ceram. Soc.*, **19** (1971) 55–76.
[141] B. C. H. Steele, *Solid State Ionics*, **129** (2000) 95–110.
[142] T. Ishihara, H. Matsuda and Y. Takita, *J. Amer. Chem. Soc.*, **116** (1994) 3801.
[143] R. Hartung and R. Maass, *Z. Chem. (Leipzig)*, **21** (1981) 337–338.

Chapter 3

Thermodynamics

Wolfgang Winkler

3.1 Introduction

A solid oxide fuel cell (SOFC) is an electrochemical device that converts chemical energy of a fuel and an oxidant gas (air) directly into electricity without irreversible oxidation. It can be treated thermodynamically in terms of the free enthalpy of the reaction of the fuel with oxidant. Hydrogen and oxygen are used to illustrate the simplest case in the early part (Section 3.2) of this chapter. This treatment allows the calculation of the reversible work at equilibrium for the reversible reaction. Heat must also be transferred reversibly to the surrounding environment in this instance.

During operation of a SOFC, described in Section 3.3, two effects intervene to reduce the electrical power available from an ideal cell; the first is ohmic resistance which generates heat; the second is the irreversible mixing of gases which causes the voltage to fall as progressively more fuel is used in the reaction. Essentially, this means that a SOFC cannot realistically use all the fuel. Some fuel, typically about 10%, must be left in the spent fuel stream which exits from the cell.

The losses in SOFC appear as heat, so it is necessary to consider a SOFC system as a heat generator as well as an electricity source. In effect, the whole SOFC system can be treated as a power generating burner, as in Section 3.4.

In a real engineering device, heat is exchanged within the SOFC in several ways including fuel processing, air preheating, flue gas cooling, etc. Excess air is normally required to prevent overheating, while the conversion of hydrocarbons into hydrogen and carbon monoxide often absorbs heat. The complex heat pathways are described in Section 3.5.

Ultimately, the heat output from the SOFC can be used to drive a heat engine such as a piston engine or gas turbine. These combined SOFC/heat engine cycles are analysed in Section 3.6.

First it is essential to list the symbols and concepts used in this chapter.

List of terms

Symbols

C_P	temperature-dependent heat capacity
e	elementary charge
e_i	specific exergy of the component i
F	Faraday constant
$\Delta^r G$	Gibbs enthalpy or free enthalpy of the reaction
H	enthalpy
$\dot{H}$	enthalpy flow
$\Delta^r H$	reaction enthalpy
h^*	specific enthalpy related to the standard state
I	electrical current
K	equilibrium constant
LHV	lower heating value
$\dot{m}$	mass flow
N_A	Avogadro constant
$\dot{n}^*$	constant molar flow at the fuel cells anode
n^{el}	quantity of released electrons related on the utilised fuel
$\dot{n}_{el}$	molar flow of electrons
n_F	molar quantity of the supplied fuel
$\dot{n}$	molar flow
n	molar quantity
p	total pressure
p_0	standard pressure
P_{el}	electrical power
p_{el}	specific electrical power
p_i	partial pressure of the component i
P_{loss}	power loss
P_{rev}	reversible electrical power
Q	heat
q	specific heat
R	electric or ohmic resistance
R_m	universal gas constant
s^*	specific entropy related to the standard state
S	total entropy
$\Delta^r S$	reaction entropy
Δs	entropy production
T	absolute temperature
U_f	fuel utilisation
V	voltage or potential
V_N	Nernst voltage
ΔV	voltage loss
v	specific volume
$-W_t$	technical work

$-w_t$	specific technical work
y	molar concentration
η	efficiency
θ	celsius temperature
λ	excess air value
μ	fuel-related specific mass
ν	fuel-related quantity
ζ	exergetic efficiency

Indices and abbreviations

0	stoichiometric value
A	air
aB	after burner outlet
AH	air heater
AFC	air at thermodynamic state of the fuel cell
An	anode
Ca	cathode
CC	Carnot cycle
CHP	combined heat and power generation
ECO	economiser
EXCO	external cooling
F	fuel
FC	fuel cell
FFC	fuel at thermodynamic state of the fuel cell
FGC	flue gas cooler
FH	fuel heater
G	flue gas
GT	gas turbine
H_2, H_2O	hydrogen, water
HEG	reversible heat engine (flue gas)
HEX	heat exchangers
HP	high pressure
HPA	heat pump (air)
HPF	heat pump (fuel)
I	inlet
i, j	components
INEX	intermediate expansion
irr	irreversible
LP	low pressure
O	outlet
O_2	oxygen
PH	product heater
ref	reformer
rev	reversible
RG	reaction product gas
RH	reheater

SH   superheater
ST   steam turbine
syst   system

3.2 The Ideal Reversible SOFC

The use of the first and the second laws of thermodynamics allows a simple description of a reversible fuel cell. The fuel and the air enter the fuel cell as non-mixed flows of the different components and the flue gas leaves the fuel cell as a non-mixed flow as well if we assume a reversible operating fuel cell. The non-mixed reactants deliver the total enthalpy $\Sigma n_i H_i$ to the fuel cell and the total enthalpy $\Sigma n_j H_j$ leaves the cell with the non-mixed products. Furthermore the heat Q_{FCrev} must be extracted reversibly from the fuel cell and transported reversibly to the environment. This can be done, for example, if the fuel cell and the environment have the same thermodynamic state. Q_{FCrev} is defined as a positive number if it is transported to the fuel cell. The reversible work $-W_{tFCrev}$ is delivered by the fuel cell. An idealised description of this model is given in Figure 3.1.

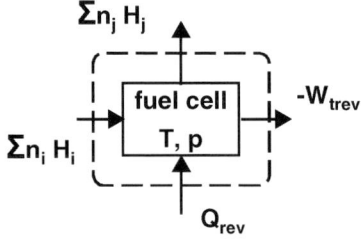

Figure 3.1   The reversible fuel cell, its energy balance and its system boundary.

Usually specific mass or mol related figures are used and the fuel quantity is the reference. The first law of the thermodynamics gives, with Figure 3.1

$$q_{FCrev} + w_{tFCrev} = \Delta^r H. \qquad (1)$$

The reaction enthalpy $\Delta^r H$ of the oxidation covers the production of the reversible work and heat. The second law of thermodynamics gives

$$\oint dS = 0. \qquad (2)$$

The reaction entropy $\Delta^r S$ is a result of the reaction itself and must be compensated by the transport of the reversible heat q_{FCrev} to the environment and Eq. (2) gives

$$\Delta^r S - \frac{q_{FCrev}}{T_{FC}} = 0. \tag{3}$$

Equations (1) and (3) give the reversible work w_{tFCrev}

$$w_{tFCrev} = \Delta^r H - T_{FC} \cdot \Delta^r S. \tag{4}$$

The reversible work w_{tFCrev} of the reaction is equal to the free or Gibbs enthalpy $\Delta^r G$ of the reaction

$$w_{tFCrev} = \Delta^r G = \Delta^r H - T_{FC} \cdot \Delta^r S. \tag{5}$$

The reversible efficiency η_{FCrev} of the fuel cell is the ratio of the Gibbs enthalpy $\Delta^r G$ and the reaction enthalpy $\Delta^r H$ at the thermodynamic state of the fuel cell and Eq. (5) gives

$$\eta_{FCrev} = \frac{\Delta^r G}{\Delta^r H} = \frac{\Delta^r H - T_{FC} \cdot \Delta^r S}{\Delta^r H}. \tag{6}$$

The process environment of a SOFC cannot exist near the ambient state and it is thus an artificial model only. A general reversible SOFC system operation is only possible within a system connecting reversibly the process environment with the ambient state. This is the approach in Section 3.5.

An SOFC can also be described as an electrical device where the electrical effects are explained by thermodynamics. Figure 3.2 shows the transport processes within a SOFC connecting the thermodynamic and the electrical effects using the example of hydrogen oxidation.

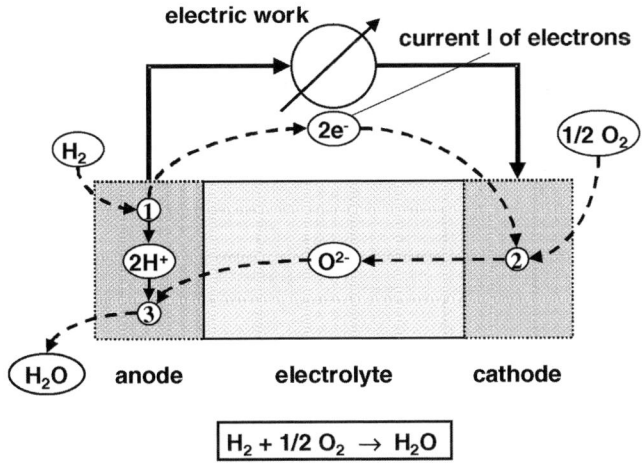

Figure 3.2 Transport processes within a SOFC.

The oxidation of hydrogen follows the equation

$$H_2 + \frac{1}{2}O_2 \to H_2O. \tag{7}$$

This equation is independent of the process itself. The reaction path in a SOFC depends on the anode and the cathode reactions. Hydrogen is adsorbed at the anode, ionised and the electrons are removed by the connection to the electrical load where the electrical work is used. Oxygen is adsorbed at the cathode connected with the load and ionised by the arriving electrons. The oxide ion is conducted by the electrolyte to the anode. The hydrogen ions (protons) and the oxide ion form a molecule of water. The first reaction ① on the anode is

$$H_2 \to 2H^+ + 2e^-. \tag{8}$$

The reaction ② on the cathode is

$$\frac{1}{2}O_2 + 2e^- \to O^{2-}. \tag{9}$$

The oxide ion O^{2-} is conducted through the electrolyte and arrives at the anode. At the anode, water forms ③ according to the reaction

$$2H^+ + O^{2-} \to H_2O. \tag{10}$$

As Figure 3.2 shows, the product H_2O is mixed with the anode gas and its concentration increases with increasing fuel utilisation U_f. The fuel utilisation U_f is the ratio of the spent fuel flow and the inlet fuel flow and is defined by

$$U_f = 1 - \frac{\dot{m}_{FAnO}}{\dot{m}_{FI}}. \tag{11}$$

where $\dot{m}_{FI}$ is the fuel mass flow at the cell's inlet and $\dot{m}_{FAnO}$ is the fuel mass flow at the outlet of the anode. A similar definition can be made with the molar flow. Because these mixing effects are irreversible processes they produce entropy and a reversible SOFC operation is only possible as the limiting process of the real process with $U_f \to 0$. Equation (8) shows that the molar flow of the electrons is twice that of the molar flow of hydrogen, thus

$$\dot{n}_{el} = 2\dot{n}_{H2}. \tag{12}$$

The electric current I is a linear function of the molar flow $\dot{n}_{el}$ of the electrons or the molar flow of the spent fuel – in this example the molar flow of $\dot{n}_{H2}$ the spent hydrogen

$$I = \dot{n}_{el} \cdot (-e) \cdot N_A = -\dot{n}_{el} \cdot F = -2\dot{n}_{H2} \cdot F. \tag{13}$$

In Eq. (13) we introduced the elementary charge e

$$e = (1,60217733 \pm 0,00000049) \cdot 10^{-19} C. \tag{14}$$

and the Faraday constant F

$$F = e \cdot N_A = (96485,309 \pm 0,029) C/mol \tag{15}$$

as the product of the elementary charge and the Avogadro constant N_A. Equations (13) and (15) show that the electric current I is a measure of the fuel spent. Thus a current measurement is a very simple method to measure the fuel spent. The matching between thermodynamic and electrical quantities can be done by the power but not by the work. The reversible power can be written as a product of the reversible voltage V_{FCrev} and the current I as well as a product of the molar flow of the fuel $\dot{n}_{H2}$ and the free enthalpy $\Delta^r G$ of the oxidation reaction

$$P_{FCrev} = V_{FCrev} \cdot I = \dot{n}_{H2} \cdot w_{tFCrev} = \dot{n}_{H2} \cdot \Delta^r G. \tag{16}$$

The reversible voltage V_{FCrev} results from Eqs. (13) and (16)

$$V_{FCrev} = \frac{-\dot{n}_{H2} \cdot \Delta^r G}{\dot{n}_{el} \cdot F}. \tag{17}$$

Equation (12) shows the ratio between the molar flow of the electrons and the spent hydrogen as 2. This can be generalised and n^{el} is the number of the electrons that are released during the ionisation process of one utilised fuel molecule, related on the molar flows we get with Eq. (11)

$$n^{el} = \frac{\dot{n}_{el}}{U_f \cdot \dot{n}_{FI}}, \tag{18}$$

and finally for the reversible voltage V_{FCrev} of the oxidation of any fuel gas

$$V_{FCrev} = \frac{-\Delta^r G}{n^{el} \cdot F}. \tag{19}$$

It has already been mentioned that the mixing effects during fuel utilisation within a SOFC do not allow a reversible SOFC operation. These effects and the voltage reduction can be calculated by considering the fuel utilisation connected with a change of the partial pressures of the components within the system [2]. We can write Eq. (4) more precisely as

$$\Delta^r G(T,p) = \Delta^r H(T,p) - T \cdot \Delta^r S(T,p). \tag{20}$$

Using the common assumption of the ideal gas we get

$$\Delta^r G(T,p) = \Delta^r H(T) - T \cdot \Delta^r S(T,p) \tag{21}$$

and we can write, with $dS = (dH - v.dp)/T$, for the entropy S_j of any component j

$$S_j(T,p) = S_j^0 + \int_{T_0}^{T} \frac{C_{Pj}(t)}{t} dt - R_m \cdot \ln\left(\frac{p_j}{p_0}\right). \quad (22)$$

where C_{Pj} is the temperature dependent heat capacity of the component j. The pressure dependence of C_{Pj} can be neglected in this common assumption, Eq. (21). Using Eq. (22), we can write for the reaction entropy $\Delta^r S(T,p)$

$$\Delta^r S(T,p) = \Delta^r S(T) - R_m \cdot \ln(K) \quad (23)$$

with the equilibrium constant K (see, e.g., [1])

$$K = \prod_j \left(\frac{p_j}{p_0}\right)^{v_j}. \quad (24)$$

v_j is the fuel-related quantity of the component j in the equation of the oxidation reaction and p_0 is the standard pressure (1 bar):

$$p_0 = 1 \, bar. \quad (25)$$

Using Eqs. (21)–(24) we get

$$\Delta^r G(T,p) = \Delta^r G(T) + T \cdot R_m \cdot \ln(K). \quad (26)$$

This use of the assumption of an ideal gas allows one to express the Nernst potential or the Nernst voltage V_N by using Eqs. (18), (19) and (26) as

$$V_N = \frac{-\Delta^r G(T)}{n^{el} \cdot F} - \frac{R_m \cdot T \cdot \ln(K)}{n^{el} \cdot F}. \quad (27)$$

The following reversible oxidation of hydrogen (H_2), of carbon monoxide CO and of methane (CH_4) can be analysed as examples by using Eq. (27):

$$H_2 + \frac{1}{2}O_2 \rightarrow H_2O. \quad (7)$$

$$CO + \frac{1}{2}O_2 \rightarrow CO_2. \quad (28)$$

$$CH_4 + 2O_2 \rightarrow 2H_2O + CO_2. \quad (29)$$

The equations (7), (28), and (29) determine the reaction enthalpy, the reaction entropy and thus the free enthalpy and the voltage of the reversible oxidation as formulated in Eqs. (5) and (19) with the thermodynamic data of the reactions at the standard conditions 0 (25°C, 1 bar) as collected in e.g. [1,4]. A variation of the thermodynamic state of the environment of the reversible cell provides a first idea of the behaviour of real cells under changing operation conditions. It can be assumed that the reaction enthalpy and the reaction entropy depend only slightly on the temperature as a first assumption. Thus the free enthalpy of the reaction at a higher temperature can be approximated with the values of the reaction enthalpy and the reaction entropy at standard conditions and it depends on the temperature linearly and yields the reversible cell voltage, Eqs. (26) and (27). The values of the reversible cell voltage are calculated for the standard state 0, at 1000°C/1 bar and for 25°C and 1000°C at 0.1 and 10 bar. These linearised values of the free enthalpy and the reversible cell voltage at different thermodynamic states are written as $\Delta^r G^*$ and V^*. The water is always assumed to be in the gaseous phase because a SOFC operates at high temperatures. The values can be found in Table 3.1 and Figure 3.3.

Equations (7) and (28) indicate that the volumes of the products of the oxidation reactions of hydrogen and carbon monoxide are smaller than the

Table 3.1 The reversible oxidation of hydrogen, carbon monoxide and methane

Fuel	H_2	CO	CH_4
$\Delta^r H^0$ in kJ/mol	−241.82	−282.99	−802.31
$\Delta^r S^0$ in J/(mol K)	−44.37	−86.41	−5.13
$\Delta^r G^0$ in kJ/mol	−228.59	−257.23	−800.68
$\Delta^r G^*$ at 1000°C, 1 bar in kJ/mol	−185.33	−172.98	−795.68
n^{el}	2	2	8
V^0 in V	1.185	1.333	1.037
V^* at 1000°C, 1 bar in V	0.960	0.896	1.031
$\ln(K)$ at 0.1 bar	1.1513	1.1513	0
$\ln(K)$ at 10 bar	−1.1513	−1.1513	0
V^* at 25°C, 0.1 bar in V	1.170	1.318	1.037
V^* at 1000°C, 0.1 bar in V	0.897	0.833	1.031
V^* at 25°C, 10 bar in V	1.199	1.348	1.037
V^* at 1000°C, 10 bar in V	1.024	0.960	1.031

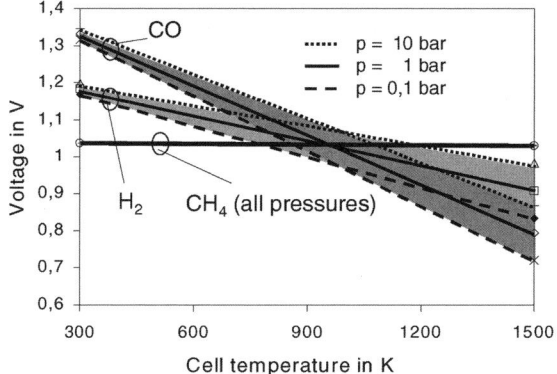

Figure 3.3 The reversible cell voltage of different fuels at different states (p,T) of the environment (linearised model and assumption of ideal gas).

cumulative volume of the reactants. But the cumulative volume of the reactants and of the products of the oxidation reaction of methane is the same. This is also apparent by the values of the activity K ($\ln(1) = 0$). Thus there is theoretically no change of the entropy in the last case and the real change of the measured values at the standard state 0 is very limited as indicated in Table 3.1. This is the reason for the very small dependence on temperature of the free enthalpy or the reversible cell voltage of the oxidation reaction of methane, the idealised pressure dependence of the entropy yields no change in the cell voltage caused by the system pressure. The reversible cell voltage of the hydrogen and the carbon monoxide oxidation decreases with increasing system temperature and increases with increasing system pressure.

3.3. Voltage Losses by Ohmic Resistance and by Mixing Effects by Fuel Utilisation

The thermodynamic relevance of the voltage can be understood by considering Eqs. (13) and (16). The voltage is a measure of the exergetic quality similar to the work of a thermodynamic process. The exergy is defined as the potential of the reversible work of a system related to the ambient state 0 [1]. Thus it is clear that the voltage loss ΔV due to the electric resistance R is connected with an additional irreversible production of entropy. We get for the voltage loss

$$\Delta V = I \cdot R \tag{30}$$

and for the power loss

$$P_{loss} = \Delta V \cdot I = I^2 \cdot R = T_{SOFC} \cdot U_f \cdot \dot{n}_{FI} \cdot \Delta s_{irr}. \tag{31}$$

Equation (31) shows that the irreversible entropy production of an ohmic loss P_{loss} in a SOFC is smaller than that in another fuel cell operating at a lower temperature and a generalised form of Eq. (13) yields the irreversible entropy production

$$\Delta s_{irr} = \frac{U_f \cdot \dot{n}_{FI} \cdot R \cdot (n^{el} \cdot F)^2}{T_{SOFC}}. \tag{32}$$

Equation (13) was generalised here as

$$I = -n^{el} \cdot U_f \cdot \dot{n}_{FI} \cdot F. \tag{33}$$

It has already been mentioned that the mixing effects during fuel utilisation within a SOFC do not allow a reversible SOFC operation. These influences and the voltage reduction can be easily calculated by considering the fuel utilisation connected with a change of the partial pressures of the components within the system [2].

It should be mentioned that any reactant must have the same thermodynamic state in the case of the reversible cell. This is, for example, not the case if we use air as the oxidant gas. We can calculate cases like this with Eq. (27) of the Nernst voltage, but however we use the ideal process the total process is not reversible any more. The oxidation of hydrogen (Eq. (7)) is a good example to illustrate this. Using p_i as the partial pressure of the component i we get

$$p_i = y_i \cdot p, \tag{34}$$

writing y_i for the molar concentration of the component i and p for the total pressure of the system. Using Eq. (11) we can write

$$U_f = \frac{y_{FI} \cdot \dot{n}_{AnI} - y_{FO} \cdot \dot{n}_{AnO}}{y_{FI} \cdot \dot{n}_{AnI}} \tag{35}$$

if we consider the molar flow of the fuel F as the product of the molar concentration y and the total molar flow at the inlet I and the outlet O of the anode side An. U_f will be used as a variable thus the outlet O can be interpreted as a space variable along the axis of the parallel flowing fuel and air defined by a certain U_f to be obtained. The local Nernst voltage $V_N(U_f)$ depends on the local gas concentration. The molar flow on the anode side is constant in our example of the hydrogen oxidation and we get

$$\dot{n}_{AnI} = \dot{n}_{AnO} = \dot{n}^* \tag{36}$$

and Eq. (35) yields

$$U_{fH2} = 1 - \frac{y_{H2,O}}{y_{H2,I}}. \tag{37}$$

The equation of the reaction (7) shows that the molar flow of the utilised fuel is equal to the molar flow of the produced water at the outlet O

$$\dot{n}_{H2,U} = \dot{n}_{H2O,O} \tag{38}$$

if the used hydrogen is dry ($y_{H2,I} = 1$). This yields

$$U_{fH2} = \frac{\dot{n}_{H2,U}}{\dot{n}^*} = \frac{\dot{n}_{H2O,O}}{\dot{n}^*} = y_{H2O,O}. \tag{39}$$

Following Eq. (7) we can write for the cathode side

$$\dot{n}_{O2,U} = \frac{1}{2} \cdot \dot{n}_{H2,U}. \tag{40}$$

Practical SOFC systems operate with air instead of oxygen and with an excess air $\lambda > 1$. The incoming air flow is defined by the inlet flow of the cathode

$$\dot{n}_{CaI} = \frac{1}{2} \cdot \lambda \cdot \frac{\dot{n}^*}{0,21}. \tag{41}$$

The outlet flow of the cathode can be calculated by

$$\dot{n}_{CaO} = \frac{1}{2} \cdot \lambda \cdot \frac{\dot{n}^*}{0,21} - \frac{1}{2} \cdot \dot{n}_{H2,U}. \tag{42}$$

The molar oxygen flow at the inlet is

$$\dot{n}_{O2,I} = \frac{1}{2} \cdot \lambda \cdot \dot{n}^* \tag{43}$$

and the molar oxygen flow at the outlet is

$$\dot{n}_{O2,O} = \frac{1}{2} \cdot (\lambda \cdot \dot{n}^* - \dot{n}_{H2,U}). \tag{44}$$

Equations (43) and (44) can be written as a function of U_f

$$\dot{n}_{CaO} = \frac{1}{2} \cdot \dot{n}^* \cdot \left(\frac{\lambda}{0,21} - \frac{\dot{n}_{H2,U}}{\dot{n}^*} \right) = \frac{1}{2} \cdot \dot{n}^* \cdot \left(\frac{\lambda}{0,21} - U_{fH2} \right) \tag{45}$$

and

$$\dot{n}_{O2,O} = \frac{1}{2} \cdot \dot{n}^* \cdot \left(\lambda - \frac{\dot{n}_{H2,U}}{\dot{n}^*} \right) = \frac{1}{2} \cdot \dot{n}^* \cdot (\lambda - U_{fH2}). \tag{46}$$

Now we can express y_i as a function of the fuel utilisation U_f

$$y_{H2,O} = 1 - U_{fH2} \tag{47}$$

$$y_{H2O,O} = U_{fH2} \tag{48}$$

and

$$y_{O2,O} = \frac{\lambda - U_{fH2}}{\lambda/0,21 - U_{fH2}} = \frac{\dot{n}_{O2,O}}{\dot{n}_{CaO}}. \tag{49}$$

Using Eqs. (24), (27) and (34) we get for the equilibrium constant K

$$K = \frac{U_{fH2} \cdot \left(\lambda/0,21 - U_{fH2} \right)^{1/2}}{(1 - U_{fH2}) \cdot [(\lambda - U_{fH2}) \cdot p]^{1/2}}. \tag{50}$$

We can use Eqs. (50) and (27) to calculate the ideal Nernst voltage V_N as a function of the fuel utilisation U_f. Using Eq. (33) we see that the current I is proportional to the fuel utilisation U_f

$$U_f = \frac{I}{n^{el} \cdot \dot{n}_{FI} \cdot F}. \tag{51}$$

The number of electrons n^{el} depends on the fuel (2 for H_2), the Faraday constant F is a constant value and the fuel inlet flow is the only variable that influences the relation between fuel utilisation U_f and current I. Any curve of the voltage V depending on the current I can be expressed by any curve of V depending on the fuel utilisation U_f at the same fuel inlet flow.

Considering Eqs. (50) and (27) we see that $V \rightarrow +\infty$ for $U_f \rightarrow 0$ and $V \rightarrow -\infty$ for $U_f \rightarrow 1$ respectively. But the model of the ideal gas gives a good approximation for $0 < U_f < 1$ in the regime of the real SOFC operation. This model allows one to evaluate the principal influences of the different parameters, system pressure p, SOFC temperature θ_{SOFC}, excess air λ and fuel utilisation U_f, on the Nernst voltage V_N.

Figure 3.4 shows the Nernst voltage V_N as a function of the fuel utilisation U_f in a SOFC with H_2 fuel and with the system pressure p as a parameter. The excess air and the SOFC temperature are fixed. The interesting area between $U_f = 0.1$ and $U_f = 0.9$ can be well approximated with the model of the ideal gas. The dotted line shows the adoption of the model. The irreversible mixing within the SOFC reduces V_N between $U_f = 0.1$ and $U_f = 0.9$ by about more than 200 mV. An increase of the system pressure from 1 to 10 bar increases V_N by about 70 mV. An increasing SOFC temperature decreases V_N as shown by Eqs. (27) and (50).

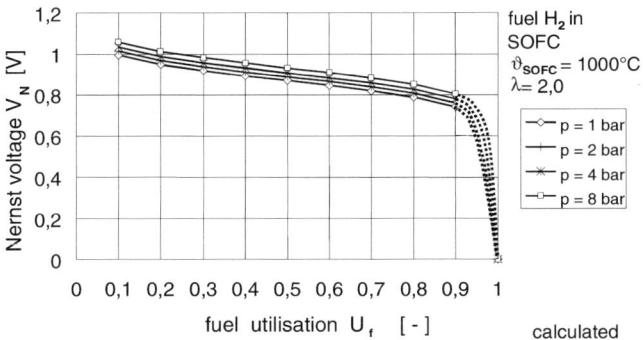

Figure 3.4   The calculated Nernst voltage V_N as a function of the fuel utilisation U_f.

The excess air λ is a very important process parameter for the design of the total system as shown later. Figure 3.5 shows the Nernst voltage V_N as a function of the excess air λ and the system pressure p as a parameter.

An increasing excess air λ increases V_N slightly. But this influence of the excess air λ on V_N decreases with an increasing excess air. An increase of the excess air λ at values > 2 does not really influence V_N any more. In the range $1 < \lambda < 2$ the voltage increase is ~30 mV. The calculation shows that a certain fuel utilisation U_f leads to a certain V_N. The maximum power P_{elmax} of one cell is determined by the Nernst voltage V_{NO} and the corresponding current I_O

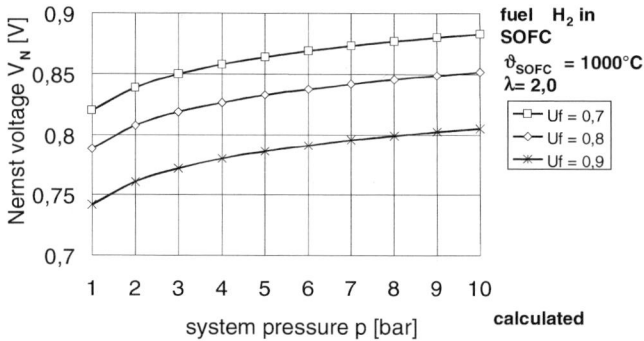

Figure 3.5   The calculated Nernst voltage V_N as a function of the excess air λ.

depending on the fuel utilisation U_{fO} at the outlet O according to Eq. (51) and can be written as

$$P_{el\max} = V_{NO} \cdot I_O. \tag{52}$$

Figure 3.6 (left) illustrates this. It shows the power from a cell as given by the curve of the Nernst voltage V_N as a function of fuel utilisation U_f. The current I is proportional to U_f, see Eq. (51), for a given substance flow and I is thus marked in parentheses in Figure 3.6. An electrical serial connection of a number of cells allows an integration of the curve of V_N as shown at the right-hand side of Figure 3.6. This integration leads to different voltages in every cell but the current or the fuel utilisation in every cell must be equal. This cascading of cells allows an increase in power and efficiency compared with a single cell.

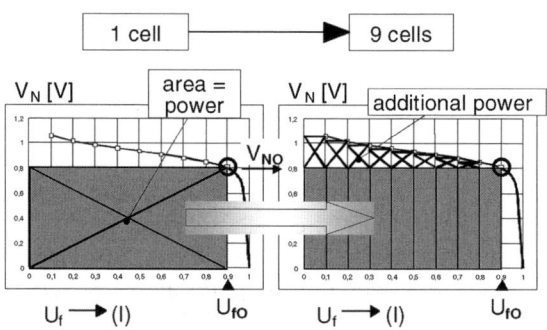

Figure 3.6   The increase in efficiency by cascading single cells.

3.4 Thermodynamic Definition of a Fuel Cell Producing Electricity and Heat

In designing practical SOFC systems with associated components such as fans, heat exchangers, etc., modelling of a SOFC as a power generating burner is very helpful (see Figure 3.7). The system is defined as a module consisting of SOFC cells connected in electrical parallel into stacks supplying a common burner with

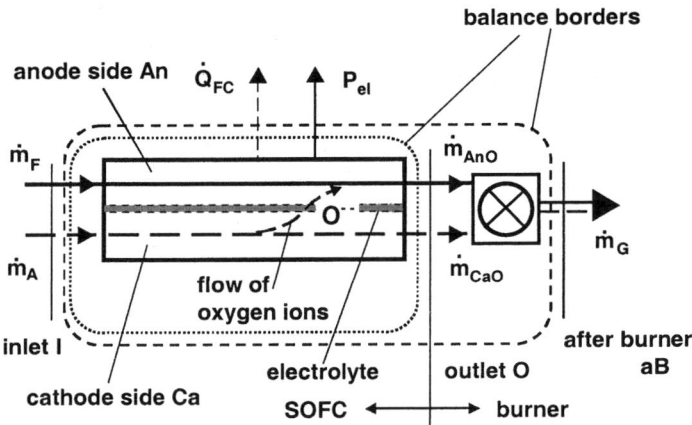

Figure 3.7 The power generating burner model of a SOFC module.

the depleted fuel. The energy balance of the stack provides the necessary requirements for cooling and excess air simultaneously [2].

There are two possible descriptions. The most simple approach is a balance border around the complete module including all stacks and the joint burner from the inlet I of the fuel F and the air A to the outlet aB of the flue gas G after the burner. The more detailed approach is a balance border surrounding all stacks from the inlet I to the outlets O of the anode side AnO and of the cathode side CaO. The calculation of this 'power generating burner' is very similar to the calculation of a combustor of a gas turbine or of a furnace of a boiler. The calculation of the mass flows of the module does not even differ from any calculation of a conventional oxidation. The energy balance of this simpler approach (from I to aB) delivers

$$\dot{H}_{FI} + \dot{H}_{AI} = \dot{Q}_{FC} + P_{el} + \dot{H}_{GaB}. \tag{53}$$

The total enthalpy flow $\dot{H}_{FI}$ of the fuel includes the reaction enthalpy (or in technical terms the LHV) as well. The enthalpy flow of the incoming air is $\dot{H}_{AI}$. Both these enthalpy flows have to cover the energy output of the SOFC module consisting of the produced power P_{el}, the generated heat $\dot{Q}_{FC}$ and the enthalpy flow of the flue gas $\dot{H}_{GaB}$. We get with Eq. (53) the respective mass flows and the respective related enthalpies h^*

$$\dot{m}_{FI} \cdot \left(LHV + h^*_{FI}\right) + \dot{m}_{AI} \cdot h^*_{AI} = \dot{Q}_{FC} + P_{el} + \dot{m}_{GaB} \cdot h^*_{GaB}. \tag{54}$$

The use of the related enthalpies is necessary to match all enthalpies with the LHV related on the chemical standard state (1 bar, 25°C). The related enthalpy is defined by

$$h^* = h(p, \vartheta) - h_0(1\,bar, 25°C). \tag{55}$$

These equations are sufficient to calculate the necessary excess air λ for a given heat extraction $-\dot{Q}_{FC}$ of the module, or vice versa to calculate the necessary SOFC cooling by the heat extraction $-\dot{Q}_{FC}$ for a defined excess air λ as shown below.

The consideration of the stacks only allows a more detailed modelling and the energy balance delivers

$$\dot{H}_{FI} + \dot{H}_{AI} = \dot{Q}_{FC} + P_{el} + \dot{H}_{AnO} + \dot{H}_{CaO}. \tag{56}$$

This equation is identical with Eq. (53) if all the fuel is used in the stacks. The enthalpy flow of the incoming fuel is

$$\dot{H}_{FI} = \dot{m}_{FI} \cdot \left(LHV + h^*_{FI}\right) \tag{57}$$

and the enthalpy flow of the incoming air is

$$\dot{H}_{AI} = \dot{m}_{AI} \cdot h^*_{AI} = \dot{m}_{FI} \cdot \lambda \cdot \mu_{AO} \cdot h^*_{AI}. \tag{58}$$

The stoichiometric specific air demand μ_{AO} is defined by the relation of the stoichiometric air mass flow and the corresponding fuel mass flow. The relation of all terms of the energy balance on the mass flow $\dot{m}_{FI}$ of the incoming fuel allows a generalised consideration. The generated heat is

$$\dot{Q}_{FC} = \dot{m}_{FI} \cdot q_{FC}. \tag{59}$$

The produced power is

$$P_{el} = \dot{m}_{FI} \cdot p_{el}. \tag{60}$$

With the fuel utilisation U_f

$$U_f = 1 - \frac{\dot{m}_{FAnO}}{\dot{m}_{FI}} \tag{11}$$

the enthalpy flow at the anode outlet is

$$\dot{H}_{AnO} = \dot{m}_{FI} \cdot \left(1 - U_f\right) \cdot \left(LHV + h^*_{FAnO}\right) + \dot{m}_{RG} \cdot h^*_{RGAnO}. \tag{61}$$

The flow of the reaction product gas RG is

$$\dot{m}_{RG} = \dot{m}_{FI} \cdot U_f + \dot{m}_{O2} = \dot{m}_{FI} \cdot U_f \cdot (1 + \mu_{O20}), \tag{62}$$

with

$$\dot{m}_{O2} = U_f \cdot \dot{m}_{FI} \cdot \mu_{O20}. \tag{63}$$

The mass flow at the outlet of the anode side consists of the not utilised fuel and of RG. RG consists of the reaction products CO_2 and H_2O. Its mass flow is equal to the mass flow of the utilised fuel and of the transferred oxygen by the ion conduction through the electrolyte. The stoichiometric demand on oxygen related on the inlet fuel mass flow is given by the figure μ_{O2O}. We finally get for the enthalpy flow at the anode outlet

$$\dot{H}_{AnO} = \dot{m}_{FI} \cdot \left[(1 - U_f) \cdot (LHV + h^*_{FAnO}) + U_f \cdot (1 + \mu_{O2O}) \cdot h^*_{RGAnO}\right]. \quad (64)$$

The enthalpy flow at the cathode outlet can be calculated by the difference of the enthalpy flow of the non-depleted air and the enthalpy flow of the oxygen being transferred to the anode both with the thermodynamic state at the cathode outlet

$$\dot{H}_{CaO} = \dot{m}_{AI} \cdot h^*_{ACaO} - \dot{m}_{O2} \cdot h^*_{O2CaO}. \quad (65)$$

Equations (58) and (63) yield with Eq. (65)

$$\dot{H}_{CaO} = \dot{m}_{FI} \cdot \left[\lambda \cdot \mu_{AO} \cdot h^*_{ACaO} - U_f \cdot \mu_{O2O} \cdot h^*_{O2CaO}\right]. \quad (66)$$

The specific generated heat q_{FC} results from Eqs. (56)–(66) as

$$\begin{aligned} q_{FC} = &\; U_f \cdot LHV + h^*_{FI} - (1 - U_f) \cdot h^*_{FAnO} - p_{el} - \\ &- U_f \cdot \left[\mu_{O2O} \cdot (h^*_{RGAnO} - h^*_{O2CaO}) + h^*_{RGAnO}\right] \\ &+ \mu_{LO} \cdot \left[\lambda \cdot h^*_{AI} - \lambda \cdot h^*_{ACaO}\right]. \end{aligned} \quad (67)$$

The use of Eq. (67) depends on a constant excess air λ as probably regulated by a oxidation control by an O_2 measurement after the burner. The necessary excess air λ to cool the cell for a fixed heat extraction q_{FC} can be calculated by rearranging Eq. (67) as

$$\lambda = \frac{U_f \cdot \left[LHV - \mu_{O2O} \cdot (h^*_{RGAnO} - h^*_{O2CaO}) - h^*_{RGAnO}\right] + h^*_{FI}}{\mu_{LO} \cdot (h^*_{ACaO} - h^*_{AI})} + \frac{-q_{FC} - p_{el} - (1 - U_f) \cdot h^*_{FAnO}}{\mu_{LO} \cdot (h^*_{ACaO} - h^*_{AI})}. \quad (68)$$

3.5 Thermodynamic Theory of SOFC Hybrid Systems

The process environment of the cell model in Figure 3.1 must be related reversibly to the ambient state to define the reversible system. As mentioned above we assume $U_f \to 0$ and the flows consist of unmixed components to assure a reversible process. Figure 3.8 shows the reversible fuel cell–heat engine system that fulfills these requirements. The reactants air and fuel in the ambient state T_0, p_0 are brought to the thermodynamic state of the cell T, p by the reversible heat pumps HPA (for air) and HPF (for fuel). The necessary

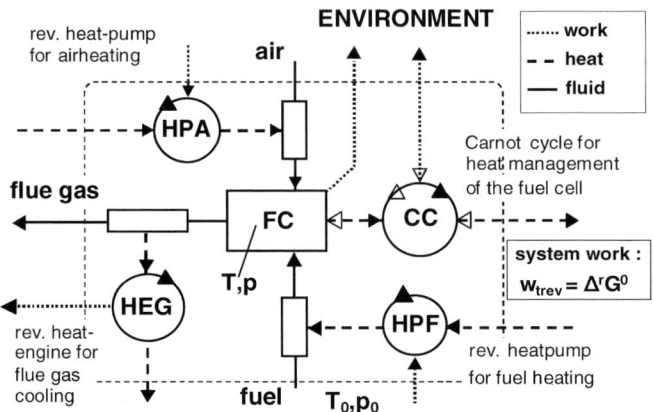

Figure 3.8 *The reversible fuel cell–heat engine hybrid system.*

heat consists of the energy from the environment (anergy) and of the exergy being supplied by the reversible working heat pumps. The fuel cell FC delivers the flue gas, the work and the heat as given in Eq. (5). The unmixed product flue gas is brought from the state T,p of FC to the ambient state T_0,p_0 by the reversible heat engine HEG. The reversible work delivered by HEG is the exergy of the flue gas with the state T,p. Finally we use the Carnot cycle CC for the heat management between FC and the environment. FC can exchange heat reversibly with CC as with the process environment before. CC operates as a heat engine for $\Delta^r S < 0$.

The fuel cell FC delivers the reversible work w_{tFCrev}

$$w_{tFCrev} = \Delta^r G = \Delta^r H - T_{FC} \cdot \Delta^r S \tag{5}$$

and the reversible heat q_{FCrev}

$$q_{FCrev} = T_{FC} \cdot \Delta^r S. \tag{69}$$

FC is the heat source of the Carnot cycle CC and delivers reversible heat q_{FCrev}. The reversible work w_{tCCrev} of CC is defined by

$$w_{tCCrev} = q_{FCrev} \cdot \left(1 - \frac{T_0}{T_{FC}}\right) = T_{FC} \cdot \Delta^r S \cdot \left(1 - \frac{T_0}{T_{FC}}\right) \tag{70}$$

and the reversible heat q_{FCrev}

$$q_{CCrev} = q_{FCrev} \cdot \frac{T_0}{T_{FC}} = T_0 \cdot \Delta^r S. \tag{71}$$

The heat pump HPF has to deliver the reversible heat q_{HPFrev} to heat the fuel

$$q_{HFrev} = h^*_{FFC} = w_{tHPFrev} + q_{HPFrev}. \tag{72}$$

HPF has to be supplied reversibly with the work $w_{tHPFrev}$ that is equal to the exergy e_{FFC} of the fuel with the thermodynamic state of the fuel cell and the heat q_{HPFrev} from the environment

$$w_{tHPFrev} = e_{FFC} = h^*_{FFC} - T_0 \cdot s^*_{FFC} = (h_{FFC} - h_{F0}) - T_0 \cdot (s_{FFC} - s_{F0}) \tag{73}$$

$$q_{HPFrev} = T_0 \cdot s^*_{FFC}. \tag{74}$$

In the right-hand part of Eq. (73) the definition of the exergy of the fuel with the thermodynamic state of the fuel cell is now worked out in more detail. Similar processes are used for the reversible heating of the air and the reversible cooling of the flue gas. The reversible air heating needs the work w_{tHPA}

$$w_{tHPArev} = \mu_A \cdot e_{AFC} = \mu_A \cdot \left(h^*_{AFC} - T_0 \cdot s^*_{AFC}\right) \tag{75}$$

and the reversible heat engine HEG for the cooling of the flue gas G delivers the work $w_{tHEGrev}$

$$w_{tHEGrev} = -(\mu_A + 1) \cdot e_{GFC} = -(\mu_A + 1) \cdot \left(h^*_{GFC} - T_0 \cdot s^*_{GFC}\right). \tag{76}$$

The total work of the reversible fuel cell–heat engine system is defined by

$$w_{tsystrev} = w_{tFCrev} + w_{tCCrev} + w_{tHPFrev} + w_{tHPArev} + w_{tHEGrev} \tag{77}$$

Using Eqs. (5), (70), (73) and (75)–(77) we get

$$w_{tsystrev} = \Delta^r H^0 - T_0 \cdot \Delta^r S^0 = \Delta^r G^0 \tag{78}$$

The reversible work $w_{tsystrev}$ of any fuel cell–heat engine system is independent of the state of the cell and is equal to the free enthalpy of the reaction $\Delta^r G^0$ at the ambient state [3]. The standard condition is assumed to be the ambient state to keep the argument simple.

It is useful to define a simplified process for further analysis, because the three reversible heat engines HPA, HPF and HEG do really nothing else than to heat fuel and air by cooling the flue gas – their total reversible work is negligible. Thus the simplified process uses a heat exchanger system for the heat recovery instead of HPA, HPF and HEG, as shown in Figure 3.9. But this simplified reference cycle is generally not reversible [5]. This is caused by the changes of the specific heat capacities of the different substances with the reaction temperature T that change the reaction enthalpy $\Delta^r H(T,p)$. A (small) amount of the waste heat of FC must be used to heat the reactants completely. We lose this amount of heat for further reversible use in the Carnot cycle CC.

The simplified fuel cell–heat engine hybrid cycle as a reference cycle fits the reversible system well; the deviation at 1000°C is −0.76% only for the hydrogen oxidation. Figure 3.10 shows the applications of this cycle. The left-hand side of

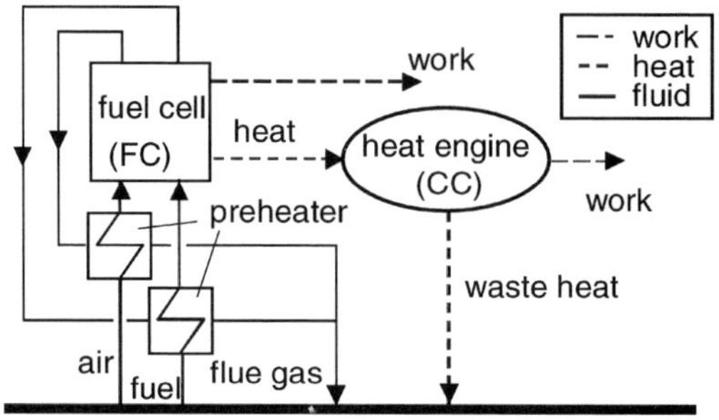

Figure 3.9 Simplified fuel cell–heat engine hybrid system as a reference cycle.

Figure 3.10 shows the efficiency of the reference cycle. The system efficiency η_{syst} is defined as

$$\eta_{syst} = \frac{\sum w_t}{LHV}. \tag{79}$$

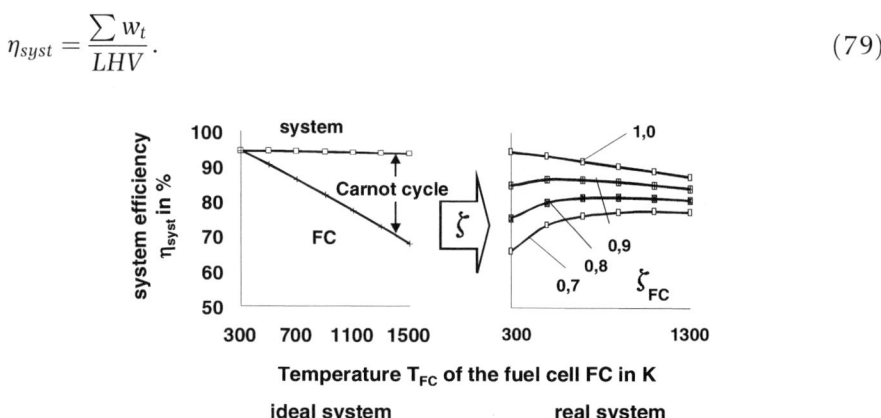

Figure 3.10 The system efficiency of the ideal and the real fuel cell–heat engine hybrid system with an exergetic efficiency $\zeta_{HE} = 0.7$ and the oxidation of hydrogen.

The cell temperature T_{FC} is equal to the temperature T of the process environment but more illustrative for further discussion. The work w_{tCC} produced by the Carnot cycle CC increases with increasing T_{FC} and the work w_{tFCrev} produced by FC decreases with increasing T_{FC} as expected. There is a compensation of both effects and the work w_{tsyst} of the system is independent of T_{FC} (or nearly independent in the case of the simplified process). FC operates reversibly in both cases but CC does not operate completely reversible in the simplified process due to the fact that a small part of the waste heat of FC is used to heat air and fuel. The practical use of the simplified combined fuel cell–heat cycle is the opportunity to use exergetic efficiencies to describe the operational conditions of real cycles by using this very simple model. The exergetic efficiency ζ is defined as

$$\zeta = \frac{w_{treal}}{w_{trev}}. \tag{80}$$

It is well known from heat engines that ζ is between about 0.7 and 0.8. All types of real cells have efficiencies between 55 and 65% but there is no significant difference caused by the cell temperature T_{FC} as we would expect from the thermodynamic considerations [2]. Thus it is obvious that real cells operating at lower temperatures do not use their potential work w_{tFCrev} properly. Real high-temperature fuel cells use their potential work reasonably well. The exergetic efficiency of fuel cells is described in [5] for the H_2 fuel. It can be shown easily that the SOFC has the best exergetic efficiency here. The exergetic efficiency ζ_{FC} (Eq. (80)) can be related to the total fuel feed if we assume that the non-utilised fuel can be burnt in an isothermal combustor. We can define the fuel cell and the isothermal combustor as one unit in that case (see Figure 3.7).

The system efficiency η_{syst} of any real hydrogen-fuelled combined fuel cell–heat cycle is plotted against the cell temperature T_{FC} in Figure 3.10 (right). The exergetic efficiency of the fuel cell ζ_{FC} is the parameter (0.7; 0.8; 0.9; 1.0) and the exergetic efficiency of the heat engine ζ_{HE} is kept constant at 0.7.

The system efficiency η_{syst} increases with an increasing cell temperature T_{FC} for all exergetic efficiencies $\zeta_{FC} < 1.0$ until a maximum is reached. The location of the maximum of η_{syst} shifts with increasing exergetic efficiencies ζ_{FC} to decreasing T_{FC}. But there are no big changes in the region of the maximum if we change T_{FC} at constant ζ_{FC}. The influence of the Carnot cycle dominates at lower temperatures T_{FC} and lower exergetic efficiencies ζ_{FC}.

The main result of these considerations is the possibility of designing hybrid fuel cell–heat cycles with efficiencies of about 80% [5]. This efficiency value is a target of the US Department of Energy since 1999 [6]. It appears useful to operate the cell at the lowest possible cell temperature T_{FC} in the region of the maximum of η_{syst} for reducing material costs of the heat engine and the heat exchangers. An increase in T_{FC} leads to only a negligible increase in the system efficiency η_{syst}.

The use of natural gas or other hydrocarbons changes the system design because of the processing of the fuel before its use in the SOFC. The following investigations will be done for methane as the main component of natural gas to keep the calculations simple. A very common fuel processing for hydrocarbons is the endothermic steam reforming process as shown for methane in Eq. (81) with the heat demand of Eq. (82)

$$CH_4 + H_2O \rightarrow 3H_2 + CO, \tag{81}$$

$$\Delta^r H(750°C)_{ref} = +14065, 1 kJ/kg_{CH4}. \tag{82}$$

Heat is also needed to evaporate the feed water. It is useful to use the waste heat of the cell for these purposes. A general model of a methane fired combined SOFC cycle based on the reference cycle of Figure 3.9 is shown in Figure 3.11 to describe the thermodynamic influences on the system's behaviour as simply as possible [2,7,8].

The SOFC can be modelled as one unit of two parallel operating SOFCs fuelled with hydrogen and carbon monoxide. All irreversible effects including mixing is

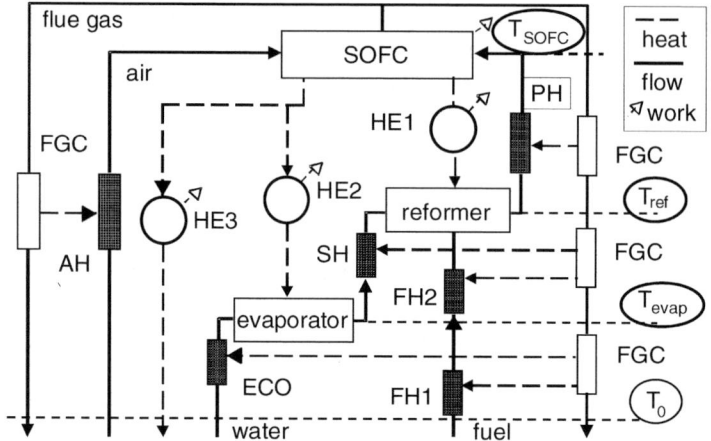

Figure 3.11   Process model for integrated reforming in SOFC systems.

be described by $\zeta_{FC} < 1$. The reasons for these irreversibilities of the SOFC and other components are not important for understanding the system's behaviour if they are considered properly in the system. The important relation here is the relation between work and heat within the single components and the temperatures of the heat sources and the heat sinks. The SOFC is the heat source of the fuel processing, i.e. reforming and evaporation. If we look at the necessary temperature levels we find generally

$$T_{FC} \geqslant T_{ref} > T_{evap}. \tag{83}$$

This can be assumed for any SOFC system.

A reversible heat transport between these different temperature levels is possible for a SOFC as the heat source of the two Carnot cycles and for the reformer and the evaporator as their heat sinks. The real engines are: the heat engine HE1, operating between the SOFC and the reformer, and the heat engine HE2, operating between the SOFC and the evaporator. The exergetic efficiencies ζ_{HE} (Eq. (80)) describe the irreversibilities in the principal reversible processes. Finally a third heat engine HE3 must operate between the SOFC and the ambient state if the waste heat of the SOFC is not used completely in the system. The flue gas flow is divided into two streams. The air is heated in the air heater AH by cooling of the major stream of the flue gas (FGC). The reactant (of reforming) feed water is heated in the economiser from the ambient temperature T_0 to the evaporator temperature T_{evap} and the saturated steam is superheated from T_{evap} to the reformer temperature T_{ref}. The reactant (of reforming) methane is heated in the fuel heaters FH1 and FH2 from T_0 to T_{ref} and finally the products (of reforming) hydrogen and carbon monoxide (+ steam) are heated from T_{ref} to the T_{SOFC} in the product heater PH. The required heat is supplied by the cooling of the second pass of the flue gas (FGC) from T_{SOFC} to a waste gas temperature $> T_0$, by the SOFC directly (for PH), by the waste heat of HE1 (for $T < T_{ref}$) and by the waste heat of HE2 (for $T < T_{evap}$). An auxiliary burner must be used for the reforming process if the waste heat of the SOFC cannot cover the heat requirement of the reformer and the evaporator. This auxiliary burner is not shown in Figure 3.11. It is possible to use efficiencies η

$$\eta = \frac{\dot{Q}_{used}}{\dot{Q}_{supplied}} \tag{84}$$

to describe any real heating process (heat exchanger, burner). The internal reforming in the SOFC is included in this modelling of the system ($T_{SOFC} = T_{ref}$). The external reforming is included in this model as well if the heat engine HE1 is replaced by a burner. The parameters listed in Table 3.2 have been used for analysis of both the systems. The water surplus was fixed at 2 to avoid coke formation [9].

Table 3.2 Standard parameters for analysis of SOFC–heat engine hybrid cycles

SOFC temperature T_{SOFC}	900°C
Reformer temperature T_{ref}	750°C
Evaporator temperature T_{evap}	200°C
Ambient temperature T_0	25°C
Excess air λ	2
Water surplus n_W	2
Exergetic efficiency SOFC ζ_{SOFC}	0.60
Exergetic efficiency heat engine ζ_{HE}	0.70
Efficiency of air heater η_{AH}	0.90
Efficiency of heat exchangers η_{HEX}	0.98

The system efficiencies η_{syst} of combined SOFC cycles with integrated and external reforming have been calculated and compared [7,8]. The possible system efficiencies η_{syst} of systems with external reforming are about 5–7% lower than of a system with integrated reforming. The differences between the processes with external or integrated reforming are caused by the utilisation of the waste heat of the SOFC within the system. External reforming systems use an external burner with an additional entropy production. This increases the usable heat of the heat engine HE3 operating with the heat source SOFC and the heat sink environment and thus the waste heat of the system increases. An external reformer cannot be used as a heat sink of the waste heat of the SOFC and the entropy cannot be recycled as within an integrated reforming. An internal reforming in the SOFC has no temperature difference available for a power generation during the heat transport. This leads to a slight decrease of the system efficiency η_{syst} of the internal reforming compared with the integrated reforming.

An important influence on the performance of combined SOFC cycles with an integrated reforming is caused by the operation of the air heater under real conditions. The excess air λ and the efficiency η_{AH} of the air heater (Eq. (83))

are the main operational and design parameters. The system efficiency η_{syst} is plotted against excess air λ in Figure 3.12 with η_{AH} as a parameter. The basic chemical thermodynamics shows that $\Delta^r G$ – and therefore w_{tFCrev} (Eq. (5)) and $w_{tsysrev}$ (Eq. (78)) – is independent of the excess air λ. This can be used to prove the model because η_{syst} is independent of λ for $\eta_{AH} = 1$ as expected. η_{syst} decreases with increasing λ for all $\eta_{AH} < 1$. The influence of η_{AH} increases with increasing λ. The behaviour of the system for $\eta_{AH} = 0.85$ is shown Figure 3.12.

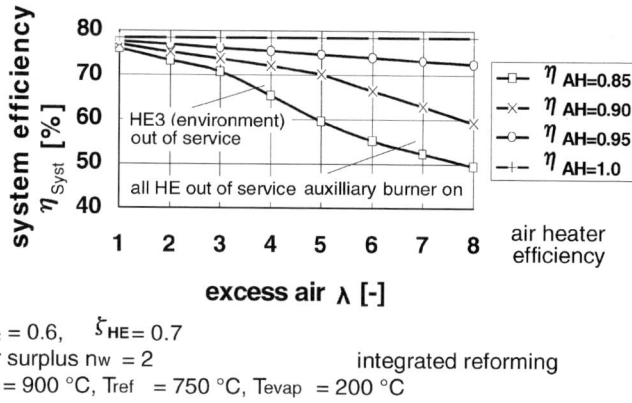

Figure 3.12 *The influence of the excess air λ and the efficiency η_{AH} of the heat transfer in the air heater on the system efficiency η_{syst} of the SOFC–heat engine hybrid cycle.*

First η_{syst} decreases slightly with an increasing λ, because the work of the heat engine HE3 decreases by compensating increasing heat losses. The other heat engines are operating at full load. The decrease of η_{syst} becomes sharper for an excess air $\lambda \approx 3$ because the heat engine HE3 goes out of service by a lack of available heat. The total waste heat of the SOFC must be used to supply the heat engines HE1 and HE2 which operate between the SOFC and the reformer and the evaporator respectively and to compensate the increasing heat loss of the air heater. This causes the sharper decrease of the system efficiency η_{syst} with an increasing λ in the region $3 < \lambda < 6$ by a decreasing supply of work by HE1 and HE2. In the region $\lambda > 6$ there is no heat engine in operation. All further heat losses (increasing with increasing excess air λ) must be compensated by the auxiliary burner. η_{syst} drops to values lower than 50% as shown in Figure 3.12. These results show that it is important to assure a good heat recovery in the air heater system and to avoid a very high excess air λ.

The system efficiency η_{syst} is influenced by the exergetic efficiency of the heat engines (HE1, HE2, HE3) ζ_{HE1}, ζ_{HE2}, ζ_{HE3} as well. η_{syst} is plotted against ζ_{HE} for each of the three heat engines in Figure 3.13. The maximum difference of about 9% in η_{syst} occurs if ζ_{HE3} (heat sink environment) is varied between 0 and 1. The difference is only about 2% if ζ_{HE1} (heat sink reformer) is varied. The variation of ζ_{HE2} (heat sink evaporator) leads to differences in η_{syst} of about 8%. The order of magnitude of these differences corresponds to the difference of the respective

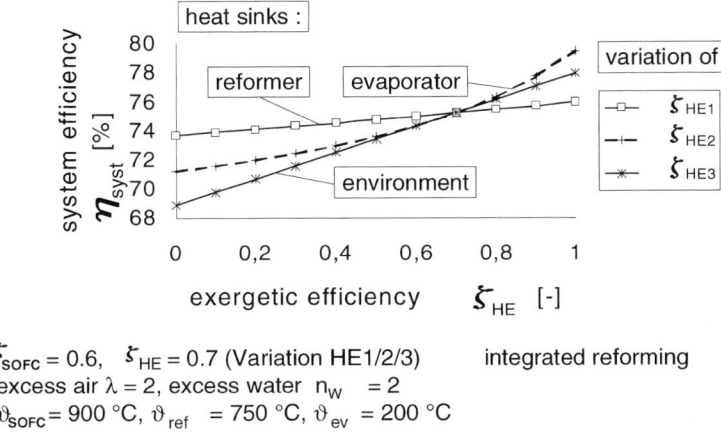

Figure 3.13 The influence of the heat engine design on the system efficiency η_{syst} of SOFC–heat engine hybrid cycles.

temperature differences between the SOFC and the heat sinks. This result confirms that entropy recycling by integrated reforming is more important to achieve high efficiencies than the power generation in the heat engine HE1. But the heat engine between the SOFC and the evaporator is important especially with a good exergetic efficiency ζ_{HE2}. The shape of η_{syst} of the variation of ζ_{HE2} seems to be unexpected compared with the variations of ζ_{HE1} and ζ_{HE3}.

But it can be explained easily if we look at the conditions that lead to the required amount of the SOFC's waste heat to supply the evaporator or the reformer. We obtain for the necessary waste heat Q_{SOFC} to operate the heat engines (HE1, HE2 = HEprocess) and to supply the processes (reformer or evaporator) with the necessary heat $Q_{process}$

$$Q_{SOFC} = \frac{Q_{process}}{1 - \zeta_{HEprocess} + \zeta_{HEprocess} \cdot T_{process}/T_{SOFC}} \tag{85}$$

This result is governed by the exergetic efficiency of the heat engine $\zeta_{HEprocess}$ and the relation of the temperature of the heat sink $T_{process}$ and the temperature of the heat source T_{SOFC} (T in K). Table 3.2 shows that the relation $T_{process}/T_{SOFC}$ is about 0.4 in the case of the evaporator and about 0.9 in the case of the reformer. Recycling of the anode outlet flow is another option to supply the reformer with steam.

3.6 Design Principles of SOFC Hybrid Systems

The process models as given in Figures 3.9 and 3.11 can be realised by different heat engines; Figure 3.14 gives an example. The heat source can be the flue gas

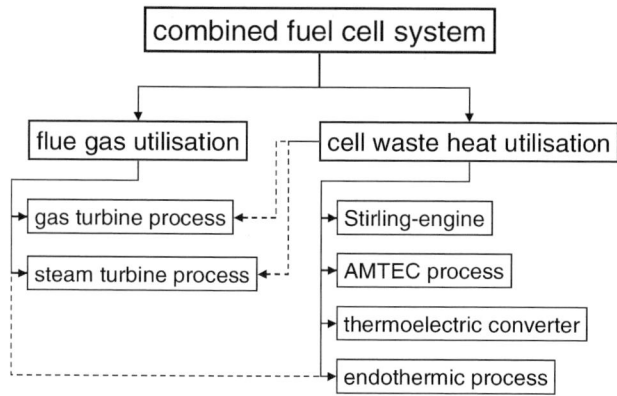

Figure 3.14   Possibilities of system integration in SOFC–heat engine hybrid cycles.

or the stack heat. Obviously a gas turbine (GT) or a waste heat boiler of a steam cycle can utilise the heat from the flue gas. But the total system integration may utilise the waste heat of the cell directly in both cases, as expressed by the dotted line in Figure 3.14. The different cycles based on a Carnot cycle with a separate process flow are other options for direct stack cooling. The use of a Stirling engine might be one option as the latest developments indicate [10]. A further option might be the conversion of heat to electricity by an AMTEC process [11]. The thermoelectric conversion might be a possibility to extract heat for electricity generation in smaller units as, for example, for defence applications [12]. The direct power generation in the last two options might be of specific interest for the electric system integration. Finally there is a further option to use the cell entropy in the sense of the second law of thermodynamics. Any endothermic process needs a transfer of heat at a certain temperature, and thus a certain supply of entropy [13]. This amount of entropy is a thermodynamic process requirement different from, for example, a heat supply for room heating that can be clearly reduced by a better heat recovery and a better insulation. However CHP for room heating might be the better commercial solution.

The SOFC-GT system is very interesting for high-efficiency power generation [14–16]. Any successful cooling strategy for SOFC of a SOFC-GT system must avoid high excess air at the system's outlet as shown above (see Figure 3.12). Figure 3.15 shows the possible strategies. The SOFC module can be divided in sub-modules and the heat of the SOFC module is extracted by cooling the waste air of the first sub-module to the inlet temperature of the cathode of the following sub-module by the power generation by a GT. This intermediate expansion (INEX) can be carried on until the last GT delivers the waste gas for the heat exchangers (HEX) to heat the air and the fuel.

The other strategy is the SOFC cooling by an external cooler (EXCO) fed with the flue gas that has been cooled by the heating of air and fuel. The SOFC module is the heat source for the GT cycle and the air is heated by the flue gas as in the generalised model. The integrated gas heater can be heated by radiation and allows an optimisation of the temperature level of the SOFC cooling together with

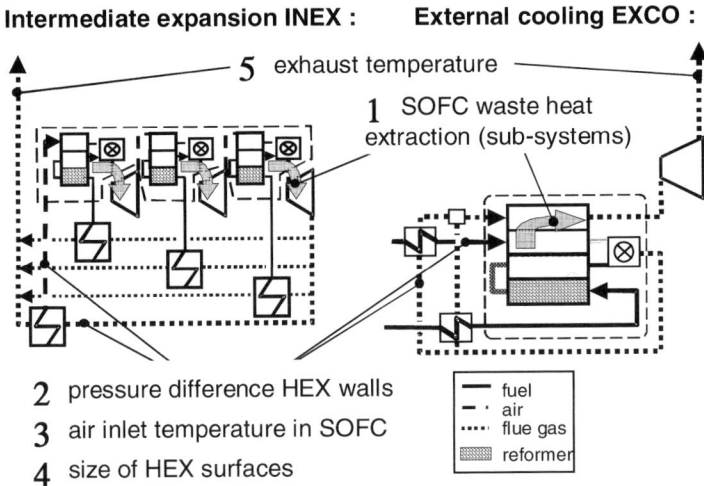

Figure 3.15 Cooling strategies for SOFC modules by GT cycles.

an integrated air heater and avoids unacceptable thermal stresses. The main differences between INEX and EXCO are indicated in Figure 3.15 and compared in Table 3.3 [17,18].

Table 3.3 Comparison of the INEX and the EXCO design

		INEX	EXCO
1	SOFC waste heat extraction (sub-systems)	2–n pressure levels (systems)	1 pressure level (system)
2	Pressure difference HEX walls	Maximal pressure difference	Only pressure loss
3	Limit for air inlet temperature in SOFC	Gas turbine outlet temperature	SOFC temperature
4	Size of HEX surfaces	Min. 1/2.5 of ambient system	Min. 1/7 of ambient system
5	Exhaust temperature	~200°C	500–600°C

1 The waste heat is extracted at one pressure level in the EXCO design and at up to n pressure levels in the INEX design, depending on the allowable temperature difference of the cathode. The number of pressure levels is equal to the number of the pressurised subsystems.

2 The pressure difference on the HEX walls of the air heater is the maximum pressure in the INEX design and the only pressure loss in the module in the EXCO design.

3 The air inlet temperature of the SOFC module is limited by the SOFC (module outlet) temperature in the EXCO design and by the lower GT outlet temperature in the INEX design.

4 The size of the HEX of an on one side pressurised INEX design is about 2.5 times smaller than under ambient conditions, but the on both sides pressurised EXCO design has up to 7 times smaller HEX surfaces.

5   The electric efficiency of an INEX design with two turbines is about 70% [19], similar to the EXCO design. The exhaust temperature in the INEX design is about 200°C and that of the EXCO design is about 500–600°C depending on the individual parameters.

The EXCO design has thus the potential for a combination with a steam turbine cycle (ST) that could be, for example, a Cheng cycle. This leads to an electric efficiency of about 75% [20]. The first studies [20] of the EXCO design included a reheat cycle with an additional heat exchanger within the SOFC module. This design seemed to be too complicated. But a comparison of both designs shows that the benefit of the EXCO design to reduce the excess air in one process step at one pressure level with small HEXs can be combined with the benefit of the INEX design to allow a simple cascading of GT cycles as needed for a reheat GT cycle. This led to the proposal of the reheat SOFC-GT cycle combined with a steam turbine (ST) cycle which reaches slightly more than 80% as the calculated efficiency [21].

3.7 Summary

Thermodynamic considerations are used to understand the processes of energy conversion in SOFCs. Such theoretical studies of the behaviour of the reversible processes have a high practical value in helping to understand complex systems. The reversible work of a fuel cell is defined by the free or Gibbs enthalpy of the reaction. If we use the assumption of the ideal gas we immediately get the equation of the Nernst voltage from the Gibbs enthalpy of the reaction. The consideration of the electrical effects shows that the molar flow of the spent fuel is proportional to the electric current and the reversible work is proportional to the reversible voltage. A coupling between the thermodynamic data and the electrical data is only possible using the quantities power or heat flow and not by using work and heat. This is caused by the fact that we use a mass or substance transport as the basis for thermodynamic considerations and we use a charge transport to describe electrical phenomena.

Irreversible losses cause a difference in the efficiency of reversible and real processes. These losses can be described and quantified by their irreversible entropy production. The consideration of the ohmic losses shows that the irreversible entropy production in a SOFC is smaller than in another low-temperature fuel cell. This is caused by the lower irreversible entropy production of the heat dissipated at a higher temperature. The effects of the irreversible mixing of reactants and products lead to an irreversible entropy production as well that reduce the cell voltage. The changes in the Nernst voltage can be understood by the analysis of the fuel utilisation.

Because all the fuel cannot be fully reacted in practice within the fuel cell, the SOFC stack can be treated like a power generating burner so as to integrate it easily into a system model. The stack cooling depends on the amount of excess air.

The combination of a SOFC with a heat engine allows an extremely high electric efficiency. Any real combination of a SOFC and a heat engine is based on a reversible system but a simplified version can be used to analyse the principles of the design of a combined SOFC–heat engine. It is important that the cell itself and not the flue gas is considered as the heat source. Integration of fuel processing is another important factor in achieving a high efficiency since the embedded fuel processor can be supplied with a certain amount of entropy from the cell heat. This entropy is used for the reforming reaction and need not be transported to the environment as an entropy loss.

References

[1]   K. Wark, *Advanced Thermodynamics for Engineers*, McGraw-Hill, New York (1995).
[2]   W. Winkler, *Brennstoffzellenanlagen*, Springer Verlag, Berlin (2002).
[3]   W. Winkler, *Brennstoff-Wärme-Kraft*, **46** (1994) Heft 7/8, pp. 334–340.
[4]   U. Bossel, Facts and figures, Final report on SOFC Data, IEA Programme of R, D&D on Advanced Fuel Cells, Annex II: Modelling & Evaluation of Advanced SOFC, Swiss Federal Office of Energy; Operating Agent Task II, Berne, April 1992.
[5]   W. Winkler, *Brennstoff-Wärme-Kraft*, **45** (1993) Heft 6, pp. 302–307.
[6]   M. C. Williams, in *Solid Oxide Fuel Cells VI*, eds. S. C. Singhal and M. Dokiya, The Electrochemical Society Proceedings, Pennington, NJ, PV99-19, 1999, pp. 3–9.
[7]   W. Winkler, in *Proceedings of the First European SOFC Forum*, ed. Ulf Bossel, Switzerland, 1994, pp. 821–848.
[8]   W. Winkler, in *Proceedings 2nd International Fuel Cell Conference*, Kobe, Japan, 1996, pp. 397–400.
[9]   A. Gubner, *Grundlagen der Modellbildung für die Methanbildung in Hochtemperaturbrennstoffzellen*, Thesis, University of Applied Sciences, Hamburg (1992).
[10]   10th International Stirling Engine Conference 2001 (10th ISEC), 24–26 September 2001, Osnabrück, VDI Gesellschaft für Energietechnik.
[11]   J.-M. Tournier, M. S. El-Genk and L. Huang, *Experimental investigations, modeling, and analysis of high-temperature devices for space applications*. Final Report, AFRL-VS-PS-TR-1998-1108. (US) Air Force Research Laboratory, Kirkland Air Force Base, January 1999.
[12]   R. J. Nowak, *A DARPA Perspective on Small Fuel Cells for the Military*. Solid State Energy Conversion Alliance (SECA) Workshop, Arlington, VA, 29 March 2001.
[13]   R. Kikuchi, K. Sasaki and K. Eguchi, in *Solid Oxide Fuel Cells VII*, eds. H. Yokokawa and S. C. Singhal, The Electrochemical Society Proceedings, Pennington, NJ, PV2001-16, 2001, pp. 214–223.

[14] M. C. Williams and C. M. Zeh, *Workshop on Very High Efficiency Fuel Cell/Gas Turbine Power Cycles*. US Department of Energy – Office of Fossil Energy, Morgantown Energy Technology Center, 1995.
[15] R. A. George, *Journal of Power Sources*, **86** (2000) 134–139.
[16] M. Koyama, H. Komiyama, K. Tanaka and K. Yamada, in *Solid Oxide Fuel Cells VII*, eds. H. Yokokawa and S. C. Singhal, The Electrochemical Society Proceedings, Pennington, NJ, PV2001-16, 2001, pp. 234–243.
[17] W. Winkler, in *Proceedings of the 3rd European SOFC Forum*, ed. P. Stevens. Switzerland, 1998, pp. 525–534.
[18] W. Winkler, in *Solid Oxide Fuel Cells VI*, eds. S. C. Singhal and M. Dokiya, The Electrochemical Society Proceedings, Pennington, NJ, 1999, pp. 1150–1159.
[19] W. G. Parker and F. P. Bevc, in *Proceedings 2nd International Fuel Cell Conference*, Kobe, Japan, 1996, pp. 275–278.
[20] W. Winkler, *Brennstoff-Wärme-Kraft*, **44** (1992) Heft 12, pp. 533–538.
[22] W. Winkler and H. Lorenz, in *Proceedings of the 4th European SOFC Forum*, ed. A. J. McEvoy, Switzerland, 2000, pp. 413–420.

Chapter 4

Electrolytes

Tatsumi Ishihara, Nigel M. Sammes and Osamu Yamamoto

4.1 Introduction

The electrolyte for solid oxide fuel cells (SOFCs) must be stable in both reducing and oxidising environments, and must have sufficiently high ionic with low electronic conductivity at the cell operating temperature. In addition, the material must be able to be formed into a thin, strong film with no gas leaks. Until now, stabilised zirconia, especially yttria-stabilised zirconia, possessing the fluorite structure, has been the most favoured electrolyte for SOFCs. Other fluorite structured oxide ion conductors, such as doped ceria, have also been proposed as the electrolyte materials for SOFCs, especially for reduced temperature operation (600–800°C). More recently, a number of other materials, including perovskites, brownmillerites and hexagonal structured oxides, have also been found to possess good ionic conductivity. This chapter first describes fundamental and practical aspects of fluorite structured electrolytes, and then it proceeds to discuss the structure and properties of perovskites and other ion conductors.

4.2 Fluorite-Structured Electrolytes

Oxide ion conductivity was first observed in ZrO_2 containing 15 wt% Y_2O_3 (yttria-stabilised zirconia or YSZ) by Nernst [1] in the 1890s. In 1937, Baur and Preis [2] constructed the first solid oxide fuel cell using this electrolyte. Since that time, many oxide systems have been examined as potential electrolytes for SOFCs. An excellent review of solid oxide electrolytes was presented by Etsell and Flengas in 1970 [3], while more recent conductivity data are summarised by Minh and Takahashi [4]. Figure 4.1 shows the temperature dependence of the ionic conductivity for several oxides, indicating that YSZ is by no means the best oxide ion conductor.

Bismuth oxide compositions [5] show the highest conductivity and several other formulations are also superior to YSZ, particularly at temperatures below

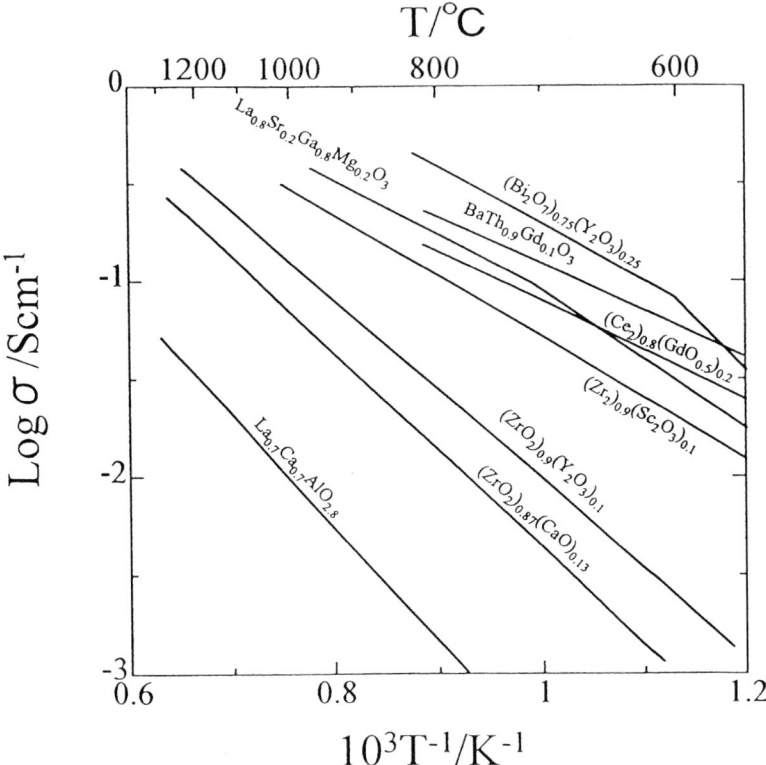

Figure 4.1   Temperature dependence of electrical conductivity for selected oxide ion conductors.

600°C [6]. However, these other oxides have disadvantages such as electronic conductivity, high cost, or difficulties in processing. The candidate electrolytes can be divided into two major structures, the fluorite structure like YSZ discussed here and the perovskite structure like lanthanum gallate discussed in Section 4.6.

The fluorite structure is a face-centred cubic arrangement of cations with anions occupying all the tetrahedral sites, leading to a large number of octahedral interstitial voids. Thus this structure is a rather open one and rapid ion diffusion might be expected. At high temperatures, zirconia has the fluorite structure, stabilised by addition of divalent or trivalent (i.e. aliovalent) cations such as Ca or Y at lower temperatures. Pure ceria also has the fluorite structure. Oxide ion conduction is provided by oxide ion vacancies and interstitial oxide ions. Intrinsic defects are fixed by thermodynamic equilibrium in pure compounds, while extrinsic defects are established by the presence of aliovalent dopants. To maintain electroneutrality, a soluble aliovalent ion in an ionic compound is compensated by an increase in the concentration of an ionic defect [7]. In the case of pure ZrO_2 and CeO_2, electrical conductivity is quite low because the concentration of the oxide ion vacancies and interstitial oxide ions is low. However, as dopants such as yttria are added, the conductivity increases.

The dissolution of yttria into the fluorite phase of ZrO_2 can be written by the following defect equation in Kroger–Vink notation [8]:

$$Y_2O_3(ZrO_2) \rightarrow 2Y'_{Zr} + 3O^x_o + V_o^{\cdot\cdot} \tag{1}$$

Each additional yttria molecule creates one oxygen vacancy. The concentration of the vacancies is given simply by the electrical neutrality condition, for this case, $2[Y'_{Zr}] = [V_o^{\cdot\cdot}]$, implying that the vacancy concentration is linearly dependent on the dopant level. The ionic conductivity, σ, can be expressed by

$$\sigma = en\mu \tag{2}$$

where n is the number of mobile oxide ion vacancies, μ their mobility, and e the charge. In the case of oxide ion conductors such as doped zirconia and ceria, Eq. (2) gives Eq. (3) with the fraction of mobile oxide ion vacancies, $[V_o^{\cdot\cdot}]$, and the fraction of unoccupied oxide ion vacancies, $[V_o^{\cdot\cdot}]^{-1}$. To move through the crystal, the ions must be able to move into an unoccupied equivalent site with a minimum of hindrance, thus

$$\sigma = A/T[V_o^{\cdot\cdot}]([V_o^{\cdot\cdot}]^{-1})\exp(-E/RT) \tag{3}$$

where E is the activation energy for conduction, R the gas constant, T absolute temperature, and A the pre-exponential factor [9]. The conductivity of doped zirconia and doped ceria varies as a function of dopant concentration, and shows a maximum at a specific concentration. However, this maximum occurs at a much lower concentration than that expected from Eq. (3). An example of this behaviour is shown in Figure 4.2 as reported by Arachi et al. [10] for the ZrO_2–M_2O_3 (M=Sc, Yb, Y, Dy, Gd or Eu) systems.

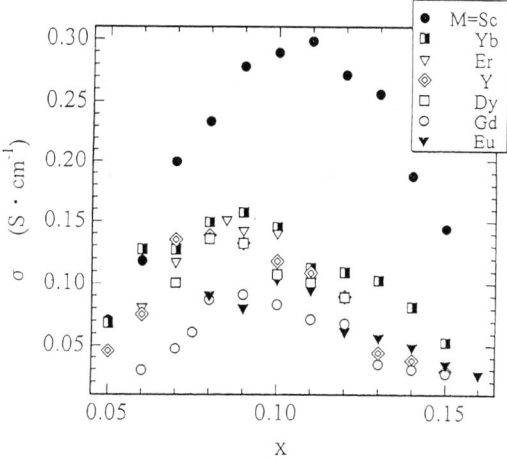

Figure 4.2 Composition dependence of the electrical conductivity at 1000°C for ZrO_2–M_2O_3 compositions.

Attempts to explain this conductivity behaviour have been made by Baker et al. [11] and Hohnke [12], involving clusters in the first and second coordination shells, and by Carter and Roth [13] based on structural effects. The relation between the dopant concentration with the highest conductivity at 1000°C and dopant ionic radius in the ZrO_2–M_2O_3 system [10] is shown in Figure 4.3. The content of dopant with the highest conductivity decreases with increasing radius of dopant ion. The dopants, Dy^{3+} and Gd^{3+}, with higher ionic radii show a limiting value of 8 mol%. The dopant Sc^{3+}, which has the closest ion radius to the host ion, Zr^{4+}, shows the highest conductivity and the highest dopant content at the maximum conductivity. Similar conductivity dependence on the dopant level was observed in the CeO_2 system. The highest conductivity was found at 10 mol% for Sm_2O_3 and at 4 mol% for Y_2O_3 dopants. The diffusion of oxide ion vacancies is affected by the elastic strain energy, which is related to the size mismatch between the host and dopant cations [14].

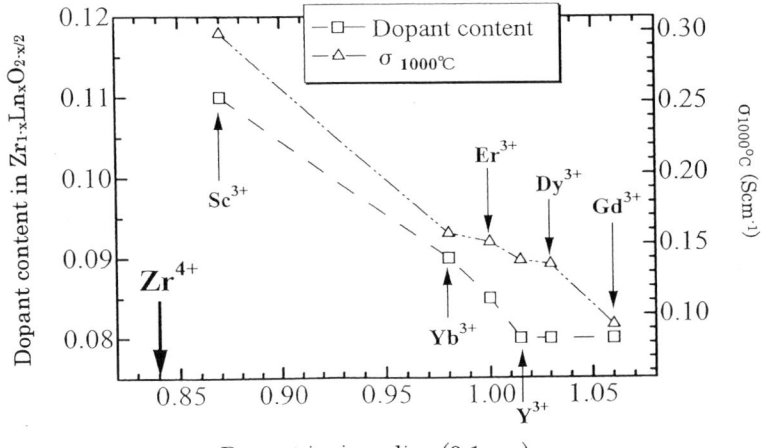

Figure 4.3 Dopant concentration exhibiting the highest conductivity dependence on dopant cation radius in ZrO_2–M_2O_3 systems.

Based on the comments of Nowick [15] and Kilner and Steele [16] who emphasised the importance of defect pairs formed due to interaction between the oxide ion vacancies, $V_o^{\cdot\cdot}$, and aliovalent cations, M_{Ce}', in CeO_2:

$$V_o^{\cdot\cdot} + M_{Ce}' = (V_o^{\cdot\cdot} M_{Ce}') \tag{4}$$

and

$$V_o^{\cdot\cdot} + 2M_{Ce}' = (V_o^{\cdot\cdot} 2M_{Ce}') \tag{5}$$

Manning et al. [17] suggested that $(V_o^{\cdot\cdot} M_{Ce}')$ is more likely to occur because of the expected random distribution of M_{Ce}'. Kilner and Brook [14] have shown that the effect of the binding enthalpy of an associated ion can significantly affect the population of free vacancies at low temperatures. At lower temperatures, association is almost complete, so that

$$[V_o^{\cdot\cdot}M_{Ce}']^{\cdot} \gg [V_o^{\cdot\cdot}] \tag{6}$$

and

$$[V_o^{\cdot\cdot}] = (B/T)\exp(-E_a/RT) \tag{7}$$

where E_a is the association binding energy and B a constant. Kilner [18] has pointed out that both calculated and experimental data for the association enthalpies could be correlated with the ionic radius of the dopant. Experimental [19] and calculated [20] E_a values in CeO_2–M_2O_3 are shown in Figure 4.4 The minimum association enthalpy occurred when the ionic radii of the host and the dopant were close to each other such that the lattice elastic strain was minimised. Kilner further postulated that this was a universal effect, common to all acceptor-doped oxides.

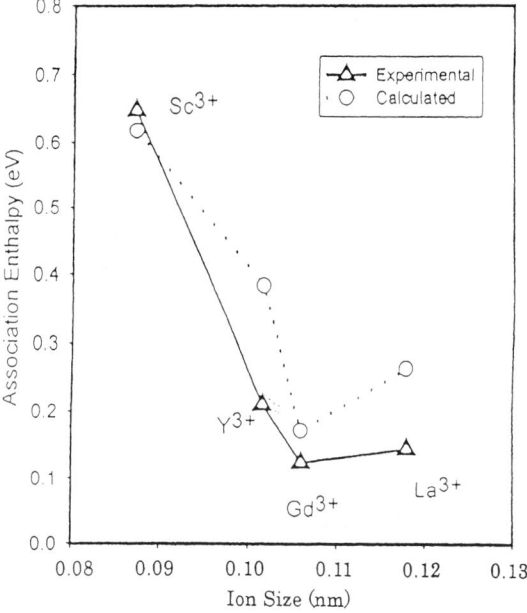

Figure 4.4 Calculated and experimental vacancy association enthalpy for oxygen conductivity in doped ceria plotted against the ionic radii of the dopant ions.

Arachi et al. [10] reported dependence of the association enthalpy on the dopant ion radius in ZrO_2–M_2O_3 systems. The activation energy for conduction, E, in Eq. (3) is expressed as the sum of the enthalpy for motion, E_m, and the association binding energy, E_a, thus

$$E = E_m + E_a \tag{8}$$

At higher temperatures the complex $(V_o^{\cdot\cdot}\text{-}M_{Zr}')^{\cdot}$ dissociates completely to free $V_o^{\cdot\cdot}$ and M_{Zr}'. The concentration of $V_o^{\cdot\cdot}$ is independent of the temperature and equal to the total concentration of dopant M^{3+}. Therefore, the migration enthalpy, E_m, could be estimated from the slope of the temperature dependence for conduction in the higher temperature range. The association enthalpy could be calculated from the difference in the slopes at the lower and the higher temperature ranges. In Figure 4.5 the dependence of ion migration enthalpy and association enthalpy on the dopant ion radius is shown along with the electrical conductivity at 1000°C.

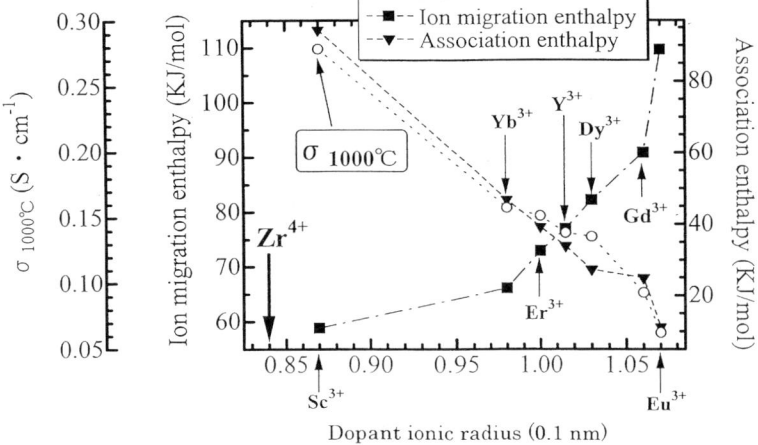

Figure 4.5 Ion migration enthalpy and association enthalpy versus dopant ionic radius.

The Sc^{3+}-doped zirconia shows the lowest ion migration enthalpy and highest association enthalpy because Sc^{3+} has the closest ion radius to Zr^{4+}. The migration enthalpy increases with increasing dopant ion radius. The high ion migration enthalpy with a dopant with different ion size than that of the host Zr^{4+} is explained by the elastic strain energy induced in the cation lattices by the size difference. On the other hand, the association enthalpy between the oxide ion vacancies and dopant cation decreases with increasing dopant cation radius. Butler et al. [21] have calculated the association binding energy for doped zirconia, defined with respect to the total energy of isolated defects that enter the associate. The calculated values were 27 kJ/mol for $(V_o^{\cdot\cdot}Y_{Zr}')$ and 16 kJ/mol for $(V_o^{\cdot\cdot}Gd_{Zr}')$. The experimental results were in good agreement with the calculated values in both ZrO_2–M_2O_3 and CeO_2–M_2O_3 systems, with the maximum oxide ion conductivity being found in the solid solution of M^{3+} having the ionic radius closest to that of the host cation.

Properties and fabrication of two of the most common fluorite structured electrolyte materials, zirconia based and ceria based, are discussed below.

4.3 Zirconia-Based Oxide Ion Conductors

Zirconia exhibits three polymorphs. It has monoclinic structure at room temperature, changing to tetragonal above 1170°C and to the cubic fluorite structure above 2370°C. The addition of a dopant such as yttria stabilises the fluorite and tetragonal phases down to room temperature, leading to an increase in the oxide vacancy concentration. Yashima et al. [22] surveyed the phase diagrams of doped zirconia systems, including the ZrO_2-Y_2O_3 [23] (shown in Figure 4.6) and ZrO_2-Sc_2O_3 [24] systems.

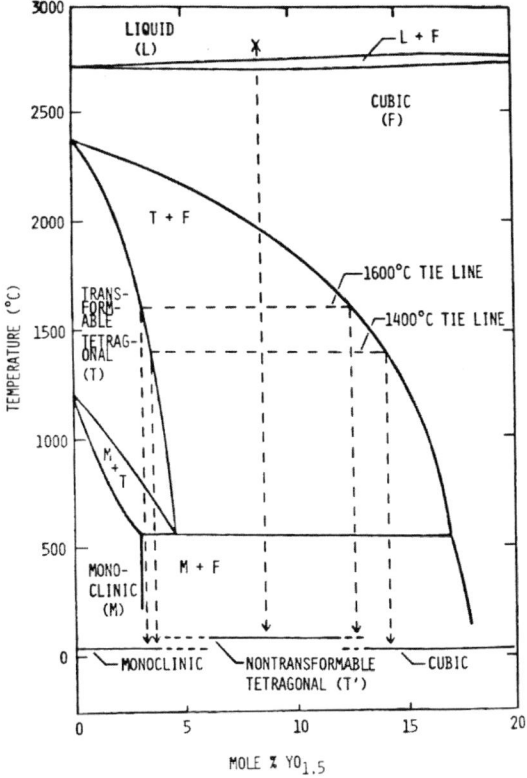

Figure 4.6 Phase diagram of ZrO_2-Y_2O_3 system.

The temperature dependence of the electrical conductivity for zirconia based oxides [25] is shown in Figure 4.7. Yttria-doped zirconia (YSZ), which has been used extensively in SOFCs, shows a conductivity of 0.14 S/cm at 1000°C. Scandia-doped zirconia (SSZ) has a higher conductivity and at 780°C its value corresponds to that of YSZ at 1000°C.

In Figure 4.7, the thickness of the electrolyte possessing a resistance of 0.2 Ω cm^2 is shown on the right-hand side of the vertical axis. The 0.2 Ω cm^2 resistance corresponds to 0.1 V lost due to the electrolyte resistance at 0.5 mA/cm^2, where it is assumed that the cell voltage should be 0.7 V to maintain a total energy

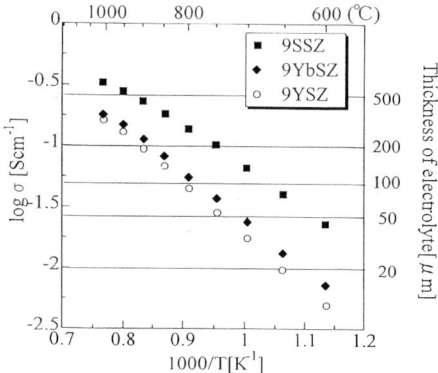

Figure 4.7 Temperature dependence of the electrical conductivity in the $ZrO-M_2O_3$ systems. 9SSZ, 9 mol% $Sc_2O_3-ZrO_2$; 9YbSZ, 9 mol% $Yb_2O_3-ZrO_2$; 9YSZ, 9 mol% $Y_2O_3-ZrO_2$.

efficiency of greater than 50%, and the cathode, the anode, and the electrolyte overvoltages contribute equally to the cell voltage drop. The relationship between the temperature and the electrolyte thickness suggests that at an operation temperature of 700°C, the thickness of the electrolyte should be less than 50 μm in the YSZ and ytterbia-doped zirconia (YbSZ) systems, while an electrolyte as thick as 150 μm could be used in the SSZ system. A 50 μm thick electrolyte is difficult to use in an electrolyte-supported cell configuration, because of the fragile nature of such a thin electrolyte sheet. As shown in the latter part of this chapter, dense YSZ films of several tens of microns thickness have been fabricated on tubular and planar supports by vapour deposition and tape casting methods.

Scandia-doped zirconia is also attractive as the electrolyte for SOFCs, especially for intermediate temperature (600–800°C) SOFCs. However, it undergoes aging on long-term exposure at high temperatures [26,27]. ZrO_2 with 8 mol% Sc_2O_3 exhibited a significant aging effect with annealing at 1000°C. Its conductivity of 0.3 S/cm at 1000°C (as sintered) decreased to 0.12 S/cm after aging at 1000°C for 1000 h. This conductivity value after aging is comparable to that of ZrO_2 with 9 mol% Y_2O_3. On the other hand, ZrO_2 with 11 mol% Sc_2O_3 showed no aging effect on annealing at 1000°C for more than 6000 h. ZrO_2 with 11 mol% Sc_2O_3 shows a phase transition from the rhombohedral structure (low-temperature phase) to the cubic structure (high-temperature phase) at 600°C with an accompanying small volume change. The cubic phase is stabilised at room temperature by the addition of a small amount of CeO_2 [28] and Al_2O_3 [29]. The conductivity of SSZ with CeO_2 and Al_2O_3 is slightly lower than that of the undoped SSZ. Similar aging effects have been observed in other zirconia-based oxide ion conductors. In Table 4.1, the conductivity changes in the $ZrO_2-M_2O_3$ system by annealing at 1000°C for 1000 h are summarised.

For electrolytes, high electrolyte strength and toughness are also desirable in addition to high electrical conductivity, especially in planar cell configurations. The bending strengths of zirconia-based electrolytes along with their thermal expansion coefficients are also shown in Table 4.1. SSZ shows as good a set of

Table 4.1 Electrical conductivity, bending strength, and thermal expansion coefficient of zirconia-based electrolytes

Electrolyte	Conductivity at 1000°C (S/cm)		Bending strength (MPa)	Thermal expan. coeff. (1/K × 10^6)
	As sintered	After annealing		
ZrO_2–3 mol%Y_2O_3	0.059	0.050	1200	10.8
ZrO_2–3 mol%Yb_2O_3	0.063	0.09		
ZrO_2–2.9 mol%Sc_2O_3	0.090	0.063		
ZrO_2–8 mol%Y_2O_3	0.13	0.09	230	10.5
ZrO_2–9 mol% Y_2O_3	0.13	0.12		
ZrO_2–8 mol%Yb_2O_3	0.20	0.15		
ZrO_2–10 mol%Yb_2O_3	0.15	0.15		
ZrO_2–8 mol%Sc_2O_3	0.30	0.12	270	10.7
ZrO_2–11 mol%Sc_2O_3	0.30	0.30	255	10.0
ZrO_2–11 mol%Sc_2O_3–1 wt% Al_2O_3	0.26	0.26	250	

mechanical properties as YSZ. ZrO_2 with 11 mol% Sc_2O_3 and 1 wt% Al_2O_3 appears as one of the best candidates for an intermediate temperature SOFC because of its high oxide ion conductivity, phase stability and excellent mechanical properties.

SOFC electrolytes should of course be stable under fuel conditions, such that the electronic conductivity remains negligible compared to the ionic conductivity. Figure 4.8 shows the dependence of both electronic and ionic conductivities on oxygen partial pressure for YSZ [30]. At oxygen partial pressures of 10^{-30} atm, the electronic conductivity can be comparable to the ionic contribution, but this oxygen pressure is far below the normal SOFC operating range of 0.21–10^{-20} atm.

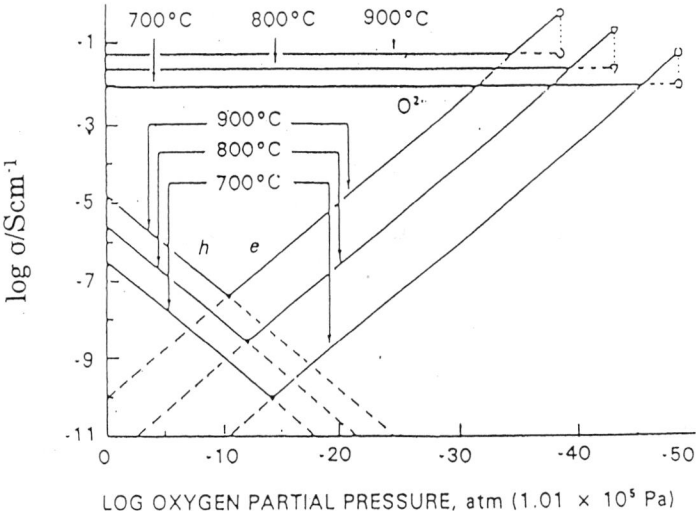

Figure 4.8 Electrical conductivity of YSZ as a function of oxygen partial pressure.

4.4 Ceria-Based Oxide Ion Conductors

Doped ceria has been suggested as an alternative electrolyte for low temperature SOFCs [6, 31, 32]. Reviews on the electrical conductivity and conduction mechanism in ceria-based electrolytes have been presented by Mogensen et al. [33] and Steele [34]. Ceria possesses the same fluorite structure as the stabilised zirconia. Mobile oxygen vacancies are introduced by substituting Ce^{4+} with trivalent rare earth ions as shown in Eq. (1). The conductivity of doped ceria systems depends on the kind of dopant and its concentration. A typical dopant concentration dependence of the electrical conductivity in the $(CeO_2)_{1-x}(Sm_2O_3)_x$ system as reported by Yahiro et al. [35] is shown in Figure 4.9.

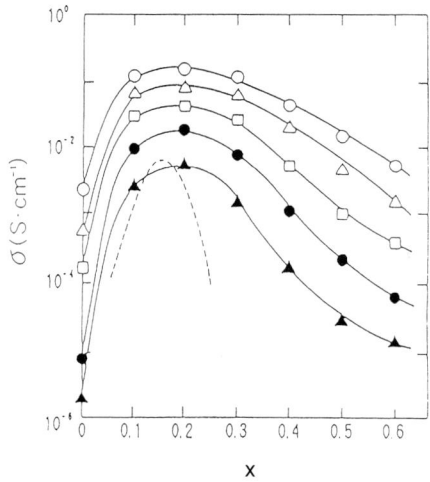

Figure 4.9 Concentration dependence of electrical conductivity for CeO_2–Sm_2O_3. (○) 900°C; (△) 800°C; (□) 700°C; (●) 600°C; (▲) 500°C; (-----) $(ZrO_2)_{1-x}(CaO)_x$ at 900°C.

The maximum conductivity is observed at around 10 mol% Sm_2O_3. The conductivity of the CeO_2–Ln_2O_3 system depends on the dopant (Ln) ionic radius, and is summarised in Figure 4.10 [36]. The binding energy calculated by Butler et al. [20] shows a close relationship to the conductivity as also illustrated in this figure, the dopant with low binding energy exhibiting higher conductivity.

In Table 4.2, the conductivity data for doped ceria are summarised. CeO_2–Gd_2O_3 and CeO_2–Sm_2O_3 show an ionic conductivity as high as 5×10^{-3} S/cm at 500°C, corresponding to 0.2 Ω cm^2 ohmic loss for an electrolyte of 10 µm thickness. These compositions are attractive for low temperature SOFCs and have been extensively examined.

Ceria-based oxide ion conductors are reported to have purely ionic conductivity at high oxygen partial pressures. At lower oxygen partial pressures, as prevalent on the anode side of an SOFC, these materials become partially reduced. This leads to electronic conductivity in a large volume fraction of the electrolyte extending from the anode side. When a cell is constructed with such

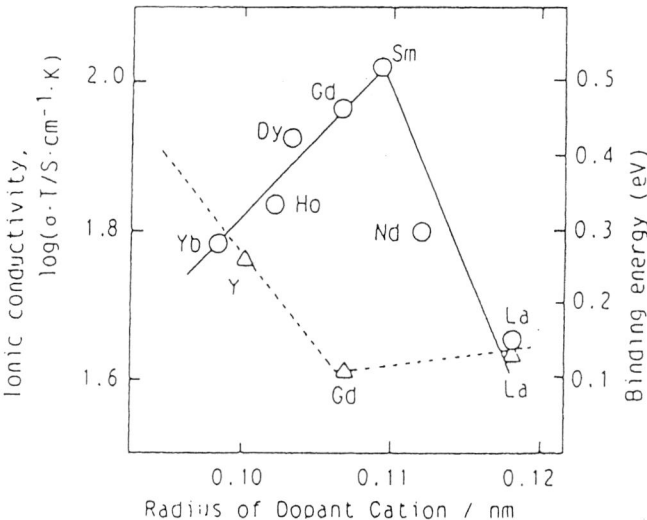

Figure 4.10 Dependence of ionic conductivity for $(CeO_2)_{0.8}(Ln_2O_3)_{0.2}$ at 800°C on ionic radius of Ln^{3+}.

Table 4.2 Electrical conductivity data for CeO_2–Ln_2O_3

Ln_2O_3	Mol%	Conductivity (S/cm)		Activation energy (kJ/mol)	Reference
		700°C	500°C		
Sm_2O_3	10	3.5×10^{-2}	2.9×10^{-3}	68	33
	10	4.0×10^{-2}	5.0×10^{-3}	75	31
Gd_2O_3	10	3.6×10^{-2}	3.8×10^{-3}	70	33
Y_2O_3	10	1.0×10^{-2}	0.21×10^{-3}	95	31
CaO	5	2.0×10^{-2}	1.5×10^{-3}	80	33

an electrolyte with electronic conduction, electronic current flows through the electrolyte even at open circuit, and the terminal voltage is somewhat lower than the theoretical value. In Figure 4.11, total electrical conductivity (ionic and electronic) of $Ce_{0.8}Sm_{0.2}O_{1.9-\delta}$ at different temperatures is shown as a function of oxygen partial pressure.

Godickemeier and Gauckler [37,38] analysed the efficiency of cells with $Ce_{0.8}Sm_{0.2}O_{1.9}$ by consideration of the electronic conduction. The maximum efficiency based on Gibbs free energy was 50% at 800°C and 60% at 600°C. An SOFC with $Ce_{0.8}Sm_{0.2}O_2$ electrolyte should be operated at temperatures below about 600°C to avoid such efficiency loss due to electronic leakage. For example, ceria based electrolytes have been used in SOFCs operating at 550°C and lower. If higher temperature operation is required, then the electronic conduction can be prevented by protecting the ceria electrolyte with a thin coating of YSZ on the anode side [39]. However, interdiffusion at the YSZ/ceria interface could then be an issue.

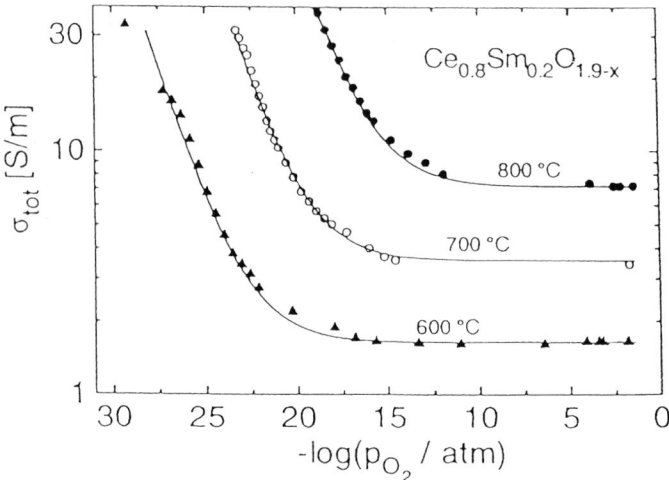

Figure 4.11 Total electrical conductivity of $Ce_{0.8}Sm_{0.2}O_{1.9-\delta}$ as a function of oxygen partial pressure.

4.5 Fabrication of ZrO_2- and CeO_2-Based Electrolyte Films

The electrolyte film can be fabricated by a number of processes depending upon the configuration of the cells. For tubular SOFCs, an electrochemical vapour deposition (EVD) technique was developed by Westinghouse Electric Corporation (now Siemens Westinghouse Power Corporation) in 1977 [40] to fabricate gas-tight thin layers of doped zirconia. This EVD process involved growing a dense oxide layer on a porous substrate at elevated temperatures and reduced pressures, as described in Figure 4.12 [41].

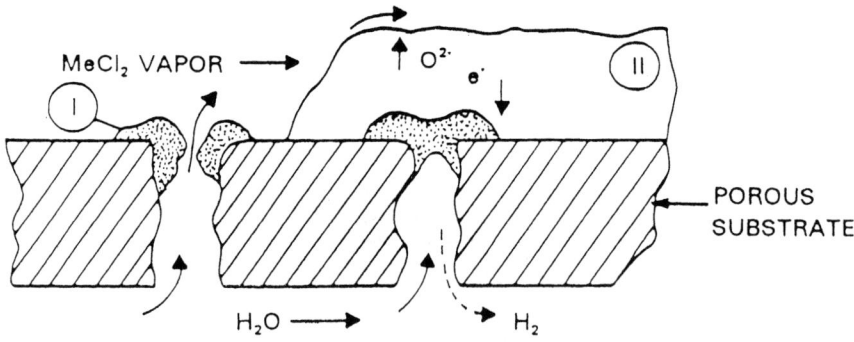

Figure 4.12 Principle of electrochemical vapour deposition (EVD).

Under the operating conditions of the EVD process, the oxide exhibits both oxide ion and electronic conductivity. Thus, the oxide ion flux during the oxide growth is balanced by an electron flux, thereby preserving the electroneutrality of the oxide. The growth rate of the oxide is commonly described by the classical parabolic rate law as in the Wagner oxidation process. Thus, the film thickness, L, is related to the deposition time, t, by the following equation:

$$L^2 = 2k_p t \tag{9}$$

where k_p is the parabolic constant. At the deposition temperatures of 1000–1200°C, the parabolic rate constant for the deposition of YSZ ranges from 1.1×10^{-5} to 3.8×10^{-3} cm²/s. The thickness of the YSZ film in a typical tubular SOFC is about 40 μm and it takes about 40 minutes to make this film at 1000°C [41]. The EVD process has been successfully used by Westinghouse for the production of thousands of tubular cells for several multikilowatt power generation systems.

More conventional slurry dipping/sintering techniques have also been used to prepare electrolyte films for tubular SOFCs, starting with a porous support which can be of either anode, cathode or an inert material. Toto Ltd in Japan prepared YSZ films on large size porous (La,Sr)MnO₃ tubes, 22 mm in diameter and 900 mm in length [42] by a slurry dipping/sintering method. Thickness and gas tightness of the YSZ layer depended on the speed of tube withdrawal from the slurry, the number of dippings, and the viscosity of the slurry. Song et al. reported that a dense YSZ film was obtained on a porous Ni-YSZ tube of effective area 20 cm² at a withdrawal speed of 22 mm/s. A YSZ film of approximately 20 μm in thickness was obtained from two slurry coats [43].

Electrolytes for planar SOFCs are often prepared by conventional tape casting, shown schematically in Figure 4.13 [4]. Tape casting slurries for YSZ electrolyte tape are prepared by dispersing YSZ powder in solvents such as 2-butanone/ ethanol, after which binders such as polyvinyl butyral, plasticisers such as polyethylene glycol and a deflocculant/wetting agent such as glycerol trioleate are added. Flat YSZ plates 50–250 μm thick have been fabricated using this

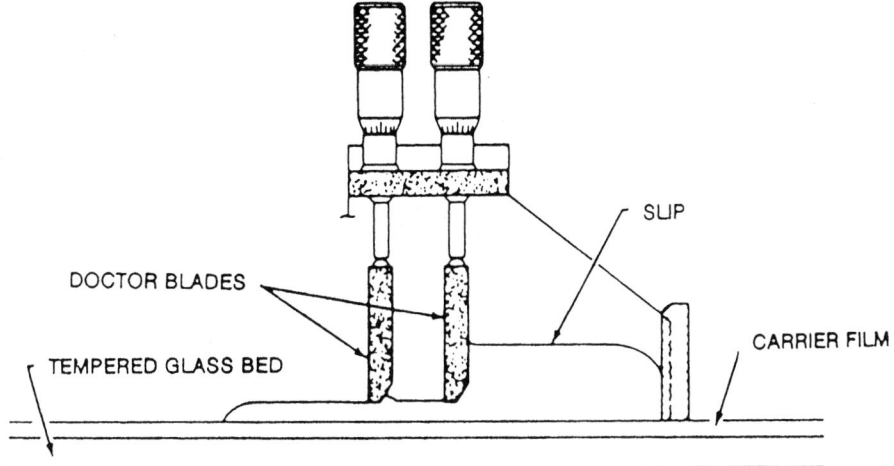

Figure 4.13 Schematic of the tape-casting process.

method. Recently, electrode-supported cell designs have been extensively studied because the electrolyte thickness can be much lower in these designs, typically 5–20 μm, giving a much lower ohmic resistance than that of an electrolyte supported cell. Thus, the electrode-supported cell design, in principle, is better suited for operation at lower temperatures [44]. Anode-supported cells are being developed using ~10 μm thick tape cast YSZ electrolyte laminated onto ~600 μm thick tape cast Ni/YSZ anode and co-sintered at about 1350°C for 1 h.

Conventional screen printing techniques have also been used to prepare thin electrolyte films in electrode-supported designs [45]. Cells with electrolyte thicknesses between 3 and 30 μm have been fabricated. The total cell resistance with a 4 μm thick electrolyte was 0.105 Ω cm^2 at 700°C which corresponds to a 10^{-2} S/cm conductivity value of YSZ at 700°C. In Table 4.3, the preparation methods for SOFC electrolytes used by several organisations are summarised.

Table 4.3 Preparation methods for SOFC electrolytes

Organisation	Method	Substrate	Electrolyte	Thickness (μm)	Reference
Siemens Westinghouse	EVD	(La,Sr)MnO$_3$	YSZ	40	[41]
AlliedSignal	Tape casting	Ni-YSZ	YSZ	1–10	[4]
Argonne National Laboratory	Tape casting	Ni-CGO	CGO	30	[46]
ECN	Screen printing	Ni-YSZ	YSZ	3–30	[45]
Toto	Slurry coating	(La,Sr)MnO$_3$	YSZ		[42]
Toho Gas	Slurry coating	Ni-YSZ	ScSZ	20	[47]
KIER	Slurry coating	Ni-YSZ	YSZ	20–30	[43]
Pacific Northwest National Laboratory	Tape casting	Ni-YSZ	YSZ	7	[48]

4.6 Perovskite-Structured Electrolytes

In addition to fluorite structure electrolytes such as stabilised zirconia and ceria, there are many non-fluorite structure oxides which are potentially attractive for SOFC electrolyte application. These include perovskites like lanthanum gallate and to a lesser degree calcium titanate. Alternative oxides are the pyrochlores such as yttrium zirconate (YZr$_2$O$_7$) and gadolinium titanate (Gd$_2$Ti$_2$O$_7$) [49,50], but these are only suitable in very limited oxygen pressure ranges. Therefore, the main discussion here focuses on the perovskites.

The perovskites based on the general formula ABO$_3$ comprise a rich family of compounds with important applications in solid oxide fuel cells, ferroelectrics, superconducting materials and oxidation catalysts [51] because the total charges on A and B (+6) can be achieved by the combinations of 1 + 5, 2 + 4, and 3 + 3, and also in more complex ways as in Pb(B$'_{1/2}$B$''_{1/2}$)O$_3$, where B$'$ = Sc or Fe and B$''$ = Nd or Ta, or A$'_{1/2}$A$''_{1/2}$TiO$_3$ where A$'$ = Li, Na and A$''$ = La, Pr, etc. Due to the high stability of the crystal structure and the variety of cations which can be accommodated within it, perovskites display a wide variety of properties. Many display both ionic and electronic conductivity and so are useful as electrodes in

SOFCs [52]. Only few perovskites are purely ionic in their conduction behaviour. Properties of selected ionic-conducting perovskites are discussed below.

4.6.1 LaAlO$_3$

Over the past 35 years, Takahashi and Iwahara have measured oxide ion conductivity in many different perovskites [53]. They first reported fast ion conductivity in Ti- and Al-based compositions and continued to measure a range of perovskite formulations as shown in Figure 4.14. From this figure, it is clear that Al or Mg doped CaTiO$_3$ exhibits the highest conductivity.

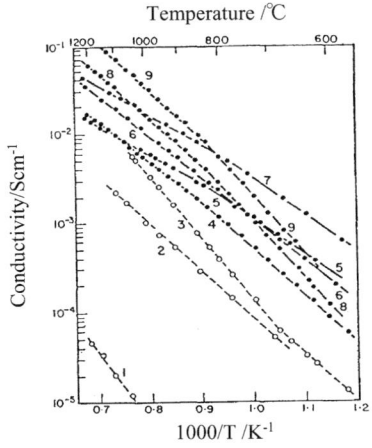

Figure 4.14 Arrhenius plots of the oxide ion conductivity in perovskites in air (numbers in figure corresponds to the following materials: (1) LaAlO$_3$, (2) CaTiO$_3$, (3) SrTiO$_3$, (4) La$_{0.7}$Ca$_{0.3}$AlO$_3$, (5) La$_{0.9}$Ba$_{0.1}$AlO$_3$, (6) SrTi$_{0.9}$Al$_{0.1}$O$_3$, (7) CaTi$_{0.95}$Mg$_{0.05}$O$_3$, (8) CaTi$_{0.5}$Al$_{0.5}$O$_3$, (9) CaTi$_{0.9}$Al$_{0.3}$O$_3$).

Figure 4.15 shows the transport number of oxide ions estimated from experiments with H$_2$–O$_2$ gas concentration cells. Although a high transport number is obtained in CaTi$_{0.95}$Mg$_{0.05}$O$_3$ at intermediate temperatures, in the range from 500 to 800°C, Ca-doped LaAlO$_3$ is suggested as an attractive SOFC electrolyte, because it displays a high transport number, always higher than 0.9 in the temperature range shown, and gives no electronic conduction in reducing atmospheres.

Since the initial work of Takahashi and Iwahara, many other researchers have investigated the oxide ion conductivity in LaAlO$_3$ based materials. For example, Mizusaki et al. [54] reported the oxide ion conductivity and defect chemistry of La$_{1-x}$Ca$_x$AlO$_3$ single-crystal samples with $x = 0.0027$–0.008, prepared by a floating zone technique. They reported that the nonstoichiometry in the oxygen content with P$_{O2}$ and temperature was negligible. This suggests that in this material, oxygen vacancies are the major defects and electron holes the minor. The conductivity due to electron holes increases with P$_{O2}^{1/4}$ and the activation energy for migration of oxide ions is 0.74 ± 0.05 eV. Perovskite-structured LaScO$_3$ has also been reported to be an oxide ion conductor [55–57]. Although

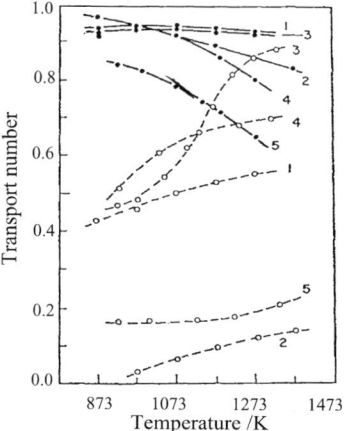

Figure 4.15 Transport number of oxide ion in perovskites (numbers in figure correspond to the same ones as in Figure 4.14 and broken line and solid line are the data obtained by O_2–air and O_2–humidified H_2 cells respectively).

the unit lattice volume of $LaScO_3$ is similar to that of $LaGaO_3$, hole conductivity is significant at high oxygen partial pressures. However, oxide ion conductivity in all reported perovskite oxides is lower than that of YSZ, since solid solubility of the dopant in these oxides is limited.

In ABO_3 perovskites, it was originally believed that the electric or dielectric properties were strongly dependent on B site cations. However, migrating oxide ions have to pass through the triangular space consisting of two large A and one small B site cations in the crystal lattice. Theoretical calculations now suggest that the enlargement in size of this triangular space is important for improving the migration of oxide ions in the crystal lattice [58]. Therefore, the ionic size of the A site cation is also significant in determining the oxide ion conductivity. This effect [59] is illustrated in Figure 4.16 which shows the effect of atomic radii

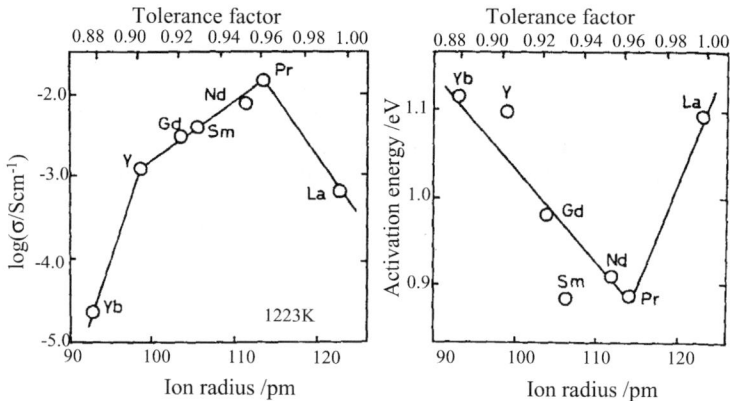

Figure 4.16 Oxide ion conductivity at 1223 K and activation energy for conduction as a function of ionic radii in A site of Al-based perovskite oxide, $Ln_{0.9}Ca_{0.1}AlO_3$.

of A site cations on the oxide ion conductivity and its apparent activation energy in $Ln_{0.9}Ca_{0.1}AlO_3$. It is clear that the ionic radii of A site cations, dictating the size of the triangular space, have a strong affect on the mobility of oxide ions. Oxide ion conductivity and activation energy increase and decrease respectively with increasing ionic radius, and $PrAlO_3$ exhibits the highest conductivity among the Al-based perovskites. These results suggest that enlargement of the size of the triangular space in the lattice can be used to increase the ionic conduction. From these results, higher ion conductivity is to be expected in perovskites with larger unit lattice dimensions [60].

4.6.2 LaGaO$_3$ Doped with Ca, Sr and Mg

Arrhenius plots of the electrical conductivity [61] of Ca-doped $LnGaO_3$ (Ln = La, Pr, Nd, Sm) are shown in Figure 4.17.

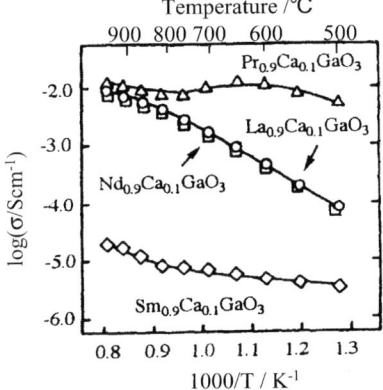

Figure 4.17 Arrhenius plots of the electrical conductivity of Ca-doped $LnGaO_3$ (Ln = La, Pr, Nd, Sm).

The oxide ion conductivity strongly depends on the particular cation on the A site, and increases in the following order, Pr > La > Nd > Sm. Electrical conductivity of all Ga-based perovskites is almost independent of the oxygen partial pressure, indicating that oxide ion conduction is dominant in these materials.

Doping with smaller valence cations generally causes oxygen vacancies to form in order to maintain electrical neutrality. Consequently, oxide ion conductivity increases. This is illustrated in Figure 4.18 which shows the effect of replacing La with various other cations [62].

Electrical conductivity depends strongly on the particular alkaline earth cation doped onto the La site and increases in the order Sr > Ba > Ca. Therefore strontium appears to be the most suitable dopant for $LaGaO_3$. Theoretically, increasing the amount of Sr increases the number of oxygen vacancies and oxide ion conductivity should therefore increase. However, the solid solubility of Sr on the La site of $LaGaO_3$ is low and the secondary phases, $SrGaO_3$ or La_4SrO_7, form when Sr is higher than 10 mol%. Thus there is only a limited gain in conductivity from increasing the dopant level.

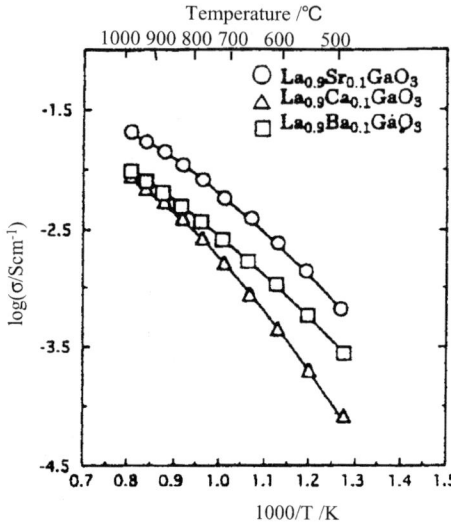

Figure 4.18 Effects of alkaline earth cation (M) in La site of $La_{0.9}M_{0.1}GaO_3$.

Oxygen vacancies can also be formed by doping an aliovalent cation into the Ga site in addition to the La site. Doping Mg into the Ga site increases conductivity substantially. The oxide ion conductivity attains a maximum at 20 mol% Mg doped on the Ga site. The lattice parameter also increases by doping Mg onto the Ga site, the ionic radius of Mg being larger than that of Ga. The solid solubility of Sr in the $LaGaO_3$ lattice is around 10 mol% without Mg; however, it increases up to 20 mol% with Mg on the Ga site. Such increase in Sr solid solubility, brought about by the enlarged crystal lattice, has also been reported by Majewski et al. [63]. It is now confirmed [64] that the highest oxide ion conductivity in $LaGaO_3$ based oxides is obtained with the composition $La_{0.8}Sr_{0.2}Ga_{0.8}Mg_{0.2}O_3$ (LSGM).

Since the initial work [62], $LaGaO_3$ based electrolytes have been studied by various groups and the various cation dopants have been investigated [65]. P. N. Huang et al. [66] reported the highest oxide ion conductivity at the composition $La_{0.8}Sr_{0.2}Ga_{0.85}Mg_{0.15}O_3$. On the other hand, K. Huang et al. [67] reported the highest conductivity of 0.17 S/cm at Sr = 0.2, Mg = 0.17. The composition at the highest ionic conductivity found by the three groups [64, 66, 67] was between $y = 0.15$ and 0.2 in $La_{0.8}Sr_{0.2}Ga_{1-y}Mg_yO_3$.

The oxide ion conductivity of Sr and Mg doped $LaGaO_3$ is higher than that of typical YSZ or ceria based materials and somewhat lower than Bi_2O_3-based oxides. However, electronic conduction and thermal instability are problems for Bi based electrolytes. Doubly doped $LaGaO_3$ formulations are very promising electrolytes for SOFCs in terms of ionic conductivity.

However, to complicate matters, formation of a secondary phase is always observed by X-ray diffraction analysis of doubly doped $LaGaO_3$. Although the crystal structure of such a secondary phase is not confirmed, it appears to be $LaSrGaO_4$. Some additional phases such as $LaSrGa_3O_7$ have also been reported.

Majewski et al. [63] and Hrovat et al. [68] have investigated the phase diagram in the system La_2O_3–SrO–MgO–Ga_2O_3 at 1623–1673 K in air. Figure 4.19 shows the phase diagram in the plane $LaGaO_3$–$SrGaO_3$–$LaMgO_3$. The system La_2O_3–SrO–MgO–Ga_2O_3 exhibits complex phase relations. Compared to the ternary La_2O_3–SrO–Ga_2O_3 system, the combined solid solubility of Sr and Mg in $LaGaO_3$ is significantly enhanced in the quaternary La_2O_3–SrO–MgO–Ga_2O_3 system. In order to obtain a single phase LSGM, great attention is required in its preparation. Huang et al. have investigated the synthesis of $La_{0.9}Sr_{0.1}Ga_{0.8}Mg_{0.2}O_3$ by the alkoxide method [69] and reported that almost single phase LSGM can be obtained at a temperature as low as 1270°C by this method.

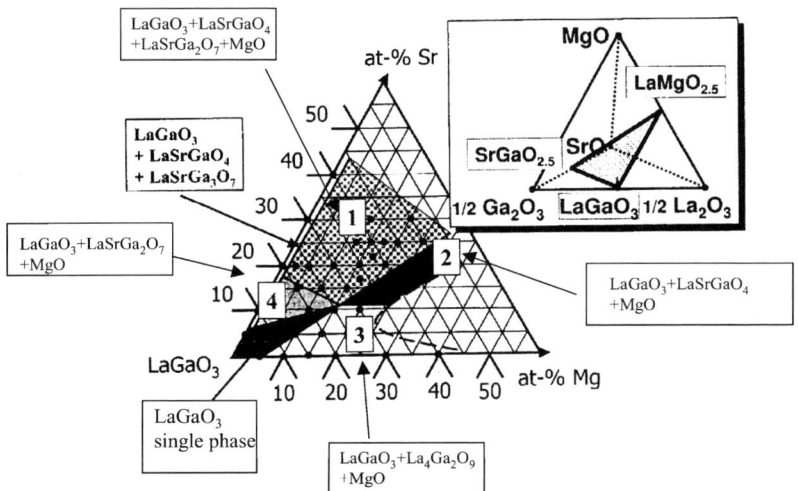

Figure 4.19 *Contour plots of the conductivity at 1073 K in $LaGaO_3$ doped with Sr and Mg for La and Ga site, respectively.*

Drennan et al. [70] and Du et al. [71] have investigated the oxide ion conductivity and the mechanical strength of $LaGaO_3$ based electrolytes. They pointed out that this $LaGaO_3$ material is slightly weaker in bending strength compared to YSZ. Mechanical property measurements for $La_{0.9}Sr_{0.1}Ga_{0.8}Mg_{0.2}O_3$ at room temperature and at 1173 K give average strengths of 162 ± 14 MPa and 55 ± 11 MPa, respectively, which are similar to strengths of CeO_2-based electrolytes [70]. Creep resistance of LSGM has been reported [72] as being worse than that of YSZ. As a result, additives such as Al_2O_3, which do not decrease the electrical conductivity, have been added to improve the mechanical properties of LSGM for application as electrolyte in practical SOFCs. Yasuda et al. [73] reported that the mechanical strength of LSGM is greatly improved by adding Al_2O_3 at around 2 wt%. Effects of the additives on mechanical strength have also been investigated by Yamada et al. [74].

Hole and electron conduction (minor carrier) and transport number of oxide ions in $LaGaO_3$-based oxides have been reported by Baker et al. [75], Yamaji et al. [76] and Kim et al. [77] based on the polarisation method. Figure 4.20 shows the

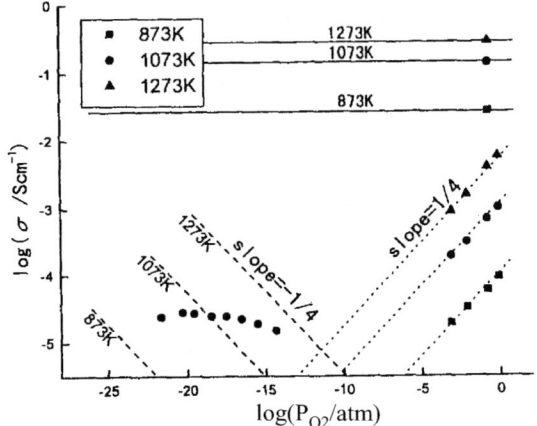

Figure 4.20 Estimated hole and electron conductivity and oxide ion conduction in LSGM as a function of oxygen partial pressure.

estimated hole and electron conduction in LSGM as a function of oxygen partial pressure [76]. It is seen that P_{O2} dependence of hole and electron conductivities is $P_{O2}^{1/4}$ and $P_{O2}^{-1/4}$, respectively, obeying the Hebb–Wagner theory. The polarisation method clearly suggests that $LaGaO_3$ exhibits almost pure oxide ion conduction over a wide range of oxygen partial pressures ($10^5 > P_{O2} > 10^{-25}$ atm).

Kim et al. [77] also investigated the temperature dependence of hole and electronic conductivity in $La_{0.9}Sr_{0.1}Ga_{0.8}Mg_{0.2}O_3$ using the polarisation method. Figure 4.21 shows the evaluated boundaries of the electrolytic domain of $La_{0.9}Sr_{0.1}Ga_{0.8}Mg_{0.2}O_3$ in the plane of $\log(P_{O2}/atm)$ versus reciprocal temperature. The lower boundary of the electrolytic domain (defined as $t_{ion} > 0.99$) for LSGM is 10^{-23} atm at 1000°C further indicating its suitability as SOFC electrolyte.

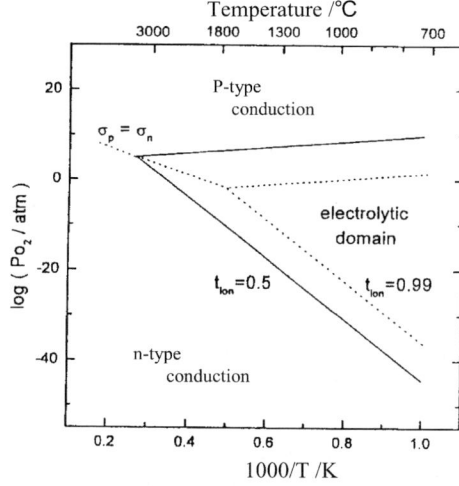

Figure 4.21 Evaluated electrolytic domain ($t_i > 0.99$) of LSGM with polarisation method.

Yokokawa et al. estimated the electrolyte efficiency of LSGM used as the electrolyte in an SOFC [78]. Electrolyte efficiency was given by a function of fuel utilisation and internal resistance [76]. When the thickness of the electrolyte was too small, chemical leakage of oxygen due to the electronic conduction became significant and the extra fuel consumption resulted in decreased electrolyte efficiency. On the other hand, with increasing thickness of electrolyte, internal resistance of the cell increased also resulting in decreased electrolyte efficiency. Consequently, an optimum thickness exists for each electrolyte material at a given temperature and current density. For example YSZ operating at 700°C has an optimum thickness of about 10 μm. The benefit of LSGM is that its highest efficiency with a thickness near 5 μm is achieved around 450°C. Thus LSGM appears to be the best electrolyte discovered so far to operate at low temperatures.

Several groups have investigated the electrochemical performance of cells with LSGM electrolyte. Figure 4.22 shows the temperature dependence of the maximum power density and the open circuit voltage (OCV) of a cell with $Sm_{0.5}Sr_{0.5}CoO_3$ cathode and Ni anode [79]. The OCV increased with decreasing temperature and was in good agreement with theoretical values estimated from the Nernst equation. The maximum power density was greater than 1.0 W/cm² at 1000°C and about 0.1 W/cm² at 600°C with 0.5 mm thick LSGM electrolyte. In other investigations with $LaGaO_3$ based electrolyte [80–82] similar large power densities have been reported at intermediate temperatures with $La_{0.6}Sr_{0.4}CoO_3$ cathode and $Ni-CeO_2$ doped with La cermet anode.

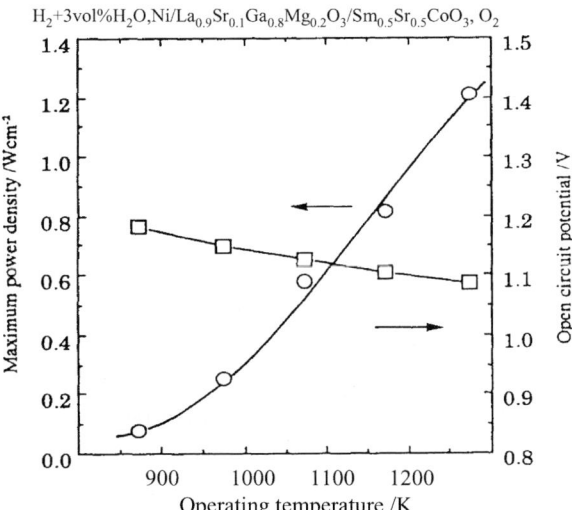

Figure 4.22 Temperature dependence of the maximum power density and the open circuit potential of the cell using LSGM for electrolyte.

Reactivity of $LaGaO_3$ with electrode materials has also been investigated [81,82]. Platinum seems to be easily reacted with gallium oxide to reduce Ga^{3+} to Ga^{2+} which is volatile. As for the stability of $LaGaO_3$-based electrolyte under reducing conditions, Yamaji et al. [83] found by SIMS analysis that the Ga content decreased at the surface of LSGM owing to the high vapour pressure of GaO. However, the vapour pressure of GaO decreases exponentially with decreasing temperature and is negligible at 600°C. Therefore, evaporation of GaO does not appear to be a problem when LSGM is used as an electrolyte in SOFCs at intermediate temperatures.

Hayashi et al. [84] and Ishihara et al. [79] investigated thermal expansion of LSGM and showed that it increases with increasing dopant content. The estimated average thermal expansion coefficient was around 11.5×10^{-6}/K in the temperature range from room temperature to 1000°C. This is slightly larger than that of YSZ but slightly smaller than that of than CGO.

The diffusivity of oxide ions in LSGM was studied with ^{18}O tracer diffusion measurements [85]. LSGM exhibits a large diffusion coefficient because of the larger mobility of oxide ions compared to that in the fluorite structured oxides (Table 4.4). The perovskite structure has a large free volume in its lattice and this gives a high diffusivity of oxide ions, resulting in high conductivity.

Table 4.4 Comparison of mobility of oxide ion in selected fluorite and LSGM oxides at 1073 K [85]

	D_t (cm^2/s)	E_a (eV)	δ	$[V_o^{\cdot\cdot}]$ (cm^{-3})	D (cm^2/s)	μ (cm^2/Vs)
$Zr_{0.81}Y_{0.19}O_2$	6.2×10^{-8}	1.0	0.10	2.95×10^{21}	1.31×10^{-6}	1.41×10^{-5}
$Zr_{0.858}Ca_{0.142}O_2$	7.54×10^{-9}	1.53	0.142	4.19×10^{21}	1.06×10^{-7}	1.15×10^{-6}
$Zr_{0.85}Ca_{0.15}O_2$	1.87×10^{-8}	1.22	0.15	4.43×10^{21}	2.49×10^{-7}	2.69×10^{-6}
$Ce_{0.9}Gd_{0.1}O_2$	2.70×10^{-8}	0.9	0.05	1.26×10^{21}	1.08×10^{-6}	1.17×10^{-5}
LSGM(9182)	3.24×10^{-7}	0.74	0.15	2.52×10^{21}	6.40×10^{-6}	6.93×10^{-5}
LSGM(8282)	4.13×10^{-7}	0.63	0.20	3.34×10^{21}	6.12×10^{-6}	6.62×10^{-5}

D_t, tracer diffusion coefficient; E_a, activation energy; δ, oxygen deficiency; [VÖ], oxygen vacancy concentration; D, self-diffusion coefficient; μ, mobility.

4.6.3 LaGaO$_3$ Doped with Transition Elements

LSGM has been modified by doping in several ways. Kim et al. [86] investigated the effects of Ba and Mg doping rather than Sr and Mg. These gave similar conductivities. Larger effects were observed when transition metals such as Co were used at levels below 10 mol% [87]. Baker et al. [75] investigated the effects of Cr and Fe on the oxide ion conductivity of LaGaO$_3$. Doping Cr or Fe on the Ga site induced hole conduction in the LaGaO$_3$, resulting in decreased stability against reduction. On the other hand, Ishihara et al. [87] found that doping with small amounts of transition metals, particularly Co or Ni, increased the oxide ion conductivity in LSGM.

Figure 4.23 shows an Arrhenius plot of electrical conductivities for LaGaO$_3$ doped with various transition metal cations on the Ga site. The conductivity increased by doping with Co, Ni and Fe, and decreases by doping with Cu and Mn. n-Type conduction is greatly enhanced by doping with Mn and Ni, and p-type conduction is increased by doping with Cu. Kharton et al. [88] also found

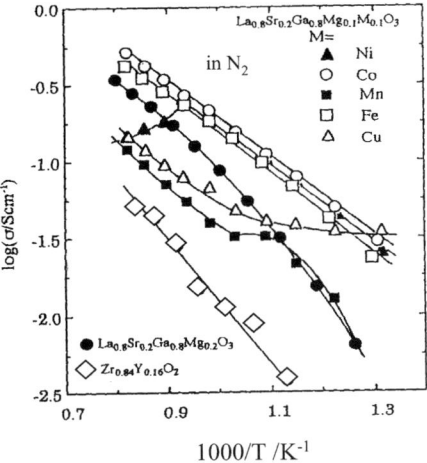

Figure 4.23 Arrhenius plots of electrical conductivity of the LaGaO$_3$-based oxide doped with some transition metal cations for Ga site.

that in LaGa$_{0.8}$Mg$_{0.2}$O$_3$, doping with Mn and Cr decreases the oxide ion conductivity, even though the amount of doped transition metal was much larger, 40 mol% on the Ga site. Effects of other transition metal doping have also been reported [89,90] but Co is the most interesting dopant found up to now for SOFC electrolytes.

Figure 4.24 shows the OCV and the maximum power density at 800°C as a function of Co content in a cell with LSGM electrolyte [91]. The OCV decreased

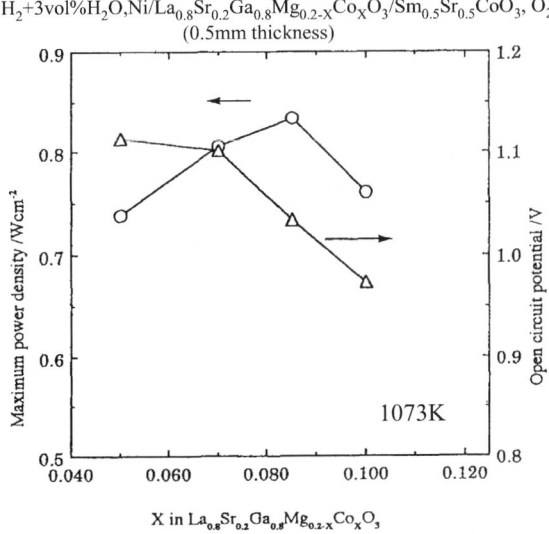

Figure 4.24 Open circuit potential as well as the maximum power density at 1073 K as a function of Co content in LaGaO$_3$ electrolyte (0.5 mm thickness). (Ni and Sm$_{0.5}$Sr$_{0.5}$CoO$_3$ are used as electrodes.)

monotonically with increasing content of Co. In particular, the OCV drop was significant for Co concentrations above 10 mol%. This was caused by the onset of hole conduction. But power density increased with Co concentration and attained a maximum value at 8.5 mol% Co. This was explained by the improved ionic conductivity of the doped electrolyte. However, current leakage became dominant at higher Co levels. When the electrolyte thickness was reduced, the power density further increased and at 180 μm thickness the maximum power density was 1.58 W/cm² at 800°C and 0.5 W/cm² at 600°C. Larger cells of 150 mm diameter using $La_{0.8}Sr_{0.2}Ga_{0.8}Mg_{0.15}Co_{0.05}O_3$ electrolyte have also been investigated recently [92].

4.7 Oxides with Other Structures

4.7.1 Brownmillerites (e.g. $Ba_2In_2O_6$)

Anther perovskite-related structure, which is interesting from the viewpoint of oxide ion conduction, is brownmillerite with a general formula of $A_2B'B''O_5$ or $A_2B_2O_5$. This structure can be viewed as a perovskite with oxygen vacancies ordered along the [101] direction in alternate layers. Such vacancy ordering results in an increased unit cell relative to the perovskite. In other words, lattice parameters of the a and c axis of the ideal brownmillerite oxide are larger than those of the ideal perovskite oxide by 2 and the b axis of brownmillerite is the same as that of perovskite. In some cases the oxygen vacancies do not order, which results in a perovskite structure with a statistical distribution of oxygen vacancies on the oxygen sites. Therefore, high oxide ion conductivity is also expected in brownmillerites.

Goodenough *et al.* [93] have reported the high oxide ion conductivity in several brownmillerite oxides. A listing [94] is given in Table 4.5. All these brownmillerites exhibit oxide ion conductivity and the conductivities are rather

Table 4.5 Oxide ion conductivity for selected brownmillerite compounds [94]

Compound	T (K)	σ (S/cm)	Compound	T (K)	σ (S/cm)
$Ba_2In_2O_5$	973	5×10^{-3}	$Ba_3In_2HfO_8$	673	1.0×10^{-3}
	1223	1×10^{-1}	$Sr_3In_2HfO_8$	973	1×10^{-4}
$BaZrO_3$	973	1×10^{-6}	$Ba_3Sc_2ZrO_8$	973	7×10^{-3}
$BaZr_{0.5}In_{0.5}O_{2.75}$	973	1×10^{-2}	$Ba_2GdIn_{0.8}Ga_{0.2}O_5$	873	5×10^{-3}
$Ba_3In_2ZrO_8$	973	5×10^{-3}	$Ba_2GdIn_{0.6}Ga_{0.4}O_5$	873	5×10^{-3}
$Ba_3In_{1.7}Zr_{1.3}O_{8.15}$	973	5×10^{-2}	$Ca_2Cr_2O_5$	973	5×10^{-3}
$Ba_2In_{1.33}Zr_{0.67}O_{5.33}$	973	1×10^{-3}	Sr_2ScAlO_5	973	1×10^{-5}
$Ba_2In_{1.75}Ce_{0.25}O_{5.125}$	973	9×10^{-3}	$Sr_2Sc_{1.3}Al_{0.7}O_5$	973	1×10^{-3}
	1223	6×10^{-2}	$Sr_2ScAl_{0.8}Mg_{0.2}O_{4.9}$	973	5×10^{-4}
$Ba_3In_2TiO_8$	973	7×10^{-4}	$Sr_2ScAl_{0.8}Zn_{0.2}O_{4.9}$	973	2×10^{-4}
$Ba_3In_2ZrO_8$	673	6.8×10^{-3}	$Sr_2Sc_{0.8}Y_{0.2}AlO_5$	973	1×10^{-4}
$Ba_3In_2CeO_8$	673	1.5×10^{-3}	$Sr_{1.8}Ba_{0.2}ScAlO_5$	973	1×10^{-4}

high compared with those of fluorite-structured oxides. Among these, $Ba_2In_2O_5$ is attracting much interest from the viewpoint of SOFC electrolyte.

Figure 4.25 shows a comparison of oxide ion conductivities in $Ba_2In_2O_5$ and Ce-doped $Ba_2In_2O_5$ in an Arrhenius plot. The P_{O2} dependence of $Ba_2In_2O_5$ suggests that hole conduction is dominant in this oxide at P_{O2} higher than 10^{-3} atm whereas oxide ion conductivity becomes dominant with decreasing P_{O2}. $Ba_2In_2O_5$ was characterised by a large jump in conductivity around 900°C. This can be explained by an order/disorder transition of the oxygen vacancy structure, namely, from oxygen vacancy ordered brownmillerite to oxygen-disordered pseudo-perovskite. A similar jump has been observed in Bi_2O_3 electrolytes. Oxide ion conductivity in $Ba_2In_2O_5$ is comparable with that of YSZ at temperatures above 900°C.

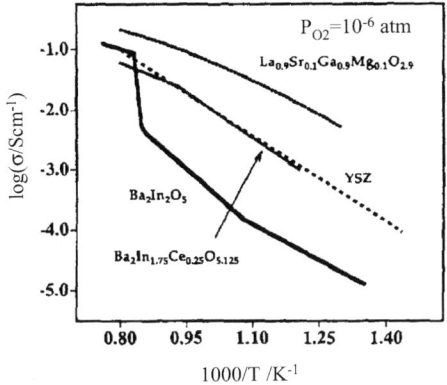

Figure 4.25 Arrhenius plots of the oxide ion conductivity in $Ba_2In_2O_5$ and Ce-doped $Ba_2In_2O_5$.

In an attempt to eliminate this order/disorder transition and to stabilise the structure down to lower temperatures, the effects of dopants on $Ba_2In_2O_5$ have been studied by several groups. For example, Figure 4.26 shows the temperature dependence of the conductivity of Ce-doped $Ba_2In_2O_5$. The large jump around 900°C was suppressed by doping with Ce, but the conductivity at high temperature was reduced. Attempts to stabilise the disordered phase at lower temperatures by selective doping of A and/or B sites have met with some success and rather high oxide ion conductivity has been observed in some phases as shown by Kendall et al. [94]. For $Ba_2In_2O_5$, La doping on the Ba site is the most promising for increasing oxide ion conductivity. The Arrhenius plots [95] of the ionic conductivity in $(Ba_{1-x}La_x)_2In_2O_5$ indicate an effect similar to Ce doping, the jump in the Arrhenius plots shifting to lower temperature with increasing amount of La and disappearing around $x = 0.2$. Since the amount of lattice oxygen decreases with increasing La, the oxygen disorder structure may be stabilised by doping with La. The electrical conductivity at high temperatures further increases by doping La on the Ba site. The highest ion conductivity is achieved at $x = 0.6$ in this system and this conductivity is higher than that of 8 mol% Y_2O_3-stabilised ZrO_2.

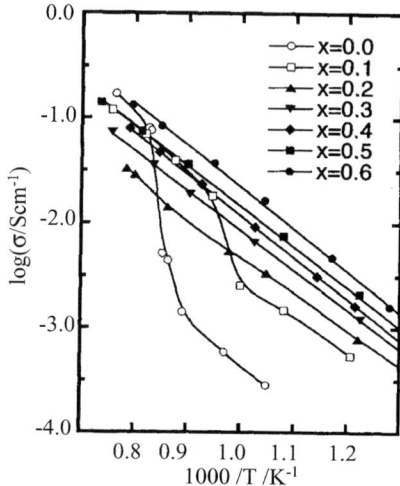

Figure 4.26 Arrhenius plots of the conductivity in $(Ba_{1-x}Ce_x)_2In_2O_5$.

The simulations of Fisher et al. [96] on the defect energies of $Ba_2In_2O_5$ by atomic modelling techniques show that oxygen Frenkel pairs are the dominant intrinsic defects in the low-temperature phase, with a preference for oxide ion diffusion via these defects in the [001] direction. $Ba_2In_2O_5$ is also predicted to be more stable against reduction than oxidation, with reduction involving removal of oxygen from octahedral layers and oxidation involving insertion of oxide ions into the interstitial sites. On the other hand, simulations suggest that this oxide exhibits proton conductivity. In fact, Zhang et al. [97] and Schober et al. [98] reported that $Ba_2In_2O_5$ exhibits proton conduction at temperatures lower than 400°C. The proton conduction in this oxide has been confirmed by several groups showing that proton conduction occurs in the crystal phase of $Ba_2In_2O_5 \cdot H_2O$.

4.7.2 Non-cubic Oxides

To a large degree, the known fast ion conductors possess either cubic or pseudo-cubic crystal lattice structures; even lanthanum gallate and barium indiate are not exceptions because they are pseudo-cubic. Therefore, it is generally believed that the high symmetry in the lattice is an essential requirement for fast ion conduction. So far there have been no reports in the literature for a notable oxide ion conductivity in non-cubic oxides. Among few exceptions, the oxide ion conductivity in hexagonal apatites, $La_{10}Si_6O_{27}$ and $Nd_{10}Si_6O_{27}$, reported by Nakayama et al. [99, 100] is interesting.

Figure 4.27 shows a comparison of oxide ion conductivity in $La_{10}Si_6O_{27}$ with that of doped bismuth oxide and YSZ. The electrical conductivity of this oxide at temperatures higher than 600°C is not high compared to that of YSZ. However, at lower temperatures, $La_{10}Si_6O_{27}$ exhibits higher oxide ion conductivity than that of conventional oxide ion conductors. Sansom et al. [101] studied the

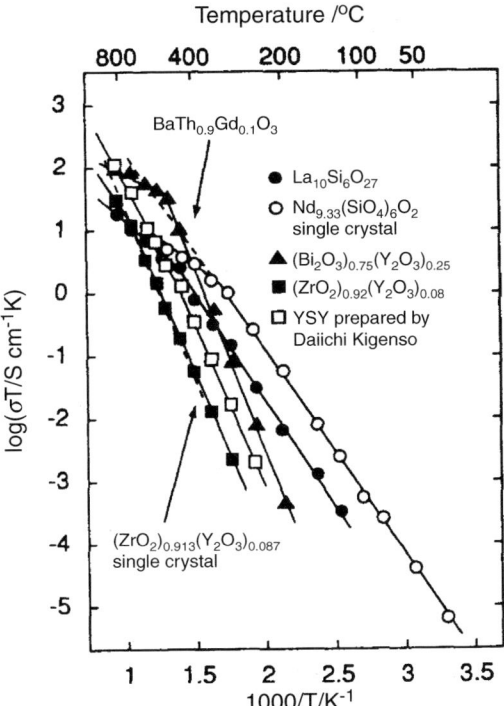

Figure 4.27 Comparison of the oxide ion conductivity in $La_{10}Si_6O_{27}$ and $Nd_{10}Si_6O_{27}$-based oxide with conventional oxide ion conductors.

relationship between crystal structure and oxide ion conductivity in this oxide. The refined crystal structure of $La_{10}Si_6O_{26}$ belongs to the hexagonal space group P-3 (no. 147), with $a = b = 972.48$ pm, $c = 718.95$ pm. This suggests that the $La_{10}Si_6O_{26}$ has a unique oxygen channelling structure, such that the high oxygen conductivity could be assigned to disorder of these channel sites [101]. On the other hand, bismuth-based oxide, so called BIMEVOX, has also been reported as a high oxide ion conductor with non-cubic structure, but only in a limited P_{O2} range [102] of limited interest to SOFCs.

La-deficient La_2GeO_5 also exhibits fast oxide ion conductivity over a wide range of oxygen partial pressures [103]. La_2GeO_5 has the monoclinic crystal structure with P21/c space group. It consists of two types of oxygen; one is covalently bonded to Ge to form a GeO_4 tetragonal unit and the other is bridged between La and GeO_4. Considering the strength of the chemical bonds in each case, the bridged oxygen is most likely to be the mobile site. The conductivity increases with La deficiency and the maximum value is attained at $x = 0.39$ in $La_{2-x}GeO_{5-\delta}$. The oxide ion transport number in La_2GeO_5 was estimated to be unity from H_2–O_2 and N_2–O_2 gas concentration cell measurements. The comparison shown in Figure 4.28 clearly reveals that the oxide ion conductivity of $La_{1.61}GeO_{5-\delta}$ is much higher than that of Y_2O_3-stabilised ZrO_2 and almost the same as that of $Gd_{0.15}Ce_{0.85}O_2$ or $La_{0.9}Sr_{0.1}Ga_{0.8}Mg_{0.2}O_3$ at temperatures above

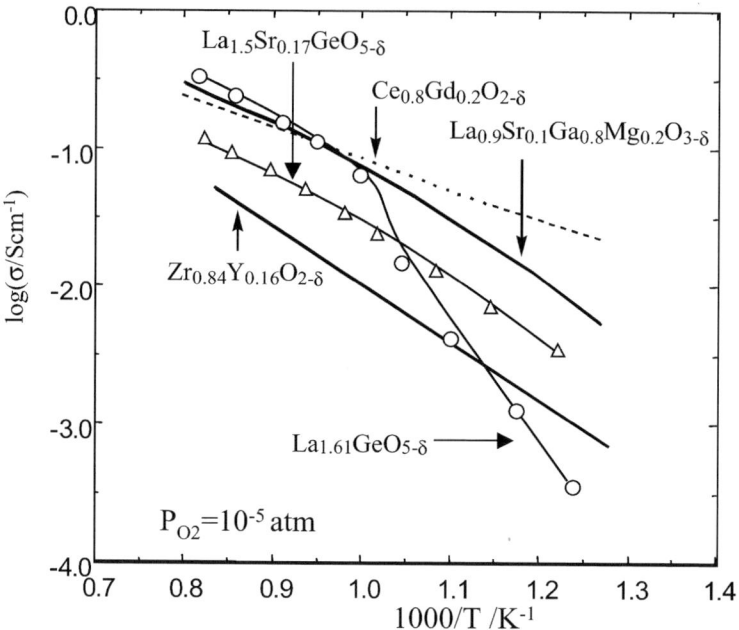

Figure 4.28 Comparison of the oxide ion conductivity in La_2GeO_5-based oxide with those of the conventional oxide ion conductors.

700°C. However, at low temperatures, the oxide ion conductivity of $La_{1.61}GeO_{5-\delta}$ becomes much smaller because of the change in activation energy. This can be explained by the order-disorder transition in oxygen vacancy structure. Such a high value of electrical conductivity at these temperatures makes $La_{1.61}GeO_{5-\delta}$ attractive for consideration as an SOFC electrolyte.

4.8 Proton-Conducting Oxides

Proton-conducting oxides are also possible electrolytes for intermediate temperature SOFCs. In this section, high-temperature proton-conduction in perovskites is briefly examined. Since the proton is the smallest positive ion, its mobility is high and good ionic conductivity may be obtained at low temperature in certain materials. Proton conductivity in oxide electrolytes at high temperatures was first found by Iwahara et al. using $BaCeO_3$-based compositions [104], with high conductivity obtained by doping $BaCeO_3$ with rare earth cations on the Ce sites. However, proton conductivity in doped $BaCeO_3$ is still smaller than oxide ion conductivity in $LaGaO_3$ or Sm-doped CeO_2 and the chemical stability of $BaCeO_3$ formulations, particularly in CO_2, is poor.

Mother compounds like $SrCeO_3$ or $BaCeO_3$ are not good conductors in themselves. However, after doping with aliovalent cations, mainly rare earth cations such as Y or Yb, electron hole conductivity appears. For example, electronic hole conductivity of $SrCe_{0.95}Yb_{0.05}O_3$ in dry air is 0.01 S/cm at 800°C. When this oxide is exposed to humidified air, protons as charge carriers are generated in the lattice by the following reaction

$$H_2 + 2h^{\cdot} = 2H^+$$

and the electrical conductivity is increased due to proton conduction. The conductivity of $SrCe_{0.95}Yb_{0.05}O_3$ in humidified atmospheres was around 0.002 S/cm. Figure 4.29 shows the comparison of proton conductivity in selected perovskites [105]. In general, proton conduction increased in the order $BaCeO_3$ > $SrCeO_3$ > $SrZrO_3$ > $CaZrO_3$. On the other hand, chemical stability deteriorated in the opposite order, $SrZrO_3$ exhibiting the highest stability. Consequently, for fuel cell application, $SrZrO_3$ would appear to be the most stable option.

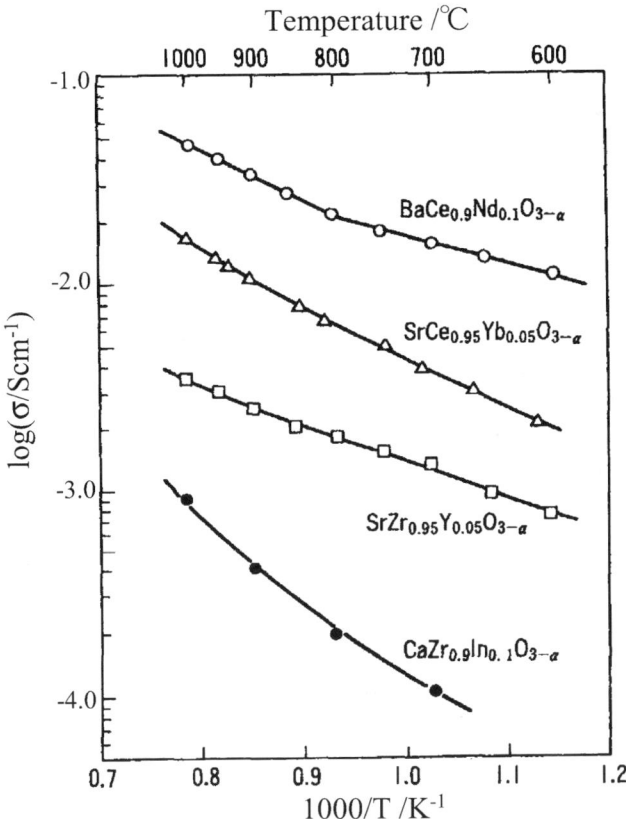

Figure 4.29 Comparison of proton conduction in perovskite oxides.

$BaThO_3$ and $BaTbO_3$ doped with Gd are also reported as possible proton-conducting electrolytes for intermediate temperature SOFCs [106, 107]. Sammells et al. reported that $BaTh_{0.9}Ga_{0.1}O_3$, $Sr_2Gd_2O_5$, and $Sr_2Dy_2O_5$ were possible fast proton conductors [108]. Although detailed measurements of proton conduction in these oxides have not been reported, proton conductivity up to 8.7×10^{-2} S/cm at 550°C has been estimated from the electrical resistance in power density curves. But improvement of chemical stability is the key requirement for application of all proton-conducting ceramics as SOFC electrolytes.

4.9 Summary

During the past decades, many oxide formulations have been extensively examined in the search for candidate SOFC electrolyte materials. Zirconia-based compositions are still the best electrolytes at present owing to their good stability under reducing atmospheres, low electronic conductivity, and acceptable oxide ion conductivity above 800°C. The recent trend of SOFC development is to operate at lower temperatures. The lowest operation temperature limit of the cell, for thin YSZ electrolytes, is estimated to be about 700°C from YSZ conductivity and mechanical property data. Scandia-doped zirconia, which shows a higher conductivity than that of YSZ, could be preferred at temperatures below 700°C, if the cost of scandia was acceptable.

Ceria-based electrolytes could be used at 550°C or less. To operate at higher temperatures, a dual layer electrolyte, with a thin YSZ layer on CGO, has been proposed to avoid the electronic current leakage. The interdiffusion issues at the interface are important in this case for long-life electrolytes.

Another possibility is to use perovskite compositions. The most promising candidate at this time is $LaGaO_3$ doped with Sr and Mg. Other possible perovskites are $Ba_2In_2O_5$ doped with Ce or La. Other ionic conducting oxides have also been found including $La_{10}Si_6O_{26}$ composition. Proton conductors such as $SrCeO_3$, $SrZrO_3$ or $BaCeO_3$ doped with Y or Yb may also be effective electrolytes, but reaction with CO_2 has to be resolved first.

References

[1] W. Nernst, Z. Elektrochem., **6** (1899) 41.
[2] E. Baur and H. Preis, Z. Electrochem., **43** (1937) 727.
[3] T. H. Etsell and S. N. Flengas, Chem. Rev., **70** (1970) 339.
[4] N. Q. Minh and T. Takahashi, in *Science and Technology of Ceramic Fuel Cells*, Elsevier, Amsterdam (1995).
[5] T. Takahashi, T. Esaka and H. Iwahara, J. Appl. Electrochem. **7** (1977) 299.
[6] B. C. H. Steele, in *High Conductivity Solid Ionic Conductors*, ed. T. Takahashi, World Scientific, Singapore (1989).
[7] E. Koch and C. Wagner, Z. Phys. Chem., **B38** (1937) 295.

[8] F. Kroger and H. J. Vink, in *Solid State Physics*, Vol. 3, eds. F. Seitz and D. Turnbull, Academic Press, New York, 1965, p. 304.
[9] J. A. Kilner, *Solid State Ionics*, **129** (2000) 13.
[10] Y. Arachi, H. Sakai, O. Yamamoto, Y. Takeda and N. Imanishi, *Solid State Ionics*, **121** (1999) 133.
[11] W. W. Baker and O. Knop, *Proc. Br. Ceram. Soc.*, **19** (1971) 15.
[12] D. K. Hohnke, in *Fast Ion Transport in Solid*, eds. P. Vashista, J. N. Mundy and G. K. Shenoy, North Holland, Amsterdam, 1968, p. 669.
[13] R. E. Carter and W. L. Roth, in *Electromotive Force Measurements in High Temperature Systems*, ed. C. B. Alcock, IMM, London, 1968, p. 653.
[14] J. A. Kilner and R. J. Brook, *Solid State Ionics*, **6** (1982) 237.
[15] S. Nowick, *Comments Solid State Phys.*, **9** (1979) 85.
[16] J. A. Kilner and B. C. H. Steele, in *Nonstoichiometric Oxides*, ed. O. T. Sorensen, Academic Press, London, 1981, p. 233.
[17] P. S. Manning, J. D. Sirman, R. A. De Souza and J. A. Kilner, *Solid State Ionics*, **100** (1997) 1.
[18] J. A. Kilner, *Solid State Ionics*, **8** (1983) 201.
[19] R. G. Anderson and A. S. Nowick, *Solid State Ionics*, **5** (1981) 547.
[20] V. Butler, C. R. A. Catlow, B. E. F. Fender and J. H. Harding, *Solid State Ionics*, **8** (1983) 109.
[21] V. Butler, C. R. A. Catlow, B. E. F. Fender and J. H. Harding, *Solid State Ionics*, **5** (1981) 539.
[22] M Yashima, M. Kakihana and M Yoshimura, *Solid State Ionics*, **86/88** (1996) 1131.
[23] R. A. Miller, J. L. Smialek and R. G. Garlik, in *Science and Technology of Zirconia*, eds. A. H. Heuer and L. W. Hobbs, Advances in Ceramics, vol. 3, American Ceramics Society, Columbus, OH, 1981, p. 241.
[24] R. Ruh, H. J. Garnett, R. F. Donagala and V. A. Patel, *J. Am. Ceram. Soc.*, **60** (1977) 399.
[25] O. Yamamoto, *Electrochemi. Acta*, **45** (2000) 2423.
[26] H. Uchida, H. Suzuki and M. Watanabe, *J. Electrochem. Soc.*, **146** (1998) 615.
[27] O. Yamamoto, Y. Arachi, Y. Takeda, N. Imanishi, Y. Mizutani, M. Kawai and Y. Nakamura, *Solid State Ionics*, **79** (1995) 137.
[28] Y. Arachi, T. Ashai, O. Yamamoto, Y. Takeda, N. Imanishi, K. Kawada and C. Tamakoshi, *J. Electrochem. Soc.*, **148** (2001) A520.
[29] Y. Mizutani, M. Kawai, K. Nomura, Y. Nakamura and O. Yamamoto, in *Solid Oxide Fuel Cells V*, eds. U. Stimming, S. C. Singhal, H. Tagawa and W. Lehnert, The Electrochemical Society Proceedings, Pennington, NJ, PV97-40, 1997, p. 37.
[30] W Weppner, *J. Solid State Chem.*, **20** (1977) 305.
[31] T. Kudo and Y. Obayashi, *J. Electrochem. Soc.*, **123** (1976) 419.
[32] H. L. Tuller and A. S. Nowick, *J. Electrochem. Soc.*, **122** (1975) 255.
[33] M. Mogensen, N. M. Sammes and G. A. Tompsett, *Solid State Ionics*, **129** (2000) 63.

[34] B. C. H. Steele, *Solid State Ionics*, **129** (2000) 95.
[35] H. Yahiro, Y. Eguchi, K. Eguchi and H. Arai, *J. Appl. Electrochem.*, **18** (1988) 527.
[36] H. Yahiro, K. Eguchi and H. Arai, *Solid State Ionics*, **36** (1989) 71.
[37] M. Godickemeier, K. Sasaki and L. J. Gauckler, *J. Electrochem. Soc.*, **144** (1997) 1635.
[38] M. Godickemeier and L. J. Gauckler, *J. Electrochem. Soc.*, **145** (1998) 414.
[39] T. Tsai and S. A. Barnett, in *Solid Oxide Fuel Cells V*, eds. U. Stimming, S. C. Singhal, H. Tagawa and W. Lehnert, The Electrochemical Society Proceedings, Pennington, NJ, PV97-40, 1997, p. 274.
[40] A. O. Isenberg, in *Proceedings of the Symposium on Electrode Materials and Processes for Energy Conversion and Storage*, May 1977, eds. D. E. McIntyre, S. Srinivasan and F. G. Will, The Electrochemical Society Proceedings, Pennington, NJ, PV77-6, 1977, p. 572.
[41] U. B. Pal and S. C. Singhal, *J. Electrochem. Soc.* **137** (1980) 2937.
[42] M. Kuroishi, S. Furuya, K. Hiwatashi, K. Omoshiki, A. Ueno and M. Aizawa, in *Solid Oxide Fuel Cells VII*, eds. H. Yokokawa and S. C. Singhal, The Electrochemical Society, Pennington, NJ, PV2001-16, 2001, p. 88.
[43] R.-H. Song, K.-S. Song, Y.-E. Ihm and H. Yokokawa, in *Solid Oxide Fuel Cells VII*, eds. H. Yokokawa and S. C. Singhal, The Electrochemical Society Proceedings, Pennington, NJ, PV2001-16, 2001, p. 1073.
[44] S. C. Singhal, in *Solid Oxide Fuel Cells VII*, eds. H. Yokokawa and S. C. Singhal, The Electrochemical Society Proceedings, Pennington, NJ, PV2001-16, 2001, p. 166.
[45] J. P. Ouweltjes, F. P. F. van Berkel, P. Nammensma and G. M. Christie, in *Solid Oxide Fuel Cells VI*, eds. S. C. Singhal and M. Dokiya, The Electrochemical Society Proceedings, Pennington, NJ, PV99-19, 1999, p. 803.
[46] R. Doshi, V. L. Richards, J. D. Carter, X. Wang and M. Krumpelt, *J. Electrochem. Soc.*, **146** (1999) 1273.
[47] Y. Mizutani, M. Kawai, K. Nomura, Y. Nakamura and O. Yamamoto, in *Solid Oxide Fuel Cells VI*, eds. S. C. Singhal and M. Dokiya, The Electrochemical Society Proceedings, Pennington, NJ, PV99-19, 1999, p. 185.
[48] S. P. Simner, J. W. Stevenson, K. D. Meinhardt and N. J. Canfield, in *Solid Oxide Fuel Cells VII*, eds. H. Yokokawa and S. C. Singhal, The Electrochemical Society Proceedings, Pennington, NJ, PV2001-16, 2001, p. 1051.
[49] M. Pirzada, R. W. Grimes, L. Minervini, J. F. Maguire and K. E. Sickafus, *Solid State Ionics*, **140** (2001) 201.
[50] S. A. Kramer and H. L. Tuller, *Solid State Ionics*, **82** (1995) 15.
[51] M. O'Connell, A. K. Norman, C. F. Hüttermann and M. A. Morris, *Catalysis Today*, **47** (1999) 123.
[52] R. A. De Souza and J. A. Kilner, *Solid State Ionics*, **106** (1998) 175.
[53] T. Takahashi and H. Iwahara, *Energy Conversion*, **11** (1971) 105.
[54] J. Mizusaki, I. Yasuda, J. Shimoyama, S. Yamauchi and K. Fueki, *J. Electrochem. Soc.*, **140** (1993) 467.
[55] D. Lybye and N. Bonanos, *Solid State Ionics*, **125** (1999) 339.

[56] D. Lybye, F. W. Poulsen and M. Mogensen, *Solid State Ionics*, **128** (2000) 91.
[57] K. Nomura and S. Tanabe, *Solid State Ionics*, **98** (1997) 229.
[58] M. Cherry, M. S. Islam and C. R. A. Catlow, *J. Solid State Chem.*, **118** (1995) 125.
[59] T. Ishihara, H. Matsuda and Y. Takita, *J. Electrochem. Soc.*, **141** (1994) 3444.
[60] T. Ishihara, H. Matsuda, Y. Mizuhara and Y. Takita, *Solid State Ionics*, **70/71** (1994) 234.
[61] T. Ishihara, H. Matsuda and Y. Takita, in *2nd Ionic and Mixed Conducting Ceramics*, eds. T. A. Ramanarayanan, W. L. Worrell and H. L. Tuller, The Electrochemical Society Proceedings, Pennington, NJ, PV92-12, 1994, p. 85.
[62] T. Ishihara, H. Matsuda and Y. Takita, *J. Am. Chem. Soc.*, **116** (1994) 3801.
[63] P. Majewski, M. Rozumek and F. Aldinger, *J. Alloys Compounds*, **329** (2001) 253.
[64] T. Ishihara, H. Matsuda and Y. Takita, *Solid State Ionics*, **79** (1995) 147.
[65] M. Feng and J. B. Goodenough, *Eur. J. Solid State Inorg. Chem.*, **31** (1994) 663.
[66] P. N. Huang and P. Petric, *J. Electrochem. Soc.*, **143** (1996) 1644.
[67] K. Huang, R. Tichy and J. B. Goodenough, *J. Am. Ceram. Soc.*, **81** (1998) 2565.
[68] M. Hrovat, Z. Samardzija, J. Holc and S. Bernik, *J. Mater. Res.*, **14** (1999) 4460.
[69] K. Huang, M. Feng and J. B. Goodenough, *J. Am. Ceram. Soc.*, **79** (1996) 1100.
[70] J. Drennan, V. Zelizko, D. Hay, F. T. Ciacchi, S. Rajendran and S. P. S. Badwal, *J. Mater. Chem.*, **7** (1997) 79.
[71] Y. Du and N. M. Sammes, *J. Eur. Ceram. Soc.*, **21** (2001) 727.
[72] J. Wolfenstine, *Electrochem. Solid-State Lett.*, **2** (1999) 210.
[73] I. Yasuda, Y. Matsuzaki, T. Yamakawa and T. Koyama, *Solid State Ionics*, **135** (2000) 381.
[74] T. Yamada, J. Akikusa, K. Adachi, K. Hoshino, T. Ishihara and Y. Takita, *Extended Abstract of 9th Symposium on Solid Oxide Fuel Cell in Japan*, 2000, p. 25.
[75] R. T. Baker, B. Gharbage and F. M. B. Marques, *J. Electrochem. Soc.*, **144** (1997) 3130.
[76] K. Yamaji, T. Horita, M. Ishikawa, N. Sakai, H. Yokokawa and M. Dokiya, in *Solid Oxide Fuel Cells V*, eds. U. Stimming, S. C. Singhal, H. Tagawa and W. Lehnert, The Electrochemical Society Proceedings, Pennington, NJ, PV97-40, 1997, p. 301.
[77] J. H. Kim and H. I. Yoo, *Solid State Ionics*, **140** (2001) 105.
[78] H. Yokokawa, N. Sakai, T. Horita and K. Yamaji, *Fuel Cell*, **1** (2001) 117.

[79] T. Ishihara, M. Honda, T. Shibayama, H. Nishiguchi and Y. Takita, *J. Electrochem. Soc.*, **145** (1998) 3177.
[80] K. Huang, J. H. Wan and J. B. Goodenough, *J. Electrochem. Soc.*, **148** (2001) A788.
[81] K. Huang, M. Feng, J. B. Goodenough and M. Schmerling, *J. Electrochem. Soc.*, **143** (1996) 3630.
[82] K. Yamaji, T. Horita, M. Ishikawa, N. Sakai, H. Yokokawa, *Solid State Ionics*, **108** (1998) 415.
[83] K. Yamaji, H. Negishi, T. Horita, N. Sakai and H. Yokokwa, *Solid State Ionics*, **135** (2000) 389.
[84] H. Hayashi, M. Suzuki and H. Inaba, *Solid State Ionics*, **128** (2000) 131.
[85] T. Ishihara, J. A. Kilner, M. Honda and Y. Takita, *J. Am. Chem. Soc.*, **119** (1997) 2747.
[86] S. Kim, M. C. Chun, K. T. Lee and H. L. Lee, *J. Power Sources*, **93** (2001) 279.
[87] T. Ishihara, H. Furutani, M. Honda, T. Yamada, T. Shibayama, T. Akbay, N. Sakai, H. Yokokawa and Y. Takita, *Chem. Mater.*, **11** (1999) 2081.
[88] V. V. Kharton, A. A. Yaremchenko, A. V. Kovalevsky, A. P. Viskup, E. N. Naumovich and P. E. Kerko, *J. Memb. Sci.*, **163** (1999) 307.
[89] V. Thangadurai, A. K. Shukla and J. Gopalakrishnan, *Chem. Commun.*, (1998) 2647.
[90] T. Ishihara, T. Shibayama, M. Honda, H. Nishiguchi and Y. Takita, *J. Electrochem. Soc.*, **147** (2000) 1322.
[91] T. Ishihara, T. Shibayama, M. Honda, H. Nishiguchi and Y. Takita, *Chem. Commun.*, (1999) 1227.
[92] J. Akikusa, K. Adachi, K. Hoshino, T. Ishihara and Y. Takita, *J. Electrochem. Soc.*, **148** (2001) A1275.
[93] J. B. Goodenough, J. E. Ruiz-Diaz and Y. S. Zhen, *Solid State Ionics*, **44** (1990) 21.
[94] K. R. Kendall, C. Navas, J. K. Thomas and H. C. zur Loye, *Solid State Ionics*, **82** (1995) 215.
[95] K. Kakinuma, H. Yamamura, H. Haneda and T. Atake, *Solid State Ionics*, **140** (2001) 301.
[96] C. A. J. Fisher and M. S. Islam, *Solid State Ionics*, **118** (1999) 355.
[97] G. B. Zhang and D. M. Smyth, *Solid State Ionics*, **82** (1995) 153.
[98] T. Schober, J. Friedrich and F. Krug, *Solid State Ionics*, **99** (1997) 9.
[99] S. Nakayama and M. Sakamoto, *J. Eur. Ceram. Soc.*, **18** (1998) 1413.
[100] S. Nakayama, *Mater. Integration*, **12** (1999) 57.
[101] J. E. Sansom, D. Richings and P. R. Slater, *Solid State Ionics*, **139** (2001) 205.
[102] F. Abraham, J. C. Boivin, G. Mairesse and G. Nowogrocki, *Solid State Ionics*, **40–41** (1990) 934.
[103] T. Ishihara, H. Arikawa, T. Akbay, H. Nishiguchi and Y. Takita, *J. Am. Chem. Soc.*, **123** (2001) 203.

[104] H. Iwahara, H. Uchida, K. Ono and K. Ogaki, *J. Electrochem. Soc.*, **135** (1988) 529.
[105] H. Iwahara, *Bull. Ceram. Soc. Jpn.*, **27** (1992) 112.
[106] R. L. Cook, R. C. MacDuff and A. F. Sammells, *J. Electrochem. Soc.*, **137** (1990) 3309.
[107] R. L. Cook, J. J. Osborne, J. H. White, R. C. MacDuff and A. F. Sammells, *J. Electrochem. Soc.*, **139** (1992) L19.
[108] A. F. Sammells, R. L. Cook, J. H. White, J. J. Osborne and R. C. MacDuff, *Solid State Ionics*, **52** (1992) 111.

Chapter 5

Cathodes

Harumi Yokokawa and Teruhisa Horita

5.1 Introduction

Cathodes for solid oxide fuel cells (SOFCs) have to possess many properties including high electrical conductivity, high catalytic activity for oxygen reduction, and compatibility with other cell components. In the earliest stages of SOFC development, platinum was used as cathode since other appropriate materials were not available. However, platinum is expensive and its use in cost-effective commercial SOFCs for power generation is not practical. Less expensive perovskites [1] also possess the required properties and have consequently attracted much interest. In 1969, $LaCoO_3$ was tested by Tedmon et al. [2] and its initial performance in cells was good. However, severe degradation occurred with increasing time of operation due to reactions with yttria-stabilised zirconia (YSZ) electrolyte. Investigations on cathodes then moved to lanthanum manganite ($LaMnO_3$)-based materials. Although degradation of lanthanum manganite cathodes was not as severe, some potential reactions with YSZ, particularly at higher cell fabrication temperatures, were recognised [3].

Success of the seal-less tubular cells with electrolyte fabricated using electrochemical vapour deposition method [4] stimulated investigations of fabricating SOFC components by a more cost-effective slurry/sintering process [5]. To utilise this process successfully, any chemical interactions between cathode and YSZ electrolyte had to be avoided during cell fabrication without sacrificing cathode performance. For this purpose, the reactivity of lanthanum manganites with YSZ was investigated using thermodynamic considerations [6], and to avoid $La_2Zr_2O_7$ formation, A-site (La)-deficient $LaMnO_3$ was proposed for the cathode. Its use inhibited $La_2Zr_2O_7$ formation and resulted in better cathode performance, as confirmed by Dokiya et al. [7] on test cells and by Ipponmatsu et al. [8] on tubular cells in a 1 kW system. Aizawa et al. [9] also used the A-site deficient lanthanum manganite cathode to fabricate tubular cells by a wet slurry/sintering process and achieved about the same cell performance as cells fabricated by using electrochemical vapour deposition for the YSZ electrolyte [4].

For intermediate temperature SOFCs, a composite cathode consisting of strontium-doped lanthanum manganite (LSM) and YSZ has shown good performance [10]. For use with ceria-based electrolytes [11], a (La,Sr)(Co,Fe)O_3 (LSCF)-based cathode has been developed [12].

In this chapter, the physical and physicochemical properties of perovskites which are important in understanding electrochemical performance are given in detail, emphasising the effect of valence stability on oxygen nonstoichiometry, crystal structure, electrical conductivity, thermal expansion and chemical reactivity with YSZ electrolyte. Then, practical aspects of the lanthanum manganite cathodes are described with a focus on how to avoid $La_2Zr_2O_7$ formation. Cathodes for the intermediate temperature SOFCs are discussed in relation to the conventional $LaMnO_3$-based cathodes. Compatibility of cathode materials with oxide ceramic interconnects and metallic interconnects is also examined with emphasis on Cr poisoning. Finally, the fabrication methods for cathodes are briefly described.

5.2 Physical and Physicochemical Properties of Perovskite Cathode Materials

5.2.1 Lattice Structure, Oxygen Nonstoichiometry, and Valence Stability

The lattice structure of perovskites, ABO_3, is shown in Figure 5.1. This oxide consists of three elements, namely the large cations, A^{n+}, the small cations,

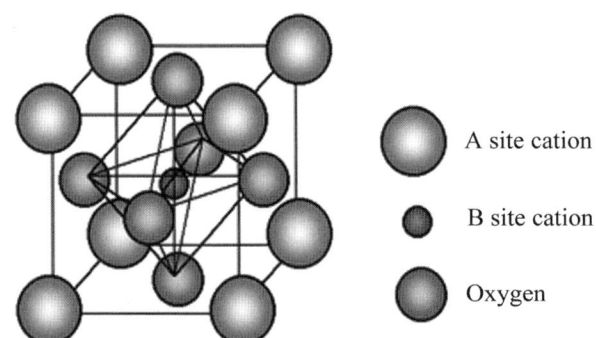

Figure 5.1 Schematic representation of lattice structure of perovskite, ABO_3.

$B^{(6-n)+}$, and the oxide ions, O^{2-}, where n is the positive charge on the A ions. Cations $B^{(6-n)+}$ are surrounded by 6 oxide ions, whereas cations A^{n+} have 12 oxide ion coordinates; the sites with 12 coordinates are often called the A sites, and the sites with 6 coordinates the B sites. The geometrical fitting of cations and anions to this structure is measured with the Goldschmidt tolerance factor defined as follows:

$$t = (r_A + r_O)/[2^{1/2}(r_B + r_O)] \tag{1}$$

where r_A, r_B, and r_O are the effective ionic radii of A, B, and O ions, respectively.

Usually, this factor is evaluated from Shannon's ionic radii [13] for respective coordination numbers. When the tolerance factor is near unity, the structure is the ideal cubic one. In other cases, orthorhombic or rhombohedral distortions appear.

Undoped stoichiometric $LaMnO_3$ shows orthorhombic structure at 25°C and rhombohedral above 600°C [14]. The crystallographic transformation temperature strongly depends on the oxygen stoichiometry, and hence on the mean manganese valence. With increasing oxygen content, the transformation temperature decreases rapidly. Substitution of La with lower valence cations (such as Sr^{2+} and Ca^{2+}) or A-site deficiency increases the concentration of Mn^{4+} in the $LaMnO_3$ lattice. This eventually decreases the transformation temperature. Although the cubic structure does not appear in pure $LaMnO_3$, it appears around 1000°C in $(La_{0.7}Sr_{0.3})MnO_3$.

The most interesting features of $LaMnO_3$-based perovskites are their oxygen nonstoichiometry and related defect structure [15–24]; that is, in addition to the oxygen deficient region, the 'oxygen-excess' region [15] appears as illustrated in Figure 5.2a. Figures 5.2b–d show oxygen contents of $La_{1-x}MnO_3$ and $La_{1-x}Sr_xMnO_3$ as a function of oxygen partial pressure [15]. For $La_{1-x}MnO_3$

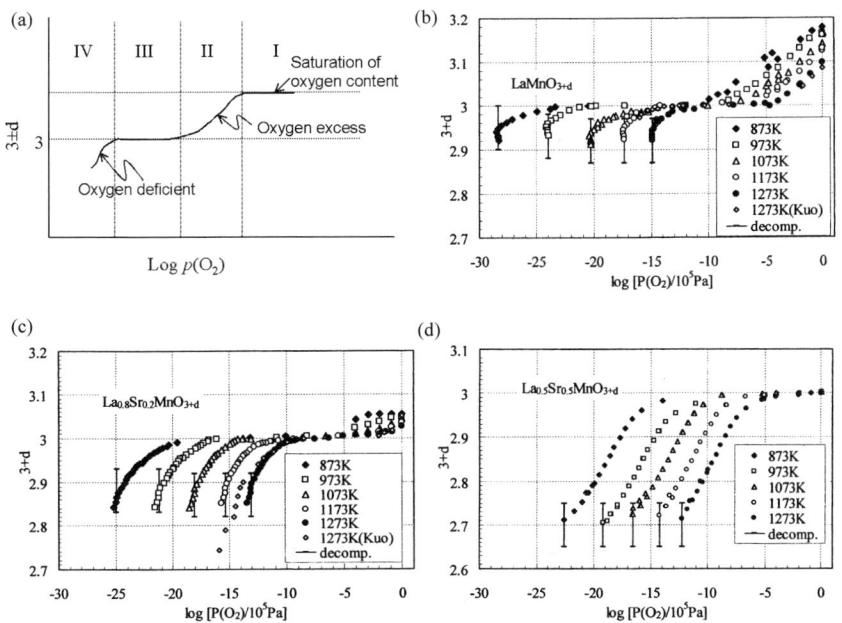

Figure 5.2 Oxygen nonstoichiometry of $LaMnO_3$ as a function of oxygen partial pressure [15]. (a) Schematic representation of oxygen nonstoichiometry. (b) Oxygen content of $LaMnO_{3\pm d}$. (c) Oxygen content of $La_{0.8}Sr_{0.2}MnO_{3\pm d}$. (d) Oxygen content of $La_{0.5}Sr_{0.5}MnO_{3\pm d}$.

and $x < 0.4$ in $La_{1-x}Sr_xMnO_3$, the oxygen content exhibits two plateaus in its oxygen potential dependence; one is around the oxygen-excess (3 + d) at high oxygen partial pressures (Region I in Figure 5.2a), the other being the stoichiometric one (around 3, Region III in Figure 5.2a) at the lower oxygen partial pressures. Particular interest has focused on the oxygen-excess region because the perovskite lattice does not allow interstitial oxygen and therefore it is a challenge in defect chemistry to explain the 'oxygen excess' region.

Many attempts have been made to take into account the thermogravimetric results by considering the formation of metal vacancies in the A- and the B-sites. Sometimes, the 'oxygen excess' can be discussed together with the A-site deficiency in the structure, because there is an expectation that the metal vacancies can be easily formed on the A sites compared with the B sites. The validity of defect models is usually checked by examining the reproducibility of thermogravimetric results. Here, however, care must be taken; good reproducibility does not necessarily mean that the adopted defect model is appropriate. For example, Roosmalen and Cordfunke [17–20] tried to interpret the defect structure of the oxygen-excess composition by the following assumptions: $[Mn^{\bullet}_{Mn}](=[Mn^{4+}])$ is constant independent of the Sr content, and oxygen nonstoichiometry and oxygen incorporation in the oxygen-excess region involves the oxidation of $Mn'_{Mn}(= Mn^{2+})$ to $Mn^{\times}_{Mn}(=Mn^{3+})$:

$$\frac{3}{2}O_2 + 6Mn'_{Mn} \rightarrow LaMnO_3 + V'''_{La} + V'''_{Mn} + 6Mn^{\times}_{Mn} \qquad (2)$$

$$K_{OX}p(O_2)^{3/2} = \frac{[V'''_{La}][V'''_{Mn}][Mn^{\times}_{Mm}]^6}{[Mn'_{Mn}]^6} \qquad (3)$$

Results of their calculations can reproduce the thermogravimetric results fairly well. However, their assumption that $[Mn^{\bullet}_{Mn}](=[Mn^{4+}])$ is constant may not be accurate. Their assumption implies that with decreasing oxygen potential, the concentration of Mn^{3+} becomes lower than that of Mn^{4+}. This does not happen in a system in which equilibration proceeds reasonably. This is therefore due to their inappropriate consideration of possible reactions among defects. Recently, rapid progress has been made in computer calculations so that complicated chemical equilibria calculations can be made without difficulty. In fact, Yokokawa *et al.* [25] made an attempt to explain the oxygen nonstoichiometry of $La_{1-x}MnO_{3-d}$ with different lanthanum deficits as a function of oxygen potential and succeeded in reproducing those thermogravimetric results by a single model. The concentrations of the respective manganese ions derived from this model show a reasonable oxygen potential dependence as expected in accordance with normal chemical understanding. Recently, Mizusaki *et al.* [15] have proposed a defect clustering model to explain the fact that the oxygen excess region disappears for $x > 0.4$ in $La_{1-x}Sr_xMnO_{3+d}$. In addition to metal vacancies, they propose a 'vacancy excluding space' that is needed for each cation vacancy to exist stably without having the vacancy in its neighbouring space; for the metal vacancy (V'''_{La}), this space covers the space of nine unit cells, while for the Sr'_{La}, it covers three unit cells. The maximum oxygen excess can be determined from the maximum number of vacancy excluding spaces available in the lattice. For $x > 0.4$, there is no room for the vacancy excluding space around the metal vacancies and this model can provide a good explanation for the disappearance of the oxygen excess. However, even in Mizusaki's model, the inappropriate assumption of constant $[Mn^{\bullet}_{Mn}](=[Mn^{4+}])$ was not corrected.

With decreasing oxygen partial pressure, $La_{1-x}Sr_xMnO_{3+d}$ oxides become stoichiometric as shown in the second plateau in Figure 5.2a. Even in the lower oxygen partial pressure region (region IV in Figure 5.2a), these oxides show deficiency from the stoichiometric composition, and oxygen vacancies are formed. According to Mizusaki [15], the formation of oxygen vacancies can be represented as:

$$2Mn^{\bullet}_{Mn} + O^{x}_{O} = 2Mn^{x}_{Mn} + V^{\bullet\bullet}_{O} + \frac{1}{2}O_2 \tag{4}$$

Here again the reduction from Mn^{4+} to Mn^{2+} appears to be improper.

5.2.2 Electrical Conductivity

$LaMnO_3$-based perovskites exhibit intrinsic p-type conductivity due to changes in the Mn valence. The electrical conductivity of these materials is greater than 10 S cm^{-1} at 700°C. The electrical conductivity is enhanced by replacing La^{3+} with lower valence cations (such as Ca^{2+}, Sr^{2+}) or doping with other cations (Mg^{2+}, Co^{3+}, etc.) for application as a cathode material [21–23, 26].

In particular, calcium and strontium doping of $LaMnO_3$ has been examined to improve its electrical conductivity because such doped materials have high electrical conductivity as well as thermal expansion and chemical properties compatible with other SOFC component materials. When a La^{3+} ion is replaced by a Sr^{2+} ion, an electric hole is formed on the Mn^{3+} site to maintain electroneutrality and this leads to an increase in electrical conductivity:

$$LaMnO_3 \xrightarrow{xSrO} La^{3+}_{1-x}Sr^{2+}_{x}Mn^{3+}_{1-x}Mn^{4+}_{x}O_3 \tag{5}$$

Figure 5.3 shows temperature dependence of the electrical conductivity [26]. Straight lines in this plot suggest the small polaron hopping conduction, which is generally expressed as follows:

$$\sigma T = (\sigma T)^{\circ} \exp\left(-\frac{E_a}{kT}\right) = A\left(\frac{h\nu^{\circ}}{\kappa}\right) c(1-c) \exp\left(-\frac{E_a}{kT}\right) \tag{6}$$

where $(\sigma T)^{\circ}$ and E_a are the pre-exponential constant and the activation energy, respectively. A constant, c, is the carrier occupancy on the sites and therefore $c(1-c)$ indicates the probability of hopping from the carrier occupied site to the unoccupied sites.

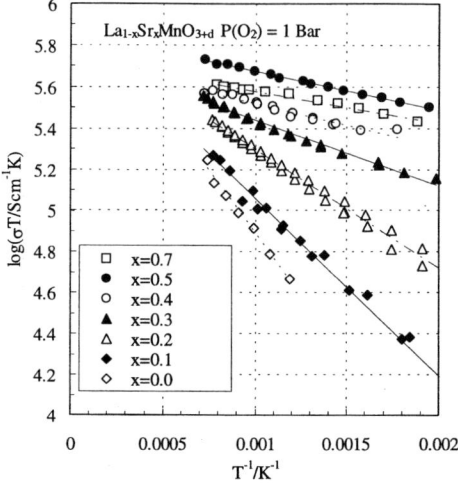

Figure 5.3   Electrical conductivity of Sr-doped LaMnO$_{3\pm d}$ as a function of inverse temperature [26].

The electronic band structure has been employed to explain the relationship between hopping conduction and Mn mean valence. In the oxygen-excess La$_{1-x}$Sr$_x$MnO$_{3+d}$ ($d > 0$), the conductivity is about the same as that of the material with stoichiometric oxygen content, $d = 0$. In the oxygen-deficient La$_{1-x}$Sr$_x$MnO$_{3+d}$ ($d < 0$), the conductivity is essentially determined by the mean Mn valence and temperature. The predominant electrical conduction was found to take place by the electron hopping on the $e_g\uparrow$ level of Mn [26].

Any A-site deficiency from the stoichiometric composition (La$_{1-x}$MnO$_{3+d}$, $0 < x < 0.1$) also affects the conductivity. According to Mizusaki et al. [27], La-deficient LaMnO$_3$ exhibits lower electrical conductivity. Figure 5.4 shows

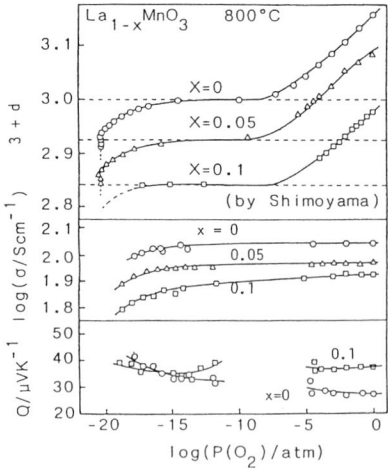

Figure 5.4   Nonstoichiometry, electrical conductivity, and Seebeck coefficient of La$_{1-x}$MnO$_{3+d}$ (x = 0, 0.05, 0.1) as a function of oxygen partial pressure [27].

nonstoichiometry, electrical conductivity, and Seebeck coefficient of $La_{1-x}MnO_{3+d}$ (x = 0, 0.05, 0.1) as a function of oxygen partial pressure. In the oxygen excess region, the conductivity (σ) and the Seebeck coefficient (Q) are essentially constant irrespective of oxygen content or metal vacancies. Therefore, the concentration of carrier in the oxygen-excess region is about the same as that in the stoichiometric composition, $LaMnO_3$. In the oxygen-deficient region, the conductivity decreases and the Seebeck coefficient increases with decreasing oxygen partial pressure. This suggests that the carrier concentration decreases with oxygen partial pressure. The decrease in conductivity in the La-deficient $LaMnO_3$ is due to a decrease in the mobility of holes.

5.2.3 Thermal Expansion

To minimise stresses during cell fabrication and cell operation, thermal expansion of the cathode should be matched with other SOFC component materials, especially electrolyte and interconnect. The thermal expansion coefficient (TEC) of undoped $LaMnO_3$ is $11.2 \pm 0.3 \times 10^{-6}$ K^{-1} in the temperature range 35–1000°C [28]. Table 5.1 summarises the TECs of undoped and doped $LaMnO_3$ [28,29].

Table 5.1 Thermal expansion coefficients of several $LaMnO_3$-based perovskites [28,29]

Composition	Thermal expansion coefficient (10^{-6} K^{-1})
$La_{0.99}MnO_3$	11.2
$La_{0.94}Sr_{0.05}MnO_3$	11.7
$La_{0.89}Sr_{0.10}MnO_3$	12.0
$La_{0.79}Sr_{0.20}MnO_3$	12.4
$La_{0.69}Sr_{0.30}MnO_3$	12.8
$LaMnO_3$	12.5
$La_{0.9}Sr_{0.1}MnO_3$	11.2
$La_{0.8}Sr_{0.2}MnO_3$	11.3
$La_{0.7}Sr_{0.3}MnO_3$	11.8, 12.0
$La_{0.6}Sr_{0.4}MnO_3$	12.6
$La_{0.9}Ca_{0.1}MnO_3$	10.6
$La_{0.8}Ca_{0.2}MnO_3$	10.0
$La_{0.7}Ca_{0.3}MnO_3$	10.5
$La_{0.6}Ca_{0.4}MnO_3$	11.7

In the slightly A-site-deficient $LaMnO_3$ ($La_{0.99}MnO_3$), the TEC values are lower than in the stoichiometric composition. This is due to a crystal structure change caused by the A-site deficiency. With Sr doping in $La_{0.99}MnO_3$, the TEC values increase with increase in the concentration of Sr.

Recently, Mori et al. [29] have observed a thermal expansion behaviour which exhibits some anomalous dependence on dopant concentration; that is, there is a minimum in TEC around a dopant concentration of 0.1–0.2. The reported minimum TEC values are about 10×10^{-6} K^{-1} and 11×10^{-6} K^{-1} for $La_{0.8}Ca_{0.2}MnO_3$ and $La_{0.9}Sr_{0.1}MnO_3$, respectively. They also observed a hysteresis in thermal expansion curves caused by thermal cycling. It should be noted that their experiments were made on dense bars, while electrodes used in SOFCs are porous in which no geometrical change has been observed on thermal cycling. This is apparently due to a difference in the relaxation time for oxygen stoichiometry to reach a new equilibrium value on changing temperature between dense and porous samples. Particularly in the oxygen-excess region, longer time is required for equilibration.

LaCoO$_3$-based cathodes show TEC values of about 20×10^{-6} K^{-1} which are too high compared to that of YSZ. Attempts have been made to reduce TEC by doping with Sr and other alkaline earth ions.

5.2.4 Surface Reaction Rate and Oxide Ion Conductivity

Oxygen incorporation is a very important process for perovskite cathode materials because the oxygen stoichiometry can change during cell operation and also on thermal cycling. The oxygen incorporation rate and the oxygen flux through the materials can be characterised in terms of two parameters: the oxygen diffusion coefficient and the oxygen surface exchange coefficient. The ^{16}O/^{18}O isotope exchange technique provides very meaningful data on these parameters [30–37]. During ^{18}O isotope annealing, the net isotope flux crossing a O$_2$/solid surface is directly proportional to the difference in isotope fractions between the gas and the solid. This flux is equal to the ^{18}O flux diffusing away from the surface into the solid. This leads to the following boundary condition:

$$-D^* \frac{\partial C}{\partial x}\bigg|_{x=0} = k^*(C_g - C_s) \tag{7}$$

where D^* and k^* are the ^{18}O diffusion coefficient and surface exchange coefficient, respectively. C_g and C_s are the ^{18}O fractions in the gas and at the surface, respectively. The solution for a semi-infinite medium with the above boundary condition has been given by Crank in the following equation [38]:

$$C'(x,t) = \frac{C(x,t) - C_{bg}}{C_g - C_{bg}} = \mathrm{erfc}\left[\frac{x}{2\sqrt{D^*t}}\right] - \left[\exp(hx + h^2 D^* t) \times \mathrm{erfc}\left(\frac{x}{2\sqrt{D^*t}} + h\sqrt{D^*t}\right)\right] \tag{8}$$

where $C'(x,t)$ is the ^{18}O fraction after being corrected for the natural isotope background level of ^{18}O ($C_b = 0.2\%$) and for the isotope enrichment of the gas ($C_g = 95$–98%); t is the corrected time of the isotope exchange, and h is a parameter, $h = k^*/D^*$. The labelled stable ^{18}O isotope is analysed by secondary ion mass spectrometry (SIMS) in the diffusion profiles or in the secondary ion images. Kilner and co-workers have collected many k^* and D^* data for perovskites, and derived a correlation between these two parameters [30]. Figure 5.5 shows the relation between D^* and k^*, a so-called h-plane plot. A linear regression of the (logarithmic) data gives a slope near 0.5. This correlation is valid over a wide temperature range for various compositions of perovskite materials. Since these materials are electronically conducting, there are many mobile electrons. On oxygen incorporation, therefore, the concentration of oxide ion vacancies plays an important role both for surface oxygen exchange and oxide ion diffusion.

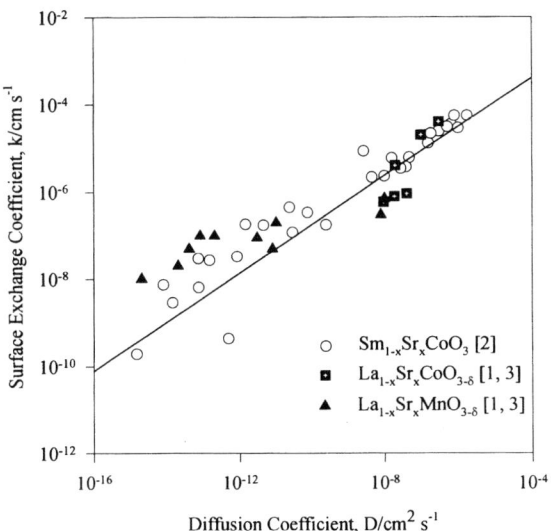

Figure 5.5 Relationship between isotope oxygen diffusion coefficient (D) and surface oxygen exchange rates (k*) (h-plane plots) for selected perovskites [30].*

The measured diffusion coefficients and surface oxygen exchange coefficients for $LaMnO_3$ and selected other perovskites are listed in Table 5.2 [33–37]. The oxygen diffusion coefficients for $LaMnO_3$-based perovskites are scattered over one or two orders of magnitude, and the values range in the orders of 10^{-13} to 10^{-11} cm^2 s^{-1} at 1000°C at $p(O_2) = 1$ atm. These values for diffusivity are low compared with the values required for mixed conducting oxide cathodes. This implies that when doped $LaMnO_3$ is used for a cathode, oxygen will be transported through the gaseous phase or on the surface of the $LaMnO_3$ to the major electrochemical reaction sites at the electrolyte/cathode/gas three-phase boundary as will be described in the next section. In the high overpotential region, however, some authors have pointed out that oxide ion diffusion can take place inside doped $LaMnO_3$ because the oxide ion vacancies can be formed at the lower oxygen potentials. This oxide ion flow could contribute to the electrochemical reaction [39–41]. Such oxygen diffusion inside the lanthanum manganite was confirmed with ^{18}O labelling and SIMS analysis [42]. $LaCoO_3$-based oxides show higher oxide ion diffusivity and larger surface exchange coefficients than the $LaMnO_3$-based ones as listed in Table 5.2.

Although a detailed description of cathode reactions and polarisations is given in Chapter 9, a brief discussion of the cathode reaction mechanisms is included here to elucidate several aspects of materials issues in cathode development. As discussed in the previous section, the cathode reduces oxygen molecules to

Table 5.2 Oxygen diffusion coefficients (D^*) of LaMnO$_3$-based perovskites [32,35]

Composition	Temperature (°C)	$P(O_2)$ (atm)	Isotope oxygen diffusion coefficient, D^* (cm^2 s^{-1})	Oxygen surface exchange coefficient, k^* (cm s^{-1})	Ref.
La$_{0.95}$Sr$_{0.05}$MnO$_3$	900	1	2.44×10^{-13}	–	32
La$_{0.90}$Sr$_{0.10}$MnO$_3$	1000	1	4.78×10^{-12}	–	32
La$_{0.80}$Sr$_{0.20}$MnO$_3$	900	1	1.27×10^{-12}	–	32
La$_{0.80}$Sr$_{0.20}$MnO$_3$	1000	1	1.33×10^{-11}	–	32
La$_{0.80}$Sr$_{0.20}$MnO$_3$	1000	1	6.60×10^{-13}	5.62×10^{-8}	35
La$_{0.80}$Sr$_{0.20}$MnO$_3$	900	1	1.60×10^{-13}	1.78×10^{-8}	35
La$_{0.80}$Sr$_{0.20}$MnO$_3$	800	1	4.00×10^{-15}	5.62×10^{-9}	35
La$_{0.80}$Sr$_{0.20}$MnO$_3$	700	1	3.10×10^{-16}	1.01×10^{-9}	35
La$_{0.80}$Sr$_{0.20}$CoO$_3$	1000	1	9.01×10^{-8}	5.64×10^{-6}	35
La$_{0.80}$Sr$_{0.20}$CoO$_3$	900	1	1.03×10^{-8}	2.02×10^{-6}	35
La$_{0.80}$Sr$_{0.20}$CoO$_3$	800	1	9.87×10^{-10}	6.31×10^{-7}	35
La$_{0.80}$Sr$_{0.20}$CoO$_3$	700	1	1.04×10^{-10}	1.58×10^{-7}	35

oxide ions around the cathode/electrolyte interface. The elemental steps for cathode reaction are: (i) oxygen molecule adsorption and dissociation into oxygen atoms at the cathode surface, (ii) surface diffusion of adsorbed oxygen, (iii) incorporation and subsequent bulk diffusion of oxygen inside the oxide lattice, (iv) incorporation of adsorbed oxygen in the O$_2$/cathode/electrolyte three-phase boundary, and (v) transport of oxide ions in the solid electrolyte. The charge transfer reaction can take place in steps (i), (iii), or (iv). Any of these elemental steps can limit the rate of the cathodic reaction.

Isotope oxygen (^{18}O$_2$) labelling is one of the effective methods to analyse the reaction mechanism, as previously mentioned. Figure 5.6 shows depth diffusion profiles of the oxygen isotope and other metal elements from a dense (La,Sr)CoO$_3$

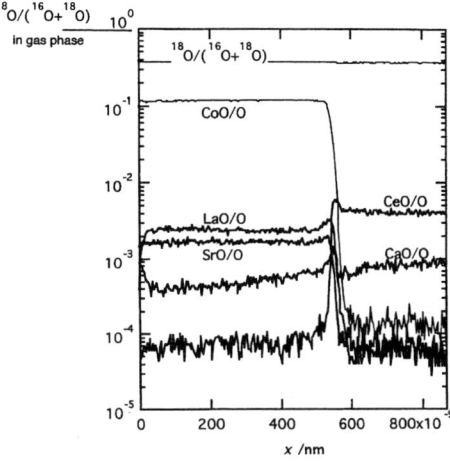

Figure 5.6 Oxygen isotope and elemental diffusion profiles from a dense LaCoO$_3$ cathode layer surface into CeO$_2$ electrolyte. ^{18}O$_2$ annealing for 240 s at 800°C, p(^{18}O$_2$) = 0.01 bar [43].

(LSC) layer into a (Ce,Ca)O$_2$-based electrolyte layer [43]. Although some cation diffusion is observed across the interface, the LSC/CeO$_2$ interface is very clear in the range of 100 nm from the interface. The oxygen isotope concentration shows no gap at the interface from LSC to CeO$_2$, while the isotope oxygen concentration at the gas phase and the LSC surface is considerably different. This suggests that the rate-determining step for cathode reaction is the surface reaction process at the O$_2$/LSC surface; that is, step (i) is the slowest among the elemental steps.

For the LaMnO$_3$/YSZ system, the oxygen transport is different than in the LSC/doped-CeO$_2$ system. Figure 5.7 shows depth diffusion profiles of the oxygen isotope and other elements around a dense La$_{0.9}$Sr$_{0.1}$MnO$_3$ layer/YSZ interface [44]. For isotope oxygen diffusion, a small decrease of isotope oxygen concentration is observed between the gas phase and the dense La$_{0.9}$Sr$_{0.1}$MnO$_3$ surface. A gradual decrease of isotope oxygen concentration is observed in the dense La$_{0.9}$Sr$_{0.1}$MnO$_3$ layer followed by a flat profile at the interface. This suggests that the oxygen diffusion in the La$_{0.9}$Sr$_{0.1}$MnO$_3$ layer is quite slow. Since the step (iii) cannot be expected to occur to a significant degree, the three-phase boundaries become the sites where the charge transfer takes place [44]. This suggests that step (*ii*) or (iv) will be rate limiting.

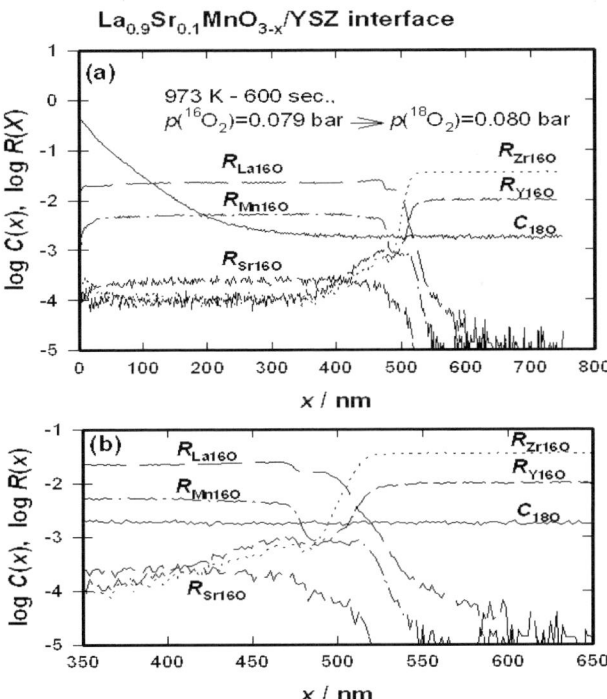

Figure 5.7 Oxygen isotope and elemental diffusion profiles at a dense La$_{0.9}$Sr$_{0.1}$MnO$_3$ layer/YSZ interface [44] (a): wide view of the LSM/YSZ interface, (b): interface region of LSM/YSZ. Samples were annealed at 973 K. The concentration of ^{18}O is defined as the ratio of secondary ion signal counts: C$_{18O}$(x)=I(^{18}O$^-$)/{I(^{18}O$^-$)+I(^{16}O$^-$)} at a depth of x, and the ratio of the other coupling metal secondary ions at a depth of x were defined as R$_{M16O}$(x)=I(M^{16}O$^-$)/{I(^{18}O$^-$)+I(^{16}O$^-$)}.

5.3 Reactivity of Perovskite Cathodes with ZrO_2

5.3.1 Thermodynamic Considerations

The perovskite lattice provides a strong stabilisation effect on the transition metal ions in B sites. For example, the trivalent Co^{3+} ions are well stabilised in the perovskite structure, although this valence state is not fully stable in other crystal structures. This stabilisation is due to the geometrical matching of the A-site and the B-site ions (see Figure 5.1). When the tolerance factor (Eq. (1)) derived from the ionic sizes is close to unity, large stabilisation energy is achieved [6,45]. For the rare earth transition metal perovskites, the tolerance factor is less than unity suggesting that rare earth ions are *small* as A-site cations. Among rare earth ions, the *largest* ions, La^{3+}, therefore exhibit the largest stabilisation. When comparison is made between Ca and Sr, the stabilisation energy of SrO-based perovskites is generally larger because of better geometrical fitting.

Yttria-stabilised zirconia also exhibits strong stabilisation on forming solid solutions in the fluorite structure [46]. Pure zirconia has the stable monoclinic phase in which the zirconium ions are coordinated with 7 oxide ions, whereas the cubic phase with 8 coordinates becomes stable only at high temperatures. On doping with Y_2O_3, the oxide ion vacancies are formed preferentially around the zirconium ions, which leads to stabilisation of zirconia in the cubic phase.

Perovskite cathodes and yttria-stabilised zirconia (YSZ) electrolyte can react in several ways as discussed below.

5.3.1.1 Reaction of Perovskites with the Zirconia Component in YSZ

The La_2O_3 component in perovskite can react with the zirconia component in YSZ to form lanthanum zirconate, $La_2Zr_2O_7$. On reaction, the valence state of the transition metal ions, M^{n+}, may change as a result of the formation of other transition metal binary oxides:

$$LaMO_3 + ZrO_2 = 0.5La_2Zr_2O_7 + MO + 0.25O_2(g) \qquad (9)$$

$$LaMO_3 + ZrO_2 = 0.5La_2Zr_2O_7 + 0.5M_2O_3 \qquad (10)$$

$$LaMO_3 + ZrO_2 + 0.25O_2(g) = 0.5La_2Zr_2O_7 + MO_2 \qquad (11)$$

where M is the transition metal and MO_n is its binary oxide. During cell fabrication at high temperatures, reaction (9) becomes important because of its large entropy change. During cell operation, oxygen potential dependence becomes important. Since YSZ is an oxide ion conductor, the oxygen potential in the vicinity of the electrode/electrolyte interface is important in determining how electrolyte/electrode chemical reactions proceed under cell operation.

5.3.1.2 Reaction of perovskite with the yttria (dopant) component in YSZ

Since the yttria component has a large stabilisation on forming solid solution, it is rare for it to react with other oxides. One exceptional case is the reaction with vanadium oxide:

$$0.2\text{VO}_{2.5} + (\text{ZrO}_2)_{0.8}(\text{YO}_{1.5})_{0.2} = 0.8\text{ZrO}_2(\text{monoclinic}) + 0.2\text{YVO}_4 \quad (12)$$

The driving force for this reaction is the large stabilisation energy of YVO$_4$. This can be categorised as the salt formation between an acidic oxide (V$_2$O$_5$) and a basic oxide (Y$_2$O$_3$). Since yttria, which is the stabiliser of the cubic phase, is extracted, this reaction leads to serious destabilisation of YSZ.

5.3.1.3 Interdiffusion between Perovskite and Fluorite Oxides
Both perovskite and fluorite oxides can form solid solutions with the common oxide components allowing interdiffusion to take place. For example, manganese ions can diffuse from the perovskite to the fluorite oxide:

$$\begin{aligned}\text{Mn}^{n+}(\text{in perovskite}) + n/2\text{O}^{2-} &= \text{Mn}^{m+}(\text{in fluorite}) \\ &+ m/2\text{O}^{2-} + (n-m)/2\text{O}_2(\text{g})\end{aligned} \quad (13)$$

When alkali earth substituted lanthanum transition metal oxides are used as cathodes, their compatibility with YSZ can be predicted from thermodynamic considerations [6,45,46]. Such considerations show that LaCoO$_3$ ($>$1173 K) and LaNiO$_3$ are unstable against reaction (9). Similarly, alkali earth transition metal oxides (ACoO$_3$, ACrO$_3$, AFeO$_3$, A = Ca, Sr, Ba) are unstable against the reductive reactions. Also, although the thermodynamic data show that LaMnO$_3$ is stable against the reactions (9), (10), and (11), experimental results do show some reactions. This can be accounted for in terms of the lanthanum nonstoichiometry in LaMnO$_3$. The reactivity with YSZ can be represented by a composition diagram as shown in Figure 5.8. The thermodynamic analysis

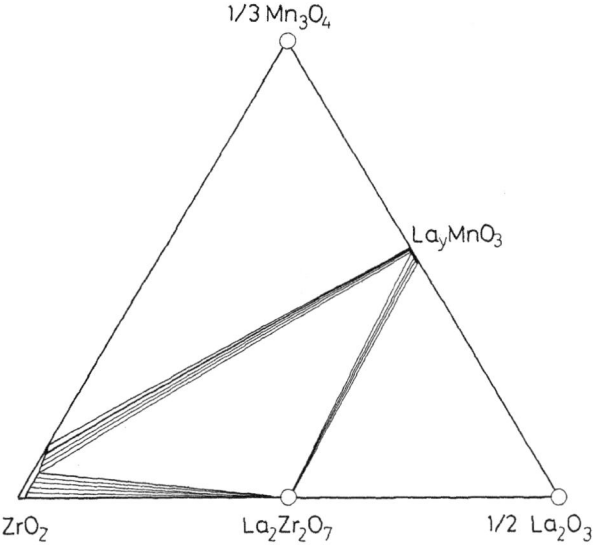

Figure 5.8   Compositional diagram for the La–Mn–Zr–O system.

without consideration of La nonstoichiometry indicates that LaMnO$_3$ and ZrO$_2$ do not react and can coexist. This is shown by the tie line between LaMnO$_3$ and ZrO$_2$. On the other hand, LaMnO$_3$ also has a tie line with La$_2$Zr$_2$O$_7$. When LaMnO$_3$ has La nonstoichiometry, these tie lines change to a three-phase tie line and other two-phase tie lines. The same feature can be shown in the chemical potential diagram as given in Figure 5.9 [6]. These diagrams show that the stoichiometric LaMnO$_3$ reacts with ZrO$_2$ to form the three-phase combination among ZrO$_2$, La$_2$Zr$_2$O$_7$ and an A-site-deficient La$_{1-x}$MnO$_3$, whereas the A-site-deficient La$_{1-x}$MnO$_3$ does not react with YSZ. In other words, the reaction of non-A-site-deficient LaMnO$_3$ with ZrO$_2$ can be written as follows:

$$\text{LaMnO}_3 + x(\text{ZrO}_2 + 0.25\text{O}_2) = 0.5x\text{La}_2\text{Zr}_2\text{O}_7 + \text{La}_{1-x}\text{MnO}_2 \quad (14)$$

In this reaction, manganese ions are oxidised and oxygen is involved as a reactant. For the A-site-deficient La$_{1-x}$MnO$_3$, the manganese solubility in YSZ increases. However, the manganese dissolution does not change the conduction characteristics very much [47].

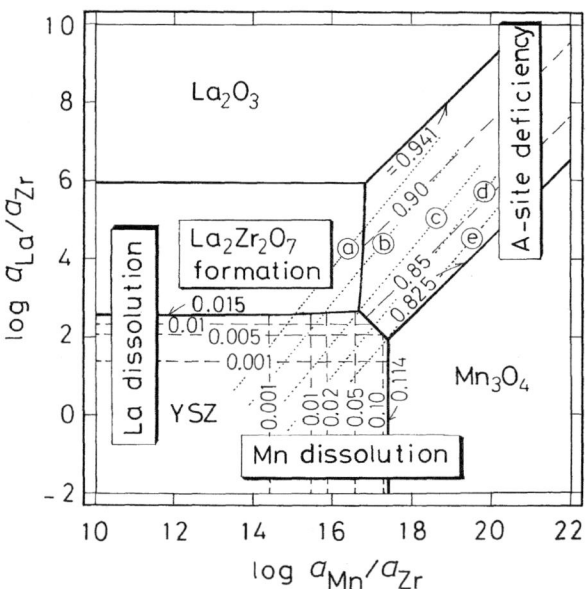

Figure 5.9   Chemical potential diagram for the La–Mn–Zr–O system.

5.3.2 Experimental Efforts

Experimental verification of the reactivity of lanthanum strontium manganite (LSM) with YSZ has been made by many researchers. Lau and Singhal [3] confirmed that LSM and YSZ can react with each other to form lanthanum zirconate, La$_2$Zr$_2$O$_7$, at high temperatures. In addition, they also found the interdiffusion of manganese, lanthanum and strontium ions across the interface into the YSZ electrolyte. Tricker and Stobbs [48] confirmed, by examining the electrode/electrolyte interface with transmission electron microscopy (TEM), that during high-temperature treatment, $La_2Zr_2O_7$ is formed between LSM and YSZ, and that after 24 h the $La_2Zr_2O_7$ formed at the interface moves towards LSM, leading to narrowing of the interface with the YSZ. The latter fact is important in understanding the chemical nature of the $La_2Zr_2O_7$ formation [49].

To minimise reactions between the cathode and the electrolyte, in Japan, most research efforts have focused on the A-site-deficient lanthanum manganite. In the USA and Europe, however, efforts [50] have been made to seek alternative cathodes, but with only limited success. Perhaps the most significant finding has been the use of composite cathodes in contact with the YSZ electrolyte. These composite cathodes minimise cathode/electrolyte interaction by mixing LSM and YSZ powders and laying down a thin layer of this mixture on the electrolyte [10]. Another step forward has been the use of an activation process to reduce the polarisation loss at the electrode [51].

To check the thermodynamic predictions of the cathode/electrolyte interaction, effects of the A-site deficiency in lanthanum manganite on the cathode performance were investigated by Dokiya *et al.* [7] using the non-A-site-deficient and the A-site-deficient lanthanum manganites. It was found that the overpotential is smaller when the A-site-deficient lanthanum manganite is applied at temperatures below 1473 K. Figure 5.10 shows a cell performance with the A-site-deficient lanthanum manganite cathode; the observed cathode overpotential (open circles) is compared with the evaluated resistivity (dashed line) originating from the electrolyte. The difference between two values gives the overpotential from the cathode reaction, which is less than 30 mV at a current density of $1400\ mA\ cm^{-2}$.

Yoshida *et al.* made systematic investigations on the effect of using partially stabilised zirconia (PSZ, Y_2O_3 content = 3 mol%) on the electrochemical performance of cathodes with different dopants and their concentrations [52]. The overpotential of the LSM/PSZ is always and systematically higher than that of the same LSM with fully stabilised YSZ. This is apparently due to the chemical interaction between LSM and PSZ; that is, the tetragonal phase (on the surface of the PSZ electrolyte) is transformed into the cubic phase after manganese dissolution into the tetragonal phase [53]. This increases the overpotential of the LSM cathode.

Although the A-site-deficient LSM can avoid reaction with YSZ, its performance as cathode depends sensitively on its heat treatment temperature [7,54]. After heat treatments at lower temperatures below 1473 K, these cathodes exhibit excellent performance (see Figure 5.10, heat treated at 1423 K), whereas heat treatment above this temperature gives rise to an increase in overpotential. Since no chemical reaction is expected as described above, this is due to the diffusion of La and Mn and related sintering behaviour. The sinterability is enhanced in the A-site-deficient LSM. It appears that diffusion inside the A-site-deficient manganite can be accelerated by increasing the number of oxide ion vacancies at high temperatures. In addition, Zr and Y

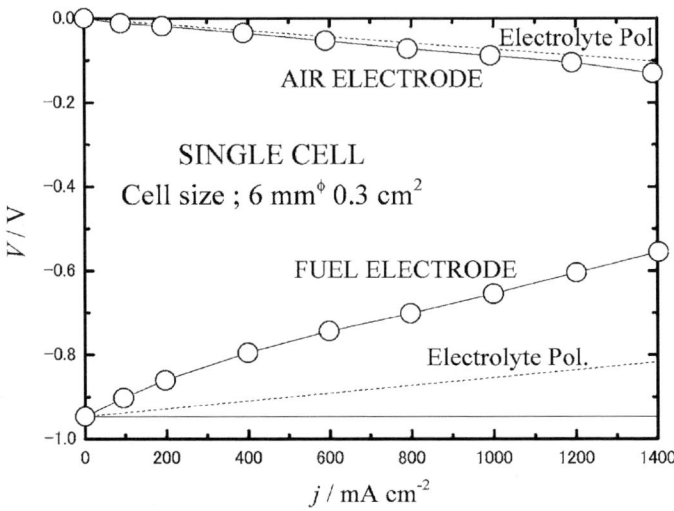

Figure 5.10 Performance of a cell prepared with a doctor-bladed electrolyte with an A-site-deficient lanthanum manganite, $(La,Sr)_{0.9}MnO_3$ cathode at 1423 K [7].

diffusion into LSM is also enhanced. These phenomena decrease the length of the three-phase boundaries at the interface and the porosity inside LSM. The fact that this is not a chemical reaction suggests that appropriate ways can be devised to avoid this degradation due to the decrease in three-phase boundary length and porosity of LSM.

5.3.3 Cathode/Electrolyte Reactions and Cell Performance

For both catalytic activity and compatibility with YSZ electrolyte, the lanthanum manganite-based perovskites are currently the best cathode materials for SOFCs. The most important issue associated with these materials is the optimisation of their composition. Initially, $(La_{0.84}Sr_{0.16})MnO_3$ was developed as a cathode material for SOFCs, whereas $(La_{0.5}Ca_{0.5})MnO_3$ was developed for water electrolysers. These initial selections can be discussed from the compatibility point of view. Thermodynamic analysis [55] predicts, as shown in Figure 5.11, that for $(La_{0.5}Ca_{0.5})MnO_3$, the zirconate formation can be avoided, whereas some zirconate formation will be expected for $(La_{0.84}Sr_{0.16})MnO_3$. This difference can be explained from thermodynamic considerations in zirconate formation. As given in Eq. (14), the $La_2Zr_2O_7$ formation can be related to the oxidation of manganese ions in the perovskites. Also, the electrode reaction mechanism on lanthanum manganite electrodes suggests that the overpotential associated with the manganite electrode can be attributed to the oxygen potential difference in the gas and in the oxygen atoms at the three-phase boundaries [56]. These considerations lead to the conclusion that $La_2Zr_2O_7$ formed at the interface will disappear from the three-phase boundaries on cathodic polarisation because of the shift of the oxygen potential

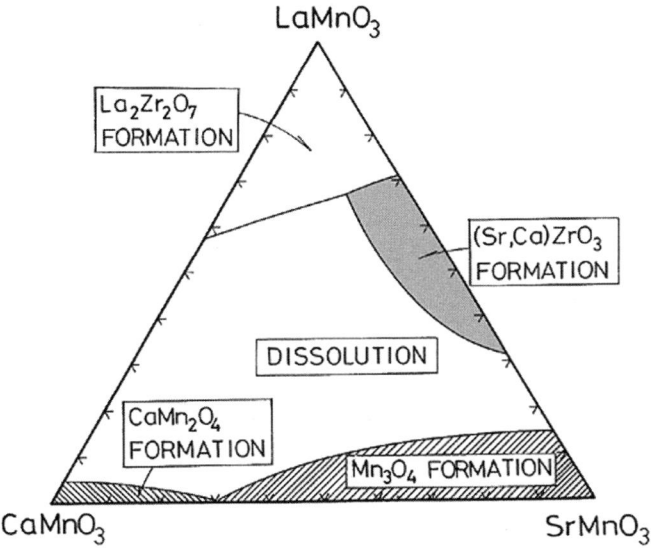

Figure 5.11 Composition diagram for (La,Sr,Ca)MnO$_3$ cathode [55]. Westinghouse developed (La$_{0.84}$Sr$_{0.16}$)MnO$_3$ for fuel cells, while Dörnier developed (La$_{0.5}$Ca$_{0.5}$)MnO$_3$ for water electrolysers.

to the more reducing side. On the other hand, on anodic polarisation (e.g. in water electrolysers), the La$_2$Zr$_2$O$_7$ formation is enhanced. Note that the (La$_{0.5}$Ca$_{0.5}$)MnO$_3$ electrode is just outside of the composition region for the La$_2$Zr$_2$O$_7$ formation as shown in Figure 5.11.

When La$_2$Zr$_2$O$_7$ phase is avoided, better initial performance is expected in SOFCs. With the use of A-site-deficient lanthanum manganite cathode (and therefore no La$_2$Zr$_2$O$_7$ formation), the cathode overpotential is extremely small as seen in Figure 5.12 (about 12 mV of polarisation at a current density of 1.5 A cm^{-2}) [8]. In this case, the LSM/YSZ interface was fabricated by electrochemical vapour deposition (EVD) of YSZ on a porous LSM substrate so that morphologically stable and long three-phase boundaries were formed. Also, the interfacial resistivity remains low even at lower temperatures down to 1073 K (Figure 5.12). This cathode clearly indicates that lanthanum manganite on YSZ provides excellent performance if the microstructure is well tailored. Furthermore, the difference in cathode overpotential in air and in pure oxygen (Figure 5.12) shows the importance of gaseous diffusion in the gas channels in cathodes.

A technological key issue for cathodes for 1000°C cell operation is therefore how to fabricate the cathode/electrolyte interface with a fine microstructure by cost-effective methods. It should also be pointed out that the cathode used for Figure 5.12 was fabricated in a reducing atmosphere and around 1200°C; this means that it was free from any degradation that may be caused by heat treatment above 1200°C in air. After many attempts, similar good behaviour has been observed in cells fabricated by the wet slurry/sintering method [9].

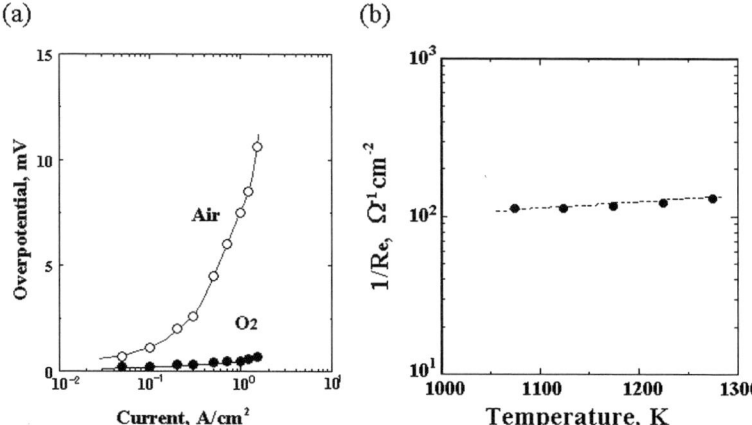

Figure 5.12 Performance of a cell with a thin film of EVD YSZ electrolyte on A-site-deficient cathode [8]: (a) the cathode overpotential for air and pure oxygen; (b) the reciprocal interface resistivity as a function of temperature.

In addition to A-site deficiency, improvement in cathode performance can also be obtained by using composite cathodes consisting of LSM and YSZ [10,57–59]. Since oxide ion conductivity is small in lanthanum manganite, enlargement of the electrochemical reaction sites can be expected from introduction of an oxide ion path (through YSZ) in the cathode layer. This composite cathode is also effective in minimising the formation of $La_2Zr_2O_7$. During the fabrication of the composite cathode, some $La_2Zr_2O_7$ may form shifting the composition of LSM to where no further formation of $La_2Zr_2O_7$ occurs. When such a composite cathode is placed on a YSZ electrolyte, no $La_2Zr_2O_7$ is formed at the interface of LSM particles in the composite cathode and the YSZ electrolyte. This helps in obtaining a LSM/YSZ interface with long three-phase boundaries.

5.3.4 Cathodes for Intermediate Temperature SOFCs

At lower temperatures, the catalytic activity of oxide cathodes for oxygen reduction decreases and the selection of cathodes becomes critical for obtaining high performance in SOFCs. There are several strategies for utilising oxide cathodes successfully in the intermediate temperature (600–800°C) region. One option is to use the same materials as those for 1000°C cell operation as in Figure 5.12, which indicates little temperature dependence. Attempts have also been made to use a composite (LSM + YSZ) cathode for intermediate temperature SOFCs with some success. Even though the interfacial resistivity of the cathode in Figure 5.12 shows very small temperature dependence, typical composite cathodes do exhibit some temperature dependence and lowering of activity with decreasing temperature. Although it is well recognised that it is difficult to obtain reproducibility of cell performance with composite cathodes, their main features can be summarised as follows:

(i) Some composite cathodes exhibit good performance even at low temperatures such as 800°C.
(ii) The main electrochemical active sites are located at the interface between the electrolyte and the composite cathode, although three-phase boundaries are spread throughout the cathode layer. The presence of YSZ particles in the cathode layer makes it easier to have longer three-phase boundaries at the interface.
(iii) The presence of YSZ in the cathode layer enhances oxygen permeability through the cathode layer. However, the presence of YSZ does not contribute to nitrogen removal from the electrochemically active sites when air is used as oxidant.

Properties (ii) and (iii) strongly depend on the microstructure of the composite cathodes making it difficult to quantitatively characterise such composite electrodes and also to reproduce their electrode activities. Attempts to utilise doped ceria instead of YSZ as a component oxide of composite cathodes have shown some success, although it is hard to say which property dominates in enhancing the cathode activity. Introducing a fine microstructure at the LSM/YSZ interface by depositing a metal organic layer (several 10 nm thick) of LSM, which spontaneously forms micropores in nm size on the passage of current, provides fine three-phase boundaries which are stable during long term operation [67].

Another option is to find alternative cathodes with higher catalytic activity at lower temperatures, particularly cobaltite-based cathodes. As given in Table 5.2, the high catalytic activity for oxygen incorporation reaction and the high oxide ion conductivity of cobaltites make them superior to lanthanum manganite. Furthermore, the driving force of reactions of $LaCoO_3$ with YSZ disappears below 1173 K. This makes it attractive to use cobaltites at low temperatures. However, there are several other issues regarding their use in YSZ electrolyte cells including their high thermal expansion coefficient and high reactivity of dopant oxide with YSZ. To reduce thermal expansion, alkaline earth substitution works to some extent; however, it also enhances reactions with YSZ to form alkaline earth zirconates. So far, no cobaltite cathode has been used successfully in YSZ electrolyte cells below 800°C.

However, attempts have been made to use cobaltite cathodes with ceria-based electrolytes. Compared with YSZ, ceria has less reactivity with perovskite cathode materials; this is because of the less acidic nature of CeO_2 compared with ZrO_2; La_2O_3 or SrO component can be regarded as basic oxides so that the interaction with ZrO_2 or CeO_2 can be judged from their acidity. The same trend can also be explained from other physicochemical properties; that is, compared with Zr ions, Ce ions are too big to form the perovskites with La or Sr. On this basis, it is possible to use lanthanum strontium cobaltite-based cathode with doped ceria electrolyte in intermediate temperature SOFCs. Even so, $(La,Sr)CoO_3$ still has a high thermal expansion coefficient ($16-22 \times 10^{-6}$ K^{-1}) compared with doped ceria (about 12×10^{-6} K^{-1}). Consequently, $(La,Sr)CoO_3$ cathode is not used even though the oxide ion conductivity is high. Steele and co-workers have focused on development of new cathodes appropriate for intermediate temperature SOFCs and (La,Sr)(Co,Fe)O_3 cathodes have been optimised for use with CeO_2-based electrolytes [11,12], a typical composition being (La$_{0.6}$Sr$_{0.4}$)(Co$_{0.8}$Fe$_{0.2}$)O_3 (LSCF). This cathode has about the same thermal expansion coefficient as doped ceria.

Ferrite-based cathodes [60] have also been explored for use with YSZ electrolyte in intermediate temperature SOFCs. However, their high-temperature phase behaviour and long-term stability remain unclarified and need further investigation. Use of interlayers between the perovskite cathode and the electrolyte has also been tried to minimise interfacial reactions. One typical interlayer consists of a thin layer of ceria based oxide between the cobaltite-based cathode and YSZ electrolyte because the chemical reactivity and thermal expansion mismatch between YSZ and cobaltites can be moderated by inserting ceria between the two materials. Another approach is to use multilayer cathodes consisting of a manganite at the interface with the electrolyte and a cobaltite on top of the manganite [68]. In this case, the electrochemical active sites are located in the manganite part, whereas the cobaltite-rich layer acts as good electrical conductor.

5.4 Compatibility of Perovskite Cathodes with Interconnects

In addition to compatibility with the electrolyte, compatibility of the cathode with the interconnect is also important. Both oxide ceramic and metallic materials are used as interconnects in SOFCs. As expected, these two types of interconnects present quite different issues in their compatibility with the cathode.

5.4.1 Compatibility of Cathodes with Oxide Interconnects

The main oxide interconnects for SOFCs are based on lanthanum chromite [5]. In this case, since both the cathode and the interconnect materials are perovskites, there are no severe chemical reactions between them and interdiffusion and precipitation of third phases are the main issues. Interdiffusion can take place in both the A and the B sites of the perovskite lattice. Ideal mixing in the A and the B sites gives rise to the driving force for interdiffusion. Usually, the cation diffusivity in perovskite oxides is faster in the A site than in the B site [61]. Thus, mixing can occur first in the A site and then in the B site. This implies that when (La,Sr)MnO_3 and (La,Ca)CrO_3 are contacted, interdiffusion among the A-site elements will take place to give a constant La or Sr content throughout the manganite and chromite phases. This is due to the driving force originating from random mixing. This can be called the 'entropy effect'. Another driving force for interdiffusion arises from the difference in stabilisation energy among the different combinations of A-site and B-site cations. When comparison is made of the valence stabilities of manganese and chromium ions in a cathode atmosphere, chromium ion tends to be trivalent, whereas the manganese ion tends to be tetravalent. This 'enthalpy effect' leads to a driving force to form $LaCrO_3$ and $AMnO_3$ (A = Ca, Sr) [62]. Thus, there are two different driving forces which promote interdiffusion in opposite directions.

Since the A-site-deficient lanthanum manganite shows better compatibility with YSZ, its compatibility with oxide interconnect is of particular interest. Nishiyama *et al.* [62] found the following interesting behaviour of the A-site-deficient manganite at the interface with $(La,Ca)CrO_3$ (LCC) that had excess CaO to enhance its air sinterability:

(i) With porous manganite cathode, the elemental distribution across the cathode/interconnection interface suggests the following replacement reaction because of the above second driving force:

$$LaMnO_3(\text{in LSM}) + CaCrO_3(\text{in LCC}) = LaCrO_3(\text{in LCC}) + CaMnO_3(\text{in LCC}) \quad (15)$$

Here, $LaCrO_3$ is formed as a dense layer next to the original LCC while $CaMnO_3$ is formed as a porous layer next to the $LaCrO_3$ layer. This reaction is triggered by the presence of calcium oxychromates in the original LCC which can be squeezed out of grain boundaries under oxygen potential gradient.

(ii) With dense manganite cathode, CaO instead of calcium oxychromate initiates the precipitation of manganese oxide at the interface, suggesting that oxygen potential distribution plays an important role in determining the mass transfer. Inside the dense manganite, the oxygen potential is lowered due to low oxide ion diffusivity.

5.4.2 Compatibility of Cathodes with Metallic Interconnects

Even though many alloys have been investigated as metallic interconnects, almost all form chromia as a protective oxide scale [63]. The main issues associated with the use of such chromia-forming alloys are chemical reactions at the interface with the cathode (and also with anode) material and chromium poisoning of the cathode. Taniguchi *et al.* [59] found that degradation by chromium poisoning occurs more severely at lower temperatures (see Figure 5.12) and that degradation measured in terms of the cathode life time is proportional to the logarithmic oxygen activity derived from the overpotential values, η, as follows:

$$\tau(\text{degradation}) \propto \Delta \log a_O (= 2\eta F/2.303RT) \quad (16)$$

Their observations on distribution of chromium in the electrolyte/cathode vicinity (as shown in Figure 5.13) indicated that although the average Cr content in the cathode layer increases with increasing temperature, this quantity is not directly related to cathode degradation. Instead, concentrated chromium deposition on the three-phase boundaries can be directly related with the cathode lifetime.

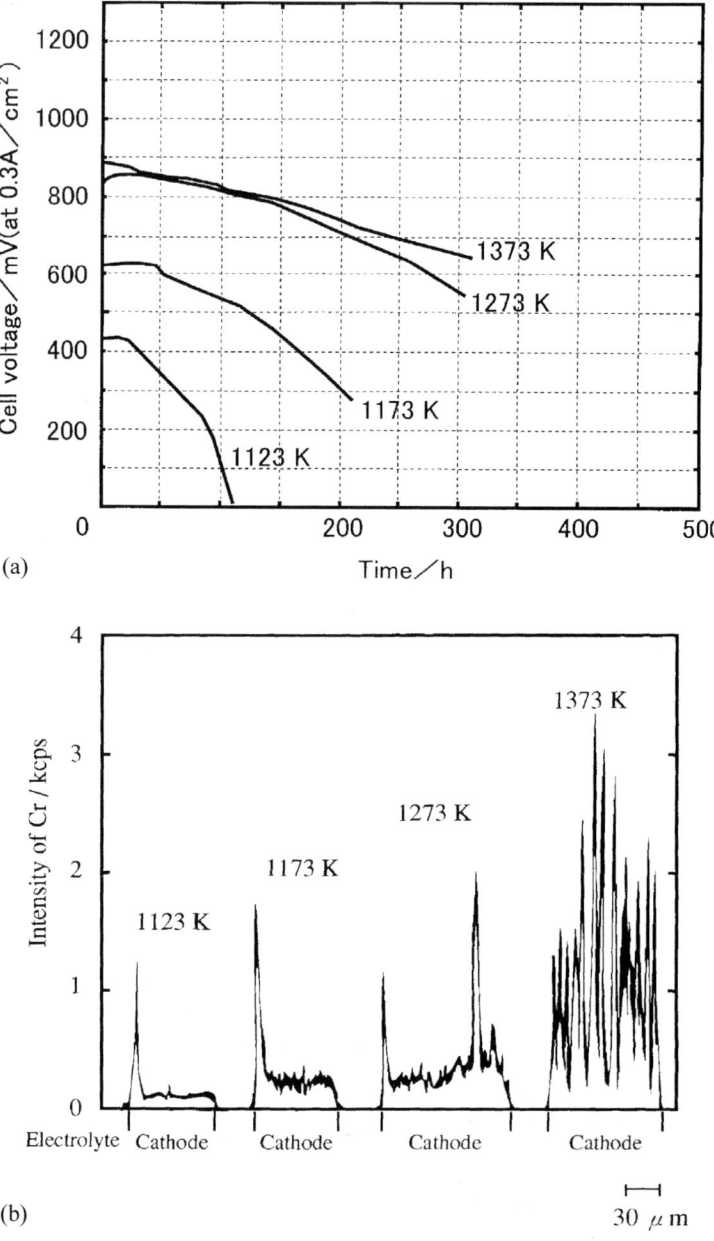

Figure 5.13 Cathode degradation due to chromium vapours from the metal interconnects [59]: (a) the cell performance (voltage) as a function of operation time at selected temperatures; (b) chromium distribution inside the cathode layer after operation; the left hand side is the interface with electrolyte.

It is well known that oxidation/ corrosion of metallic materials is enhanced [63] in the presence of water vapour. Chromium poisoning can also be enhanced in the presence of water vapour. The vapour pressure of chromium containing species increases with increasing water vapour pressure, because $CrO_2(OH)_2$ is relatively stable species in the presence of water vapour [64]. Its vapour pressure shows only a small temperature dependence so that even at low temperatures its vapour pressure is high. Taniguchi *et al.* found that the degradation increases with water vapour pressure consistent with the increasing vapour pressure of Cr-containing species.

Chromium poisoning was found to depend strongly on the particular combination of electrolyte and cathode by Matsuzaki *et al.* [65]. In their investigations, the most severe degradation occurred in LSM/YSZ combination, whereas no apparent degradation occurred in LSCF/SDC (samaria-doped ceria) combination as seen in Figure 5.14. These results appear to be consistent with Taniguchi's results [59]. As described previously, the cathode reaction mechanism is quite different in the LSM and the LSCF cathodes. In LSM, the oxide ion diffusivity is low and only the three-phase boundaries are the electrochemically active sites. The oxygen potential gradient appears near these sites. This gives rise to the driving force for chromium-containing species to attack the electrochemically active sites resulting in the deposition of chromia-containing oxides that are responsible for performance degradation. On the other hand, in LSCF, the rate-determining step is the surface reaction rate. Thus, the electrochemically active sites are widely distributed on the surfaces of the LSCF cathode. The oxygen potential difference corresponding to the overpotential appears on the surface, leading to no driving force for chromium-containing species to attack any electrochemically active sites. Matsuzaki *et al.* [65] also observed that degradation is different between the LSM/YSZ and the LSM/SDC combinations. Although it is difficult to ascertain the proper reasons for this difference, a possible explanation can be derived from the differences in water

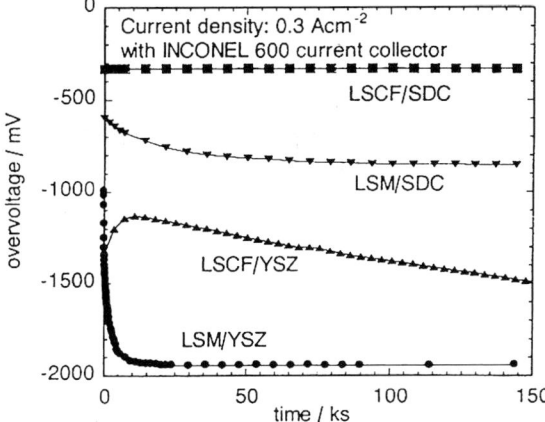

Figure 5.14 Chromium poisoning for different electrode/electrolyte combinations [65].

solubilities of YSZ and SDC and enhancement effects of water vapour on the cathode reaction [66].

To minimise/avoid Cr poisoning, several approaches have been tried. These include coating a dense, electrically conducting oxide, such as cathode-related perovskites (LSM) or lanthanum chromite-based oxides, on metallic interconnects, and utilisation of Cr getters. Cr containing species easily react with other oxides (particularly basic oxides) so that such basic oxides can act as Cr getters. La_2O_3 can be a strong Cr getter, but this material is hard to handle because of its hygroscopic nature. Lanthanum cobaltite can also be a Cr getter because the stabilisation energy of $LaCrO_3$ is larger than $LaCoO_3$.

5.5 Fabrication of Cathodes

In general, cathodes are made by powder processing routes. Cathode material powders are either made by solid state reaction of constituent oxides or high surface area powders are precipitated from nitrate and other solutions as a gel product, which is dried, calcined and comminuted to give crystalline particles in the 1–10 µm size range.

Fabrication methods for cathodes largely depend on the cell design. For cathode-supported tubular cells, porous cathode tubes are first extruded and then sintered at high temperatures [9,69,70]. After sintering, appropriate porosity and strength develop. State-of-the-art cathode tubes are more than 1.5 m in length. The most important issue when depositing other cell component layers on the cathode tubes is the adherence of the dense electrolyte and interconnect. When electrochemical vapour deposition (EVD) process is used to fabricate a thin and dense YSZ electrolyte as well as the dense $LaCrO_3$ interconnect [69], a highly adherent and reliable interface is formed without any chemical reaction or change in cathode microstructure [69]. A more cost-effective non-EVD process, a wet slurry/sintering process has also been developed [9,70] which has been very successful in fabricating reliable cells at a lower cost than the EVD process.

For planar-type SOFCs, particularly for anode-supported SOFCs, the cathode layer is usually deposited after preparing the anode/electrolyte assembly. This makes it possible to use various processes including slurry coating, screen printing, tape casting, and wet powder spraying for cathode deposition [71–74]. After deposition of the cathode slurry, it is dried followed by sintering. In many cases, the sintering temperature can be lower for anode-supported cells than for the cathode-supported type, giving higher surface area cathodes. Since the sintering of lanthanum manganite-based cathode with YSZ at higher than 1473 K gives rise to a drastic change in microstructure, lower sintering temperatures are advantageous.

Physical processes have also been used to make cathodes. For example, vacuum plasma spraying has been tried for fabricating entire anode/electrolyte/cathode assembly [75]. In this approach, cathode layers are deposited onto a porous metallic felt substrate.

5.6 Summary

The most important properties of cathodes are their catalytic activity for oxygen reduction and their compatibility with the electrolyte (including thermal expansion match and chemical non-reactivity). From the viewpoint of catalytic activity, many perovskites have been considered and investigated. Although the lanthanum manganite-based materials are not the best because of their low oxide ion diffusivity and resulting limited electrochemical activity, these are the most common perovskites that are used with YSZ electrolyte at 1000°C because of their superior chemical stability. For lower cell operation temperatures also, even though chemical reactivity at the cathode/electrolyte interface is less important, composite cathodes made from LSM/YSZ are used down to 700°C. Less severe conditions for the electrolyte/electrode chemical reactions at temperatures below 800°C make it attractive to use more catalytically active perovskites as cathodes. Sr- and Co-doped lanthanum ferrite $(La,Sr)(Co,Fe)O_3$ has been widely investigated for intermediate temperature SOFCs. Recent attempts have focused on $(La,Sr)FeO_3$ since it has lower area specific resistivity; however, fundamental phase relations and related high-temperature behaviour associated with this material still need clarification. Interactions with the interconnect can also be significant because poisoning of the cathode can occur, for example by Cr contamination. The cathode performance depends substantially on its surface area, porosity and microstructure, and therefore the processing method used is very important in determining cathode performance.

References

[1] J. B. Goodenough and J. M. Longo, in *Magnetic Oxides and Related Oxides*, Vol. 4a, Ch. 3, pp. 126–314, Landolt-Bernstein (1970); S. Nomura, in *Magnetic Oxides and Related Oxides*, Vol. 4a, pp. 368–520, Landolt-Bernstein (1978).

[2] C. S. Tedmon, Jr., H. S. Spacil and S. P. Mitoff, *J. Electrochem. Soc.*, **116** (1969) 1170.

[3] S. K. Lau and S. C. Singhal, Potential electrode/electrolyte interactions in solid oxide fuel cells, *Corrosion*, **85** (1985) 1–9.

[4] S. C. Singhal, in *Solid Oxide Fuel Cells VI*, eds. S. C. Singhal and M. Dokiya, The Electrochemical Society Proceedings, Pennington, NJ, PV 99-19, 1999, pp. 39–51.

[5] H. Yokokawa, *Fuel Cells–From Fundamentals to Systems*, **1**(2), (2001) 1–15; see also H. Yokokawa and N. Sakai, History of high temperature fuel cell development, in *Handbook for Fuel Cell Technology*, Wiley, Vol. 1, Ch. 13, 2003, pp. 219–266.

[6] H. Yokokawa, N. Sakai, T. Kawada and M. Dokiya, *J. Electrochem. Soc.*, **138**(9) (1991) 2719–2727.

[7] M. Dokiya, N. Sakai, T. Kawada, H. Yokokawa, T. Iwata and M. Mori, in *Solid Oxide Fuel Cells I*, ed. S. C. Singhal, The Electrochemical Society Proceedings, Pennington, NJ, PV 89-11, 1989, pp. 325–336.

[8] M. Suzuki, H. Sasaki, S. Otoshi, A. Kajimura, N. Sugiura and M. Ipponmatsu, *J. Electrochem. Soc.*, **141**(7) (1994) 1928–1931.

[9] H. Takeuchi, H. Nishiyama, A. Ueno, S. Aikawa, M. Aizawa, H. Tajiri, T. Nakayama, S. Suehiro and K. Shukuri, in *Solid Oxide Fuel Cells VI*, eds. S. C. Singhal and M. Dokiya, The Electrochemical Society Proceedings, Pennington, NJ, PV 99-19, 1999, pp. 879–884.

[10] M. Juhl, S. Primdahl, C. Manon and M. Mogensen, *J. Power Sources*, **61** (1996) 173.

[11] B. C. H. Steele, *Solid State Ionics*, **129** (2000) 95–110; see also M. Mogensen, N. M. Sammes and G. A. Tompsett, *Solid State Ionics*, **129** (2000) 63–94.

[12] B. C. H. Steele and J.-M. Bae, *Solid State Ionics*, **106** (1998) 255–261.

[13] R. D. Shannon, *Acta Cryst.*, **A32** (1976) 751–768.

[14] B. C. Tofield and W. R. Scott, *J. Solid State Chem.*, **10** (1974) 183.

[15] J. Mizusaki, N. Mori, H. Takai, Y. Yonemura, H. Minamiue, H. Tagawa, M. Dokiya, H. Inaba, K. Naraya, T. Sasamoto and T. Hashimoto, *Solid State Ionics*, **129** (2000) 163–177.

[16] J. A. M. Van Roosmalen and E. H. P. Cordfunke, *J. Solid State Chem.*, **93** (1993) 212–223.

[17] J. A. M. Van Roosmalen, E. H. P. Cordfunke and R. B. Helmholdt, *J. Solid State Chem.*, **110** (1994) 100–105.

[18] J. A. M. Van Roosmalen and E. H. P. Cordfunke, *J. Solid State Chem.*, **110** (1994) 106–108.

[19] J. A. M. Van Roosmalen and E. H. P. Cordfunke, *J. Solid State Chem.*, **110** (1994) 109–112.

[20] J. A. M. Van Roosmalen and E. H. P. Cordfunke, *J. Solid State Chem.*, **110** (1994) 113–117.

[21] J. Nowotny and M. Rekas, *J. Am. Ceram. Soc.*, **81**(1) (1998) 67–80.

[22] J. H. Kuo, H. U. Anderson and D. M. Sparlin, *J. Solid State Chem.*, **83** (1989) 52–60.

[23] J. H. Kuo, H. U. Anderson and D. M. Sparlin, *J. Solid State Chem.*, **87** (1990) 55–63.

[24] F. W. Poulsen, *Solid State Ionics*, **129** (2000) 145–162.

[25] H. Yokokawa, T. Horita, N. Sakai and M. Dokiya, *Solid State Ionics*, **86–88** (1996) 1161–1165.

[26] J. Mizusaki, N. Mori, H. Takai, Y. Yonemura, H. Minamiue, H. Tagawa, M. Dokiya, H. Inaba, K. Naraya, T. Sasamoto and T. Hashimoto, *Solid State Ionics*, **132** (2000) 167–180.

[27] J. Mizusaki, H. Tagawa, Y. Yonemura, H. Minamiue and H. Nambu, in *Proc. 2nd Ionic and Mixed Conducting Ceramics*, eds. T. A. Ramanarayanan, W. L. Worrell and H. L. Tuller, The Electrochemical Society, Pennington, NJ, PV 94-12, 1994, pp. 402–411.

[28] S. Srilomsak, D. P. Schlling and H. U. Anderson, in *Solid Oxide Fuel Cell I*, ed. S. C. Singhal, The Electrochemical Society Proceedings, Pennington, NJ, PV 89-11, 1989, p. 129.

[29] M. Mori and Y. Hiei, in *Solid Oxide Fuel Cell VI*, eds. S. C. Singhal and M. Dokiya, The Electrochemical Society Proceedings, Pennington, NJ, PV 99-19, 1999, pp. 347–354.

[30] J. A. Kilner, R. A. De Souza and I. C. Fullarton, *Solid State Ionics*, **86–88** (1996) 703–709.

[31] B. C. H. Steele, *Solid State Ionics*, **86–88** (1996) 123–1234.

[32] S. Adler, J. A. Lane and, B. C. H. Steele, *J. Electrochem. Soc.*, **143** (1996) 3554.

[33] I. Yasuda, K. Ogasawara, M. Hishinuma, T. Kawada and M. Dokiya, *Solid State Ionics*, **86–88** (1996) 1197–1201.

[34] R. A. De Souza and J. A. Kilner, *Solid State Ionics*, **106** (1998) 175–187.

[35] R. A. De Souza, J. A. Kilner and J. F. Walker, *Mater. Lett.*, **43** (2000) 43–52.

[36] R. A. De Souza and J. A. Kilner, *Solid State Ionics*, **126** (1999) 153–161.

[37] A. V. Berenov, J. L. MacManus-Driscoll and J. A. Kilner, *Solid State Ionics*, **122** (1999) 41–49.

[38] J. Crank, *The Mathematics of Diffusion*, Oxford University Press, Oxford, 1975.

[39] A. Hammouche, E. Siebert, A. Hammou and M. Kleitz, *J. Electrochem. Soc.*, **138** (1991) 1212.

[40] E. Siebert, A. Hammouche and M. Kleitz, *Electrochim. Acta*, **40**(11) (1995) 1212.

[41] J. Van herle, A. J. McEvoy and K. R. Thampi, *Electrochim. Acta*, **41** (1996) 1447.

[42] T Horita, K. Yamaji, N. Sakai, H. Yokokawa and T. Kato, *J. Electrochem. Soc.*, **148**(5) (2001) J25–J30.

[43] T. Kawada, K. Masuda, J. Suzuki, A. Kaimai, K. Kawamura, Y. Nigara, J. Mizusaki, H. Yugami, H. Arashi, N. Sakai and H. Yokokawa, *Solid State Ionics*, **121** (1999) 271–279.

[44] T. Horita, K. Yamaji, N. Sakai, M. Ishikawa, H. Yokokawa, T. Kawada and T. Kato, *J. Electrochem. Soc.*, **145**(9) (1998) 3196–3202.

[45] T. Kawada and H. Yokokawa, in *Electrical Properties of Ionic Solids*, eds. J. Nowotny and C. C. Sorrell, Trans Tech Publications, 1997, pp. 187–248.

[46] H. Yokokawa, in *Zirconia Engineering Ceramics: Old Challenges–New Ideas*, ed. Erich Kisi, Trans Tech Publications, 1998, pp. 37–74.

[47] T. Kawada, N. Sakai, H. Yokokawa and M. Dokiya, *Solid State Ionics*, **53–56** (1992) 418–425.

[48] D. M. Tricker and W. M. Stobbs, in *High Temperature Electrochemical Behavior of Fast Ions and Mixed Conductors*, eds. F. W. Poulsen, J. J. Bentzen, T. Jacobsen, E. Skou and M. J. L. Østergård, Risø National Laboratory, Roskilde, 1993, pp. 453–460; D. M. Tricker, *The Microstructure of Solid Oxide Fuel Cells and Related Metal/Oxide Interfaces*, Thesis, University of Cambridge, 1993.

[49] H. Yokokawa, T. Horita, N. Sakai, T. Kawada and M. Dokiya, in *Proceedings of the First European SOFC Forum*, ed. U. Bossel, Switzerland, 1994, pp. 425–434.

[50] For example, L. Kindermann, D. Das, C. C. Appel, F. W. Poulsen, H. Nickel, R. Weiß and K. Hilpert, *J. Electrochem. Soc.*, **144** (1997) 717–720; L. Kindermann, D. Das, R. Bahadur, R. Weiß, H. Nickel, K. Hilpert, *J. Am. Ceram. Soc.*, **80** (1997) 909–914.

[51] A. Müller, H. Schichlein, M. Feuerstein, A. Weber, A. Krügel and E. Ivers-Tiffée, in *Solid Oxide Fuel Cells VI*, eds. S. C. Singhal and M. Dokiya, The Electrochemical Society Proceedings, Pennington, NJ, PV 99-19, 1999, pp. 925–931.

[52] T. Yoshida, Y. Someya, H. Koide, I. Mukaisawa and M. Andoh, in Extended abstracts for *2nd Symposium on Solid Oxide Fuel Cells in Japan*, Solid Oxide Fuel Cell Society, Tokyo, Japan, 1993, pp. 41–45.

[53] J.-P. Zhang, S.-P. Jiang and K. Föger, in *Solid Oxide Fuel Cells IV*, eds. S. C. Singhal and M. Dokiya, The Electrochemical Society Proceedings, Pennington, NJ, PV 99-19, 1999, pp. 962–971.

[54] M. Mori, N. Sakai, T. Kawada, H. Yokokawa and M. Dokiya, *Denki Kagaku*, **58** (1990) 528.

[55] H. Yokokawa, N. Sakai, T. Kawada and M. Dokiya, *Solid State Ionics*, **40/41** (1990) 398.

[56] J. Mizusaki and H. Tagawa, in *Proceedings of High Temperature Electrode Materials and Characterization*, eds. D. D. Macdonald and A. C. Khandkar, The Electrochemical Society, Pennington, NJ, PV91-6, 1991, pp. 75–87.

[57] A. V. Virkar, J. Chen, C. W. Tanner and J.-W. Kim, *Solid State Ionics*, **131** (2000) 189–198.

[58] D. Ghosh, E. Tang, M. Perry, D. Prediger, M. Pastula and R. Boersma, in *Solid Oxide Fuel Cells VII*, eds. H. Yokokawa and S. C. Singhal, The Electrochemical Society Proceedings, Pennington, NJ, PV 2001-16, 2001, pp. 100–110.

[59] S. Taniguchi, M. Kadowaki, H. Kawamura, T. Yasuo, Y. Akiyama, Y. Miyake and T. Saitoh, *J. Power Sources*, **55** (1995) 73–79.

[60] J. M. Ralph, J. T. Vaughey and M. Krumpelt, in *Solid Oxide Fuel Cells VII*, eds. H. Yokokawa and S. C. Singhal, The Electrochemical Society Proceedings, Pennington, NJ, PV 2001-16, 2001, pp. 466–475.

[61] N. Sakai, T. Tsunoda, Isao Kojima, Katsuhiko Yamaji, Teruhisa Horita, Harumi Yokokawa, Tatsuya Kawada and Masayuki Dokiya, *Ceramic Interfaces 2*, IOM Communications Ltd., 2001, pp. 135–156.

[62] H. Nishiyama, M. Aizawa, H. Yokokawa, T. Horita, N. Sakai, M. Dokiya and T. Kawada, *J. Electrochem. Soc.*, **142** (1996) 2332.

[63] P. Kofstad and R. Bredesen, *Solid State Ionics*, **52** (1992) 69–75.

[64] K. Hilpert, D. Das, M. Miller, D. H. Peck and R. Weiß, *J. Electrochem. Soc.*, **143** (1996) 4013.

[65] Y. Matsuzaki and I. Yasuda, *J. Electrochem. Soc.*, **148** (2001) A126.

[66] N. Sakai, K. Yamaji, H. Negishi, T. Horita, H. Yokokawa, Y.-P. Xiong and M. B. Phillipps, *Electrochemistry (former Denki Kagaku)*, **68**(6) (2000)

499–503; see also N. Sakai, K. Yamaji, H. Negishi, T. Horita, Y.-P. Xiong and H. Yokokawa, in *Solid Oxide Fuel Cells VII*, eds. H. Yokokawa and S. C. Singhal, The Electrochemical Society Proceedings, Pennington, NJ, PV 2001-16, 2001, pp. 511–520.

[67] D. Herbstritt, C. Warga, A. Weber and E. Ivers-Tiffée, in *Solid Oxide Fuel Cells VII*, eds. H. Yokokawa and S. C. Singhal, The Electrochemical Society Proceedings, Pennington, NJ, PV 2001-16, 2001, pp. 349–357.

[68] M. Cassidy, C. Bagger, N. Brandon and M. Day, in *Proc. of the 4th European SOFC Forum*, ed. A. J. McEvoy, Switzerland, 2000, pp. 637–646.

[69] S. C. Singhal, *Solid State Ionics*, **135** (2000) 305–313.

[70] M. Kuroishi, S. Furuya, K. Hiwatashi, K. Tsujimoto, Y. Uchida, H. Yoshinaga and K. Shukuri, in *Solid Oxide Fuel Cells VII*, eds. H. Yokokawa and S. C. Singhal, The Electrochemical Society Proceedings, Pennington, NJ, PV 2001-16, 2001, p. 88.

[71] G. Rietveld, P. Nammensma and J. P. Ouweltjes, in *Solid Oxide Fuel Cells VII*, eds. Yokokawa and S. C. Singhal, The Electrochemical Society Proceedings, Pennington, NJ, PV 2001-16, 2001, p. 125.

[72] L. G. J. de Haart, I. C. Vinke, A. Janke, H. Ringel and F. Tietz, in *Solid Oxide Fuel Cells VII*, eds. H. Yokokawa, S. C. Singhal, The Electrochemical Society Proceedings, Pennington, NJ, PV 2001-16, 2001, p. 111.

[73] I. Yasuda, Y. Baba, T. Ogiwara and H. Yakabe, in *Solid Oxide Fuel Cells VII*, eds. H. Yokokawa and S. C. Singhal, The Electrochemical Society Proceedings, Pennington, NJ, PV 2001-16, 2001, p. 131.

[74] M. Hattori, A. Nakanishi and M. Iio, in *Solid Oxide Fuel Cells VII*, eds. H. Yokokawa and S. C. Singhal, The Electrochemical Society Proceedings, Pennington, NJ, PV 2001-16, 2001, p. 1061.

[75] G. Schiller, T. Franco, R. Henne, M. Lang, R. Ruckdäschel, P. Otschik and K. Eicher, in *Solid Oxide Fuel Cells VII*, eds. H. Yokokawa and S. C. Singhal, The Electrochemical Society Proceedings, Pennington, NJ, PV 2001-16, 2001, p. 895.

Chapter 6

Anodes

Augustin McEvoy

6.1 Introduction

Like the cathode, the anode must combine catalytic activity for fuel oxidation with electrical conductivity. Catalytic properties of the anode are necessary for the kinetics of the fuel oxidation with the oxide ions coming through the solid electrolyte. Ionic conductivity allows the anode to spread the oxide ions across a broader region of anode/electrolyte interface, and electronic conductivity is necessary to convey the electrons resulting from the electrode reaction out into the external circuit.

Early in the twentieth century, many candidate anode materials were tested, including precious metals like platinum and gold, and transition metals such as iron and nickel, as described in Chapter 2. But platinum does not last long in an operating solid oxide fuel cell (SOFC), peeling off after a few hours, and nickel aggregates at high temperatures inhibiting access of the fuel. Spacil [1] first recognised that the nickel aggregation problem could be solved by mixing yttria-stabilised zirconia (YSZ) electrolyte particles in with the nickel matrix to form a composite anode. Such nickel cermet anodes can provide adequate performance under certain conditions but do exhibit problems such as carbon fouling from carbonaceous fuels. However, nickel cermet is the material which has been most successful in SOFC development until now and so is emphasised in this chapter.

This chapter first considers the complex mix of attributes required of SOFC anodes, including matching of thermal expansion coefficients, chemical compatibility with the electrolyte and the interconnect, porous structure to allow gas permeation, and corrosion resistance to the fuel and impurities therein. Then the nickel cermet anode is described in detail, especially its fabrication processes. Steady-state anode reactions of hydrogen and carbon monoxide are analysed, followed by a description of transient effects. Finally, behaviour under current load and operation on different fuels are discussed. The details of the anode reactions and polarisations are described in Chapter 9.

6.2 Requirements for an Anode

The role of an anode in a solid oxide fuel cell is to provide the sites for the fuel gas to react with the oxide ions delivered by the electrolyte, within a structure which also facilitates the necessary charge neutralisation by its electronic conductivity. These functional considerations together with the operating environment of the anode are the key factors in the materials selection for the anode. It must evidently be refractory, given the high cell operating temperature to be sustained over a commercially useful lifetime, and be compliant with thermal cycling to ambient temperature. In addition, the equilibrium between fuel gas and oxidation products within the anode compartment results in an oxygen partial pressure which is very low, but variable over several orders of magnitude depending on the precise reactant and product conditions. Chemical and physical stability despite such variations is essential since certain metallic components of the anode could suffer corrosion by fuel oxidation products, while electrical properties and lattice geometry of oxide components of the anode could change by variation of stoichiometry.

For simplicity and reliability of operation, including start-up and shut-down as well as tolerance of transients, redox stability is a further desirable attribute of an anode material to permit brief excursions to high oxygen concentrations, even to air, without irreversible loss of structural coherence and electrochemical functionality. Stability implies the maintenance of structural integrity of the anode itself over the whole temperature range to which it is exposed, from the fabrication temperature through normal operating conditions, to the repeated cycling down to ambient temperature.

Throughout these ranges of temperature and gas environment there should also be maintained the necessary compatibility with the other materials with which the anode comes into contact, specifically the electrolyte, the interconnect and any relevant structural components. Physical compatibility requires a match of thermomechanical properties such as thermal expansion coefficient and an absence of phase-change effects which could generate stresses during temperature variations. For chemical compatibility there should be no solid-state contact reaction, interdiffusion of constituent elements of those materials or formation of reaction product layers which would increase resistive losses or otherwise interfere with anode functionality, despite the extremes of temperature. After assembly into a series connected stack, of course the same applies to the anode–interconnect interface. Compatibility must extend also to the behaviour of the material towards the ambient gases including corrosion or poisoning by trace impurities such as sulphur.

Obviously the anode material should not only be an adequate electronic conductor, but also electrocatalytically active such that a rapid charge exchange can be established. Resistive and overpotential losses are thereby minimised. However, the catalytic behaviour of anode materials should not extend to the promotion of unwanted side reactions, hydrocarbon pyrolysis followed by deposition of carbon being an example. The electrochemical reaction takes place in the region (Figure 6.1) where oxygen ions available from the electrolyte can

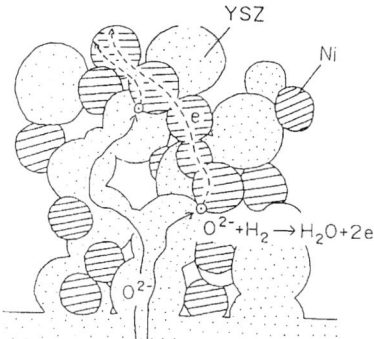

Figure 6.1 Schematic of anode cermet structure, showing interpenetrating networks of pores and conductors – nickel for electrons, yttria-stabilised zirconia for oxide ions. The reactive sites are the contact zones of the two conducting phases which are accessible to fuel through the porosity [2].

discharge electrons to the conducting anode. This requires gas phase for the fuel access, electrolyte phase for oxide ion entry and metal phase for electron output, the so-called 'three-phase boundary' zone. For efficient operation this should not be simply a linear structure at a two-dimensional interface of the solid materials, but rather distributed to provide an active 'volumetric' reaction region in three dimensions. Consequently, the fabrication of the anode is important in determining this complicated three-phase structure.

6.3 Choice of Cermet Anode Components

Given these stringent requirements, only a few metallic or ceramic candidate materials are available. After the 'Nernst mass', now known as yttria-stabilised zirconia had been identified as the favoured high-temperature ceramic electrolyte, Baur and Preis evaluated iron and graphite as anode materials [3]. Graphite, of course, is susceptible to electrochemical oxidative corrosion, so that cell life with a graphite anode is unduly short. Platinum also attracted attention due to its high-temperature stability and catalytic properties, as did other transition metals as presented in the historical review by Möbius [4]. Even platinum, however, was unsuccessful as its bond to the electrolyte tends to fail in service, with the anode layer spalling off probably due to electrochemical generation of water vapour at the interface. The transition metals also have limitations. Iron is no longer protected by the reducing activity of the fuel gas once the partial pressure of oxidation products in the anode compartment exceeds a critical value, and it then corrodes with formation of a red iron oxide. Cobalt is somewhat more stable, but also more costly. Nickel has a significant thermal expansion mismatch to stabilised zirconia, and at high temperatures the metal aggregates by grain growth, finally obstructing the porosity of the anode and eliminating the three-phase boundaries required for cell operation. As a consequence, all-metal anodes have not found acceptance in SOFC technology. Pure ceramic oxide anode technology is a very recent development and is discussed later.

As an alternative to a single-phase metallic or conductive ceramic electrode material, the accepted compromise has been the use of a porous composite of metal and ceramic, a 'cermet' (Figure 6.1). In the SOFC anode most commonly used at present, the nickel–zirconia cermet, the primary role of the zirconia is structural, to maintain the dispersion of the nickel phase and its porosity by inhibition of the aggregation and grain growth of the metal and so to achieve an adequate anode lifetime. The adhesion of the zirconia part of the cermet to the electrolyte gives a structural ruggedness able to withstand the thermal stress due to differential thermal expansion, which anyway is lowered in the composite by the ceramic volume fraction. The provision of oxide ion conductivity complementary to the electronic conductivity and electrocatalytic action of the metal is a useful secondary role of the ceramic, enhancing the electrochemical performance by delocalisation of the electrochemically active zone already mentioned. However, even as a compromise material the nickel–zirconia cermet does not fulfil all the requirements of an ideal anode. Fuel specification in particular is an important parameter. Most literature results report on hydrogen as fuel, but the commercial imperative requires hydrocarbons. These, however, are rapidly pyrolysed on nickel surfaces at high temperature, depositing a dense carbon which blocks the anode porosity and ultimately disrupts the structural integrity of the cermet. There is, however, a compensating consideration. Carbon monoxide, which is not tolerated by the platinum catalyst in low-temperature polymer electrolyte fuel cells, is a perfectly acceptable SOFC fuel, though slower to oxidise than hydrogen. Therefore the SOFC is robust with respect to fuel specifications, and can be fuelled with gas mixtures rich in hydrogen and carbon monoxide derived from hydrocarbons by partial oxidation or by reforming reactions with steam or carbon dioxide. There still remains a concern about impurities in the hydrocarbon fuels. Nickel at high temperature is sensitive to sulphur compounds at concentrations even as low as 0.1 ppm. These may occur at source in natural gas, for example, and thiophene and mercaptans are systematically added to it as odorants for safety reasons. Fortunately the sulphided nickel surface is not necessarily irrecoverable, and operation with a clean sulphur-free fuel may restore performance. Nonetheless desulphurising systems which either adsorb the contaminant on activated carbon, or react it with zinc oxide to form a solid sulphide are regarded as essential for high-temperature fuel cell operation. Dependent on the origin of the fuel, for example from coal gasification, biomass pyrolysis or fermentation, other impurities may also occur, particularly ammonia and possibly hydrochloric acid. Tolerance over 2000 h for HCl is somewhat better than that for H_2S, though still in the low ppm range. However, from all evidence, ammonia even in elevated concentrations does not pose a problem; concentrations up to 5000 ppm have had no effect on SOFC cell voltage over a 2500 h test [5]. Inorganics may also be found, entrained as dust, which can then react with the ceramic components of the cell giving, for example, silicates. It is therefore evident that adequate cell performance can only be achieved, and maintained, by careful fuel pretreatment. Despite these compromises, however, the nickel–zirconia cermet has become the most common anode in SOFC technology.

6.4 Cermet Fabrication

The development of nickel–zirconia anodes leading to their present performance in fuel cells has generally proceeded empirically since the first introduction of this combination of materials by Spacil, in 1964 [1]. Though written almost 40 years ago, his text remains a strikingly modern presentation of the anode specification, making reference to many of the concepts and procedures still current in SOFC technology. The densification of a porous metallic structure with time at temperature is the problem to which the metal–ceramic composite is the answer. Intimate anode to electrolyte bonding is a further requirement satisfied by the cermet, if necessary by adoption of a flux to aid sintering. A minimum metal proportion in the cermet is necessary for continuity of electronic conduction, while the zirconia particles are by preference also continuous. Spacil had therefore recognised that the functionality of the ceramic in the composite was essentially structural, to retain the dispersion of the metal particles and the porosity of the anode during long-term operation. Structure and elemental distribution in a typical anode cermet is shown in Figure 6.2 [6].

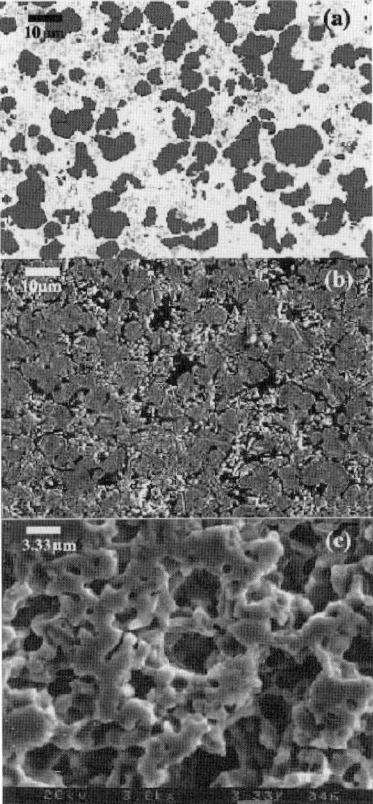

Figure 6.2 Scanning electron microscopy images of a Ni–YSZ cermet with elemental resolution, showing (a) nickel distribution, (b) overall cermet morphology, and (c) zirconia structural skeleton [6].

Spacil's fabrication procedure was essentially modern. Nickel oxide was mixed with stabilised cubic zirconia powder in an aqueous slurry of the type used for slip casting. This slurry was applied to the electrolyte surface and fired at a temperature up to 1550°C. To provide a percolation path for electrons through the anode, the nickel oxide was reduced to metal under hydrogen. The metal being more dense than the oxide, the initial volume of the NiO component was diminished by over 25% during this reduction step, and Spacil mentioned the consequent enhanced porosity as advantageous for the anode. He also presented alternative deposition procedures for the anode, such as plasma spraying. When considering the performance of fuel cells today, it is interesting to note that almost 40 years ago, he reported a power density of over 500 mW/cm^2, admittedly at the very high temperature of 1200°C.

The physical and chemical properties of both nickel and zirconia in the cermet are critical to their compatibility and functionality. An intimate bonding at the interface on the nanometre scale is necessary for the synergy of the materials and for the establishment of the three-phase electrochemically active zone. Therefore some level of physicochemical interaction or 'wetting' between metal and ceramic is necessary, though the affinity of metallic nickel for zirconia is weak, with a contact angle of 120° [7]. To promote bonding, Spacil suggested a lithium carbonate surfactant flux; in current practice the powder specifications are chosen to ensure a sufficient surface activity. Addition of metal dopants such as titanium to the zirconia ceramic is another way to engineer suitable interfaces [8]. It is useful, however, that this affinity of zirconia for nickel is limited, because this inhibits interfacial reaction or elemental interdiffusion and allows the two-phase nature of the cermet to be maintained under operating conditions. It is known that in a particularly reducing environment, close to open-circuit operation of a fuel cell with dry fuel, a nickel–zirconium intermetallic, Ni_5Zr, may form [9], but in normal operation the nickel is precipitated as the zirconium component is reoxidised. This observation confirms the very low solubility of nickel in stabilised zirconia, perhaps 2% at 1000°C [7], which has even made possible a recent synthesis of nanodispersed cermet from homogeneous solutions of nickel, zirconium and yttrium salts [10].

The key to optimisation of durable efficient anodes in the decades since Spacil lies in the improvement of materials specifications permitting a sensitive control of cermet morphology. The original cermet used a high proportion of nickel, over 50% by volume, reduced from nickel oxide of grain size around 45 μm as sieved through a 325 mesh screen, with non-connected inclusions of 10 μm zirconia after sintering. The thermal expansion was therefore unduly high, since the proportion of nickel was in excess of that required for electronic percolation conductivity, and the lack of connectivity of the ceramic component permitted long-term nickel aggregation while blocking oxygen ionic transport. With these materials a temperature of 1550°C was required to sinter the anode to the zirconia electrolyte substrate. Modern submicron ceramic powders sinter at 1400°C or lower, maintaining a higher specific surface area anode. Associated with the reduced thermal expansivity of the cermet due to the increased ceramic content, stresses during fabrication, reduction and operation are minimised, eliminating microfissuring which contributes to electrode ageing [7]. Nickel oxide of grain size around 1 μm is now used, while the zirconia component often contains a proportion of coarse powder 25 μm or larger [11] to form the anode structural skeleton and inhibit the nickel aggregation, mixed with 0.5 μm fine powder to promote sintering. The powder mixtures are applied to form the electrode layer on the YSZ electrolyte substrate, 150 μm or thicker, which also provides the structural rigidity of the cell.

A similar development strategy applies to the more recent anode-supported cells, now beginning to find favour for operation at 800°C or lower. This lower temperature relaxes materials specifications throughout the system, permitting the use of lower cost metallic structural and interconnect components. Another benefit is diminished thermomechanical stress and reaction at these lower temperatures, significantly improving durability. On the other hand, lower power output is unacceptable, so for YSZ electrolyte, its higher resistivity at temperatures under 800°C must be compensated by reduction in its thickness to 10 μm. Recent cells therefore tend to use the anode, not just as a functional component, but also structurally as the load and stress-bearing support for a thin electrolyte. As will be later noted, this structural cermet, up to 1 mm thick, can serve not only as the site of oxidation of a reformate composed principally of hydrogen and carbon monoxide, but also for the preliminary hydrocarbon fuel processing reactions. At lower temperatures also, the thermal activation of the oxidation reactions is significantly diminished, implying increased polarisation and giving added importance to considerations of electrocatalysis at the anode [12]. In this case graded anode structures are often advisable, with a high-porosity large-grain substrate bearing a finer-structured electrocatalytically active functional layer to contact with the electrolyte [13]. The micrograph in Figure 6.3 shows materials of an anode-supported cell [14].

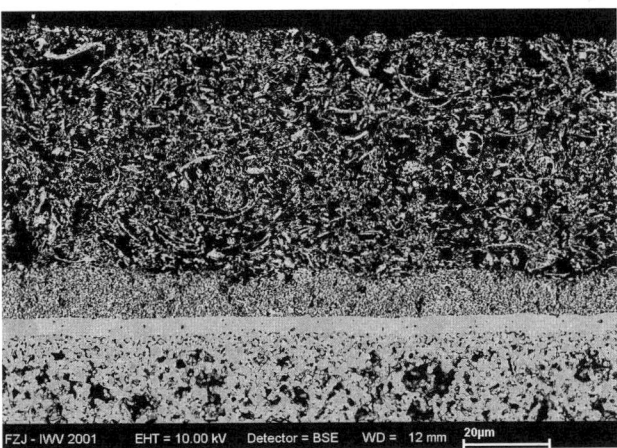

Figure 6.3 Micrograph of anode-supported thin electrolyte cell structure. On either side of the dense electrolyte is a fine-structured 'functional layer' for the electrocatalytic promotion of the electrode reactions. The full thickness of the cathode layer is imaged, but only a small section of the anode substrate (bottom). (Reproduced by courtesy of the Research Centre, Jülich (FZJ).)

6.5 Anode Behaviour Under Steady-State Conditions

The details of the electrode reactions and associated polarisations are discussed in Chapter 9. In this and the next two sections, anode behaviour under various conditions is described.

Having formed the anode onto the electrolyte, it is necessary to test its electrochemical performance and to compare it with the electrolyte, cathode and interconnect contributions to the overall cell resistance. The characteristics of the individual electrodes, anode or cathode, and their interfaces with the electrolyte are difficult to distinguish from the behaviour of the fuel cell as a whole. In standard liquid-phase electrochemistry this can be done by use of a third reference electrode, with respect to which the potential of the electrode under investigation can be determined. In the fuel cell case the geometry and location of the reference contact can introduce experimental artefacts, complicating the investigation of the electrode characteristics [15]. Also since the Nernst equation predicts an open-circuit potential difference between the electrodes dependent on the oxygen partial pressure on the anode side, it is frequently advisable to monitor potentials generally in the cell using a cathode-side reference the potential of which is therefore fixed by exposure to air. Despite these difficulties, d.c. measurement using the three-electrode configuration is the best steady-state evaluation technique which has given valuable insights into anodic processes.

The current/voltage characteristics at typical SOFC cermet anodes have been investigated in detail with different proportions of fuel (H_2, CO) and combustion product gases (H_2O, CO_2), and thus a range of oxygen partial pressures [16] as shown in Figure 6.4, and at different practical operating temperatures in Figure 6.5. In the logarithmic current density to anode overpotential relation as measured with respect to the relevant equilibrium potential, a standard

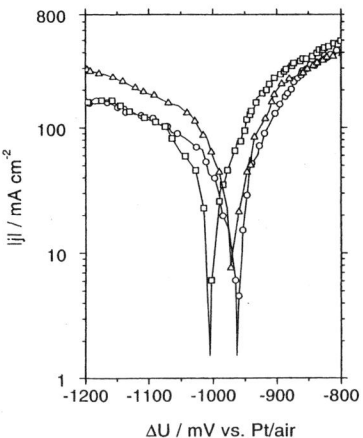

Figure 6.4 Steady-state current density (j)–potential (ΔU) characteristics of a Ni–YSZ cermet anode (after [16]), at 950°C, with variation of hydrogen partial pressure: (○) $P_{(H_2)} = 0.19$ bar, $P_{(H_2O)} = 0.05$ bar; (□) $P_{(H_2)} = 0.48$ bar, $P_{(H_2O)} = 0.05$ bar; (△) $P_{(H_2)} = 0.48$ bar, $P_{(H_2O)} = 0.12$ bar. Reference electrode Pt on air side; note displacement of equilibrium potential according to Nernst relationship.

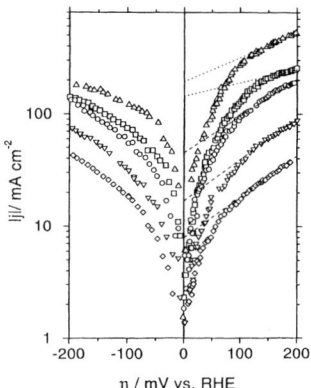

Figure 6.5 Steady-state current density (j)–overpotential (Δη) characteristics [16] under invariant Nernst gas composition conditions, $P_{(H_2)} = 0.48$ bar, $P_{(H_2O)} = 0.05$ bar, for temperatures: (◇) 725°C, (▽) 780°C, (○) 845°C, (□) 890°C, (△) 950°C.

Tafel-type characteristic is presented, once past a threshold value and dependent on temperature. The overpotentials for hydrogen oxidation, the fuel cell anodic reaction, are on the positive axis in Figure 6.5; negative values of overpotential, cathodic of the equilibrium potential, represent electrolysis of the water vapour present in the system. It is also reported that the reaction order for hydrogen, deduced from $\Delta \log j / \Delta \log P_{(H_2)}$ at a constant overpotential, where j = current density, is close to 0.5 at 725°C, a value compatible with adsorbed atomic hydrogen being involved in the charge-transfer reaction. This species is identified as a consequence of molecular hydrogen dissociation after adsorption from the gas phase, presumably on nickel surface sites. This presumption is upheld by the observed poisoning of anodes by sulphur, which chemisorbs on a nickel surface. However, the reaction order shows a decreasing trend with higher overpotentials and temperatures (Figure 6.6). This would indicate that

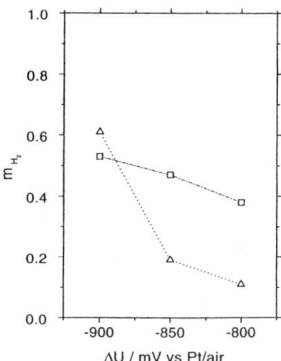

Figure 6.6 Overpotential referred to equilibrium (ΔU) determined by Nernst conditions. Reaction order m as determined for hydrogen, mH_2, with $P_{(H_2O)} = 0.05$ bar, at temperatures (□) 725°C and (△) 950°C, indicates a change of limiting reaction mechanism.

the anode surface coverage of hydrogen under those conditions becomes less sensitive to the partial pressure.

Similarly for thermal activation on the anode, Arrhenius plots can be constructed from the logarithmic dependence of current density on reciprocal temperature for the evaluation of the activation enthalpy of the oxidation reaction. Here again, within the Tafel regime, the activation enthalpy falls from 130 kJ/mol below 840°C to 110 kJ/mol at higher temperatures. While at the lower temperatures a charge-transfer controlled reaction dominates, other limiting considerations emerge as the temperature increases. This is compatible with the actual development strategy for advanced anodes in lower temperature fuel cells where enhancement of charge-transfer behaviour through electrocatalysis is the objective, a strategy less relevant at higher temperatures where anode performance is already adequate through thermal activation.

6.6 Anode Behaviour Under Transients Near Equilibrium

Complementary to steady-state characterisation by continuous current methods are the perturbation techniques, such as time-domain transients induced by current or potential steps, or frequency-domain analysis of system behaviour by electrochemical impedance spectroscopy (EIS) [17]. The methods reveal dissipative mechanisms contributing to overpotential losses on the basis of response time, and distinguish them from the purely resistive effects. Identification of the observed EIS features with specific physicochemical processes can then be effected by variation of other system parameters such as applied potential, temperature or gas environment. The physical system can then be analysed in terms of an equivalent circuit, an electronic analogue reproducing the frequency response of the actual device. Processes giving rise to impedance spectral features are therefore not necessarily electrical, but any process involving storage and dissipation effects contributing to the overall polarisation loss will appear in the equivalent circuit through a corresponding electronic model component.

At the high-frequency limit of the spectrum the intercept on the resistive axis specifies the serial ohmic component in the measured system, since this element does not introduce a phase shift. Similarly the low-frequency limit approaches the steady-state condition corresponding to the d.c. characteristics of the cell. Each spectral feature detected between these limits represents a dissipation process with the specific time dependence indicated by the inverse of the frequency at which it occurs. It should be noted, therefore that two processes with similar time constants in the anodic system will not be distinguishable by impedance spectroscopy.

In the case of high-temperature fuel cell anodes, conflicting evidence has been presented on the number and significance of the dissipative mechanisms detected as contributing to polarisation losses by impedance spectroscopy. At least three features may be distinguishable, dependent on the particular anode structure and the experimental conditions [17,19]. This variability of the impedance spectra is clearly illustrated with two cermet anodes of differing microstructures [20] operating under identical experimental conditions, at 1000°C in an environment of hydrogen with 3% H_2O. While in both results (Figure 6.7) three components with similar time constants in each case can be distinguished in the experimental EIS results, the amplitudes of the corresponding spectral features differ considerably, and hence the appearance of the overall spectra. The anodes differ in that one consisted of 0.4 µm YSZ powder with bidispersed NiO, 0.4 and 10 µm, to give after reduction a 40% Ni/60% YSZ porous cermet, the other being prepared from a submicron fine-grain plasma-processed powder to give 50% Ni/50% YSZ.

By detailed investigation of some experimental model anode systems, however, a consensus is emerging on the interpretation of anode impedance

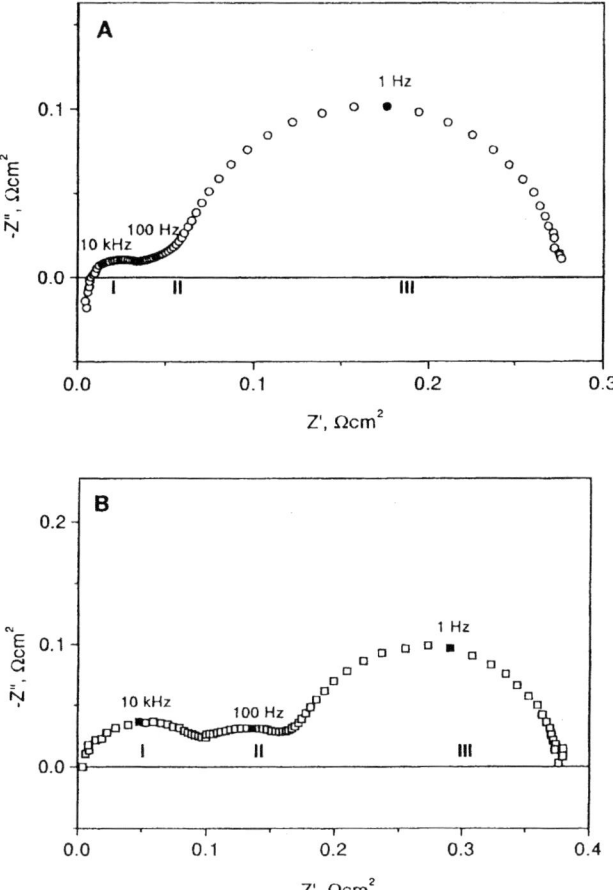

Figure 6.7 Impedance spectral variation with differing cermet microstructures, prepared from a standard anode formulation (A) and a fine-grain plasma-processed powder (B). Both results distinguish three dissipative processes with similar time constants in each case. The experimental conditions are identical, 1000°C in hydrogen, 3% H_2O (after [18]). Z' = real or resistive, and Z'' the imaginary or reactive, components of the total impedance $Z = Z' + iZ''$.

spectra. The charge transfer process is the only strictly electrical component of the system, and its time constant can be estimated. The resistive component should not exceed 1 Ω, and the reactive component, the interfacial capacitance, should be around 10 $\mu F/cm^2$ [21] implying a characteristic frequency of the order of 10 kHz. The feature I (Figure 6.7) is therefore identified with the charge transfer process. Other experimental work concurs in this identification [19], where the variation of the capacitance with temperature and with gas environment, particularly the partial pressures of hydrogen and steam indicate some dependence on adsorption in the interface region. Diffusion and mass transfer processes must therefore be associated with the slower features. A very slow mechanism may in fact be an artefact, due to variation of the Nernst potential in the vicinity of the active anode sites over the time of the modulation, of the order of 1 s (1 Hz), as measured against a reference electrode exposed to a constant gas composition, for example on the cathode side of the sample. This concentration polarisation effect was suppressed using an anode-side reference [18], clearly demonstrating that it is a function of the experimental system, not specific to the electrode. In the same work the feature II could also be suppressed if gas transport to the test anode was electrochemical, by direct contact to a second identical device exchanging reagent and reaction product, rather than gas-phase diffusion. Given that there was no gas transport out of the system, the corresponding impedance spectral feature does not appear.

6.7 Behaviour of Anodes Under Current Loading

In the discussion of impedance spectroscopy so far, it has been assumed that the modulation is applied to an anode under equilibrium conditions on each electrode with no net d.c. current transfer. Further information can be obtained by impedance analysis of the behaviour under load, the overpotential then being an independent variable. In this way a further confirmation of the identification of spectral features can be obtained, because there is a logarithmic relationship between current and overpotential (Figure 6.4) in the Tafel region. As a consequence the impedance spectroscopic feature corresponding to this reaction rapidly shrinks with increasing overpotential, reducing both resistive and reactive components and decreasing the apparent interface polarisation. At the same time a further reactance, in the inductive or opposite sense to the capacitative phenomena observed thus far, may emerge near the low-frequency limit. High-frequency inductive effects are usually artifacts of the measurement system, associated with self-inductance, whereas the low-frequency inductive feature is generally interpreted as evidence of adsorbed reaction intermediates, and may be associated with autocatalytic effects. This is illustrated in Figure 6.8. Mass transport effects may also be influenced by overpotential, for example by change of concentration gradients, increased occupation of surface sites by reaction products, or the formation of reaction intermediates. They may also be temperature influenced, for example due to thermal desorption, or even gas viscosity changes with temperature. A clear example when the anode

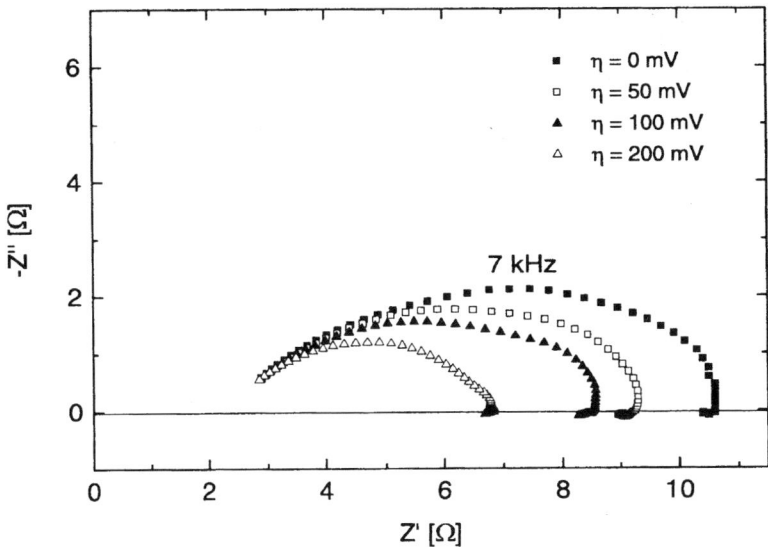

Figure 6.8 Evolution of electrochemical impedance spectra with increasing overpotential η at 700°C, $p_{(H_2)}$ = 0.13 bar, $p_{(H_2O)}$ = 5 × 10^{-4} bar, on thin screen-printed Ni/YSZ cermet anode (after [21]). Note the low-frequency inductive loop.

approaches the diffusion-limited condition is the expansion of the corresponding impedance spectral feature, indicating increasing polarisation due to fuel starvation or slow desorption and evacuation of the reaction products.

In this discussion it should be noted that water vapour is not simply a passive reaction product. It has been recognised for several years that the ratio of fuel to reaction product partial pressures modifies not only the oxygen partial pressure on the anode side, and therefore the equilibrium Nernst potential, but also the polarisation. Mogensen and Lindegaard [23] presented impedance spectra on a cermet anode at 1000°C, with $P_{(H_2)}$ = 1.0 bar, $P_{(H_2O)}$ = 0.03 and 0.0022 bar with corresponding values of $P_{(O_2)}$ = 4.5 × 10^{-18} and 6.5 × 10^{-21} bar (Figure 6.9). While the charge transfer high-frequency spectral feature is little changed, the low-frequency transport polarisation is an order of magnitude higher. Water, therefore, has a catalytic function at the cermet anode. Again, recent work can provide an explanation. Sakai *et al.* [24] report that oxygen isotope exchange with oxide-conducting ceramics is much faster when the isotope source is water rather than molecular dioxygen. The following reaction is suggested:

$$H_2^{18}O_{ad} + V_o'' \Longleftrightarrow 2H_i^+ + {}^{18}O_o^-$$

Here the water molecule adsorbed on a zirconia surface oxygen vacancy, for example, dissociated reversibly into an oxygen ion occupying the vacancy, plus two interstitial or adsorbed hydrogen ions. The solubility of hydrogen interstitially into zirconia is low, 2 × 10^{-5} moles of water equivalent per mole

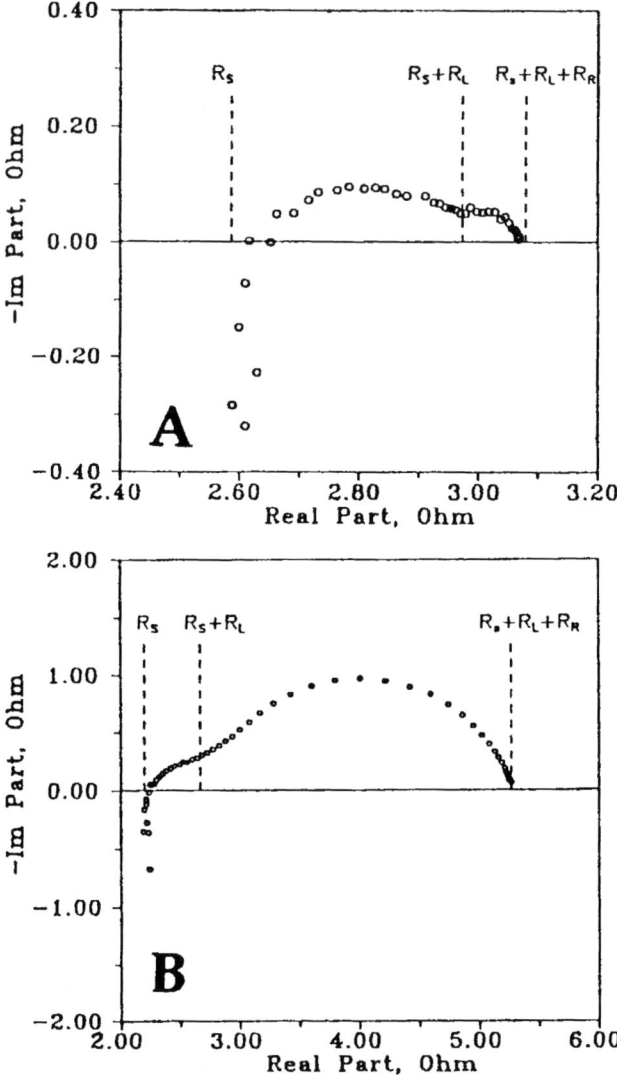

Figure 6.9 Impedance spectra on a cermet anode at 1000°C, are dependent on water partial pressure, $P_{(H_2O)}$ = 0.03 and 0.0022 bar with P_{H_2} = 1.0 bar; the oxygen partial pressures, $P_{(O_2)}$, are 4.5×10^{-18} and 6.5×10^{-21} bar for these ratios of fuel to reaction product. R_S = ohmic resistance of materials; R_L = charge transfer resistance; R_R = mass transport resistance [22].

of standard YSZ at 900°C and three orders of magnitude lower than in hydrogen ion conductors such as strontium cerate, but its effects are not negligible in a ceramic where electronic conduction is also extremely low. Raz et al. [25] presented the energetics of water adsorption: chemisorption on the ceramic surface as required for the exchange reaction above is maintained to temperatures consistent with fuel cell operation due to the high enthalpy of the process, of the order of 100 kJ/mol. The chemisorption of hydrogen on the ceramic has an even higher enthalpy, and therefore can be maintained to temperatures beyond the 700°C where hydrogen ionic conductivity has been measured. Dependence of interface polarisation on water partial pressure has been experimentally verified up to 1000°C [26]. These processes may be represented as a hydroxylated surface (Figure 6.10). Some mobility of the hydrogen ion either on the surface of the electrolyte through these chemisoption effects, or near the surface as hydrogen intercalate can therefore be postulated.

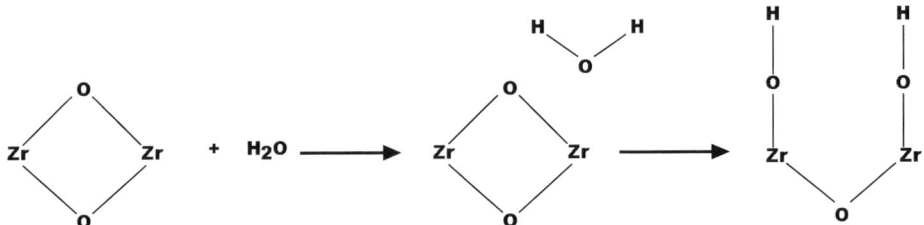

Figure 6.10 Model for hydroxylation of a zirconia surface by chemisorption of water (after [25]).

Adsorbed oxygen on the nickel component of the cermet is also discussed by Mizusaki et al. [27] as mediating hydrogen mobility on that surface. The required delocalisation of the anodic reaction from the much-discussed linear three-phase boundary, giving rise to a functionally volumetric anode, is a consequence of these transport mechanisms. On a given surface the linear three-phase boundary (TPB) structure is widened to present an active area; by electronic and ionic conductive percolation through the anode structure, an active volume is developed (Figure 6.1). With a three-phase boundary width approaching 1 µm, of the order of the grain size of the nickel in the cermet, effectively the whole surface of the grains of the cermet structure within the active electrode volume is available for the anodic reaction. A plausible model is Fick or Knudsen diffusion if the porosity is submicron, followed by dissociative adsorption of dihydrogen molecules on the nickel surface and their ionisation. Oxygen and hydrogen ions can exchange across the three-phase boundaries within the cermet giving hydroxyl sites, which can then pair and desorb water.

This model raises the issue of the effective thickness of the electrochemically active portion of the anode structure. Primdahl and Mogensen [20] found no correlation between polarisation effects and electrode thickness down to 20 µm, and in more recent work [26] a depth of 10 µm for the active zone is sustained. Mathematical modelling [29] is in accord with this experimental evidence (Figure 6.11). Beyond that thickness, the cermet can be regarded as a passive contact layer, and in anode-supported intermediate temperature fuel cells, as also having a structural and mechanical function. It is therefore available as a site for fuel reactions such as reforming. Some studies with this as objective have already been reported, such as the incorporation of ruthenium as catalyst [30].

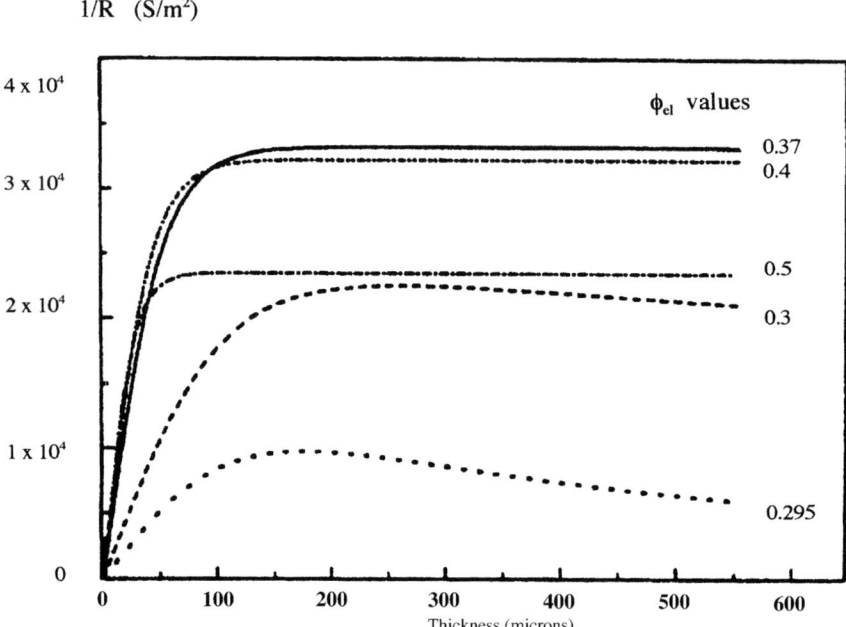

Figure 6.11 Mathematical modelling of dependence of polarisation on electrode thickness (after [29]). ϕ_{el} = volume fraction of the electronic conductor (nickel) in the cermet; exchange current density i_o = 0.1 A/cm²; conductivity of the metal component σ_{el} = 2 × 10⁶ S/m, oxygen ionic conductivity of the ceramic component σ_i = 15 S/m, grain size of both components 1 µm.

Optimisation of supporting anodes in advanced fuel cell designs to incorporate thermal control and chemical functionality into that zone, which is otherwise merely mechanical and structural, will depend on the further experimental validation of the reaction mechanisms, interpreted with the new mathematical models now appearing [31–34].

6.8 Operation of Anodes with Fuels other than Hydrogen

The study of hydrogen-fuelled cells has provided essential information on the mechanisms. In the absence of a hydrogen distribution infrastructure, however, practical engineering requires compatibility with hydrocarbon fuels. Natural gas is favoured for demonstration units, and interest in other fuels has already been noted. To avoid the growth of carbon on nickel cermet anodes exposed to hydrocarbons, reforming, a reaction of the hydrocarbon with steam to produce a hydrogen/carbon monoxide mixture as the actual cell fuel, is standard practice. Therefore the performance of the cermet as site for the electrochemical reaction of carbon monoxide is as important in practice as its kinetics for hydrogen. Verifying that the role of carbon dioxide is analogous to that of water in hydrogen-fuelled cells, Aaberg et al. [35] observed a minimum in interfacial polarisation with 55% CO_2 and CO as fuel. Evidence is now emerging [36] that while the polarisation is little influenced by CO partial pressure over a wide range, decreasing towards $P_{(CO)} = 1$ atm, the effect of carbon dioxide is uniformly to reduce polarisation with increasing partial pressure, with a reaction order of 0.5. Since oxygen partial pressure is not an independent variable given that

$$P_{(O_2)} = KP^2_{(CO_2)}/P^2_{(CO)}$$

a reaction order of 0.5 for CO_2 is equivalent to 0.25 for oxygen. The adsorbed species are then identified as oxygen – probably on oxygen vacancies near the three-phase boundary, and CO on the metal. In [36] it was also recorded that impedance spectra were difficult to obtain at low frequencies, due to instability. This is in accord with the observation of high electrical noise on electrodes exposed to CO [37]. It is tempting to consider this effect as associated with the reversible Boudouard coking reaction, $2CO \Longleftrightarrow C + CO_2$, with occupation of CO adsorption sites by carbon, followed by the reaction with CO_2 or electrochemical oxidation to remove it.

When after reaction of hydrocarbon fuel with steam, both hydrogen and CO with traces of CO_2 are admitted to an anode as reformate, the situation is even more complex. The electrochemical oxidation rate of hydrogen is several times faster than that for CO, the divergence increasing with temperature and the water-gas shift reaction being faster than either [38]. However, the concentration of CO is a fraction of that of hydrogen, 14.9% of the total gas present when the feedstock to the reformer has a steam to carbon ratio of 2 at 800°C, rising to 17.2% at 1000°C. As a consequence the electrochemical depletion rates of the two fuels are comparable, polarisation does not significantly rise, and there is no accumulation of carbon species in the system. These laboratory results are fully confirmed by the successful long-term operation of SOFC systems on reformate fuel mixtures in large-scale demonstration plants.

6.9 Anodes for Direct Oxidation of Hydrocarbons

Reforming of the hydrocarbon fuel does present a balance of plant requirement impacting on investment, maintenance and overall system efficiency, providing an incentive to develop systems and materials capable of sustaining direct oxidation on a fuel cell anode. It has been known for some time that at high current densities, with steam and carbon dioxide being formed electrochemically, and therefore with a higher $P_{(O_2)}$ over the anode, methane can oxidise on a nickel cermet without serious carbon deposition [39]. It is presumed that the oxidation products of the cell reform the incoming fuel, though to maintain a current and therefore a power density above the necessary threshold [39] may not be a practical procedure in commercial operation. Admission of some steam with the fuel, or an autothermal process by partial oxidation, can extend the regime of operation, particularly with a co-catalyst as already mentioned [30] through this distributed reforming over the anode [41] (Figure 6.12). However, it is now evident that such a direct oxidation process, without reforming, requires an innovation in anode materials, diverging from the established nickel–zirconia option.

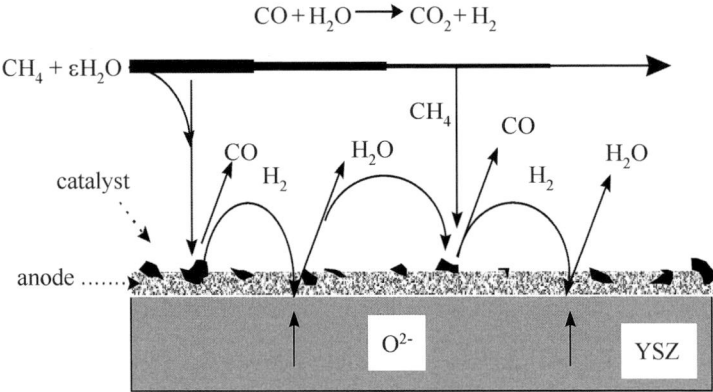

Figure 6.12 Schematic of distributed processing of humidified methane with internally generated reaction products.

Spacil's alternative transition metals for the cermet, iron and cobalt, though less active for the pyrolysis of hydrocarbons, do not have the corrosion resistance of nickel in a high-steam environment. Considering silver and copper, their oxides either decompose or melt at temperatures below the requirements for cermet sintering; neither are the metals refractory (Ag m.p. 962°C; Cu m.p. 1083°C). However, in catalytic technology the advantages of copper composites with ceria are recognised [42]. Partial reduction of copper oxide when exposed to fuel at elevated temperature, and the resulting redox properties, permit exchange of oxygen between the lattice and the gas phase, with availability for surface reactions. A copper–ceria composite anode [43,44] is a recent promising initiative. The difficulty of sintering a copper composite was avoided by forming a porous zirconia skeleton on a dense electrolyte substrate of the same material, then introduction of copper and cerium as their nitrate salts in solution, followed by drying and pyrolysis, similar to a procedure already demonstrated for anode and cathode catalysts [45]. Co-insertion of the two cations is possible since copper does not form a solid solution in ceria so the two phases remain separate as required for functionality of the electrode. In Figure 6.13 the reported performance of cells using this zirconia-supported copper–ceria composite is presented. Obviously the power density with methane fuel is significantly lower than that with hydrogen, but the synergetic catalysis by ceria is evident from the negligible power density with copper alone in the zirconia matrix. Figure 6.14 presents evidence of the stability of the composite anode, in contrast to a nickel cermet where the fuel cell operation is suppressed irreversibly within 30 min by the carbon accumulation. The ceria–copper system is now being further investigated for the direct oxidation of higher hydrocarbons [43].

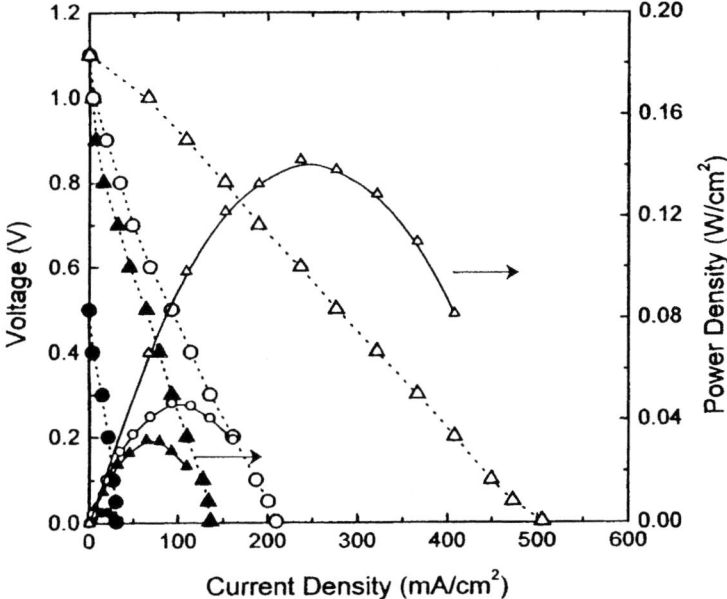

Figure 6.13 Current–potential characteristics of copper cermets at 800°C showing effect of hydrocarbon fuel and electrocatalysis by ceria: (○) hydrogen on Cu/YSZ cermet; (△) hydrogen on cermet catalyzed by ceria. Filled symbols = methane [43].

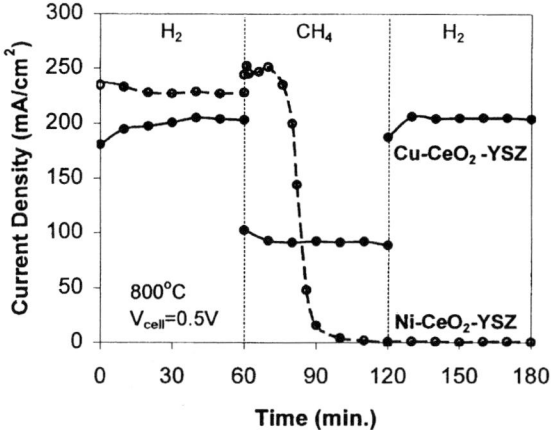

Figure 6.14 Stability of ceria-catalyzed copper cermet under different fuel conditions contrasts with rapid irreversible deactivation of nickel cermet [43].

Refractory electronic-conducting ceramics may also provide candidate materials for the direct oxidation anode. Ceria and its solid solutions with other rare earths, cerates and titanates [46] have been investigated. Given their variation in conductivity with gas environment they may have to be used with metals in composites. The perovskites, of general formula ABO_3, where A and B are two metals with a total valency of 6, form a wide variety of solid solutions by partial substitution on either the A or B sites, or both. These compounds are established in SOFC technology with the chromites used as electrical interconnects and gas separators, for example. They have the required stability over the whole range of fuel cell oxygen partial pressure from 10^{-20} bar near an open-circuit anode to atmospheric or even pressurised conditions on the cathode side, and with sufficient electrical conductivity. Lanthanum chromite is electrically conductive and stable, but unfortunately has negligible electrocatalytic activity. Baker and Metcalfe [47] therefore applied the strategy of substitution, with calcium on the A site and with nickel or cobalt on the B site. Primdahl et al. [48] used 3% vanadium on the chromium site of a lanthanum-strontium perovskite. Use of a lower valency metal ion as substituent on the A site requires compensation, either by a higher valency ion on the B site or by oxygen lattice defects, which can increase the activity of the material towards oxygen exchange and catalysis in comparison with the non-catalytic parent structure. Sauvet et al., in a review of oxide-based anodes [49], noted that to enhance the activity of a chromite substituted partially with strontium on the A site, a C–H bond breaking catalyst, specifically nickel, ruthenium or platinum, is required on the B site. The catalytic oxidation of methane over ceria and chromites promotes deep oxidation, producing CO_2 and water. However nickel substituting up to 10% of the chromium sites gives selectivity for hydrogen and CO in the temperature range 500–800°C [50]. In fuel cell operation the low surface coverage of metallic nickel avoids carbon deposition while providing the selective sites necessary for fuel activation [51]. Finally, mention should be made of exotic options, like the vanadium carbide anode for oxidation of gas-entrained solid fuel [52].

6.10 Summary

Empirical development of the nickel–zirconia anode over several decades has led to solid oxide fuel cells with adequate service life and performance, but fuel reforming is still required to operate with commercially available hydrocarbon fuels. It has become evident that the anode reactions are dominated by the 'three-phase boundary' and that the microstructure of the composite cermet anodes is pivotal. Consequently, the processing methods used for making the anode powders, and the fabrication techniques used for deposition on the electrolyte are critical in making high performance anodes. Anode-supported cells with very thin electrolyte films are becoming interesting for operation at lower temperatures.

Anode behaviour is evaluated by d.c. methods under steady state and by impedance spectroscopy under transient conditions. The reaction pathways for hydrogen have been elucidated, and mathematical modelling is providing micro- and nanoscale understanding of electrode processes. At higher current loadings, the diffusion processes have been evaluated showing that the electrochemically active anode thickness is around 10 μm. In practice, however, much thicker anode layers are used in order to provide sufficient conductivity to transport electrons out to the interconnect.

Operation of anodes on fuels other than hydrogen is commercially necessary, and challenging because of carbon deposition. In general, pre-reforming of hydrocarbons is carried out before the fuel contacts the anode. Direct internal reforming is possible with nickel cermet anodes, but carbon deposition can be a problem in long-term operation. Other possible anode metals, such as copper, can prevent carbon formation but their properties are not entirely satisfactory. One strand of current research is towards stable direct oxidation anodes, with modified cermets incorporating copper and ceramics like ceria; a further option lies in the identification of mixed conducting ceramics with sufficient electrocatalytic activity to function alone without any metal as anodes. The ultimate challenge is to produce anodes which can directly oxidise hydrocarbons. Indications are that this is possible and represents ongoing anode development.

References

[1] H. S. Spacil, US Patent 3,558,360; filed October 30, 1964, modified November 2, 1967, granted March 31, 1970.
[2] T. Ogawa, T. Ioroi, Y. Uchimoto, Z. Ogumi and Z.-I. Takehara, in *Solid Oxide Fuel Cels III*, The Electrochemical Society Proceedings, Pennington, NJ, PV 93-4, 1993, p. 479.
[3] E. Baur and H. Preis, *Z. Elektrochem.*, **43** (1937) 727.
[4] H.-H. Möbius, *J. Solid State Electrochem.*, **1** (1997) 2; this work, Chapter 2.
[5] S. C. Singhal, in *Solid Oxide Fuel Cels III*, The Electrochemical Society Proceedings, Pennington, NJ, PV 93-4, 1993, p. 665.
[6] J.-H. Lee, H. Moon, H.-W. Lee, J. Kim, J.-D. Kim and K.-H. Yoon. *Solid State Ionics*, **148** (2002) 15.
[7] S. Primdahl, B. F. Sørensen and M. Mogensen, *J. Am. Ceram. Soc.*, **83** (2000) 489.
[8] D. Skarmoutsos, F. Teitz and P. Niikolopoulos, *Fuel Cells*, **1** (2001) 243.
[9] T. Wagner, R. Kirchheim and M. Rühle. *Acta Metall. Mater.*, **40** (1992) S85.
[10] A. Ringuedé, J. A. Labrincha and J. R. Frade. *Solid State Ionics*, **141–142** (2001) 549.
[11] H. Itoh, Y. Heie, T. Yamamoto, M. Mori and T. Watanabe, in *Solid Oxide Fuel Cells VII*, eds. H. Yokokawa and S. C. Singhal, The Electrochemical Society Proceedings, Pennington, NJ, PV2001-16, 2001, p. 750.
[12] P. Holtappels, Jülich Research Centre Report 3414, 1997/Thesis, University of Bonn.
[13] R. N. Basu, G. Blass, H. P. Buchkremer, D. Stöver, F. Tietz, E. Wessel and I. C. Vinke, in *Solid Oxide Fuel Cells VII*, eds. H. Yokokawa and S. C. Singhal, The Electrochemical Society Proceedings, Pennington, NJ, PV2001-16, 2001, p. 995.
[14] Research Centre Jülich website http://www.sofc.de (2002).
[15] F. van Heuveln, Thesis, University of Twente, Netherlands, 1997, p. 167.

[16] P. Holtappels, L. G. J. de Haart and U. Stimming. *J. Electrochem. Soc.*, **146** (1999) 1620.
[17] J. R. MacDonald, *Impedance Spectroscopy*, Wiley, New York, 1987.
[18] S. Primdahl and M. Mogensen. *J. Electrochem. Soc.*, **146** (1999) 2827.
[19] P. Holtappels, L. G. J. de Haart and U. Stimming. *J. Electrochem. Soc.*, **146** (1999) 2976.
[20] S. Primdahl and M. Mogensen. *J. Electrochem. Soc.*, **144** (1997) 3409.
[21] N. L. Robertson and J. N. Michaels. *J. Electrochem. Soc.*, **138** (1991) 1494.
[22] A. Bieberle, PhD Thesis, ETH Zürich, Switzerland, 2000.
[23] M. Mogensen and T. Lindegaard, in *Solid Oxide Fuel Cells III*, The Electrochemical Society Proceedings, Pennington, NJ, PV93-4, 1993, p. 484.
[24] N. Sakai, K. Yamaji, H. Negishi, T. Horita, H. Yokokawa, Y.-P. Xiong and M. B. Phillips, *Electrochemistry*, **68** (2000) 499.
[25] S. Raz, K. Sasaki, J. Maier and I. Riess, *Solid State Ionics*, **143** (2001) 181.
[26] M. Brown, S. Primdahl and M. Mogensen, *J. Electrochem. Soc.*, **147** (2000) 475.
[27] J. Mizusaki, H. Tagawa, T. Saito, T. Yamamura, K. Kamitani, K. Hirano, S. Ehara, T. Takagi, T. Hikata, M. Ippomatsu, S. Nakagawa and K. Hashimoto, *Solid State Ionics*, **70/71** (1994) 52.
[28] J. Misusaki, H. Tagawa, T. Saito, K. Kamitani, T. Yamamura, K. Hirano, S. Ehara, T. Takagi, T. Hikata, M. Ippomatsu, S. Nakagawa and K. Hashimoto, *J. Electrochem. Soc.*, **141** (1994) 2129.
[29] P. Costamagna, P. Costa and V. Antonucci, *Electrochim. Acta*, **43** (1998) 375.
[30] P. Vernoux, PhD Thesis, INP Grenoble, France (1998).
[31] P. Costamagna, P. Costa and E. Arato, *Electrochim. Acta*, **43** (1998) 967.
[32] S. Sunde, *J. Electroceramics*, **5** (2000) 153.
[33] A. S. Ioselevich and A. A. Kornyshev, *Fuel Cells*, **1** (2001) 40.
[34] S. H. Chan and Z. T. Xia, *J. Electrochem. Soc.*, **148** (2001) A388.
[35] R. J. Aaberg, R. Tunold, S. Tjelle and R. Odegard, in *Proceedings of the 17th Risø International Symposium on Materials Science: High Temperature Electrochemistry: Ceramics and Metals*, eds. F. W. Poulsen, N. Bonanos, S. Linderoth, M. Mogensen and B. Zachau-Christiansen, Risø National Lab., Roskilde, Denmark, 1996, p. 511.
[36] F. Z. Boulenouar, K. Yashiro, M. Oishi, A. Kaimai, Y. Nigara, T. Kawada and J. Mizusaki, in *Solid Oxide Fuel Cells VII*, The Electrochemical Society Proceedings, Pennington, NJ, PV2001-16, 2001, p. 759.
[37] P. H. Middleton, International Energy Agency Adv. Fuel Cells Programme, Annex II Activity Meeting, Chexbres, Switzerland (1993) p. 54.
[38] Y. Matsuzaki and I. Yasuda, *J. Electrochem. Soc.*, **147** (2000) 1630.
[39] T. Horita, N. Sakai, T. Kawada, H. Yokokawa and M. Dokiya, *J. Electrochem. Soc.*, **143** (1996) 1161.
[40] J.-H. Koh, B.-S. Kang, H. C. Lim and Y.-S. Yoo, *Electrochem. Solid State Lett.*, **4** (2001) A12.

[41] A.-L. Sauvet and J. T. S. Irvine, *Fuel Cells*, **1** (2001) 205.
[42] W. Liu and M. Flytzani-Stephanopoulos, *J. Catal.*, **153** (1995) 304.
[43] S. Park, R. Craciun, J. M. Vohs and R. J. Gorte, *J. Electrochem. Soc.*, **146** (1999) 3603.
[44] C. Wang, W. L. Worrell, S. Park, J. M. Vohs and R. J. Gorte, *J. Electrochem. Soc.*, **148** (2001) A864.
[45] K. R. Thampi, A. J. McEvoy and J. Van Herle, in *Ionic and Mixed Conducting Ceramics II*, eds. T. A. Ramanarayanan, W. L. Worrell and H. L. Tuller, The Electrochemical Society Proceedings, Princeton, NJ, PV94-12, 1994, p. 239.
[46] J. T. Irvine, P. R. Slater, A. Kaiser, J. C. Bradley, P. Holtappels and M. Mogensen, in *Proceedings of the 4th European SOFC Forum*, ed. J. Huijmans, Switzerland, 2000, p. 471.
[47] R. T. Baker and I. S. Metcalfe, *Appl. Catal. A* **126** (1995) 319; *Solid Oxide Fuel Cells IV*, eds. M. Dokiya, O. Yamamoto, H. Tagawa and S. C. Singhal, The Electrochemical Society Proceedings, Pennington, NJ, PV95-1, 1995, p. 781.
[48] S. Primdahl, J. R. Hansen, L. Grahl-Madsen and P. H. Larsen, *J. Electrochem. Soc.*, **148** (2001) A74.
[49] A.-L. Sauvet, J. Guindet and J. Fouletier, in *Proceedings of the 4th European SOFC Forum*, ed. J. Huijmans, Switzerland, 2000, p. 567.
[50] J. Sfeir, P. A. Buffat, P. Möckli, N. Xanthopoulos, R. Vasquez, H. J. Matthieu, J. Van herle and K. R. Thampi, *J. Catal.*, **202** (2001) 229.
[51] J. Sfeir, PhD Thesis, EPFL, Lausanne, Switzerland, 2001.
[52] T. Horita, N. Sakai, T. Kawada, H. Yokokawa and M. Dokiya, US patent 6,183,896, 2001.

Chapter 7

Interconnects

Harlan U. Anderson and Frank Tietz

7.1 Introduction

Two roles of the interconnect in high-temperature solid oxide fuel cells (SOFCs) are the electrical connection between cells and the gas separation within the cell stack. The fact that the interconnect must be compatible with all of the cell components as well as be stable with respect to both oxidising and reducing gases places very stringent materials requirements on it. These requirements plus the additional constraints of cost and ease of fabrication tend to limit the possible choices to only a few materials. These materials come from either perovskite-type oxide ceramics based on rare earth chromites for operating temperatures in the 900–1000°C range or metallic alloys for lower temperature cell operation.

The properties which an interconnect must possess are rather extensive and somewhat dependent upon the particular SOFC configuration [1]. However, typical requirements are:

- High electronic conductivity with low ionic conductivity
- Chemical stability in both fuel and air
- Thermal expansion match to other cell components
- High mechanical strength
- High thermal conductivity
- Chemical stability with regard to other cell components

Depending upon the particular SOFC design, additional requirements such as the ease of fabrication to gas-tight density, the ability to make gas-tight seals with other cell components, and the material cost also play an important role.

Of the requirements listed above, the first three are crucial and tend to eliminate most candidate materials. In fact, for operation at temperatures above 800°C, the only oxides that fit these criteria are the doped rare earth chromites. In particular, compositions from the system $(La,Sr,Ca)(Cr,Mg)O_3$ are the leading interconnect materials. However, compositions from the system $(Y,Ca)CrO_3$ also have acceptable properties. These rare earth chromites satisfy most of the requirements, but have problems in fabrication and have high cost. Metallic interconnects are easier to fabricate and potentially less costly than oxide ceramics but their lifetimes under SOFC operating conditions remain to be demonstrated.

In this chapter the requirements of interconnect materials, the characteristics that the leading candidate materials possess, and how well these fulfil the requirements are discussed. The oxide ceramic materials are discussed first followed by a description of several types of metallic interconnection materials. Then, the special protective and contact materials applied as coatings on the interconnects to match them to the electrodes are described.

7.2 Ceramic Interconnects (Lanthanum and Yttrium Chromites)

7.2.1 Electrical Conductivity

The electronic conductivity for an interconnect to perform adequately should be greater than about 1 S/cm at 1000°C. For either $YCrO_3$ or $LaCrO_3$ to obtain this level of conductivity, acceptor doping is required. Tables 7.1 and 7.2 list typical conductivity values that are obtainable.

Upon exposure to reducing atmospheres, all oxides tend to lose oxygen and form oxygen vacancies. In the case of p-type oxides like Y and La chromites, the loss of oxygen results in a decrease in electrical conductivity. Figure 7.1 illustrates the behaviour that these oxides display [7]. Figures 7.2 and 7.3 show

Table 7.1 Electrical conductivity data for substituted $LaCrO_3$ (in air)

Dopant	Composition (mol%)	Electrical conductivity at 1000°C (S/cm)	Activation energy (eV) (kJ/mol)	Ref.
None	0	1	0.19 (18)	2
Mg	10	3	0.20 (19)	3
Sr	10	14	0.12 (12)	4
Ca	20	35	0.14 (13)	5
Ca, Co	20 Ca, 10 Co	34	0.15 (14)	6

Table 7.2 Electrical conductivity data for Ca-doped $YCrO_3$ (in air)

Ca content (mol%)	Electrical conductivity at 1000°C (S/cm)	Activation energy (eV) (kJ/mol)	Ref.
5	4.5	0.17 (16)	7
10	7.7	0.18 (17)	7
15	13.0	0.18 (17)	7
20	15.5	0.18 (17)	7

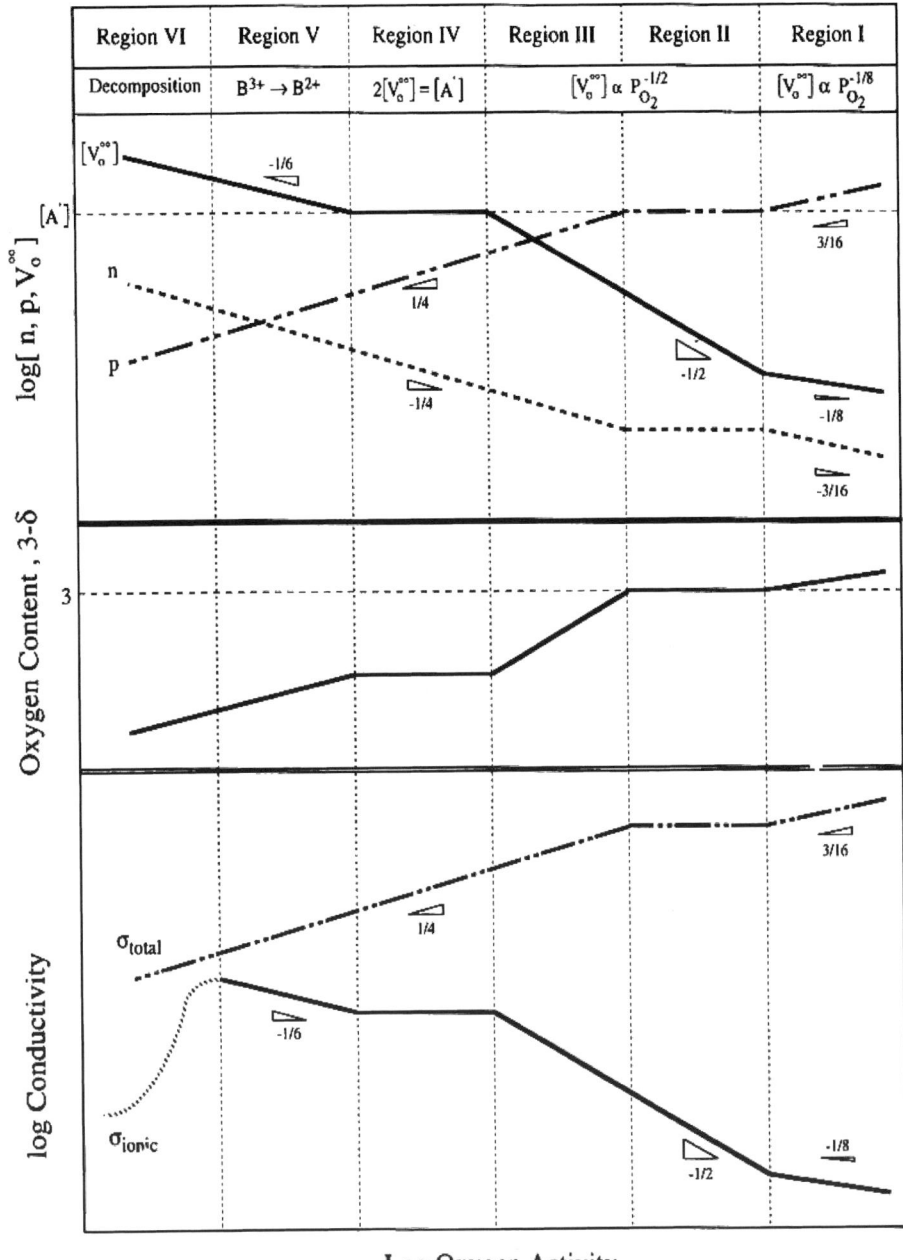

Figure 7.1 Defect concentration, oxygen content (3-δ) and electrical conductivity in acceptor-substituted $ABO_{3-\delta}$ as function of oxygen activity at constant temperature. The B site is occupied by a transition metal ion [8].

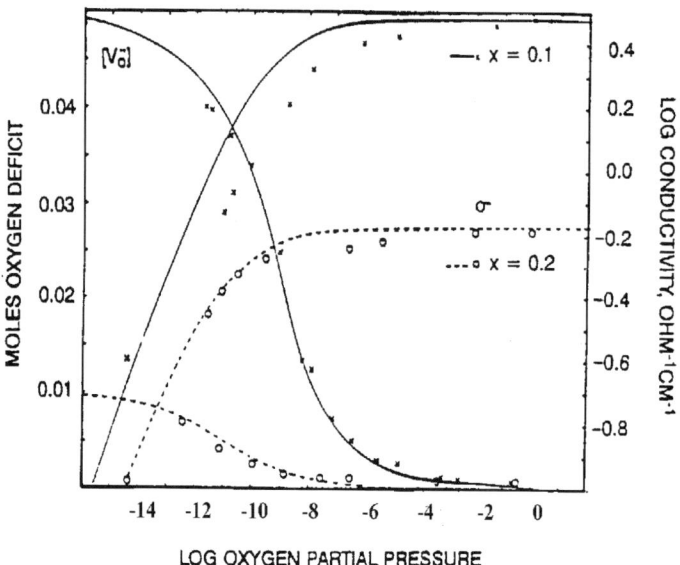

Figure 7.2 Oxygen vacancy concentration and electrical conductivity as a function of oxygen partial pressure and dopant content for $LaCr_{1-x}Mg_xO_3$ at $1200°C$ [3].

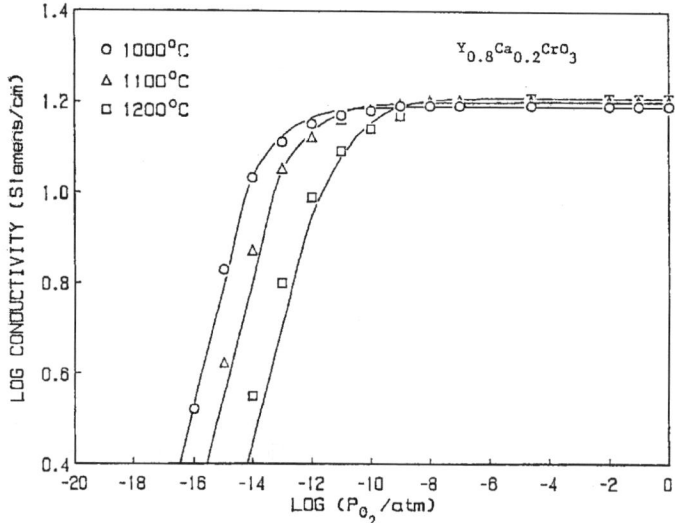

Figure 7.3 Electrical conductivity vs. oxygen partial pressure for $Y_{0.8}Ca_{0.2}CrO_3$ at various temperatures [7].

how the electrical conductivity changes with oxygen activity for both of these chromites [2, 6]. What makes the chromites useful as interconnects is that at 1000°C and 10^{-16} bar oxygen pressure, they remain single phase and do not dissociate. There are no other oxides which have these levels of electrical conductivity and can survive such reducing conditions. Thus, the chromites are quite unique and are the only oxides available for use as interconnects. The $LaCrO_3$ doped with either Ca or Sr has sufficient conductivity in fuel atmospheres to exceed 1 S/cm and therefore is preferred to Mg-doped $LaCrO_3$.

There has been some concern about the oxygen ion conductivity in $(La,SrCa)CrO_3$, particularly under reducing conditions, but studies by Yokokawa et al. [9] and Singhal [10] suggest that this is not a serious problem since at 1000°C the oxygen diffusion coefficient appears to be less than 10^{-7} cm^2/s. This would yield an ionic transport number of less than 0.01 even at the most reducing conditions. Thus, oxygen permeation through the interconnect should be minimal.

7.2.2 Thermal Expansion

It is important that the thermal expansion coefficients of all SOFC components match well. This is particularly true for the dense components, the electrolyte (most commonly yttria-stabilised zirconia, YSZ) and the interconnect. Table 7.3 compares the thermal expansion coefficients (TECs) and shows that the TECs of $LaCrO_3$ and $YCrO_3$ do not match that of YSZ, but the addition of dopants makes the match possible. Thus, thermal expansion is not a significant problem. However, the loss of oxygen in a reducing atmosphere leads to lattice expansion which has the potential of causing cracking problems [10–18]. For example, at 1000°C, when exposed to hydrogen, $LaCr_{0.85}Mg_{0.15}O_3$ and $La_{0.8}Sr_{0.2}CrO_3$ expand about 0.1% and 0.3%, respectively (Figure 7.4). The amount of expansion due to oxygen loss is directly related to the oxygen vacancy concentration. Several studies have shown that this expansion can be minimised by the addition of elements such as Al and Ti, but it is difficult to completely avoid this behaviour without the loss of other desirable properties, such as the electrical conductivity [16, 18]. Thus, it is important to allow for this expansion in the cell and stack design.

Table 7.3 Thermal expansion coefficients of $LaCrO_3$ and $YCrO_3$ [1–13]

Composition (nominal)	Thermal expansion coefficient ($\times 10^{-6}$/K)
$LaCrO_3$	9.5
$LaCr_{0.9}Mg_{0.1}O_3$	9.5
$La_{0.9}Sr_{0.1}CrO_3$	10.7
$La_{0.8}Sr_{0.2}CrO_3$	11.1
$La_{0.65}Ca_{0.35}CrO_3$	10.8
$LaCr_{0.9}Co_{0.1}O_3$	13.1
$La_{0.8}Ca_{0.2}Cr_{0.9}Co_{0.1}O_3$	11.1
$YCrO_3$	7.8
$Y_{0.9}Ca_{0.1}CrO_3$	8.9
$Y_{0.8}Ca_{0.2}CrO_3$	9.6
YSZ	9.4–11

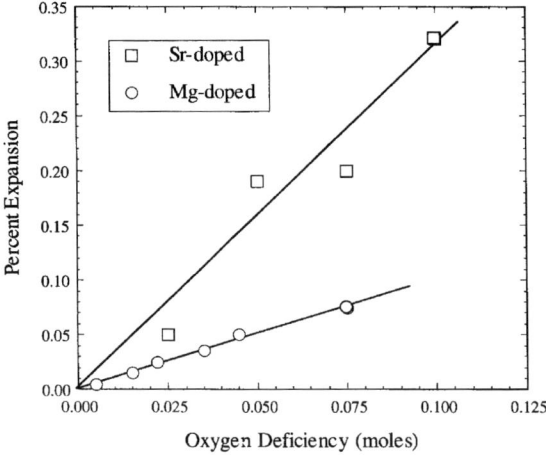

Figure 7.4   Percent expansion as function of oxygen deficiency in LaCrO$_3$ at 1000°C [11–17].

7.2.3 Thermal Conductivity

The thermal conductivities of the SOFC components are all in the order of 1.5–2 W/(mK) [3]. These are low compared to 20 W/(mK) for stainless steel and 400 W/(mK) for copper. Therefore, heat dissipation could be a problem which must be considered in the SOFC design. This is particularly true for higher power density monolithic and planar SOFCs, but is less of a problem in the tubular design.

7.2.4 Mechanical Strength

The mechanical strength of many compositions of LaCrO$_3$ is low compared to YSZ (see Table 7.4) and appears to be variable. This low strength and variability

Table 7.4   Mechanical strength of LaCrO$_3$ (MPa)

Dopant (mol%)	Temperature (°C)			Ref.
	25	1000 (air)	1000 (H$_2$)	
10% Mg	390–418			20
20% Mg	300			20
10% Sr	268	166		20
10% Sr	245	77		20
20% Sr	234			19
20% Sr		100	140	21
10% Ca	140	36		20
15% Ca		80	40	13
20% Ca	100–150	20–60	50–60	13, 22
30% Ca	100–140	60–130	20	13
YSZ		200		13

appears to be related to structural flaws due to inhomogeneities in composition, grain size and/or density and not to the $LaCrO_3$ itself. Thus, improvements in processing can alleviate this problem.

7.2.5 Processing

In general, Cr-containing oxides are difficult to sinter. The primary problem is related to the vaporisation of Cr–O species which leads to enhancement of the evaporation–condensation mechanism of sintering. This tends to suppress densification and causes coarsening of the powder [22, 23]. For $LaCrO_3$, this problem was initially addressed by Groupp and Anderson who demonstrated that densification could be achieved by sintering at temperatures exceeding 1700°C in an oxygen partial pressure in the 10^{-10}–10^{-9} bar range [24]. Under these conditions the Cr–O volatility is suppressed, thus minimising the evaporation–condensation mechanism and thereby allowing densification to occur. They showed that densification of stoichiometric $LaCrO_3$ is possible through a solid-state sintering mechanism by controlling the sintering atmosphere, but unfortunately the required conditions are so extreme, that they are not compatible with the processing of the other SOFC components and are also uneconomical.

Hot pressing has been considered as an alternative for densifying $LaCrO_3$. Several studies have shown that hot pressing at temperatures around 1500–1600°C in graphite dies achieves more than 93% of theoretical density. However, due to the low oxygen activity because of the C–O reaction, the $LaCrO_3$ dissociates to Cr metal both at the $C/LaCrO_3$ interface and in the grain boundaries. As a result, upon reoxidation cracking occurs due to Cr oxidation. Therefore, hot pressing has proven to be unsatisfactory both from a structural and an economical point of view.

The densification problem has led researchers to search for sintering aids which promote densification by suppressing the Cr–O volatility and enhancing mass transport through liquid-phase mechanisms. Perhaps the first successful demonstration of this process was made by a group at Argonne National Laboratory which was attempting to co-sinter $LaCrO_3$ with other SOFC components in monolithic SOFCs [25]. They showed that the addition of boron and fluorides of Sr and La promoted densification in air at temperatures as low as 1300°C. Owing to the volatility and the interaction of the liquid phase with other cell components, this is a difficult process to control, but it does show that liquid-phase sintering is a viable option.

Since that initial work, a number of other liquid promoters have been investigated and several systems have been rather successful. For example, Koc showed that compositions within the system $(La,Ca)(Co,Cr)O_3$ sintered well and yielded nearly theoretically dense structures at temperatures as low as 1350°C [6]. These compositions are stable in a fuel atmosphere at 1000°C and therefore are potential interconnect candidates. The main problem with these compositions is that Ca and Co tend to react with other cell components and therefore their long time stability is suspect [21].

In an effort to alleviate the Co migration problem, a number of investigators have studied the $(La,Ca)Cr_{1-y}O_3$ and $(La,Sr)Cr_{1-y}O_3$ systems [12, 17, 26–29]. What has been found is that when powders of these compositions are prepared at temperatures below 700°C, the powders tend to be multiphase with substantial quantities of either La, Ca, or Sr chromates present. These chromate phases melt incongruently in the 1000–1200°C range and the liquid promotes sintering. Unfortunately, the best sinterable compositions are Cr deficient and the excess A site components, La, Ca, or Sr tend to segregate to the grain boundaries and create hydration and cracking problems under both SOFC operating and ambient conditions. Thus, this method of sintering has not been entirely satisfactory, but until a better one is discovered, it is the one being most frequently employed in the planar SOFC configuration. The search is still on to develop a method of densifying stoichiometric $LaCrO_3$ which is both economical and yields stable interconnects.

A number of methods are used to fabricate the interconnect. The method used depends on the SOFC design [30]. For the tubular design, fabrication methods such as electrochemical vapour deposition (EVD), plasma spraying, laser ablation and slurry coating/sintering have been used, with EVD and plasma spraying being favoured. Economics is an issue with EVD while porosity and interfacial cracking are the difficulties with plasma spraying.

In the early 1980s, the monolithic SOFC design made use of tape casting, lamination and calendaring technology to produce a structure which was then sintered to produce a completed SOFC stack. On the surface, this process is attractive since it offers the potential of low cost and high power density. In practice, it is a very difficult process because it requires the simultaneous sintering of all cell components. This means that the shrinkages and shrinkage rates must be matched for all four cell components during sintering. Also, the interdiffusion between the components under the high-temperature processing conditions must be minimised. As a result, this design has been abandoned.

A variation of the monolithic design was introduced by Allied Signal [31] (that first became part of Honeywell and now a part of General Electric Power Systems). This design co-sinters the electrolyte, cathode and anode, but fabricates the interconnect separately. This design has eliminated the fabrication incompatibility problem between the interconnect and other cell components, but it does have the sealing problems of the planar cell design. The main advantage of this design is the densification of the interconnect by itself so that it gives the option of liquid-phase sintering the interconnect without inducing problems with the other cell components.

The conventional planar cell designs build the gas distribution channels into the interconnect in a bipolar structure. In this design, good electrical contact between the cell components must be maintained and the edges sealed gas-tight. These seals are made by using either glasses or cements which, when heated, give both gas-tight seals and electrical contact. In addition to interlayer seals, side seals are required which are both electrically insulating and gas-tight. A number of different schemes have been tried to provide these two seals, but at the moment they still remain a major issue with planar SOFCs.

7.3 Metallic Interconnects

The reduction of the cell operating temperature from 900–1000°C down to 600–850°C makes the use of metallic materials for the interconnect feasible and attractive. The advantages of metallic interconnects over ceramic interconnects are obvious: lower material and fabrication costs, easier and more complex shaping possible, better electrical and thermal conductivity and no deformation or failure due to different gas atmospheres across the interconnection. The interconnects can be fabricated by machining, pressing or, in the case of powder metallurgical alloys, by near-net-shape sintering. The gas distribution is usually realised by parallel channels whilst the ridges separating the channels serve as electrical contact with the electrodes.

The first reports on SOFC stacks built with metallic interconnect plates were published in the early 1990s [32,33]. Initial experiments with FeNiCr alloys showed a steady decrease in power output during single cell operation [34], and later also in stack tests [35]. This deterioration was ascribed to the release of chromium from the alloy leading to catalytic poisoning of the cathode [36, 37]. This phenomenon has been investigated intensively, is now fairly well understood, and described later in this chapter. All early attempts at using metallic materials as interconnect were not very successful, because the materials (heat-resistant steels) often contained a significant amount of Ni leading to large thermal expansion mismatch between the metallic interconnect and the ceramic SOFC components. The situation changed with the use of chromia-forming materials. Various metallic interconnect materials are discussed below.

7.3.1 Chromium-Based Alloys

After a screening of different chromium-based alloys, Metallwerke Plansee AG proposed a chromium alloy containing 5 wt% iron and 1 wt% yttria (Cr 5Fe 1Y_2O_3), the so-called Ducrolloy, for use with electrolyte-supported SOFCs [38]. In a close collaboration with Siemens AG, this alloy was used for assembling electrolyte-supported planar cells in 1–10 kW size stacks [39,40]. The alloy composition was optimised to match its thermal expansion to that of the 8 mol% yttria-stabilised zirconia (8YSZ) electrolyte to successfully thermally cycle the stacks. The good match of thermal expansion is shown in Figure 7.5. Only at temperatures above 800°C, the increased thermal expansion of the alloy leads to deviations from the thermal expansion of YSZ and the two materials differ in TEC at 1000°C by 8%.

This alloy has been investigated in detail with respect to corrosion behaviour [41,42] and contact resistance across its interfaces with the electrodes [43]. Typically, Cr 5Fe 1Y_2O_3 is a chromia former and even after long-term exposure in oxygen or air, the chromia scales are very thin. Thicker corrosion scales grow in carbon containing atmospheres (methane, coal gas) due to formation of carbides [42].

The fabrication of interconnect plates of Cr 5Fe 1Y_2O_3 is done by powder metallurgical methods and starts with the alloying of Cr flakes with Fe and Y_2O_3 by

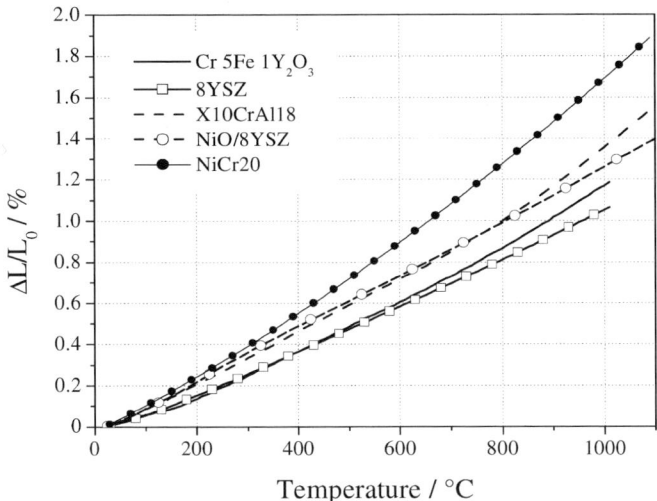

Figure 7.5 Thermal expansion curves of CrFe5Y$_2$O$_3$1, the ferritic steel X10CrAl18 and for comparison a nickel-based alloy (Ni 20Cr, VA Chromium) as well as the two mostly used supporting components in planar SOFCs, the electrolyte (8YSZ) and the anode substrate (NiO/YSZ).

high-energy milling [44]. Then pressing and sintering in hydrogen atmosphere is followed by a hot forming process like hot rolling in vacuum. For the shaping of interconnects for the Siemens stacks, electrochemical machining was applied [38,45]. Because of such sophisticated processing steps, the interconnects made of this alloy are almost as expensive as ceramic interconnects. Although a decrease in cost of about one order of magnitude from the R&D stage to mass production is estimated [45], Ducrolloy interconnects remain an expensive stack component.

Use of near-net shape processing of interconnect parts has also been tried for tubular SOFCs [46]. The aim of this near-net-shape processing is a reduction in cost by avoiding machining and by more efficient use of the chromium powder [47]. For this purpose, however, a new materials development had to be conducted with different Cr powder grades, additional alloying elements, and different oxide dispersoids to improve sinterability, pressing behaviour, resulting density, corrosion, and contact resistance with thermally sprayed protective coatings [47,48]. Such coatings are necessary on the one hand to improve the contact between the interconnect and the adjacent electrode and on the other hand to avoid fast deterioration of cell performance. The best alloy compositions were found to be Cr 5Fe 0.3Ti 0.5CeO$_2$ and Cr 5Fe 0.5CeO$_2$ resulting in a contact resistance of about 30 mΩ cm^2 after 1400 h of exposure in air and using La$_{0.8}$Sr$_{0.2}$MnO$_3$ as coating. An endurance stack test with one of these near net-shaped alloys showed very stable performance for a period of 1000 h.

7.3.2 Ferritic Steels

Compared with Cr 5Fe 1Y$_2$O$_3$, ferritic steels have the advantages of the lower cost of the material, easier processing and fabrication of components, weldability, and thermal expansion match with the anode substrate (Figure 7.5). Several SOFC stacks with ferritic steel interconnects have been fabricated and tested [49–51]. However, long-term stack tests showed large degradation in power output (a degradation rate between 2 and 25%/1000 h of operation [50,52] is typical), and the corrosion, for instance of the ferritic steel X10 CrAl 18, was not sufficient for the targeted 40,000 h operation of SOFC systems; after only 3000 h of operation, the growth of nodular corrosion products led to a partial detachment of the cathode contact layer from the cathode (Figure 7.6). Therefore, it became evident that new steel compositions having better corrosion resistance than the commercially available ferritic steels needed to be developed.

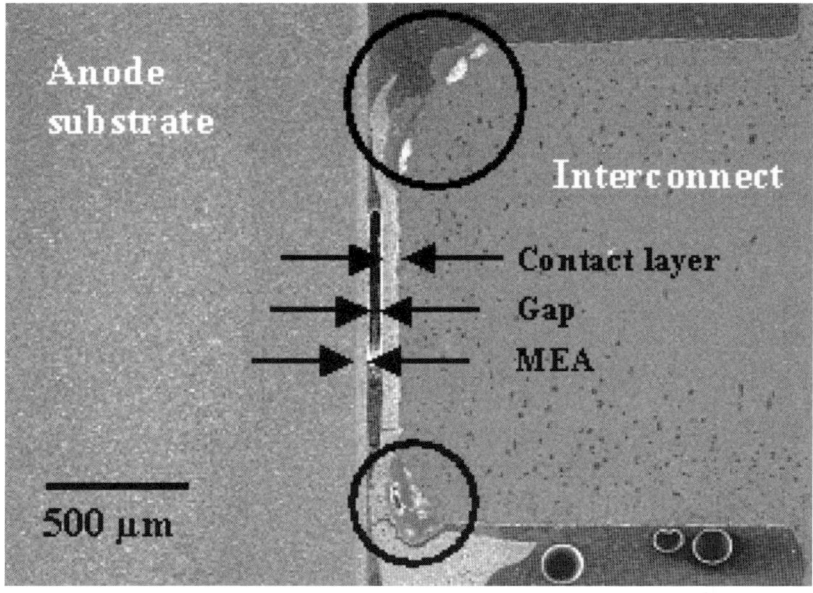

Figure 7.6 Corrosion products (circles) at the edges of an interconnect coated with LaCoO$_3$ after 3000 h of operation at 800°C and a constant power density of 0.22 W/cm^2.

Malkow et al. systematically investigated the thermal expansion and corrosion behaviour of commercial steels and model alloys [53]. The thermal expansion coefficient of ferritic steels decreases with increasing Cr content up to 20 wt% and increases with increasing Al content. By such alloying, the thermal expansion of ferritic steels can be adjusted and matched to the Ni/YSZ anode substrate, but not to the YSZ electrolyte. The oxidation of steels depends not only on the Cr content, but also on small amounts of alloyed elements, especially Al and Si. Once a compositional threshold is reached, alumina and silica layers are formed instead of a chromia layer. This leads to a reduction in oxidation rate. However, such alumina and silica layers are insulating and have to be avoided when the steel interconnect is in contact with a contact or electrode material. In a comparative study [54] of commercially available ferritic steels with chromium contents between 12 and 28%, the contact resistance against

$La_{0.6}Sr_{0.4}Fe_{0.8}Co_{0.2}O_3$ strongly increased with time for those steels containing Al and Si in the range 1–2 wt%. Lowest contact resistances were obtained with X3 CrTi 17 and X2 CrTiNb 18 steels, remaining below 10 mΩ cm^2 after 4000 h of exposure in air. In corrosion experiments, both of these steels formed scales composed of chromia and Fe–Mn–Cr spinels together with an internal oxidation of the stabilising elements Ti and Nb [54].

Further progress has been made in developing ferritic steels that form thin spinel-type corrosion scales with significant electrical conductivity and have well-adherent corrosion scales which reduce the release of volatile Cr species [55, 56]. By adding various alloying elements in the range 0.1–2.5 wt% to alloys with 17–25 wt% chromium, it was learned that:

- Ni does not support a stable and protective scale formation
- Ti leads to higher oxidation rate due to enhanced growth rate of the chromia scale and formation of internal Ti oxides
- Y, La, Ce, and Zr reduce the oxide growth rate independent of Cr content; especially La promotes very thin oxide scales
- Mn increases the oxide scale growth rate even if a lanthanide element is present, and preferentially forms a Cr–Mn spinel with low electrical resistance on top of a chromia scale

This systematic study led to an optimised steel composition – at laboratory scale – with small additions of Mn, La, and Ti but without any Al and Si. This steel forms the desired thin and electrically conductive oxide scales [55, 56], good contact resistances with ceramic coatings [57] (Figure 7.7), and reduced permeability for volatile chromium species [58].

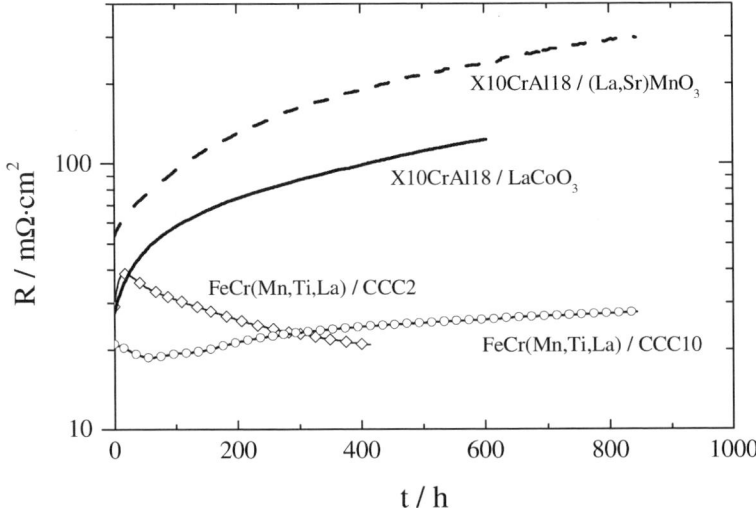

Figure 7.7 Change of electrical resistance of ferritic steel/perovskite ceramics combinations during exposure in air at 800°C.

In FeCr 18 steels containing 7–10 wt% tungsten [59], increasing tungsten content results in thermal expansion coefficients close to that of YSZ electrolyte and small additions of Ce, La or Zr lead to reduced corrosion rates. Also, in contact with lanthanum manganite, a Cr–Mn spinel/Cr_2O_3 double layer is formed.

Honegger et al. [60] investigated powder metallurgically made model steels containing 22–26% Cr and minor additions of Mo, Ti, Nb and Y_2O_3. After oxidation in air, all materials showed a double-layered oxide scale composed of chromia directly in contact with the alloy and Mn–Cr spinel at the outer surface. In humidified hydrogen (20 vol% H_2O), the corrosion experiments also revealed a Cr_2O_3/Cr–Mn spinel double layer at 700°C, but at higher temperatures a Fe–Cr spinel/Cr–Mn–Fe spinel system was formed. The contact resistance measurements performed on steel/$La_{0.6}Sr_{0.4}CoO_3$ paste/$La_{0.8}Sr_{0.2}MnO_3$ ceramic samples gave very low contact resistances at 800°C in air after 15,000 h of exposure. With Mo-containing model alloys (FeCr 22 Mo 2 Ti Y_2O_3 and FeCr 26 Mo 2 Ti Y_2O_3), a contact resistance of 20 mΩ cm^2 for 40,000 h could be deduced from the parabolic rate observed.

To achieve acceptable power densities in terms of W/kg and W/dm^3, the metallic interconnects are now made significantly thinner as listed in Table 7.5 and use of thin steel foils (0.1–1 mm) for interconnects becomes feasible [61,62]. Such foil interconnects can easily be mass produced by pressing, stamping, cutting and punching [63]. However, the corrosion behaviour of thin foils of ferritic steels can be very different compared with that of thick plates due to the effect of selective oxidation and depletion of alloyed elements leading to composition changes in the foil. As long as the metallic component is thick enough to serve as a quasi-infinite reservoir of chromium

Table 7.5 Stack developer and the interconnect materials used

Company	Interconnect material	Interconnect thickness (mm)	Method to avoid poisoning of the cathode by volatile chromium species	Ref.
Sanyo Electric	Inconel 600	5–6	Addition of La_2O_3 to cathode	35
Siemens	Cr 5Fe 1Y_2O_3	ca. 3.5	Wet coating of $LaCoO_3$	39
Siemens	Cr 5Fe 1Y_2O_3	ca. 3.5	Plasma-sprayed coating of $(La,Sr)CrO_3$	64, 65
Sulzer Hexis	Cr 5Fe 1Y_2O_3		High-velocity oxygen flame (HVOF)-sprayed coating of $(Y,Ca)MnO_3$	66
Sulzer Hexis	Cr 5Fe 1Y_2O_3		HVOF-sprayed coating of $La_{0.8}Sr_{0.2}MnO_3$	47
Forschungszentrum Jülich	Ferritic steel	6	$LaCoO_3$ coating	49
Ceramic Fuel Cells Ltd	Ferritic steel	3	Al_2O_3 coating in gas channels, conductive coating on the ribs	50

the enrichment of chromium at the surface (i.e. the formation of the chromia scale) has no significant influence on the composition of the thick steel plate. However, when thin foils are used, the amount of chromium is limited and the chromia formation can lead to compositional changes within the thin foil leading to very different corrosion behaviour. Tests over several thousand hours are required to demonstrate the reliability of these lightweight interconnect designs.

7.3.3 Other Metallic Materials

Sanyo and Fuji Electric started their SOFC stack development with metallic interconnects using nickel-based alloys such as Inconel 600 and Ni 22Cr, respectively [35,67]. In a long-term exposure experiment of 12,000 h duration, the electrical resistance of a Ni 20Cr alloy coated with $La_{0.6}Sr_{0.4}CoO_3$ did not change significantly and remained below 10 mΩ cm^2, although $SrCrO_4$ formed at the interface [68]. However, thermal cycling with these alloys led to voltage drops after each thermal cycle due to the mismatch in thermal expansion with the other cell components (Figure 7.5) leading to cracks at the interconnect/electrode interfaces [67]. Nevertheless, properties of austenitic steels and Ni-based superalloys for use in SOFC stacks continued to be explored. Linderoth et al. [69] investigated the oxidation resistance and the corrosion products of Fe–Cr–Ni steels (Haynes 230, Inconel 601), Ni–Cr steel (Inconel 657), Fe–Cr–Al steel (APM-Kanthal), and the Plansee Ducrolloy. Among the Ni-containing steels, the Haynes 230 showed the best oxidation resistance and the oxide scale composed of Cr_2O_3 and spinel might have better electronic conductivity than a pure chromia scale.

England and Virkar [70, 71] investigated thin foils of Ni-based superalloys (Inconel 625, Inconel 718, Hastelloy X, and Haynes 230) as possible interconnect materials. They also observed the slowest oxidation in air for Haynes 230 and the formation of a Cr–Mn spinel at the outer surface leading to a complete depletion of Mn in the inner part of the thin foil. Hastelloy X also formed a spinel layer at the beginning and both alloys exhibited the lowest electronic resistance of the oxide scale formed. In wet hydrogen, the oxidation resistance was also the best for Haynes 230 but the oxide scale growth was much faster than in air [71], chromia was the dominating phase in the oxide scale and hence the electronic resistance of the oxide scale was 1–2 orders of magnitude higher than after oxidation in air.

Another concept for interconnecting SOFCs is the use of FeCrAlY steels in combination with silver pins [62]. The FeCrAlY steel is used as a thin foil and quickly forms an alumina scale inhibiting the release of Cr from the steel. To avoid high resistances of the alumina scales, the steel foil is perforated with Ag pins acting as contacts between the anode of one cell to the cathode of the next cell in the stack. The use of silver is very attractive due to the low contact resistances [62,72,73]. However, problems regarding silver evaporation at operation temperatures $> 700°C$ [62] and during thermal cycling [72] need to be addressed.

7.4 Protective Coatings and Contact Materials for Metallic Interconnects

The metallic interconnects have two main disadvantages. The first is the release of volatile Cr species. In atmospheres containing water vapour, the most volatile specie is chromium acid, $H_2Cr^{(VI)}O_4$ [37, 74], which is transported with the oxidant gas through the cathode to the cathode/electrolyte interface, competes with the oxygen molecules for the electrochemically active sites and blocks them with $Cr^{(III)}$ [74]. This results in an increase in cathode polarisation [75, 76]. After the initial blocking of the electrochemically active sites, the ongoing transport and reduction of chromium species lead to decomposition of the cathode perovskite material and the formation of spinels [76,77]. The chromium transport in the cathode compartment needs to be minimised to overcome these problems, either by using 'Cr getter' materials like La_2O_3 [35] in the cathode or by applying protective coatings of lanthanum chromites [64, 65], lanthanum manganites [47] or yttrium manganites [66] to the interconnect (Table 7.5). The thermally sprayed manganite coatings have led to stable long-term performance lasting about 12,000 h with a degradation rate in cell voltage of less than 1%/1000 h [78].

Matsuzaki and Yasuda reported that $La_{0.6}Sr_{0.4}Fe_{0.8}Co_{0.2}O_3$ cathodes are much more stable against Cr poisoning than $La_{0.85}Sr_{0.15}MnO_3$ or $Pr_{0.6}Sr_{0.4}MnO_3$ cathodes [79], especially when ceria-based solid electrolytes are used, and no enrichment of chromium was found at the $La_{0.6}Sr_{0.4}Fe_{0.8}Co_{0.2}O_3$/ceria interface (after 10 h). A possible explanation might be the different overvoltages of the cathodes for oxygen and $H_2Cr^{(VI)}O_4$ reduction. While for the manganites the reduction of the chromium oxyhydroxide is the energetically preferred reaction, in the presence of the ferrites the reduction of oxygen appears to require less activation energy. Although the exposure times were short in their experiments, modification of cathodes may be a possible alternative to avoid Cr poisoning and the use of protective coatings. The extent of Cr evaporation from steels or other alloys and the effectiveness of protective coatings can be estimated before stack assembly by transpiration experiments [74,80]. Such investigations are useful in new interconnect material development and also for an understanding of the cell and stack degradation rates.

The second major disadvantage of metallic interconnects is the formation of oxide scales leading to significant ohmic losses. The interaction of the metallic interconnect with the adjacent ceramic cell components and the resulting time-dependent resistance of these material combinations is highly important. In the case of Cr 5Fe $1Y_2O_3$, the best contact material for the cathode was found to be $LaCoO_3$ [64] before protective layers of $(La,Sr)CrO_3$ were applied to avoid Cr evaporation. $LaCoO_3$ was also successfully used in combination with ferritic steel [49], although the thermal expansion coefficient of the cobaltite is higher [64,81] than the other SOFC materials [82]. Besides the lower contact resistance (Figure 7.7), $LaCoO_3$ appears to react with the released chromium species to form a $La(Co,Cr)O_3$ perovskite and therefore retains the Cr vapour.

Contact materials are used in stack assembly for better electrical contact between the interconnect and the electrodes and also for compensation of dimensional tolerances of the parts. Such contact layers have no direct role in electrochemical reactions, but they can provide a homogeneous contact over the whole area of the fuel cell and minimise the ohmic losses within the stack. The maximum assembling temperature depends on the interconnect material used. For SOFCs with only ceramic components [83, 84], the bond between the cell and the $LaCrO_3$ interconnect is realised by sintering at about 1300°C and a solid, stiff bond with good electrical contact is obtained requiring no other contact material. In the case of Cr 5Fe 1Y_2O_3 interconnects, sintering can be utilised for stack assembly providing good contact without any contact material due to the high melting point (1700°C) of the alloy [38]. However, with ferritic steel interconnects, the stack assembly temperature cannot be higher than 900–950°C due to enhanced corrosion and thus contact material is needed for good electrical contact.

Since there are no electrochemical requirements for the contact materials, they can be different from the electrode materials and be selected on the basis of their electrical conductivity and thermal expansion. Lanthanum cobaltites have high specific conductivities, up to 1700 S/cm [81]. However, the thermal expansion of these cobaltites has a large mismatch with the other cell components as mentioned previously. For these ceramic contact materials, therefore, a compromise between acceptable conductivity and tolerable mismatch in thermal expansion is generally required.

A chemical interaction between the contact layer and an electrode or the interconnect should not occur, but cannot be avoided in most cases due to the reaction of the contact material with the chromia scale formed on the interconnect. In all cases where alkaline earth-containing chromite contact materials were used, the formation of chromates was observed [57, 64, 68, 85, 86] leading to progressive decomposition of the perovskite material. The change in contact resistance (Figure 7.7) is not only due to the scale formation on the surface of the interconnect but also driven by the reaction between the oxide scale and the volatile Cr species with the contact material, by the formation of alkaline earth chromates and the steady depletion of material at the contact material/interconnect interface due to the volatility of these chromates. The latter process was demonstrated by Hou et al. [87] by applying different cathode materials – (i) Pt, (ii) $La_{0.6}Sr_{0.4}CoO_3$, (iii) $La_{0.85}Sr_{0.15}MnO_3$ + $La_{0.8}Sr_{0.2}Ga_{0.83}SMg_{0.17}O_3$ – onto an un-oxidised Fe-based alloy with composition similar to X18 CrN 28. They found that the area-specific resistance of the cobaltite specimen increased at a greater rate than for the other two material combinations although the cobaltite is more conductive than the manganite/gallate mixture. For the contact material, it is important to have not only an initial low contact resistance but also a constant resistance with time (or even a decreasing resistance as shown in Figure 7.7).

Often the corrosion of the interconnect on the anode side is not an issue because Ni meshes are used and these make good electrical contact with the interconnect. However, the Ni wires can also be corrosively attacked during

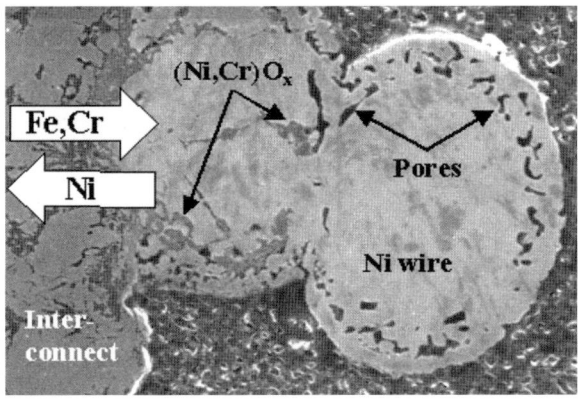

Figure 7.8 Cross-section of damaged Ni wires of a Ni mesh in contact with X10CrAl18. The stack was operated for 3700 h at 800°C. The pores along the outer part of the wires were formed during start-up of the stack due to partial oxidation of the Ni mesh and subsequent reduction of the oxide layer by the fuel gas [89].

long-term stack operation (Figure 7.8), because of the interdiffusion of the metals across the interface; Fe and Cr diffuse into the Ni wire and Ni diffuses into the interconnect [57]. Such deterioration of contacts can be minimised by an additional Ni coating on the interconnect [88]. The interdiffusion cannot be avoided, but the diffusion zone in this case is not in direct contact with the gas atmosphere and causes no internal oxidation by the formation of Ni–Cr oxides as shown in Figure 7.8.

7.5 Summary

The interconnect material is vitally important in connecting cells electrically and in separating the reactants. The requirements placed upon it are stringent and range from compatibility with electrodes and chemical stability to corrosion resistance combined with excellent electronic conductivity. The two types of materials that have been extensively used are the chromite ceramics and the chromium-based metallic alloys.

A number of issues can be listed which need to be addressed before a completely acceptable ceramic interconnect can be developed (Table 7.6). However, most of them are of secondary importance when compared to the two

Table 7.6 Current issues of ceramic interconnects

Of most importance	Of next most importance
1. Fabrication and processing costs 2. Material costs	1. Thermal expansion match to YSZ 2. Chemical compatibility to YSZ and sealing glass or cement 3. Expansion due to loss of oxygen 4. Mechanical strength and durability in reducing atmosphere 5. Electrical conductivity in reducing atmosphere

major challenges: reduction of fabrication costs and reduction of material costs. The reduction in material costs can be achieved by both the interconnection design and the quantity of interconnect required. However, reduction in fabrication costs will only come with improved processing and automation [1,31,32,63].

Lanthanum chromite has provided long lifetimes, as long as 69,000 h in Siemens Westinghouse tubular cells, at 900–1000°C. However, metallic interconnects have not yet shown equivalent lifetime performance. Improvements in metallic interconnect compositions and contact layers between cells/interconnects are still issues for materials development. In particular, the metal/ceramic interface in cells should have low corrosion, low contact resistance and low permeability of chromium species. Recent results have shown that optimised steels for SOFC applications are available and alkaline earth-free and cobalt-containing perovskites are the most suitable materials for contact layers; however, their long-term performance under fuel cell operation conditions needs to be proven.

References

[1] N. Q. Minh and T. Takahashi, in *Science and Technology of Ceramic Fuel Cells*, Elsevier, Amsterdam, 1995.

[2] D. P. Karim and A.T. Aldred, *Phys. Rev. B*, **20** (1979) 2255.

[3] B. F. Flandermeyer, M. M. Nasrallah, D. M. Sparlin and H. U. Anderson, *High Temp. Sci.*, **20** (1985) 259.

[4] W. J. Weber, C. W. Griffin and J. L. Bates, *J. Amer. Ceram. Soc.*, **70** (1987) 265.

[5] I. Yasuda and T. Hikita, in *Solid Oxide Fuel Cells II*, eds. F. Grosz, P. Zegers, S. C. Singhal and O. Yamamoto, Commission of the European Communities, Luxembourg, 1991, p. 632.

[6] R. Koc and H. U. Anderson, *J. Mater. Sci.*, **27** (1992) 5477.

[7] G. F. Carini II, H. U. Anderson, D. M. Sparlin and M. M. Nasrallah, *Solid State Ionics*, **49** (1991) 233.

[8] H. U. Anderson, C. C. Chen, L-W. Tai and M. M. Nasrallah, in *Proc. 2nd Int. Symp. Ionic and Mixed Conducting Oxides*, eds. T. A. Ramanarayanan, H. L. Tuller and W. L. Worrell, The Electrochemical Society Proceedings, Pennington, NJ, PV94-12, 1994, p. 376.

[9] H. Yokokawa, T. Horita, N. Sakai, B. A. van Hassel, T. Kawada and M. Dokiya, in *Solid Oxide Fuel Cells III*, eds. S. C. Singhal and H. Iwahara, The Electrochemical Society Proceedings, Pennington, NJ, PV93-4, 1993, p. 364.

[10] S. C. Singhal, Interconnection Material Development for Solid Oxide Fuel Cells, DOE contract DE-AC21-84MC21184, Final Report (1985).

[11] S. Srilomsak, D. P. Schilling and H. U. Anderson, in *Solid Oxide Fuel Cells I*, ed. S.C. Singhal, The Electrochemical Society Proceedings, Pennington, NJ, PV89-11, 1989, p. 129.

[12] M. M. Nasrallah, J. D. Carter, H. U. Anderson and R. Koc, in *Solid Oxide Fuel Cells II*, eds. F. Grosz, P. Zegers, S. C. Singhal and O. Yamamoto, Commission of the European Communities, Luxembourg, 1991, p. 637.

[13] M. Dokiya, T. Horita, N. Sakai, T. Kawada, H. Yokokawa, B. A. van Hassel and C. S. Montross, in *High Temperature Electrochemical Behaviour of Fast Ion and Mixed Conductors, Roskilde, Denmark*, eds. F. W. Poulsen, J. J. Bertzen, T. Jacobson, E. Skou and M. J. L. Østergård, Risø Natl. Lab., Denmark, 1993, p. 33.

[14] A. Zuev, L. Singheiser and K. Hilpert, *Solid State Ionics*, **147** (2002) 1.

[15] F. Boroomand, E. Wessel, H. Bausinger and K. Hilpert, *Solid State Ionics*, **129** (2000) 251.

[16] M. Mori and Y. Hiei, *J. Am. Ceram. Soc.*, **84** (2001) 2573.

[17] S. P. Simner, J. S. Hardy and J. W. Stevenson, *J. Electrochem. Soc.*, **148** (2001) A351.

[18] G. Pudmich, B. A. Boukamp, M. Gonzalez-Cuenca, W. Jungen, W. Zipprich and F. Tietz, *Solid State Ionics*, **135** (2000) 433.

[19] N. M. Sammes and R. Ratnaraj, in *High Temperature Electrochemical Behaviour of Fast Ion and Mixed Conductors, Roskilde, Denmark*, eds. F. W. Poulsen, J. J. Bentzen, T. Jacobson, E. Skou and M. J. L. Østergård, Risø Natl. Lab., Denmark, 1993, p. 403.

[20] M. Mori, H. Itoh, N. Mori and T. Abe, in *Solid Oxide Fuel Cells III*, eds. S. C. Singhal and H. Iwahara, The Electrochemical Society Proceedings, Pennington, NJ, PV 93-4, 1993, pp. 325–335.

[21] C. Milliken, S. Elangovan and A. Khandkar, in *Solid Oxide Fuel Cells III*, eds. S. C. Singhal and H. Iwahara, The Electrochemical Society Proceedings, Pennington, NJ, PV93-4, 1993, p. 335.

[22] S. W. Paulik, S. Baskaran and T. R. Armstrong, *J. Mater. Sci.*, **33** (1998) 2397.

[23] D-H. Peck, M. Miller and K. Hilpert, *Solid State Ionics*, **143** (2001) 391.

[24] L. Groupp and H. U. Anderson, *J. Am. Ceram. Soc.*, **59** (1976) 449.

[25] B. K. Flandermeyer, R. B. Poeppel, J. T. Dusek and H. U. Anderson, US Patent 4749632, 1988.

[26] J. D. Carter, V. Sprenkle, M. M. Nasrallah and H. U. Anderson, in *Solid Oxide Fuel Cells III*, eds. S. C. Singhal and H. Iwahara, The Electrochemical Society Proceedings, Pennington, NJ, PV93-4, 1993, p. 344.

[27] L. A. Chick, T. R. Armstrong, D. E. McCready, G. W. Coffey, G. D. Maupin and J. L. Bates, in *Solid Oxide Fuel Cells III*, eds. S. C. Singhal, H. Iwahara, The Electrochemical Society Proceedings, Pennington, NJ, PV93-4, 1993, p. 374.

[28] M. Mori, Y. Hiei and N. M. Sammes, *Solid State Ionics*, **135** (2000) 743.

[29] L. A. Chick, J. Liu, J. W. Stevenson, T. R. Armstrong, D. E. McCready, G. D. Maupin, G. W. Coffey and C. A. Coyle, *J. Am. Ceram. Soc.*, **80** (1997) 2109.

[30] S. C. Singhal, *Solid State Ionics*, **135** (2000) 305.

[31] N. Q. Minh, *J. Am. Cer. Soc.*, **76** (1993) 563.

[32] H. Tsuneizumi, E. Matsuda, T. Kadowaki, T. Shiomitsu, H. Nakagawa, Y. Watanabe and T. Yokosuka, in *Proceedings 1st International Fuel Cell Conference*, NEDO/MITI, Makuhari, Japan, 1992, p. 293.

[33] Y. Akiyama, T. Yasuo and N. Ishida, in *Solid Oxide Fuel Cells III*, eds. S. C. Singhal and H. Iwahara, The Electrochemical Society Proceedings, Pennington, NJ, PV93-4, 1993, p. 724.

[34] J. Jung, Th. Martens, H. Runge and M. Turwitt, in *Solid Oxide Fuel Cells II*, eds. F. Grosz, P. Zegers, S. C. Singhal and O. Yamamoto, Commission of the European Communities, Luxembourg, 1991, p. 144.

[35] Y. Miyake, T. Yasuo, Y. Akiyama, S. Taniguchi, M. Kadowaki, H. Kawamura and T. Saitoh, in *Solid Oxide Fuel Cells IV*, eds. M. Dokiya, O. Yamamoto, H. Tagawa and S. C. Singhal, The Electrochemical Society Proceedings, Pennington, NJ, PV95-1, 1995, p. 100.

[36] Y. Akiyama, S. Taniguchi, T. Yasuo, M. Kadowaki and T. Saitoh, *J. Power Sources*, **50** (1994) 361.

[37] D. Das, M. Miller, H. Nickel and K. Hilpert, in *Proceedings of the 1st European SOFC Forum*, ed. U. Bossel, Switzerland, 1994, p. 703.

[38] W. Köck, H.-P. Martinz, H. Greiner and M. Janousek, in *Solid Oxide Fuel Cells IV*, eds M. Dokiya, O. Yamamoto, H. Tagawa and S. C. Singhal, The Electrochemical Society Proceedings, Pennington, NJ, PV95-1, 1995, p. 841.

[39] L. Blum, W. Drenckhahn, H. Greiner and E. Ivers-Tiffée, in *Solid Oxide Fuel Cells IV*, eds. M. Dokiya, O. Yamamoto, H. Tagawa and S. C. Singhal, The Electrochemical Society Proceedings, Pennington, NJ, PV95-1, 1995, p. 163.

[40] H. J. Beie, L. Blum, W. Drenckhahn, H. Greiner, B. Rudolf and H. Schichl, in *Solid Oxide Fuel Cells V*, eds. U. Stimming, S. C. Singhal, H. Tagawa and W. Lehnert, The Electrochemical Society Proceedings, Pennington, NJ, PV97-40, 1997, p. 51.

[41] W. Quadakkers, H. Greiner and W. Köck, in *Proceedings of the 1st European SOFC Forum*, ed. U. Bossel, Oberrohrdorf, Switzerland, 1994, p. 525.

[42] H. Greiner, Th. Grögler, W. Köck and R. F. Singer, in *Solid Oxide Fuel Cells IV*, eds. M. Dokiya, O. Yamamoto, H. Tagawa and S. C. Singhal, The Electrochemical Society Proceedings, Pennington, NJ, PV95-1, 1995, p. 879.

[43] W. Thierfelder, H. Greiner and W. Köck, in *Solid Oxide Fuel Cells V*, eds. U. Stimming, S. C. Singhal, H. Tagawa and W. Lehnert, The Electrochemical Society Proceedings, Pennington, NJ, PV97-40, 1997, p. 1306.

[44] R. Eck, H.-P. Martinz and T. Sasaki, *Mater. Sci. Eng. A*, **120** (1989) 307.

[45] M. Janousek, W. Köck, M. Baumgärtner and H. Greiner, in *Solid Oxide Fuel Cells V*, eds. U. Stimming, S. C. Singhal, H. Tagawa and W. Lehnert, The Electrochemical Society Proceedings, Pennington, NJ, PV97-40, 1997, p. 1225.

[46] W. Glatz, M. Janousek, E. Batawi and K. Honegger, in *Proceedings of the 4th European SOFC Forum*, ed. A. McEvoy, Oberrohrdorf, Switzerland, 2000, p. 855.

[47] E. Batawi, W. Glatz, W. Kraussler, M. Janousek, B. Doggwiler and R. Diethelm, in *Solid Oxide Fuel Cells VI*, eds. S. C. Singhal and M. Dokiya, The Electrochemical Society Proceedings, Pennington, NJ, PV99-19, 1999, p. 731.

[48] W. Glatz, E. Batawi, M. Janousek, W. Kraussler, R. Zach and G. Zobl, in *Solid Oxide Fuel Cells VI*, eds. S. C. Singhal and M. Dokiya, The Electrochemical Society Proceedings, Pennington, NJ, PV99-19, 1999, p. 783.

[49] H. P. Buchkremer, U. Diekmann, L. G. J. de Haart, H. Kabs, U. Stimming and D. Stöver, in *Solid Oxide Fuel Cells V*, eds. U. Stimming, S. C. Singhal, H. Tagawa and W. Lehnert, The Electrochemical Society Proceedings, Pennington, NJ, PV97-40, 1997, p. 160.

[50] S. P. S. Badwal, R. Bolden and K. Föger, in *Proceedings of the 3rd European SOFC Forum*, ed. P. Stevens, Oberrohrdorf, Switzerland, 1998, p. 105.

[51] S. Taniguchi, M. Kadowaki, T. Yasuo, Y. Akiyama, Y. Miyake and K. Nishio, *Denki Kagaku*, **65** (1997) 574.

[52] H. P. Buchkremer, U. Diekmann, L. G. J. de Haart, H. Kabs, D. Stöver and I. C. Vinke, in *Proceedings of the 3rd European SOFC Forum*, ed. P. Stevens, Oberrohrdorf, Switzerland, 1998, p. 143.

[53] Th. Malkow, U. v. d. Crone, A. M. Laptev, T. Koppitz, U. Breuer and W. J. Quadakkers, in *Solid Oxide Fuel Cells V*, eds. U. Stimming, S. C. Singhal, H. Tagawa and W. Lehnert, The Electrochemical Society Proceedings, Pennington, NJ, PV97-40, 1997, p. 1245.

[54] D. Dulieu, J. Cotton, H. Greiner, K. Honegger, A. Scholten, M. Christie and T. Seguelong, in *Proceedings of the 3rd European SOFC Forum*, ed. P. Stevens, Oberrohrdorf, Switzerland, 1998, p. 447.

[55] W. J. Quadakkers, T. Malkow, J. Pirón-Abellán, U. Flesch, V. Shemet and L. Singheiser, in *Proceedings of the 4th European SOFC Forum*, ed. A. McEvoy, Oberrohrdorf, Switzerland, 2000, p. 827.

[56] J. Pirón-Abellán, V. Shemet, F. Tietz, L. Singheiser, W.J. Quadakkers and A. Gil, in *Solid Oxide Fuel Cells VII*, eds. H. Yokokawa and S. C. Singhal, The Electrochemical Society Proceedings, Pennington, NJ, PV2001-16, 2001, p. 811.

[57] O. Teller, W. A. Meulenberg, F. Tietz, E. Wessel and W. J. Quadakkers, in *Solid Oxide Fuel Cells VII*, eds. H. Yokokawa and S. C. Singhal, The Electrochemical Society Proceedings, Pennington, NJ, PV2001-16, 2001, p. 895.

[58] Ch. Gindorf, K. Hilpert and L. Singheiser, in *Solid Oxide Fuel Cells VII*, eds. H. Yokokawa and S. C. Singhal, The Electrochemical Society Proceedings, Pennington, NJ, PV2001-16, 2001, p. 793.

[59] M. Ueda and H. Taimatsu, in *Proceedings of the 4th European SOFC Forum*, ed. A. McEvoy, Oberrohrdorf, Switzerland, 2000, p. 837.

[60] K. Honegger, A. Plas, R. Diethelm and W. Glatz, in *Solid Oxide Fuel Cells VII*, eds. H. Yokokawa, S. C. Singhal, The Electrochemical Society Proceedings, Pennington, NJ, PV2001-16, 2001, p. 803.

[61] N. Q. Minh, R. Doshi, J. Guan, S. Huss, G. Lear, K. Montgomery and E. Ong, in *2000 Fuel Cell Seminar Abstracts*, Courtesy Associates, Washington, DC, 2000, p. 593.
[62] W. A. Meulenberg, O. Teller, U. Flesch, H. P. Buchkremer and D. Stöver, *J. Mater. Sci.*, **36** (2001) 3189.
[63] J. P. Allen, in *2000 Fuel Cell Seminar Abstracts*, Courtesy Associates, Washington, DC, 2000, p. 55.
[64] H. Schmidt, B. Brückner and K. Fischer, in *Solid Oxide Fuel Cells IV*, eds. M. Dokiya, O. Yamamoto, H. Tagawa and S. C. Singhal, The Electrochemical Society Proceedings, Pennington, NJ, PV95-1, 1995, p. 869.
[65] R. Ruckdäschel, R. Henne, G. Schiller and H. Greiner, in *Solid Oxide Fuel Cells V*, eds. U. Stimming, S. C. Singhal, H. Tagawa and W. Lehnert, The Electrochemical Society Proceedings, Pennington, NJ, PV97-40, 1997, p. 1273.
[66] A. Plas, E. Batawi, W. Straub, K. Honegger and R. Diethelm, in *Proceedings of the 4th European SOFC Forum*, ed. A. McEvoy, Oberrohrdorf, Switzerland, 2000, p. 889.
[67] T. Iwata, N. Kadokawa and S. Takenoiri, in *Solid Oxide Fuel Cells IV*, eds. M. Dokiya, O. Yamamoto, H. Tagawa and S. C. Singhal, The Electrochemical Society Proceedings, Pennington, NJ, PV95-1, 1995, p. 110.
[68] T. Shiomitsu, T. Kadowaki, T. Ogawa and T. Maruyama, in *Solid Oxide Fuel Cells IV*, eds. M. Dokiya, O. Yamamoto, H. Tagawa and S. C. Singhal, The Electrochemical Society Proceedings, Pennington, NJ, PV95-1, 1995, p. 850.
[69] S. Linderoth, P. V. Hendriksen and M. Mogensen, N. Langvad, *J. Mater. Sci.*, **31** (1996) 5077.
[70] D. M. England and A. V. Virkar, *J. Electrochem. Soc.*, **146** (1999) 3196.
[71] D. M. England and A. V. Virkar, *J. Electrochem. Soc.*, **148** (2001) A330.
[72] K. Föger, R. Donelson and R. Ratnaraj, in *Solid Oxide Fuel Cells VI*, eds. S. C. Singhal and M. Dokiya, The Electrochemical Society Proceedings, Pennington, NJ, PV99-19, 1999, p. 95.
[73] K. Ahmed, J. Love, R. Ratnaraj, in *Solid Oxide Fuel Cells VII*, eds. H. Yokokawa, S. C. Singhal, The Electrochemical Society Proceedings, Pennington, NJ, PV2001-16, 2001, p. 904.
[74] D. H. Peck, M. Miller, H. Nickel, D. Das and K. Hilpert, in *Solid Oxide Fuel Cells IV*, eds. M. Dokiya, O. Yamamoto, H. Tagawa and S. C. Singhal, The Electrochemical Society Proceedings, Pennington, NJ, PV95-1, 1995, p. 858.
[75] Y. Matsuzaki, M. Hishinuma and I. Yasuda, in *Solid Oxide Fuel Cells VI*, eds. S. C. Singhal and M. Dokiya, The Electrochemical Society Proceedings, Pennington, NJ, PV99-19, 1999, p. 981.
[76] S. P. S. Badwal, R. Deller, K. Föger and Y. Ramprakash, J. P. Zhang, *Solid State Ionics*, **99** (1997) 297.
[77] S. P. Jiang, J. P. Zhang, L. Apateanu and K. Föger, *J. Electrochem. Soc.*, **147** (2000) 4013.

[78] R. Diethelm, M. Schmidt, K. Honegger and E. Batawi, in *Solid Oxide Fuel Cells VI*, eds. S. C. Singhal and M. Dokiya, The Electrochemical Society Proceedings, Pennington, NJ, PV99-19, 1999, p. 60.

[79] Y. Matsuzaki and I. Yasuda, *J. Electrochem. Soc.*, **148** (2001) A126.

[80] Ch. Gindorf, L. Singheiser and K. Hilpert, *Steel Research*, **72** (2001) 528.

[81] A. Petric, P. Huang and F. Tietz, *Solid State Ionics*, **135** (2000) 719.

[82] F. Tietz, *Ionics*, **5** (1999) 129.

[83] D. Stolten, R. Späh and R. Schamm, in *Solid Oxide Fuel Cells V*, eds. U. Stimming, S. C. Singhal, H. Tagawa and W. Lehnert, The Electrochemical Society Proceedings, Pennington, NJ, PV97-40, 1997, p. 88.

[84] Y. Sakaki, Y. Esaki, M. Hattori, H. Miyamoto, T. Satake, F. Nanjo, T. Matsudaira and K. Takenubo, in *Solid Oxide Fuel Cells V*, eds. U. Stimming, S. C. Singhal, H. Tagawa and W. Lehnert, The Electrochemical Society Proceedings, Pennington, NJ, PV97-40, 1997, p. 61.

[85] W. J. Quadakkers, H. Greiner, M. Hänsel, A. Pattanaik, A. S. Khanna and W. Malléner, *Solid State Ionics*, **91** (1996) 55.

[86] Y. Larring and T. Norby, *J. Electrochem. Soc.*, **147** (2000) 3251.

[87] P. Y. Hou, K. Huang and W. T. Bakker, in *Solid Oxide Fuel Cells VI*, eds. S. C. Singhal, M. Dokiya, The Electrochemical Society Proceedings, Pennington, NJ, PV99-19, 1999, p. 737.

[88] Y. Yamazaki, T. Namikawa, T. Ide, H. Kabumoto, N. Oishi, T. Motoki and T. Yamazaki, in *Solid Oxide Fuel Cells V*, eds. U. Stimming, S. C. Singhal, H. Tagawa and W. Lehnert, The Electrochemical Society Proceedings, Pennington, NJ, PV97-40, 1997, p. 1291.

[89] D. Stöver, U. Diekmann, U. Flesch, H. Kabs, W. J. Quadakkers, F. Tietz and I. C. Vinke, in *Solid Oxide Fuel Cells VI*, eds. S. C. Singhal and M. Dokiya, The Electrochemical Society Proceedings, Pennington, NJ, PV99-19, 1999, p. 812.

Chapter 8

Cell and Stack Designs

Kevin Kendall, Nguyen Q. Minh and Subhash C. Singhal

8.1 Introduction

Over the years, many ingenious designs of solid oxide fuel cells (SOFCs) have been devised, starting from pressed thimbles and discs in the 1930s. Since the 1960s, most development has focused on planar and tubular design cells and other geometries have become less popular. This chapter describes the two main types of SOFCs, the planar and the tubular, emphasising their fabrication methods and performance characteristics. Each of these two designs has a number of interesting variants; for example, the planar SOFC may be in the form of a circular disc fed with fuel from the central axis, or it may be in the form of a square plate fed from the edges. The tubular SOFCs may be of large diameter (>15 mm), or of much smaller diameter (<5 mm). Also the tubes may be flat and joined together to give easily printable surfaces for manufacturing the electrode layers. Other designs which have fallen out of favour, for example the corrugated monolithic design [1], are not described here.

Under typical operating conditions, a single cell produces less than 1 V. To obtain high voltage and power from the SOFCs, it is necessary to stack many cells together and this can be done in a number of ways using interconnect materials which are often fabricated into complex shapes to provide for other functions such as air and fuel channelling and sealing. This chapter describes the various kinds of stack designs that have been tested by a number of manufacturers in recent years, and analyses the advantages and disadvantages of the various schemes. Planar SOFCs and stacks are described first, followed by tubular and then the microtubular SOFCs.

8.2 Planar SOFC Design

In a planar SOFC, cell components are configured as flat plates which are connected in electrical series [2, 3]. Figure 8.1 shows an example of typical

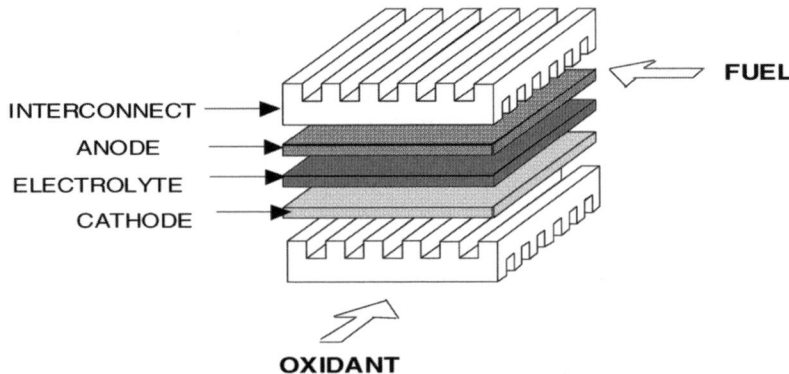

Figure 8.1   Planar SOFC design.

components of a planar SOFC. Advances in ceramic technology, especially in synthesising fine powders, engineering material compositions, tailoring composition/property/microstructure relationships, and fabricating/processing intricate structures, have contributed to the increased interest in planar SOFCs since early 1980s. Significant progress has now been made on the demonstration of fabricability, performance, and operation of planar SOFCs.

A planar SOFC, like any other cell configuration, must be designed to have the desired electrical and electrochemical performance, along with required thermal management and mechanical/structural integrity to meet operating requirements of specified power generation applications [3]. The key requirements are discussed below; this discussion is qualitative because the specific requirements depend on selected designs and intended applications.

(i) Electrical performance. This requirement means that the design must minimise ohmic losses in the stack. Thus, the current path in the components (especially those having low electrical conductivity) must be designed to be as short as possible. There must be good electrical contact and sufficient contact area between the components. The current collector must also be designed to facilitate current distribution and flow in the stack.

(ii) Electrochemical performance. This requirement means that the design must provide for full open circuit voltages and minimal polarisation losses. Thus, any significant gas leakage or cross-leakage and electrical short must be avoided. Fuel and oxidant must be distributed uniformly not only across the area of each cell but also to each cell of the stack. The gases must be able to quickly reach the reaction sites to reduce mass transport limitation.

(iii) Thermal management. This requirement means that the design must provide means for stack cooling and more uniform temperature distribution during operation. The design must permit the highest possible temperature gradient across the stack.

(iv) Mechanical/structural integrity. This requirement means that any planar SOFC stack must be designed to have adequate mechanical strength for assembly and handling. Thus, mechanical and thermal stresses must be kept to minimum to prevent cracking, delamination, or detachment of the components under the variety of operating conditions the stack is expected to experience (e.g., normal operating temperature gradients, off-design temperature gradients, thermal shock conditions such as sudden power change and cold start-up, and mechanical loading expected during installation, moving, and vibration loading).

Table 8.1 summarises the design requirements for planar SOFCs.

Table 8.1 Design requirements

	Property requirement	Design target
Electrical performance	Minimal ohmic loss	Short current path Good electrical contact and sufficient contact area Current collector design for uniform and short current path
Electrochemical performance	Full open circuit voltage	Insignificant gas leakage or cross-leakage (no or minimal sealing) No electrical short
	Low polarisation loss	Uniform gas distribution between cells and across cell Easy gas access to reaction sites
Thermal management	Cooling and uniform temperature distribution Highest possible temperature gradient across stack	Simple and efficient means for cooling Appropriate gas flow configuration Design to withstand thermal stress
Mechanical/structural integrity	Mechanical strength for assembly and handling	Minimal mechanical stress

The most important design feature of the planar SOFC relates to gas flow configuration and gas manifolding which can be arranged in several ways:

(i) Gas flow configurations. Fuel and oxidant flows in planar SOFCs can be arranged to be cross-flow, co-flow, or counter-flow. The selection of a particular flow configuration has significant effects on temperature and current distribution within the stack, depending on the precise stack design. Various flow patterns can be implemented in the different flow configurations including Z-flow, serpentine, radial, and spiral patterns (Figure 8.2). Flowfields (flow channels) are used in planar SOFCs to increase uniformity of gas distribution and to promote heat and mass transport in each cell. In addition, the flowfield is often designed to have sufficient pressure drop through the cell to promote cell-to-cell flow

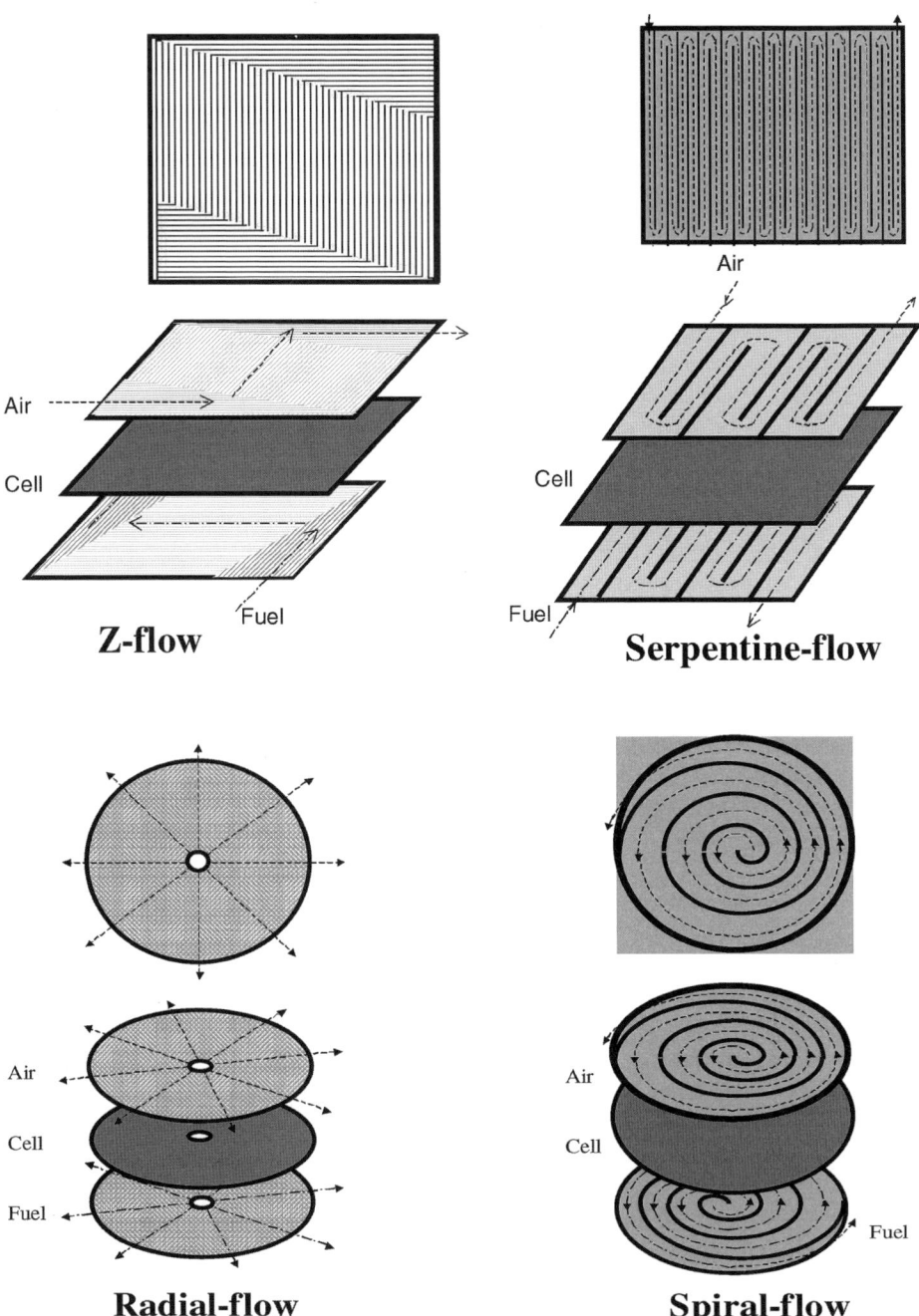

Figure 8.2 Flow patterns in planar SOFCs.

uniformity within the stack. Thus, defining the flowfield for both fuel and oxidant flows is an important aspect in designing planar SOFCs. For a specific design, the shape and arrangement of the flowfield can be varied to improve/optimise stack design. Figure 8.3 shows two examples of flowfield design used in planar SOFCs [3]. Flowfields are commonly designed as part of the interconnect although certain planar designs include the flowfield in the electrodes. Since the flowfield electrically connects the interconnect and the electrodes, contact area (between the flowfield and the electrodes) must be considered in the design to minimise contact resistance losses.

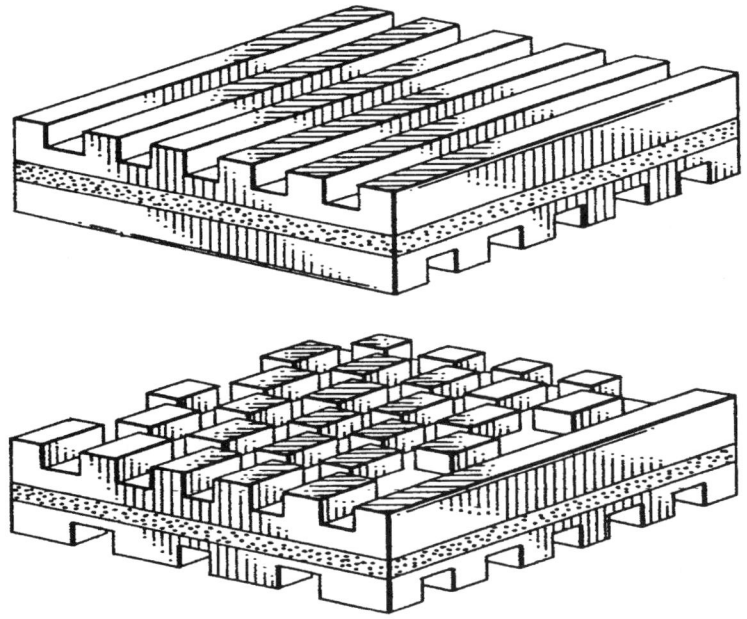

Figure 8.3 Examples of flowfield designs in planar SOFCs [3].

(ii) Gas manifolding. Any stack design must include gas manifolds for routing gases from a common supply point to each cell and removing unreacted gases and reaction products. Gas manifolds can be classified as external or integral. External manifolds are constructed separately from the cell or interconnect component of the stack. Figure 8.4 is an external manifold concept for crossflow planar SOFCs [4]. Integral manifolds are formed and designed as part of the cell or interconnect. Figure 8.5 shows several integral manifold concepts [5–7]. Depending on the design, gas manifolds often require sealing to prevent gas leakage or crossover. The manifold seal is insulating to prevent cell-to-cell electrical shorts. In principle, the manifold must be designed to have low pressure drop (relative to individual cell pressure drop) to provide uniform flow distribution to the stack.

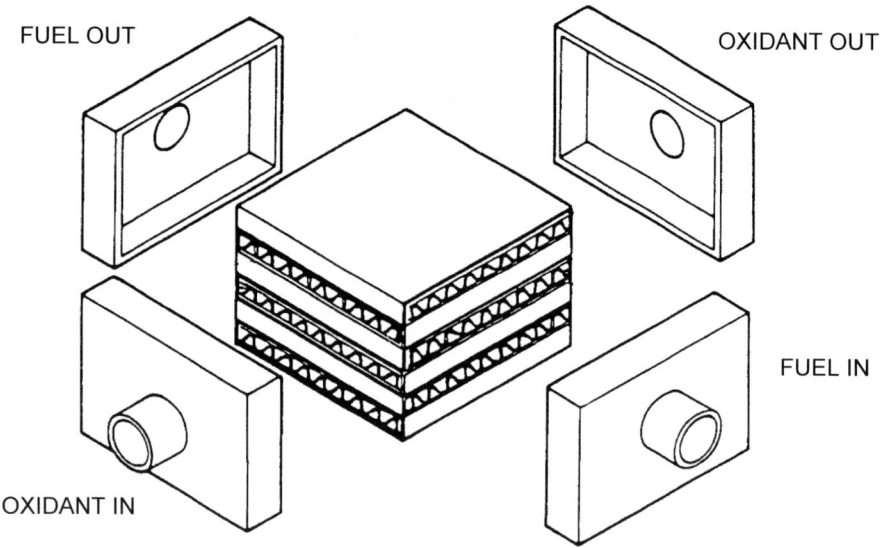

Figure 8.4   Examples of external manifolds.

Planar SOFCs employ the same materials for the single cell as other cell designs. As discussed in this book, the most common cell materials are yttria-stabilised zirconia (YSZ) for the electrolyte, lanthanum strontium manganite (LSM) for the cathode and nickel/zirconia cermet (Ni/YSZ) for the anode. Planar SOFCs can be classified into two broad categories: self-supporting and external supporting. In the self-supporting configuration, one of the cell components (often the thickest layer) acts as the cell structural support. Thus, single cells can be designed as electrolyte supported, anode supported, or cathode supported. In the external-supporting configuration, the single cell is configured as thin layers on the interconnect or a porous substrate. The various cell configurations for planar SOFCs are schematically shown in Figure 8.6. Figure 8.7 shows a micrograph of a cell on a porous metal substrate as an example of the external-supporting cell configuration [8]. The key features of each configuration are summarised in Table 8.2.

For planar SOFCs with YSZ electrolyte as the structural support, the electrolyte is typically thicker than 100 µm, and this thickness requires an operating temperature of about 900–1000°C to minimise electrolyte ohmic losses. For cell configurations with thin (5–20 µm) YSZ electrolytes (e.g., anode-supported cells), the cell can operate at reduced temperatures (< 800°C). The advantages of reduced-temperature operation for the SOFC include a wider choice of materials (especially low-cost metallic materials for the interconnect), longer cell life, reduced thermal stress, improved reliability, and potentially reduced cell cost. The main disadvantages are potential slow electrode reaction kinetics (thus high polarisations) and the reduced thermal energy that can be extracted from the hot exhaust stream by a turbine or a heat exchanger.

Cell and Stack Designs 203

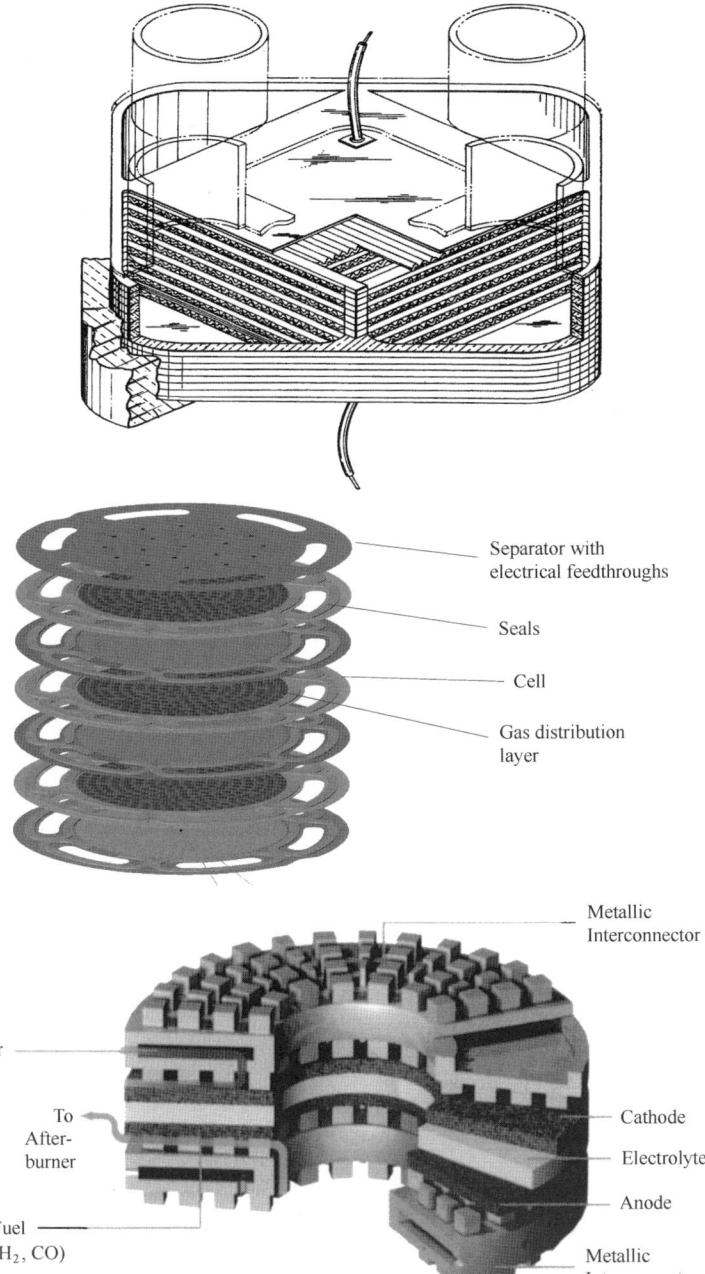

Figure 8.5   Examples of integral manifolds.

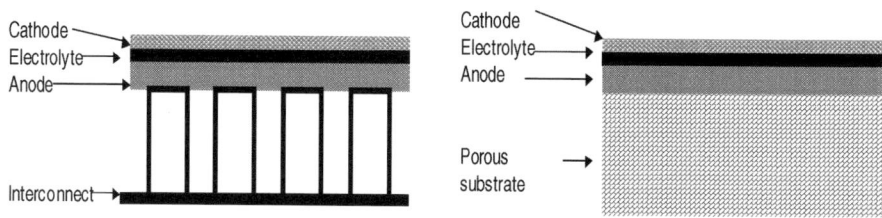

Figure 8.6　Planar cell configurations.

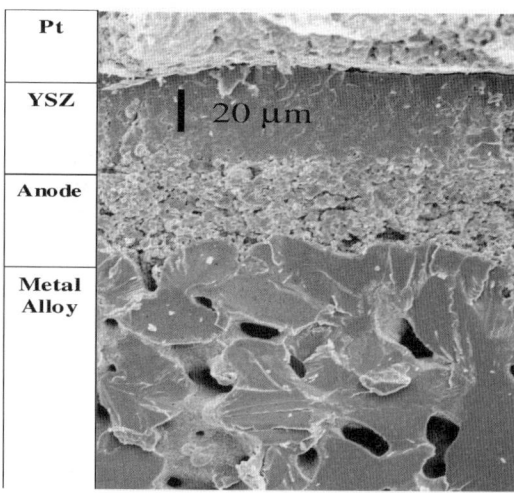

Figure 8.7　Micrograph of a planar cell on porous metal substrate [8].

Table 8.2 Features of planar single-cell configurations

Cell configuration	Advantage	Disadvantage
Self-supporting		
Electrolyte supported	Relatively strong structural support from dense electrolyte Less susceptible to failure due to anode re-oxidation	Higher resistance due to low electrolyte conductivity Higher operating temperatures required to minimise electrolyte ohmic losses
Anode supported	Highly conductive anode Lower operating temperature via use of thin electrolytes	Potential anode re-oxidation Mass transport limitation due to thick anodes
Cathode supported	No oxidation issues Lower operating temperature via use of thin electrolyte	Lower conductivity Mass transport limitation due to thick cathodes
External supporting		
Interconnect supported	Thin cell components for lower operating temperature Stronger structures from metallic interconnects	Interconnect oxidation Flowfield design limitation due to cell support requirement
Porous substrate	Thin cell components for lower operating temperature Potential for use of non-cell material for support to improve properties	Increased complexity due to addition of new materials Potential electrical shorts with porous metallic substrate due to uneven surface

8.2.1 Cell Fabrication

The fabrication processes selected for each planar SOFC cell/stack design depend on the configuration of the cells in the stack. The key step in any selected process is the fabrication of dense electrolytes. In general, ceramic fabrication processes for planar SOFCs can be classified into two groups, based on the fabrication approach for the electrolyte: the particulate approach and the deposition approach. The particulate approach involves compaction of ceramic powder into cell components and densification at elevated temperatures. Examples of the particulate approach are tape casting and tape calendering. The deposition approach involves formation of cell components on a support by a chemical or physical process. Examples of the deposition approach are chemical vapour deposition, plasma spraying, and spray pyrolysis.

8.2.1.1 Cell Fabrication Based on Particulate Approach

At present, two main particulate processes have been developed for the fabrication of planar SOFCs: tape casting [9] and tape calendering [10]. Both of these processes have been shown to be capable of making cells with electrolyte layers of various thicknesses including thin YSZ electrolytes on electrode supports.

Tape casting. Tape casting is a common method for manufacturing thin, flat sheets of ceramics and has been used to fabricate various components for planar SOFCs. The tape casting process involves making of a layer of slip (ceramic powder suspended in a liquid) using a doctor blade and drying this layer (or tape) on a temporary support. The dried layer (green tape) can be stripped from the support and fired to form a ceramic layer. Multilayer tapes can be fabricated by sequentially casting one layer on top of another.

Tape calendering. Calendering is the formation of a continuous sheet of tape of controlled size by squeezing of a softened thermoplastic material between two rolls. In the calendering of ceramic tapes, ceramic powder and organic binder are mixed in a high shear mixer to form a plastic mass. The mass is then rolled into a tape of desired thickness and the final tape is fired at elevated temperatures. To form multilayer tapes, individual layers are laminated in a second rolling operation. Figure 8.8 shows, as an example, a micrograph of a single cell with a 3 μm YSZ fabricated by the tape calendering process.

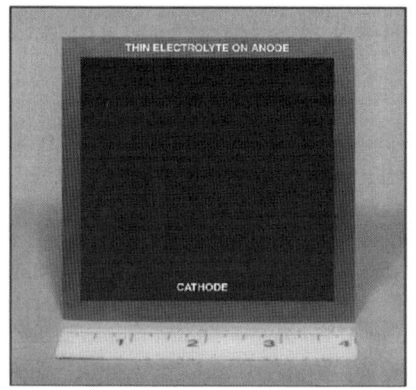
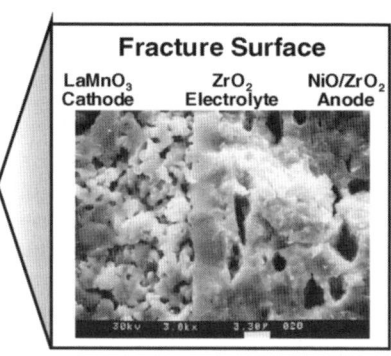

Figure 8.8   Anode-supported cell fabricated by tape calendering.

Other particulate processes such as pressing and extrusion have also been considered or developed for fabricating planar SOFC cell components.

8.2.1.2 Cell Fabrication Based on Deposition Approach

A wide range of deposition techniques have been used for fabricating planar SOFCs. Focus has been placed on developing methods for making thin (5–20 μm) YSZ electrolytes for reduced temperature operation. Selected deposition processes are described below.

Sputtering. An electrical discharge in argon/oxygen mixtures is used to deposit YSZ films (from metal targets) on substrate [11].

Dip coating. Porous substrates are immersed in YSZ slurries of colloidal size particles. Deposited films are then dried and fired [12].

Spin coating. YSZ films are produced on a dense or porous substrate by spin coating a sol–gel precursor followed by heat treatment at relatively low temperatures [13].

Spray pyrolysis. A solution consisting of powder precursor and/or particles of the final composition is sprayed onto a hot substrate followed by a sintering step to densify the deposited layer [14].

Electrophoretic deposition. YSZ particles are deposited from a suspension onto an electrode of opposite charge upon application of a DC electrical field. The deposited layer is then fired at elevated temperatures [15].

Slip casting. YSZ layers are deposited on a porous substrate by vacuum slurry coating. After deposition followed by drying, the layer is sintered at high temperatures [16].

Plasma spraying. Powders injected into a plasma jet are accelerated, melted, and deposited on the substrate [17].

Other processes investigated for planar SOFC fabrication include electrostatic-assisted vapour deposition, vapour phase electrolytic deposition, vacuum evaporation, laser spraying, transfer printing, sedimentation method, and plasma metal organic chemical vapour deposition.

As discussed in Chapter 7, the interconnect for planar SOFCs is either ceramic or metallic depending on the cell operating temperature. Ceramic interconnects are commonly used at 900–1000°C while metallic interconnects at <800°C. The most common material for ceramic interconnects is doped lanthanum chromite ($LaCrO_3$). Lanthanum chromite interconnects for planar SOFCs are often made by conventional ceramic processing methods such as pressing or tape casting followed by sintering. Flowfields are either embossed into the interconnect before firing or machined into the sintered interconnect. Lanthanum chromite is known to be difficult to densify under high oxygen activity environments; therefore, the material used for making the interconnect is generally tailored to improve its sinterability under required conditions, especially under oxidising atmospheres [3]. The most common metallic materials are chromium-based alloys and ferritic stainless steels. These materials are considered for planar SOFC interconnects because their coefficients of thermal expansion closely match those of cell components. Alloys with chromium oxide scale formation are often preferred for interconnect applications (as compared to those with alumina scales) due to the higher conductivity of the chromia scale formed on the surface of the alloy. The key technological issue with chromium-containing metallic interconnects relates to migration of chromium species into the cell, causing cell performance degradation during SOFC operation [18]. The use of metallic materials permits a variety of conventional forming methods for manufacture of the interconnect. Flowfields can be formed on metallic interconnects by machining or stamping.

The majority of planar SOFC stacks require sealing to prevent gas leakage or cross-leakage. In general, when a planar SOFC is designed, one emphasis is to minimise sealing and sealing surfaces because the seal requirements are very stringent. Two types of sealing methods have been used: compressive loads (with or without gaskets) and high-temperature sealants. Compressive seals involve use of mechanical loads to compress fuel cell components to form a seal. This type of sealing has the advantage of requiring no sealants; however, forming a gas-tight seal and minimising mechanical stress due to compression of uneven surfaces are the key issues. Gaskets can be used to improve gas tightness and provide cushion for surface unevenness. High-temperature sealants include cements, glasses, and glass–ceramics. A sealant selected for planar SOFCs must have the stability in oxidising and reducing environments, chemical compatibility with cell/stack components, and proper sealing and insulating properties. Examples of glass and glass ceramic sealants being developed for planar SOFCs are families of modified borosilicate and aluminosilicate glasses [19].

8.2.2 Cell and Stack Performance

Planar SOFCs of various sizes have been fabricated and operated under various conditions. Single cells have been shown to have extraordinarily high areal power densities. For example, power densities of up to 1.8 W/cm^2 at 800°C and 0.8 W/cm^2 at 650°C have been obtained for anode-supported planar cells with hydrogen fuel and air oxidant (Figure 8.9) [20].

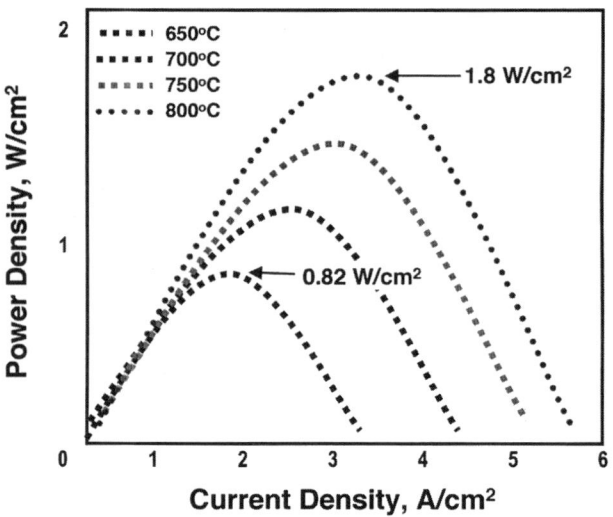

Figure 8.9   Performance of anode-supported cell with hydrogen fuel and air oxidant [20].

Figure 8.10 shows examples of planar SOFC stacks [6,21–23]. Planar stacks have been operated for thousands of hours, and operation of multi-cell stacks at multi-kW power levels has been demonstrated. An example of performance of a planar SOFC stack is given in Figure 8.11 [24].

The planar cell design offers high power density but currently has a number of significant issues such as requiring high-temperature gas seals at the edges of the cell components to isolate oxidant from fuel. Difficulties in successfully developing such high-temperature seals have slowed the development and use of planar design cells for SOFC generators. However, SOFC stacks in 1–25 kW size utilising planar cells are now beginning to be designed, fabricated, and electrically tested. Also, power systems (up to several kilowatt size) based on planar SOFCs have been assembled and tested, as described in Chapter 13.

Cell and Stack Designs 209

Figure 8.10   Examples of planar SOFC stacks.

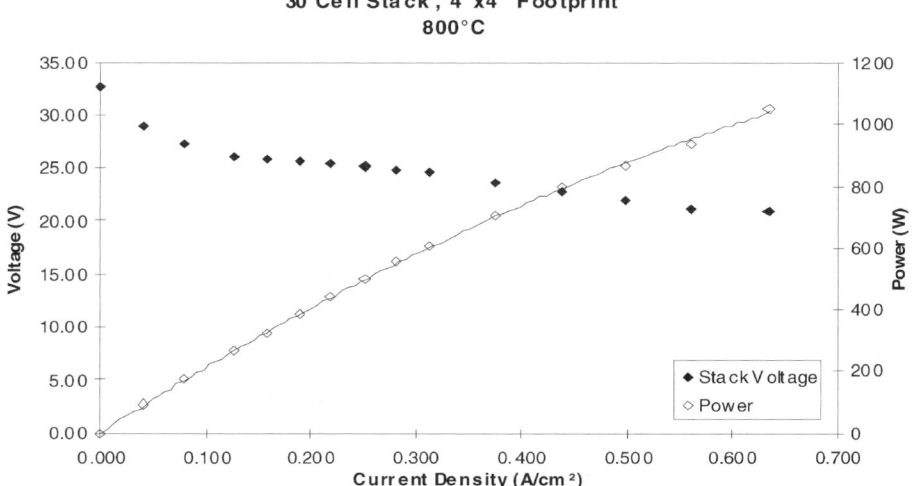

Figure 8.11   Performance of a planar SOFC stack [24].

8.3 Tubular SOFC Design

Two general types of tubular cells are currently being pursued, cells with a large diameter (>15 mm), and microtubular cells with a very small diameter (<5 mm); the microtubular cells are discussed in Section 8.4.

In the most common tubular design, pioneered by Westinghouse Electric Corporation (now Siemens Westinghouse Power Corporation), the cell components are deposited in the form of thin layers on a cylindrical tube [25]. In the early designs, this tube was made of calcia-stabilised zirconia; this porous support tube (PST) acted both as a structural member onto which the active cell components were fabricated and as a functional member to allow the passage of air to the cathode during cell operation. This porous support tube was fabricated by extrusion followed by sintering at an elevated temperature. Although sufficiently porous, this tube presented an inherent impedance to air flow toward the cathode. In order to reduce such impedance to air flow, the wall thickness of the porous support tube was first decreased from 2 mm (thick-wall PST) to 1.2 mm (thin-wall PST), and then the porous support tube was completely eliminated and replaced by a doped $LaMnO_3$ tube (air electrode-supported cell); this tube serves as the cathode onto which the other cell components are deposited. The voltage–current characteristics of these three variations of tubular cells, of similar dimensions, are compared in Figure 8.12, clearly illustrating the significantly improved performance of the design with no porous support tube [25].

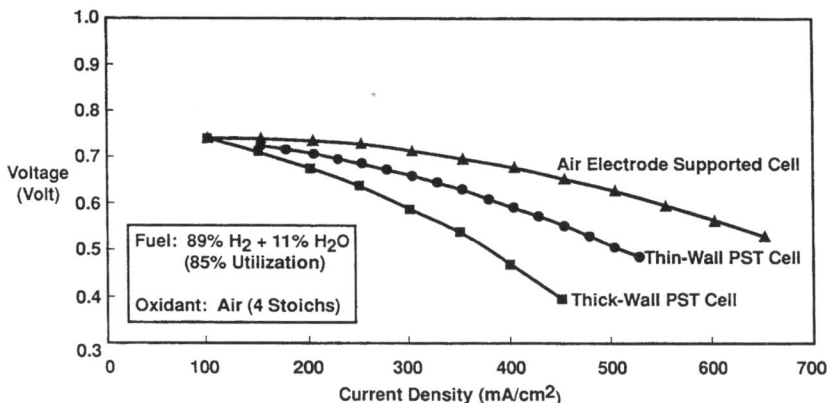

Figure 8.12   Comparison of the voltage–current characteristics of the thick-wall PST, the thin-wall PST, and the air electrode-supported tubular cells at 1000°C [25].

In addition to eliminating the porous support tube, the active length of the cells was continually increased to increase the power output per cell; a greater cell power output decreases the number of cells required in a given power size generator and thus improves power plant economics. The active length (the length of the interconnection) was increased from 30 cm for pre-1986 thick-wall PST cells to 150 cm for today's commercial prototype air electrode-supported cells. Additionally, the diameter of the cells has been increased from 1.6 cm to 2.2 cm to accommodate larger pressure drops encountered in longer length cells.

Figure 8.13 schematically illustrates the design of the latest Siemens Westinghouse tubular cell [26], and Figure 8.14 shows a photograph of an actual cell. The lanthanum manganite-based air electrode tube (2.2 cm diameter, 2.2 mm wall thickness, about 180 cm length) is fabricated by extrusion followed by sintering to obtain about 30–35% porosity. Electrolyte, zirconia doped with about 10 mol% yttria (YSZ), is deposited in the form of about 40 μm thick layer by an electrochemical vapour deposition process (EVD) [27, 28]. In this process, chlorides of zirconium and yttrium are volatilised in a predetermined ratio and passed along with hydrogen and argon over the outer surface of the porous air electrode tube. Oxygen mixed with steam is passed inside the cathode tube. In the first stage of the reaction, molecular diffusion of oxygen, steam, metal chlorides, and hydrogen occurs through the porous cathode and these react to fill the pores in the cathode with the yttria-stabilised zirconia according to the following reactions:

$$2MeCl_y + yH_2O = 2MeO_{y/2} + 2yHCl \tag{1}$$

$$4MeCl_y + yO_2 + 2yH_2 = 4MeO_{y/2} + 4yHCl \tag{2}$$

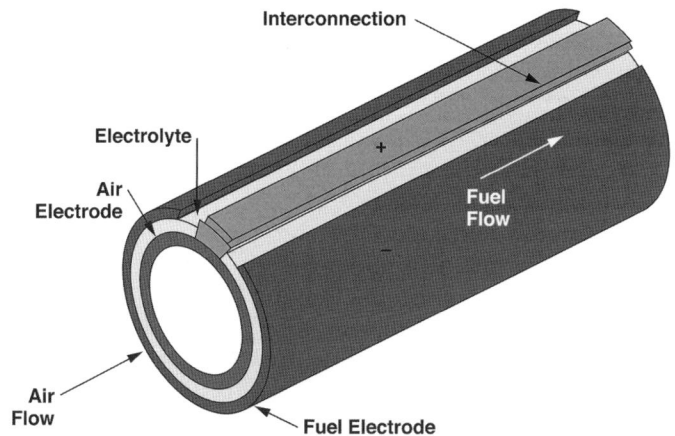

Figure 8.13 Schematic illustration of a Siemens Westinghouse tubular SOFC [26].

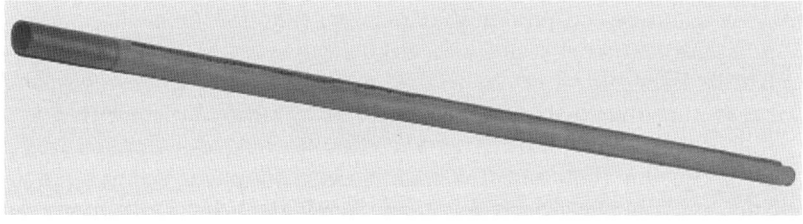

Figure 8.14 Photograph of a single tubular SOFC. (Courtesy of Siemens Westinghouse Power Corporation.)

where Me is the cation species (zirconium and yttrium); and y is the valency associated with the cation. The temperature, the pressure and the different gas flow rates are so chosen that the above reactions are thermodynamically and kinetically favoured.

During the second stage of the reaction after the pores in the air electrode are closed, electrochemical transport of oxide ions maintaining electroneutrality occurs through the already deposited yttria-stabilised zirconia in the pores from the high oxygen partial pressure side (oxygen/steam) to the low oxygen partial pressure side (chlorides). The oxide ions, upon reaching the low oxygen partial pressure side, react with the metal chlorides and the electrolyte film grows in thickness. The flows of the metal chloride vapours are maintained above a critical level to eliminate any gas-phase control of the EVD reaction. Furthermore, the ratio of yttrium chloride to zirconium chloride is so chosen that the electrolyte deposited contains about 10 mol% yttria.

The growth of the electrolyte film is parabolic with time and occurs by the oxide ions diffusing through yttria-stabilised zirconia from the oxygen/steam side to the chlorides side. The rate controlling step in this process is the electronic transport (diffusion of electrons) through the electrolyte film. The electrochemical vapour deposition process ensures the formation of a pore-free, gas-tight, uniformly thick layer of the electrolyte over porous air electrode. A representative micrograph of the electrolyte layer over porous air electrode is shown in Figure 8.15.

Figure 8.15 Representative micrograph of the electrochemically vapour deposited YSZ electrolyte over a porous air electrode.

The EVD technique to deposit the electrolyte is complex, capital-cost intensive, and requires vacuum equipment that makes scaling it up to a cost-effective, continuous manufacturing process for high volume SOFC production difficult if not impossible. Fabrication of the YSZ electrolyte films by a more cost-effective non-EVD technique such as plasma spraying followed by sintering, is being investigated to reduce cell manufacturing cost.

The Ni/YSZ anode, 100–150 μm thick, is deposited over the electrolyte by a two-step process. In the first step, nickel powder slurry is applied over the electrolyte. In the second step, YSZ is grown around the nickel particles by the same EVD process as used for depositing the electrolyte. Deposition of a

Ni/YSZ slurry over the electrolyte followed by sintering has also yielded anodes that are equivalent in performance to those fabricated by the EVD process. Deposition of the anode by a thermal spraying method is also being investigated. Use of these non-EVD processes should result in a substantial reduction in the cost of manufacturing SOFCs.

Doped lanthanum chromite interconnection is deposited in the form of about 85 µm thick, 9 mm wide strip along the air electrode tube length by plasma spraying followed by densification sintering [29].

8.3.1 Cell Operation and Performance

The cell tube is closed at one end. For cell operation, oxidant (air or oxygen) is introduced through an alumina injector tube positioned inside the cell. The oxidant is discharged near the closed end of the cell and flows through the annular space formed by the cell and the coaxial injector tube. Fuel flows on the outside of the cell from the closed end and is electrochemically oxidised while flowing to the open end of the cell generating electricity. At the open end of the cell, the oxygen-depleted air exits the cell and is combusted with the partially depleted fuel. Typically, 50–90% of the fuel is utilised in the electrochemical cell reaction. Part of the depleted fuel is recirculated in the fuel stream and the rest combusted to preheat incoming air and/or fuel. The exhaust gas from the fuel cell is at 600-900°C depending on the operating conditions.

A large number of tubular cells have been electrically tested over the years, some for times as long as 8 years. These cells perform satisfactorily for extended periods of time under a variety of operating conditions with less than 0.1% per 1000 h performance degradation. The voltage–current and power–current characteristics of a commercial prototypic 2.2 cm diameter, 150 cm active length cell at 900, 940, and 1000°C with 89% H_2 + 11% H_2O fuel (85% fuel utilisation) and air as oxidant (4 stoichs) are shown in Figure 8.16.

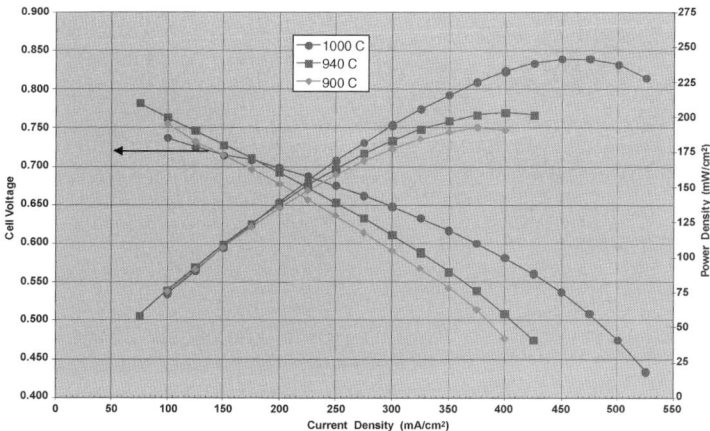

Figure 8.16   Voltage–current density and power–current density plots of a commercial prototypical tubular SOFC. (Courtesy of Siemens Westinghouse Power Corporation.)

The tubular SOFCs have also shown the ability to be thermally cycled to room temperature from 1000°C over 100 times without any mechanical damage or electrical performance loss. This ability to sustain thermal cycles is essential for any SOFC generator to be commercially viable.

The tubular SOFCs have also been tested at pressures up to 15 atm on hydrogen and natural gas fuels [26]. Figure 8.17 shows the effect of pressure on cell power output for a 2.2 cm diameter, 150 cm active length cell at 1000°C. Operation at elevated pressures yields a higher cell power at any current density due to increased Nernst potential and reduced cathode polarisation, and thereby permits higher stack efficiency and greater power output. With pressurised operation, SOFCs can be successfully used as replacements for combustors in gas turbines for SOFC/turbine hybrid systems as discussed in Chapter 13.

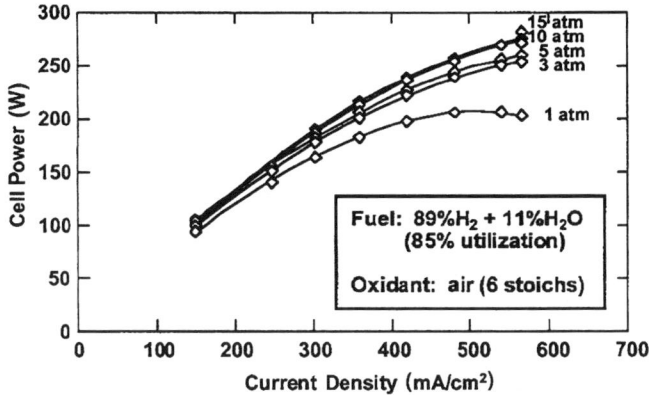

Figure 8.17 Effect of pressure on cell power at 1000°C [26].

8.3.2 Tubular Cell Stack

To construct an electric generator, individual cells are connected in both electrical parallel and series to form a semi-rigid bundle that becomes the basic building block of a generator [26]. Nickel felt, consisting of long nickel metal fibres sinter bonded to each other, is used to provide soft, mechanically compliant, low electrical resistance connections between the cells. This material bonds to the nickel particles in the fuel electrode and the nickel plating on the interconnection for the series connection, and to the two adjacent cell fuel electrodes for the parallel connection; such a series–parallel arrangement provides improved generator reliability. A three-in-parallel by eight-in-series cell bundle is shown in Figure 8.18. The individual cell bundles are arrayed in series to build voltage and form generator modules. A photograph of the cell stack for a 100 kW atmospheric power system, described in detail in Chapter 13, is shown in Figure 8.19; it consists of 48 cell bundles of 24 cells each, which are arranged in 12 rows. The cell rows are interconnected in serpentine fashion in electrical series. Between each cell row is an in-stack radiantly heated reformer. The thermal and hydraulic features of this stack are shown in Figure 8.20 [26].

Cell and Stack Designs   215

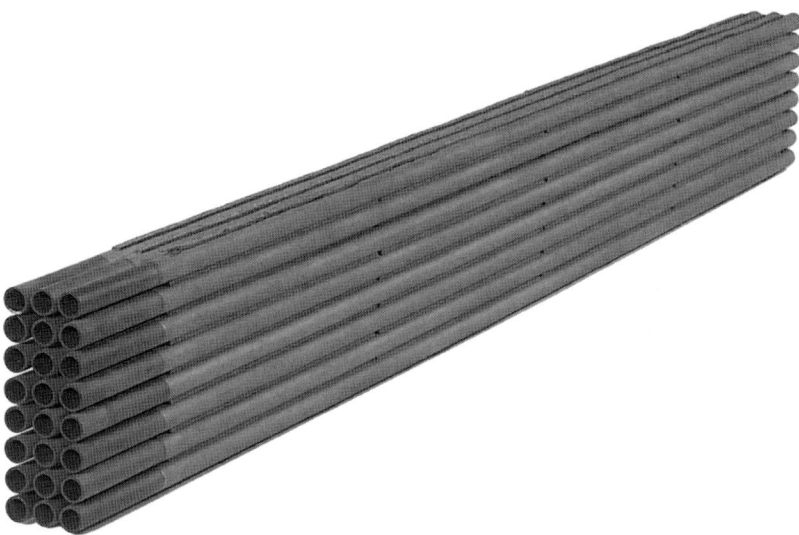

Figure 8.18   Three-in-parallel by eight-in-series tubular cell bundle [26].

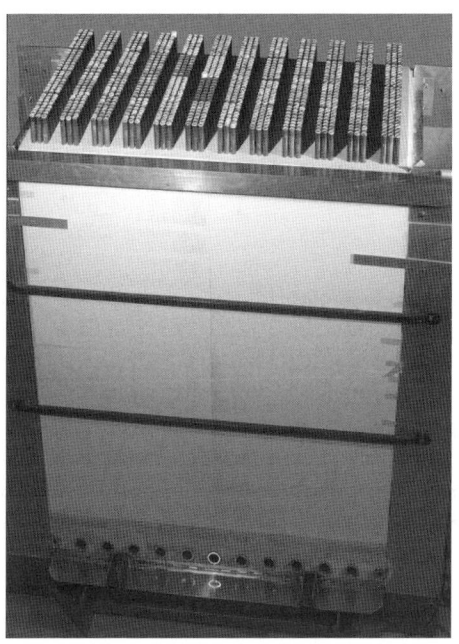

Figure 8.19   Photograph of a 100 kW tubular cell stack, showing 48 bundles of 24 cells each. (Courtesy of Siemens Westinghouse.)

The stack is partitioned in elevation by porous baffles forming a fuel distribution plenum, an active cell zone, a spent fuel plenum, a combustion zone, and an air plenum. An ejector using pressurised desulphurised natural gas as the primary fluid is used to extract a portion of the spent fuel and mix it with fresh fuel before the mixture is introduced into an adiabatic pre-reformer where the higher hydrocarbons are reformed. From the pre-reformer, the predominantly methane stream is routed to the top of the in-stack reformers. The mixture flows downward through catalyst material before exiting within the fuel plenum at the bottom of the stack. The completely reformed fuel flows upward within the stack along the exterior of cells where it is electrochemically oxidised. The stack exhaust gas departs at the combustion zone temperature, approximately 850°C. The stack is cooled with process air which enters the stack at approximately 600°C. The thermally and hydraulically integrated reformer requires no external source of water during normal operation.

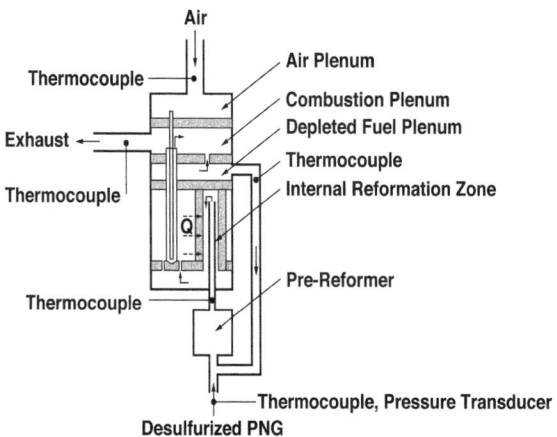

Figure 8.20   Thermal and hydraulic features of the 100 kW SOFC stack [26].

8.3.3 Alternative Tubular Cell Designs

Tubular cells, identical in design to that of Siemens Westinghouse cells, are also being developed by Toto Ltd of Japan; they use a 'wet slurry dip/sintering' method for depositing cell components on the cathode tube to reduce the manufacturing cost of the cells [30–34]. However, the performance and performance stability with time of these cells still need improvement.

Another tubular design, the so-called segmented cell-in-series design [2, 3], is being pursued by Mitsubishi Heavy Industries in Japan. In Europe, ABB and Rolls Royce Fuel Cells have been developing this system over the past 20 years. This design, shown schematically in Figure 8.21, consists of segmented cells connected in electrical and gas flow series. The cells are arranged as thin banded structure on a porous support tube, typically aluminate. The interconnection provides sealing (and electrical contact) between the anode of one cell and the cathode of the next. The fuel flows from one cell to the next inside the tubular stack of cells and the oxidant flows on the outside. The active cell components are currently deposited by plasma spraying. A photograph of such a cell stack is shown in Figure 8.22. Up to 10 kW size stacks have been built and tested,

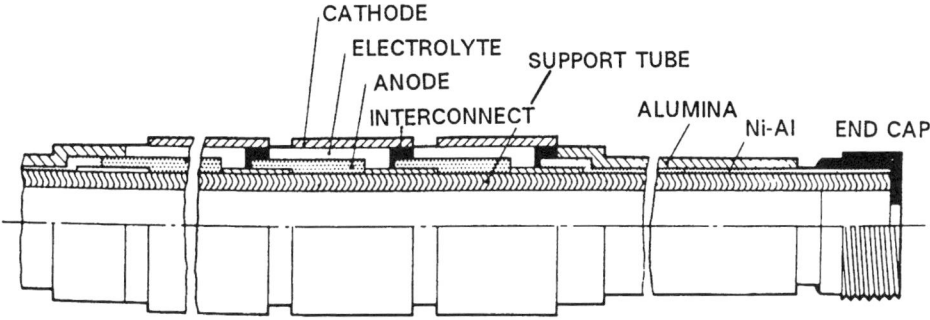

Figure 8.21   Schematic illustration of the segmented-in-series tubular cell design [3].

Figure 8.22   Photograph of the segmented-in-series cell stack. (Courtesy of Mitsubishi Heavy Industries.)

both under atmospheric and pressurised conditions, using such segmented-in-series cells [35].

The single biggest advantage of tubular cells over planar cells is that they do not require any high-temperature seals to isolate oxidant from the fuel. However, their areal power density is much lower (about 0.2 W/cm^2) compared to planar cells (from up to 2 W/cm^2 for single cells to at least 0.5 W/cm^2 for stacks) and manufacturing costs higher. The volumetric power density is also lower for tubular cells than for planar cells. For this reason, large-diameter tubular SOFCs are mainly suitable for stationary power generation applications and not very attractive for transportation and military applications.

To increase the power density and reduce the physical size and cost of tubular SOFC generators, alternate geometry cells are under development [36]. Such alternate geometry cells combine all of the advantages of the tubular SOFCs, such as not requiring high temperature seals, while providing higher power per unit length and higher volumetric power density. One new design, referred to as high power density solid oxide fuel cell (HPD-SOFC) or the flattened ribbed cell, has closed ends similar to the tubular design that provide integral air return paths for air to flow the entire length of the cell from the closed to the open end. The HPD-SOFC departs from the tubular design in that it is flattened and incorporates a number of ribs in the air electrode that act as bridges for current flow. Figure 8.23 compares the cross-sections of the tubular and the HPD-SOFCs. The ribs reduce the current path length, which in turn reduces the internal

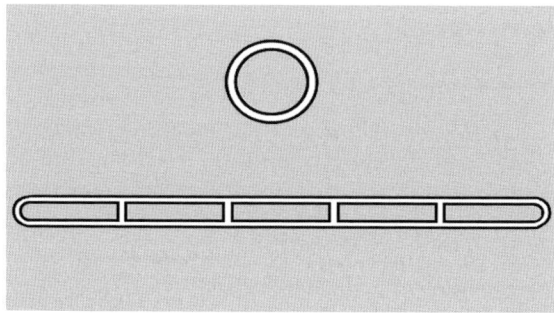

Figure 8.23 Cross-sections of tubular and flattened ribbed HPD cells [37].

resistance of the cell. The presence of the ribs, due to the decreased internal resistance of the cell, also allows use of thinner air electrodes which reduces the air electrode polarisation (a thicker air electrode results in a higher diffusion path for oxygen from the gas phase to the air electrode/electrolyte interface resulting in higher polarisation). The ribs also form air channels that eliminate the need for full-length air injector tubes. A comparison of the theoretically predicted performance of a tubular SOFC and an HPD-SOFC is shown in Figure 8.24 [37]. The higher performance of the HPD-SOFC results from the decreased resistance

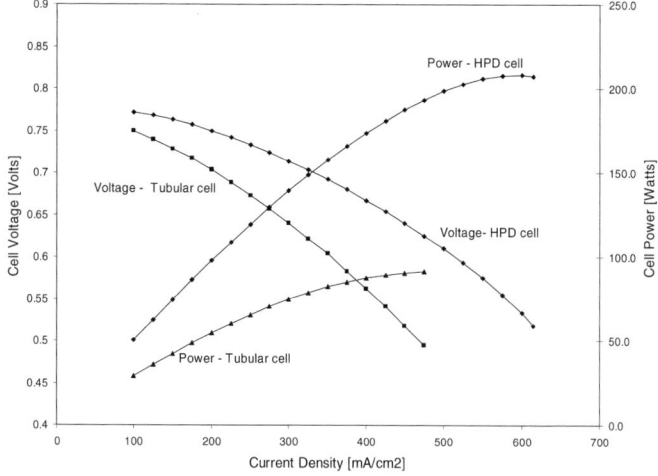

Figure 8.24 Comparison of the theoretical performance for the tubular SOFC and the HPD-SOFC [37].

of the cells compared to the tubular SOFC. When fully optimised with respect to the number of ribs and the resulting performance, such HPD-SOFCs are expected to be initially used in 5 kW residential power systems mentioned in Chapter 13.

8.4 Microtubular SOFC Design

The earliest reports of work on microtubular SOFCs were in the early 1990s when the possibility of extruding thin-walled YSZ electrolyte tubes, 1–5 mm in diameter and between 100 and 200 µm wall thickness, was demonstrated [38], and the ionic conductivity and leak tightness of such electrolyte tubes were found to be good [39, 40]. There are two major benefits of microtubular SOFCs. The first is the increase in volumetric power density when compared with the large-diameter tubular designs discussed in Section 8.3. Power density scales with the reciprocal of tube diameter. Therefore a 2 mm diameter microtubular SOFC could provide ten times more power per stack volume than a 20 mm diameter tubular cell. Another order of magnitude increase could be achieved by going to 0.2 mm diameter tubes, but this is difficult because the connections are then more numerous and problematic to apply. The most significant issue in microtubular cells is applying the electrode and connecting the metal contact inside the bore of a very small-diameter tube.

The second major benefit of the microtubular design is a high thermal shock resistance [41]. Whereas the large-diameter tubular SOFCs are prone to cracking if they are rapidly heated, the microtubular SOFCs do not crack even when heated in a blow torch to their operating temperature of about 850°C in as little as 5 s. This is a marked advantage in applications where start-up time is critical.

A typical design of a microtubular SOFC is shown in Figure 8.25. A YSZ electrolyte tube (typically 2 mm in diameter and about 150 µm wall thickness), is used as a support for the electrodes, as a gas inlet tube, and also as a combustor tube at its outlet. The overall length of the tube is between 100 and 200 mm, whereas the cell region only occupies a length of about 30 mm towards the outlet end of the tube. The Ni + YSZ anode, 30 mm long, is coated on the inner wall of

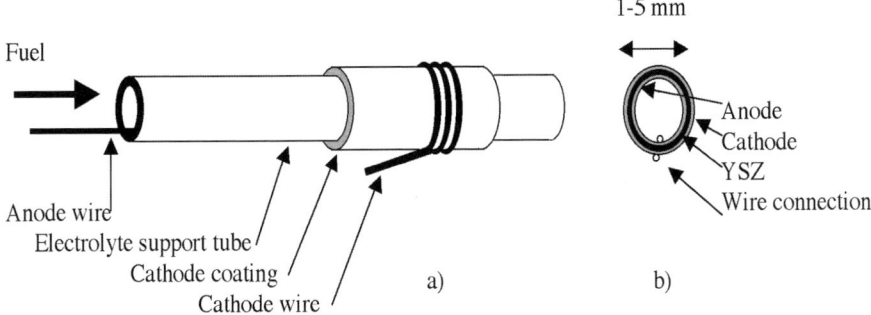

Figure 8.25 Microtubular fuel cell design: (a) arrangement of cell on electrolyte support tube; (b) cross-section of the electroded cell region.

the tube to a thickness of around 50 μm, and a nickel current collector wire is inserted and brought out from the fuel entry side of the tube. On the outside of the electrolyte tube, the lanthanum strontium manganite cathode layer, some 100 μm thick, is deposited, fired, and a silver wire is wound around it to obtain the cathode current. Figure 8.25b shows a cross-section of the cell region on the tube, illustrating the electrolyte support tube, the inner anode and nickel wire, and the outer cathode and its silver wire connector. Cell interconnection is made away from the cells in this design, so that a single interconnect material is not necessary; however, applying the anode and its current collector down inside a narrow tube is not trivial.

Such a design of microtubular SOFC allows small SOFC power generation devices to be configured. For example, a microtubular SOFC can be heated in a burner to provide a small amount of electrical output to drive an electronic device. One such arrangement is shown in Figure 8.26 with a long YSZ electrolyte tube sealed using a rubber connector to a gas inlet pipe, and extending through the thermal insulation into a hot zone at 800°C [41–44]. In this case the long YSZ support tube acts as an inlet pipe to bring fuel to the electroded cell region of the tube. The electroded cell area of the tube extends typically for 30 mm near the tube exit, the anode and cathode wires being brought out for external connection to the electrical load. Upstream of the electroded region, a catalyst layer can be coated onto the YSZ tube for fuel processing, whereas downstream, it is possible to apply a combustion catalyst to aid the reaction of the spent anode gas with the surrounding air. The advantages of this design are rapid start-up, ease of sealing, and integrability into conventional flame systems. Drawbacks are the high in-plane resistance of the cells, the long current leads, and the difficulty of connecting and stacking many small cells together.

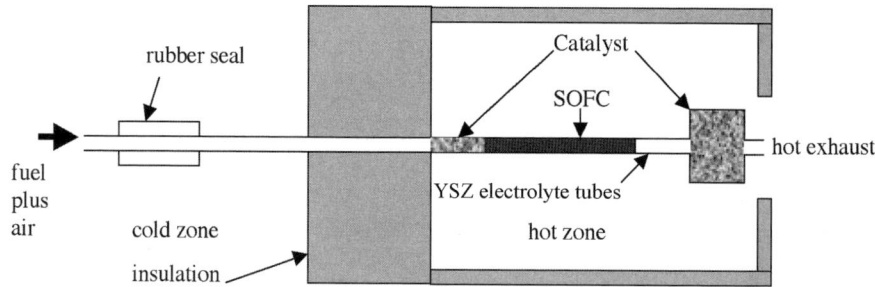

Figure 8.26 Microtubular SOFC system showing the YSZ electrolyte tube sealed with a rubber connector, with the electroded cell region in the hot zone.

This cell design illustrates several inherent features; the ease of sealing to a rubber connector in the cold zone, the high thermal shock resistance which allows the electrolyte tube to go through the thermal insulation into the hot zone, the feasibility for carrying out some fuel processing upstream of the cell region, and the ease of combustion at the exit of the tube. Typically, the gas feed is composed of a mixture of fuel and air, as in a conventional burner. Start-up is then achieved by conventional ignition of the gas/air mixture using a spark or glow plug igniter just downstream of the SOFC. This warms up the combustion catalyst which then heats the cell tube. The ignition does not damage the tubes. Also, temperature cycling can be achieved within minutes in this design.

Since the early work described above, further papers and patents on the microtubular cell design have appeared [44–50] and a number of companies (Acumentrics Corporation, Adelan Ltd) have begun developing microtubular SOFCs. An important factor facilitating the fabrication of microtubular SOFCs is the improvement in the quality of the YSZ electrolyte tubes by the ceramic extrusion process. The problem of making strong ceramics from powders has been known for many years [51–53]. Defects, such as particle aggregates, become trapped in the powder and cause premature failure of the finished ceramic, leading to poor thermal shock resistance. Strengths of ceramic parts made by powder processes are consequently an order of magnitude lower than those made by melt or vapour processes [54, 55]. An additional problem is porosity which can occur in ceramic tubes because of the presence of agglomerates which fail to sinter as the product is fired, often causing gas leakage. Usually, ball milling is used to break the hard agglomerates, producing sub-micron grains; a typical process uses ball milling of the powder in a solvent with a dispersing agent to inhibit re-aggregation [56]. However, in a novel process developed in 1996, a high surface area YSZ was bead milled in water with ammonium polyacrylate surfactant, cellulose polymer was added, filtered at 1 µm to remove any stray aggregates, and then de-watered and dried to produce an extrudable composition; this gave excellent thin-walled extruded tubes of high strength [57]. The more the particle agglomerates are broken down during the powder processing to make the microtubes, the higher the strength and reliability in the final extruded and fired cells. The aggregates are not broken down by simple mixing and need to be broken down by milling or high shear mixing to obtain microtubes with optimum performance [58].

Other possible microtubular cell designs have also been explored, including anode support and through-wall interconnect similar to that used in the large-diameter tubular SOFCs designs. For example, co-extrusion of nickel + YSZ cermet anode with a 30 µm thick YSZ electrolyte demonstrated the possibility of fabricating anode-supported microtubular cells [43]; co-extrusion of anode-supported cells can provide thinner electrolyte (and hence lower ohmic resistance) and better process economy. In an experiment [59], four layers of plastic paste with matched rheology were wrapped together, and then extruded through a tube die to give a wall thickness of 0.3 mm as shown in Figure 8.27. The outer layer was 100 µm thick YSZ electrolyte, and the innermost anode layer was 90% nickel + 10% YSZ, with intermediate anode layers containing 60 and 30% nickel, respectively. This provided improved multilayer anode structure together with thin electrolyte in a single step process. The dried tubes were co-fired at 1400°C for 2 h and this gave a product without substantial microcracking across the layers. An outer cathode of lanthanum strontium manganite was pasted on and fired, and the performance of such cells was

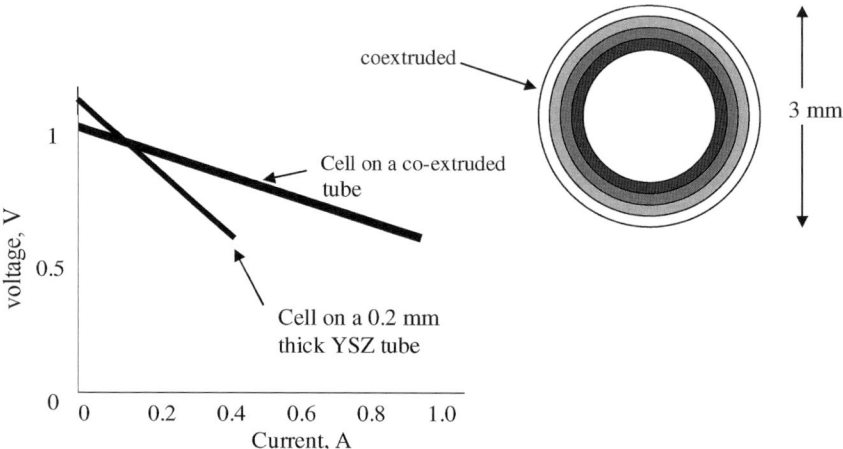

Figure 8.27 (a) Cross-section of a four-layer microtube made in a one-step co-extrusion process [59]. The inner layer is 90% nickel + YSZ and the outer layer is YSZ with the two intermediate anode layers containing 30 and 60% nickel, respectively. This was made into a cell by painting a LSM cathode ink on the outside, then connecting with wires. (b) Improvement in performance of the co-extruded multilayer cell compared with an extruded YSZ electrolyte-supported SOFC.

compared with that of microtubular cells fabricated on an extruded 0.2 mm thick YSZ electrolyte support tube with the anode applied on the inside and cathode on the outside of the tube. The results showed a factor of two improvement in power output on hydrogen fuel at 800°C, even though the open circuit voltage was slightly lower as a result of electrolyte microcracking. The anode-supported microtubular cell design thus appears feasible.

Co-extruding a strip of lanthanum chromite based interconnect along the length of a YSZ microtube has also been demonstrated [45], although a number of difficulties remain. Firstly, the tubes are much weakened by the interconnect strip, and secondly the mixing of lanthanum chromite and YSZ at the boundary of the co-extruded materials leads to a 'dead-zone' of material, about 350 μm in extent. Thus any microtubular cell design with co-extruded interconnect will require much further development to be successful.

8.4.1 Microtubular SOFC Stacks

A number of microtubular SOFC stacks have been built and demonstrated since 1993. An early stack of 20 microtubular cells was built at Keele University, UK, with a control system to introduce the fuel, ignite the gas, bring in air and control the stack temperature [60]. The control system also incorporated shut-down procedures to prevent accidental oxidation of the nickel anodes. Although warm-up was achievable in a couple of minutes, cooling down required about an hour as the heat gradually diffused through the thick ceramic fibre insulation. The same control system was later used to demonstrate a 1000-cell unit built to model a residential combined heat and power (CHP) device [61]. A cross-section of this unit is shown in Figure 8.28. The YSZ electrolyte tubes were arranged as

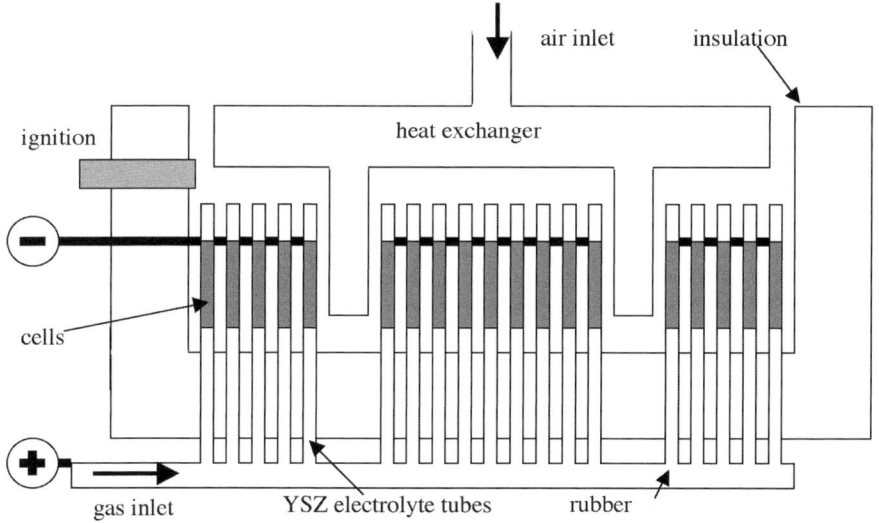

Figure 8.28   Cross-section of a 1000 microtubular cells unit.

racks on a gas inlet manifold which was outside the hot zone. Each YSZ tube was sealed to the gas inlet pipe using silicone rubber. The YSZ tubes then extended through the insulation into the hot zone where the cathode and anode layers were deposited over a 30 mm tube length. When the gas was switched on, it was ignited by a spark or glow plug, to give a flame which heated the incoming air via the heat exchanger. This hot air soon warmed the YSZ tubes to 800°C when the cells began to deliver electrical power which was fed out through metal wires. The starter flame then went out and the cells glowed red-hot as the catalytic oxidation reactions occurred. The heat output was collected using a tubular heat exchanger to provide hot water. The 1000-cell unit could operate on a 2 min cycle. This device was designed to use 20 kW of natural gas for heating, while providing base-load power of around 500 W for a household. The natural gas was premixed with air before entering the tubular cells to prevent coking of the anodes.

A smaller stack comprising 400 microtubular cells was built to power a small vehicle for a student in the Shell mileage marathon of 1996 [62]. This stack was to run on diesel fuel which required significant pre-reforming using platinum supported on ceramic fibre. Hydrogen was also necessary to preheat the reformer and the stack, and so the start-up was relatively sluggish, requiring 30 min. The stack delivered 100 W which was used to drive the vehicle at 30 km per hour around the track.

In 2000, Acumentrics Corp built a 1000-cell stack to illustrate the possibility of providing reliable power for computer systems back up. Since then, Acumentrics has designed and built several 2–5 kW systems using microtubular cells for use as back up power sources for broadband and computer systems.

Microtubular SOFCs have also been effectively used to test the operation of SOFCs on various hydrocarbon fuels. A significant benefit of the microtubular design is the ease of manifolding and introducing fuel without leaks or contamination. A microtubular cell can be made long enough to emerge at both ends of the furnace, so that rubber tubes can be attached in the cold zone to introduce fuel and to analyse the reaction products as shown in Figure 8.29. Fuel may be introduced from a gas cylinder or by means of a bubbler/saturator using a carrier gas such as helium. Other gases such as steam, carbon dioxide or air can be metered in by flow controllers to give the desired composition. The microtubular cell is maintained at a constant temperature in the furnace and oxygen flow through the electrolyte is controlled using a potentiostat. The output stream can be analysed by mass spectroscopy [63], and carbon deposition on the anode can be measured after the test by temperature-programmed oxidation [64, 65].

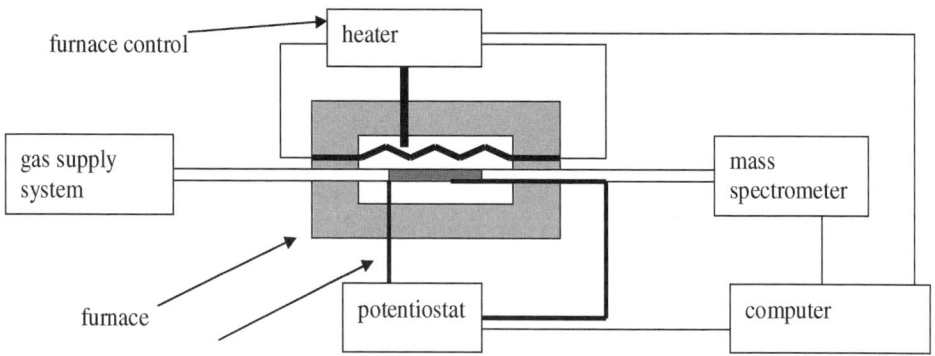

Figure 8.29 Apparatus for measuring fuel reactions in a microtubular cell.

8.5 Summary

This chapter has described the three major cell designs that are dominating research and development at the present time: these are the planar, the large-diameter tubular, and the microtubular designs. Planar SOFCs provide very high areal (W/cm^2) and volumetric (W/cm^3) power densities and can be manufactured by low-cost conventional ceramic processing techniques; however, sealing around the edges of the cells and the control of temperature gradients which can cause cell cracking remain issues to be resolved. Large-diameter tubular SOFCs have been the most successful so far. Their main advantage is the seal-less stack design; the disadvantages are the low power density, the long start-up times, and the expensive fabrication techniques. Microtubular SOFCs are especially useful for smaller systems, providing rapid start-up; the reason for this is the small diameter of the cells and the low wall thickness which prevent the build-up of damaging thermal stresses. Start-up in about a minute is possible and leaks can be prevented by bringing the microtubes through the insulation for sealing in the cold zone. On the negative side, cell interconnection and assembly issues are significant, and it seems likely that microtubular systems will mainly be applicable in small systems.

Various alternative cell designs are under development to avoid the difficulties mentioned above. For example, flattened ribbed cells have been conceived to obtain higher power densities than tubular cells yet require no seals. Another example is the seal-less planar design which is being developed to ease the sealing issues of planar cell stacks.

References

[1] N. Q. Minh, C. R. Horne, F. Liu, P. R. Staszak, T. L. Stillwagon and J. J. Van Ackeren, in *Solid Oxide Fuel Cells I*, ed. S. C. Singhal, The Electrochemical Society Proceedings, Pennington, NJ, PV89-11, 1989, pp. 307–316.
[2] N. Q. Minh, *J. Am. Ceram. Soc.*, **76** (1993) 563.
[3] N. Q. Minh and T. Takahashi, in *Science and Technology of Ceramic Fuel Cells*, Elsevier Science, Amsterdam, The Netherlands, 1995.
[4] N. Q. Minh and C. R. Horne, US Patent 5,162,167, 10 Nov. 1992.
[5] N. Q. Minh, US Patent 5,256,499, 26 Oct. 1993.
[6] B. Godfrey, K. Foger and N. Comer, in *2002 Fuel Cell Seminar Abstracts*, Courtesy Associates, Washington, DC, 2002, p. 964.
[7] H. Raak, R. Diethelm and S. Riggenbach, in *Proceedings of the Fifth European SOFC Forum*, ed. J. Huijsmans, Oberrohrdorf, Switzerland, 2002, p. 425.
[8] S. J. Visco, Lawrence Berkeley National Laboratory, Berkeley, CA, 2003.
[9] R. C. Huiberts and J. P. P. Huijsmans, in *Solid Oxide Fuel Cells VII*, eds. H. Yokokawa and S. C. Singhal, Electrochemical Society, Pennington, NJ, PV2001-16, 2001, p. 155.
[10] N. Q. Minh and C. R. Horne, in *Proceedings of the 14th Risø International Symposium on Materials Science, High Temperature Electrochemical Behaviour of Fast Ion and Mixed Conductors*, eds. F. W. Poulsen, J. J. Benzen, T. Jacobsen, E. Skou and M. J. L. Ostergard, Risø National Laboratory, Roskilde, Denmark, 1993, p. 337.
[11] L. S. Wang and S. A. Barnett, *J. Electrochem. Soc.*, **139** (1992) 1134.
[12] S. De Souza, S. J. Visco and L. C. De Jonghe, *Solid State Ionics*, **98** (1997) 57.
[13] C. C. Chen, M. M. Nasrallah and H. U. Anderson, in *1992 Fuel Cell Seminar Abstracts*, Courtesy Associates, Washington, DC, 1992, p. 515.
[14] L. Lunot and Y. Denos, in *Proceedings of the 1998 International Gas Research Conference*, Gas Research Institute, Chicago, IL 1998, p. 834.
[15] T. Ishihara, T. Kudo, H. Matsuda, Y. Mizuhara and Y. Takita, in *Solid Oxide Fuel Cells III*, eds. S. C. Singhal and H. Iwahara, The Electrochemical Society Proceedings, Pennington, NJ, PV93-4, 1993, p. 65.
[16] L. G. J. de Haart, Th. Hauber, K. Mayer and U. Stimming, in *Proceedings of the Second European SOFC Forum*, ed. B. Thorstensen, Oberrohrdorf, Switzerland, 1996, p. 229.
[17] M. Lang, T. Franco, R. Henne, S. Schaper and G. Schiller, in *Proceedings of the Fourth European SOFC Forum*, ed. A. J. McEvoy, Oberrohrdorf, Switzerland, 2000, p. 231.

[18] W. J. Quadakkers, H. Greiner and W. Kock, in *Proceedings of the First European SOFC Forum*, ed. U. Bossel, Oberrohrdorf, Switzerland, 1994, p. 525.
[19] P. H. Larsen, in *Sealing Materials for Solid Oxide Fuel Cells*, Risø National Laboratory, Roskilde, Denmark, 1999.
[20] J. W. Kim, A. V. Virkar, K. Z. Fung, K. Mehta and S. C. Singhal, *J. Electrochem. Soc.*, **146** (1999) 69.
[21] N. Komada, Paper presented at SOFC Society of Japan Meeting on Present and Future SOFC Technology and Systems, Tokyo, Japan, 24–25 September 2002.
[22] T. Ogiwara, Paper presented at SOFC Society of Japan Meeting on Present and Future SOFC Technology and Systems, Tokyo, Japan, 24–25 September 2002.
[23] D. Ghosh, in *Proceedings of the Fifth European SOFC Forum*, ed. J. Huijsmans, Oberrohrdorf, Switzerland, 2002, p. 453.
[24] N. Minh, A. Anumakonda, B. Chung, R. Doshi, J. Ferrall, J. Guan, G. Leer, K. Montgomery, E. Ong and J. Yamanis, in *Solid Oxide Fuel Cells VI*, eds. S. C. Singhal and M. Dokiya, The Electrochemical Society, Pennington, NJ, PV99-19, 1999, p. 67.
[25] S. C. Singhal, in *Solid Oxide Fuel Cells IV*, eds. M. Dokiya, O. Yamamoto, H. Tagawa and S. C. Singhal, The Electrochemical Society, Pennington, NJ, PV95-1, 1993, pp. 195–207.
[26] S. C. Singhal, *MRS Bulletin*, March 2000, pp. 16–21.
[27] A. O. Isenberg, in *Electrode Materials and Processes for Energy Conversion and Storage*, eds. J. D. E. McIntyre, S. Srinivasan and F. G. Will, The Electrochemical Society Proceedings, Princeton, NJ, PV77-6, 1977, p. 572.
[28] U. B. Pal and S. C. Singhal, *J. Electrochem. Soc.*, **137** (1990) 2937.
[29] L. J. H. Kuo, S. D. Vora and S. C. Singhal, *J. Am. Ceram. Soc.*, **80** (1997) 589.
[30] M. Aizawa, M. Kuroishi, H. Takeuchi, H. Tajiri and T. Nakayama, in *1998 Fuel Cell Seminar Abstracts*, Courtesy Associates, Washington, DC, 1998, pp. 270–274.
[31] T. Nakayama, H. Tajiri, K. Hiwatashi, H. Nishiyama, S. Kojima, M. Aizawa, K. Eguchi and H. Arai, in *Solid Oxide Fuel Cells V*, eds. U. Stimming, S. C. Singhal, H. Tagawa and W. Lehnert, The Electrochemical Society Proceedings, Pennington, NJ, PV97-40, 1997, p. 187.
[32] M. Aizawa, M. Kuroishi, A. Ueno, H. Tajiri, T. Nakayama, K. Eguchi and H. Arai, in *Solid Oxide Fuel Cells V*, eds. U. Stimming, S. C. Singhal, H. Tagawa and W. Lehnert, The Electrochemical Society Proceedings, Pennington, NJ, PV97-40, 1997, p. 330.
[33] M. Kuroishi, S. Furuya, K. Hiwatashi, K. Omoshiki, A. Ueno and M. Aizawa, in *Solid Oxide Fuel Cells VII*, eds. H. Yokokawa and S. C. Singhal, The Electrochemical Society Proceedings, Pennington, NJ, PV2001-16, 2001, p. 88.

[34] H. Takeuchi, A. Ueno, M. Kuroishi, S. Aikawa and T. Abe, in *Solid Oxide Fuel Cells VIII*, eds. S. C. Singhal and M. Dokiya, The Electrochemical Society Proceedings, Pennington, NJ, PV2003-07, 2003, p. 70.
[35] J. Iritani, K. Kougami, N. Komiyama, K. Nagata, K. Ikeda and K. Tomida, in *Solid Oxide Fuel Cells VII*, eds. H. Yokokawa and S. C. Singhal, The Electrochemical Society Proceedings, Pennington, NJ, PV2001-16, 2001, p. 63.
[36] S. C. Singhal, *Solid State Ionics*, **135** (2000) 305–313.
[37] R. A. George, in *Proceedings of the Third Annual Solid State Energy Conversion Alliance (SECA) Workshop*, Washington, DC, 2002.
[38] K. Kendall, in *Proceedings of the International Forum on Fine Ceramics*, Japan Fine Ceramics Center, Nagoya, 1992, pp. 143–148.
[39] M. Kendall, *Tubular cells: a novel SOFC design*, Final Year Project Report, Middlesex University, July 1993.
[40] K. Kendall, in *Solid Oxide Fuel Cells III*, eds. S. C. Singhal and H. Iwahara, The Electrochemical Society Proceedings, Pennington, NJ, PV93-4, 1993, pp. 813–821.
[41] K. Kendall and G. Sales, in *Proceedings of the 2nd International Conference on Ceramics in Energy Applications*, Institute of Energy, London, 1994, pp. 55–63.
[42] W. Price and K. Kendall, in *Proceedings of the 1st European SOFC Forum*, ed. U. Bossel, Switzerland, 1994, pp. 757–766.
[43] K. Kendall and M. Prica, in *Proceedings of the 1st European SOFC Forum*, ed. U. Bossel, Switzerland, 1994, pp. 163–170.
[44] K. Kendall, International Patent Publication No WO 94/22178 (1994).
[45] K. Kendall, E. Wright and A. Golds, in *Solid Oxide Fuel Cells IV*, eds. M. Dokiya, O. Yamamoto, H. Tagawa and S. C. Singhal, The Electrochemical Society Proceedings, Pennington, NJ, PV95-1, 1995, pp. 229–235.
[46] K. Kendall and I. Kilbride, International Patent Publication No WO 97/48144 (1997).
[47] V. Kozhukarov, M. Machkova, M. Ivanova and N. Brashkova, in *Solid Oxide Fuel Cells VII*, The Electrochemical Society Proceedings, Pennington, NJ, PV2001-16, 2001, pp. 244–253.
[48] H. Mobius *et al.*, US Patent 3,377,203 (1968).
[49] P. Sarkar and H. Rho, in *Solid Oxide Fuel Cells VIII*, eds. S. C. Singhal and M. Dokiya, The Electrochemical Society Proceedings, Pennington, NJ, PV2003-07, 2003, pp. 135–138.
[50] C. M. Finnerty, G. A. Tompsett, K. Kendall and R. M. Ormerod, *J. Power Sources*, **86** (2000) 459–463.
[51] W. H. Rhodes, *J. Am. Ceram. Soc.*, **64** (1981) 19.
[52] F. F. Lange, *J. Am. Ceram. Soc.*, **66** (1983) 396.
[53] N. McN. Alford, J. D. Birchall and K. Kendall, *Nature*, **330** (1987) 51.
[54] K. Kendall, *Molecular Adhesion and its Applications*, Kluwer Academic, New York, 2001.
[55] K. Kendall, *Powder Technol.*, **58** (1989) 151.

[56] C. Bagger, in *1992 Fuel Cell Seminar Abstracts*, Courtesy Associates, Washington, DC, 1992.
[57] M. Prica, K. Kendall and S. Markland, *J. Am. Ceram. Soc.*, **81** (1998) 541–548.
[58] K. Kendall, C. M. Finnerty, G. A. Tompsett, P. Windibank and N. Coe, *Electrochemistry*, **68** (2000) 403–406.
[59] Z. Liang, Coextrusion of multilayer tubes, PhD thesis, University of Birmingham (1999), Ch. 8.
[60] T. Longstaff, Masters thesis, Keele University (1996).
[61] M. Prica, T Alston and K Kendall, in *Solid Oxide Fuel Cells V*, eds. U. Stimming, S. C. Singhal, H. Tagawa and W. Lehnert, The Electrochemical Society Proceedings, Pennington, NJ, PV97-40, 1997, p. 619.
[62] K. Kendall, R. Copcutt, M. Palin and I. Kilbride, in *European Fuel Cell Group Newsletter* (1996).
[63] G. J. Saunders and K. Kendall, in *Solid Oxide Fuel Cells VIII*, eds. S. C. Singhal and M. Dokiya, The Electrochemical Society Proceedings, Pennington, NJ, PV2003-07, 2003, pp. 1305–1314.
[64] G. J. Saunders and K. Kendall, *J. Power Sources*, **106** (2002) 258.
[65] K. Kendall, C. M. Finnerty, G. Saunders and J. T. Chung, *J. Power Sources*, **106** (2002) 323.

Chapter 9

Electrode Polarisations

Ellen Ivers-Tiffée and Anil V. Virkar

List of terms

ΔG	Free energy change
p_i	Partial pressure of species i
F	Faraday constant
Φ	Electrostatic potential
N_A	Avogadro's number
j_i	Flux of species i
E	Nernst potential
η	Overpotential
R	Ideal gas constant
R_{ct}	Charge transfer resistance
R_p	Polarisation resistance
ρ_i	Resistivity of the material for the transport of species i
D_{i-j}	Binary diffusivity of $i - j$
$D_{c(eff)}$	Effective cathode gas diffusivity
$D_{a(eff)}$	Effective anode gas diffusivity
V_v	Volume fraction porosity
k_{exc}	Surface exchange coefficient
$Z(\omega)$	Impedance at angular frequency ω
$B(\omega)$	Imaginary part of admittance
$G(\omega)$	Real part of admittance
l_a	Anode thickness
l_c	Cathode thickness
l_e	Electrolyte thickness
i_o	Exchange current density
i_{as}	Anode limiting current density
i_{cs}	Cathode limiting current density
l_{TPB} or l_{tpb}	Three phase boundary length
β	Transfer coefficient of symmetry factor

k_{ads}	Rate constant for adsorption
k_{des}	Rate constant for desorption
k_{red}	Rate constant for reduction
k_{ox}	Rate constant for oxidation

9.1 Introduction

Polarisation is a voltage loss or overpotential, which is a function of current density. It can be broken down into a number of terms, originating in various phenomena that occur when a finite current flows in a cell. The three dominant polarisations are: (a) ohmic polarisation or ohmic loss; (b) concentration polarisation; and (c) activation polarisation. This chapter defines and discusses these polarisations, and describes methods to measure them.

Solid oxide fuel cells (SOFCs) generally operate above 600°C, with the typical operating range being from 800 to 1000°C. High temperature operation makes it possible to use hydrocarbon fuels once they have been processed to form a gaseous mixture of H_2 and CO, with appropriate amounts of H_2O and CO_2 present in the fuel to prevent the deposition of solid carbon. The SOFC thus can use CO as a fuel in addition to H_2. Even in the case of SOFC, however, the largest component of the fuel mixture is H_2. For this reason, and for the sake of simplicity, much of the discussion in this chapter is restricted to hydrogen as the fuel and oxygen (air) as the oxidant.

The overall reaction in an SOFC is the oxidation of H_2 to form H_2O, namely,

$$H_2(\text{gas, anode}) + \tfrac{1}{2}O_2(\text{gas, cathode}) \rightarrow H_2O(\text{gas, anode}).$$

Under open circuit conditions, with electrochemical potential of oxide ions equilibrated across the oxide-ion conducting electrolyte, a voltage difference, E, the Nernst potential, appears between the anode and the cathode. It is related to the net free energy change, ΔG, of the reaction via the following relation [1,2]

$$\Delta G = -nFE = -2FE \tag{1}$$

where n denotes the number of electrons participating in the reaction. The Nernst potential, E, is the open circuit voltage, OCV, and is given in terms of the various partial pressures by [1,2]

$$E = -\frac{\Delta G}{2F} = -\frac{\Delta G^o}{2F} - \frac{RT}{2F}\ln\left(\frac{p^a_{H_2O}}{p^a_{H_2}p^{c\,1/2}_{O_2}}\right) = E^o + \frac{RT}{4F}\ln\left(\frac{p^c_{O_2}p^{a\,2}_{H_2}}{p^{a\,2}_{H_2O}}\right) \tag{2}$$

where $p^c_{O_2}$ is the partial pressure of oxygen in the cathode gas, $p^a_{H_2}$ and $p^a_{H_2O}$ are respectively the partial pressures of H_2 and H_2O in the anode gas, R is the gas constant, F the Faraday constant, and T the absolute temperature.

In what follows, it is assumed that partial pressures of the various species, namely, $p^c_{O_2}$, $p^a_{H_2}$, and $p^a_{H_2O}$, are fixed just outside of the electrodes, regardless of the local current density. Thus, the Nernst voltage, E, is not a function of current density. This assumption is valid only if the flow rates of fuel and oxidant are sufficiently high such that the fuel and oxidant compositions just outside of the anode and cathode, respectively, are virtually fixed. If this is not the case, then the OCV itself must be treated as a function of current density. The dependence of E on current density can be estimated assuming the respective cathodic and anodic chambers as being continuously stirred tank reactors. Figure 9.1 shows a schematic voltage vs. current density polarisation curve of a typical cell with E being a function of current density.

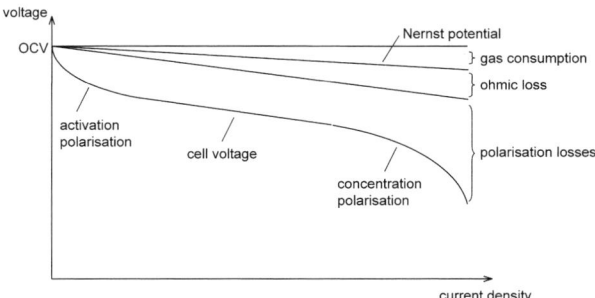

Figure 9.1 Schematic plot of voltage versus current density showing different types of polarisations: activation polarisation is usually dominant at low current densities, and concentration polarisation is dominant at high current densities when the transport of reactive species to the electrolyte/electrode interface becomes a limiting factor for the cell reaction.

Equation (1) also gives the maximum possible electrical work that can be derived, $w_{max} = 2FE = -\Delta G$ [1]. However, the rate at which this work can be realised near equilibrium is essentially zero as the current flowing through the cell at OCV is also zero. When an external load is connected, a finite, non-zero current flows through the circuit, and the process is carried out irreversibly. At any given current density, i, part of the open circuit voltage, E, is reflected as a loss, which appears as the thermal effect. If the voltage across the external load is $V(i)$, and the voltage loss is $\eta(i)$, then

$$E = V(i) + \eta(i) \tag{3}$$

assuming very high flow rates such that E is fixed. If this is not the case E will be a function of current density such that

$$E(i) = V(i) + \eta(i) \tag{4}$$

The difference $E - E(i)$ is a measure of the change in gas phase compositions just outside of the electrodes. This difference must be accounted for in the overall description of cell performance. The voltage loss term $\eta(i)$ is known as the polarisation or overpotential, and is a function of current density; it consists of a number of terms, with their origins related to various phenomena occurring in the cell, under a finite current. The different polarisations are termed: (a) ohmic polarisation, (b) concentration polarisation, and (c) activation polarisation; these are discussed below.

9.2 Ohmic Polarisation[1]

All matters (except superconductors, of course) offer a resistance to the motion of electrical charge, and this behavior, in the simplest case, can be described by Ohm's law. The assumed linear behavior between voltage drop and current density can be described by resistivity, a material property. Transport of oxide ions through the electrolyte is thus governed by the ionic resistivity of the electrolyte. Similarly, transport of electrons (or electron holes) through the electrodes (the cathode and the anode) is governed by their respective electronic resistivities (corrected for porosity and the possible existence of secondary, insulating phases). Because of these ohmic resistances, at a given current density, there is a voltage loss, η_{ohm}, given by

$$\eta_{ohm} = (\rho_e l_e + \rho_c l_c + \rho_a l_a + R_{contact})i \qquad (5)$$

where ρ_e, ρ_c, and ρ_a, are respectively electrolyte, cathode, and anode resistivities, and l_e, l_c, and l_a, are respectively electrolyte, cathode, and anode thicknesses, and $R_{contact}$ is any possible contact resistance. The ohmic polarisation can be described using an equivalent circuit comprising a simple resistor with a zero capacitance in parallel. For this reason, its response time is essentially zero, i.e. it's instantaneous. In reality, however, the response time is not zero but very, very small. Fast response allows its determination using current interruption.

In most SOFCs, the main contribution to η_{ohm} is from the electrolyte, since its (e.g. yttria-stabilised zirconia, YSZ) ionic resistivity is much greater than electronic resistivities of the cathode (e.g. Sr-doped LaMnO$_3$, LSM), and the anode (e.g. Ni + YSZ cermet). For example, the ionic resistivity of YSZ at 800°C is $\sim$50 Ωcm. By contrast, electronic resistivity of LSM is $\sim 10^{-2}$ Ωcm and that of the Ni + YSZ cermet is on the order of 10^{-4} Ωcm. Thus, the electrolyte contribution to ohmic polarisation can be large, especially in thick electrolyte-supported cells. The recent move towards electrode-supported cells, in which electrolyte is a thin film of 5 to 30 microns, reduces the ohmic polarisation. Also, the use of higher conductivity electrolyte materials such as doped ceria and lanthanum gallate lowers the ohmic polarisation.

Most of the discussion in this chapter is centered on cells made with traditional materials such as YSZ electrolyte, Ni + YSZ anode, and LSM + YSZ cathode; although its extension to other materials is essentially straightforward. The relative contributions of various polarisations vary widely among the different cell designs; anode-supported, cathode-supported, and electrolyte-supported. Ohmic contribution is the smallest in electrode-supported cells due to the thin

---

[1] The term 'ohmic polarisation' is often referred to as the 'ohmic loss', and is part of the overall loss, $\eta(i)$. As such, here it is referred to as ohmic polarisation, although both terminologies are in general use.

electrolyte (typically ~10 microns), and highest in electrolyte-supported cells. Anode-supported cells thus exhibit higher performance.

9.3 Concentration Polarisation

In fuel cells, the reacting species are gaseous; at the anode H_2 (or $H_2 + CO$), and at the cathode O_2. At the anode, H_2 (or $H_2 + CO$) must be transported from the fuel stream, through the porous anode, to (or near) the anode/electrolyte interface. Hydrogen (or $H_2 + CO$) then reacts with oxide ions transported through the electrolyte, at or near the anode/electrolyte interface, to form H_2O (or $H_2O + CO_2$), and release electrons to the anode, for their subsequent transport to the cathode, through the external circuit. The H_2O (or $H_2O + CO_2$) formed must be transported away from the electrolyte/anode interface, through the porous anode, to the fuel stream. This transport of H_2 ($H_2 + CO$) and H_2O ($H_2O + CO_2$) must be consistent with the net current flowing through the cell, adjusted for appropriate charge balance/mass balance parameters. In steady state, the following equality

$$|j_{H_2}| + |j_{CO}| = |j_{H_2O}| + |j_{CO_2}| = 2|j_{O_2}| = \frac{iN_A}{2F} \tag{6}$$

must be obeyed, where j_{H_2} and j_{CO} are respectively the fluxes of hydrogen and carbon monoxide through the porous anode to the anode/electrolyte interface, j_{H_2O} and j_{CO_2} are respectively the fluxes of water vapor and carbon dioxide through the porous anode, away from the anode/electrolyte interface, j_{O_2} is the flux of oxygen through the porous cathode, to the cathode/electrolyte interface, and N_A is the Avogadro's number.

For simplicity, the following discussion is confined to pure hydrogen as the fuel. Thus, equation (6) reduces to

$$|j_{H_2}| = |j_{H_2O}| = 2|j_{O_2}| = \frac{iN_A}{2F} \tag{7}$$

Transport of gaseous species usually occurs by binary diffusion, where the effective binary diffusivity is a function of the fundamental binary diffusivity $D_{H_2-H_2O}$, and microstructural parameters of the anode [3, 4]. In electrode microstructures with very small pore sizes, the possible effects of Knudsen diffusion, adsorption/desorption and surface diffusion may also be present. The physical 'resistance' to the transport of gaseous species through the anode at a given current density is reflected as an 'electrical voltage loss'. This polarisation loss is known as concentration polarisation, η_{conc}^a, and is a function of several parameters, given as

$$\eta_{conc}^a = f(D_{H_2-H_2O}, \text{Microstructure, Partial Pressures, Current Density}) \tag{8}$$

where $D_{H_2-H_2O}$ is the binary H_2-H_2O diffusivity. It is assumed here that the effects of Knudsen diffusion, adsorption/desorption and surface diffusion are negligible. The η_{conc}^a increases with increasing current density, but not in a linear fashion. A simplified equivalent circuit can be used to describe the process, using what is known as the Warburg element, which consists of a number of resistors and capacitors [5]. The presence of capacitors ensures that the response time or time constant is non-zero. Since the relevant time dependences are not describable by simple first order kinetics, it is not appropriate to describe response time as a time constant. Nevertheless, a characteristic time can be defined, which depends on electrode thickness, electrode microstructure and the representative diffusivity.

In terms of physically measurable parameters, analytical expressions for anodic concentration polarisation have been derived which allow its explicit determination as a function of a number of parameters. One of the important parameters is the anode-limiting current density, which is the current density at which the partial pressure of the fuel, e.g. H_2, at the anode/electrolyte interface, is near zero such that the cell is starved of fuel. If this condition is realised during operation, the voltage precipitously drops to near zero. This anode-limiting current density, i_{as}, has the following form [6]

$$i_{as} = \frac{2F p_{H_2}^a D_{a(eff)}}{RT l_a} \qquad (9)$$

where $D_{a(eff)}$ is the effective gaseous diffusivity through the anode, and l_a is the anode thickness. The effective anode diffusivity contains the binary diffusivity of the relevant species, namely H_2 and H_2O, $D_{H_2-H_2O}$, the volume fraction of porosity, $V_{v(a)}$, and the tortuosity factor, τ_a [3,4]. If the fuel contains hydrocarbons, multi-component nature of gaseous diffusion must be addressed. The tortuosity factor is a measure of the tortuous nature of the anode through which diffusion must occur. In very fine microstructures, the tortuosity as a phenomenological parameter may include effects of Knudsen diffusion, surface diffusion, and possible effects of adsorption/desorption. The anodic concentration polarisation is then of the form [6]

$$\eta_{conc}^a = -\frac{RT}{2F} \ln\left(1 - \frac{i}{i_{as}}\right) + \frac{RT}{2F} \ln\left(1 + \frac{p_{H_2}^a i}{p_{H_2O}^a i_{as}}\right) \qquad (10)$$

Note that as the current density approaches the anode limiting current density, that is when $i \rightarrow i_{as}$, the first term approaches infinity. The maximum value of η_{conc}^a is limited by the OCV. Thus, the maximum achievable current density will always be less than i_{as}. The dependence of the anodic concentration polarisation given by equation (10) on various parameters can be qualitatively described as follows: From the standpoint of physical dimensions, and microstructural parameters, the lower the volume fraction porosity, the higher the tortuosity factor, and the greater the anode thickness, the higher is η_{conc}^a. From the standpoint of fuel gas composition, the lower the partial pressure of hydrogen, $p_{H_2}^a$, the higher is the η_{conc}^a. The temperature dependence is complicated. It is seen that $i_{as} \propto T^{1/2}$, since $D_{a(eff)} \propto T^{3/2}$, which would mean η_{conc}^a increases as temperature decreases. At the same time, as seen from equation (10), η_{conc}^a is linearly dependent on temperature, which would mean η_{conc}^a

decreases with decreasing temperature. In general, the η^a_{conc} is not a very strong function of temperature.

As stated earlier, the process of gaseous transport through porous electrodes is not describable by first order kinetics; nevertheless a characteristic time constant can be approximated by:

$$t_{characteristic} \sim \frac{l_a^2}{D_{a(eff)}} \quad (11)$$

For a typical anode-supported cell, l_a is 0.5 to 1 mm, and $D_{a(eff)}$ is $\sim$0.1 to $\sim$0.5 cm^2/sec. Thus, the corresponding characteristic time is on the order of several milliseconds to a few tenths of a second. The estimated tortuosity factors, based on cell performance measurements, range between $\sim$5 or 6 to as high as 15 to 20. The estimated tortuosity factor based on geometrical path a molecule traverses is typically less than 5 or 6. High values of the tortuosity factors estimated from cell performance data thus cannot be described solely on the basis of geometric considerations; other effects such as Knudsen diffusion, adsorption and surface diffusion probably also play a role. It is to be emphasised, however, that very high tortuosity factors have indeed been measured in many other cases involving gaseous transport through porous bodies with low porosities and small pore sizes [7]. Despite the fact that a high tortuosity factor cannot be justified on geometric arguments alone, it still is a useful parameter for describing concentration polarisation.

Concentration polarisation at the cathode similarly is related to the transport of O_2 and N_2 through the porous cathode. The net flux of O_2 from the oxidant stream, through the cathode to the cathode/electrolyte interface, is linearly proportional to the net current density. In this case also, gaseous transport is a function of the fundamental binary diffusivity, $D_{O_2-N_2}$, and cathode microstructure. The physical 'resistance' to the transport of gaseous species through the cathode is reflected as an 'electrical voltage' loss. This polarisation loss is known as cathodic concentration polarisation, η^c_{conc}, and is given as

$$\eta^c_{conc} = f(D_{O_2-N_2}, \text{Microstructure, Partial Pressures, Current Density}) \quad (12)$$

The η^c_{conc} increases with increasing current density, but not in a linear fashion. The time constant or response time must be a function of diffusivity and a characteristic diffusion distance, and thus the response time is finite, non-zero. Similar to the anode, a characteristic time for the cathode may be given by:

$$t_{characteristic} \sim \frac{l_c^2}{D_{c(eff)}} \quad (13)$$

where $D_{c(eff)}$ is the effective diffusivity through the cathode, and l_c is the cathode thickness. For an anode-supported cell, for a cathode thickness of $\sim$200 microns, and effective cathode diffusivity, $D_{c(eff)}$ of $\sim$0.05 cm^2/s, the characteristic time is $\sim$8 milliseconds; that is, in the millisecond range. In terms of physically measurable parameters, analytical expressions for cathodic concentration polarisation have been derived which allow its explicit determination as a function of a number of parameters. As with the anode, one of the important parameters is the cathode-limiting current density, which is the current density at which the partial pressure of the oxidant, e.g. O_2, at the cathode/electrolyte interface is near zero such that the cell is starved of oxidant. Depending upon the contributions of the other terms, such a condition may not be realised in cell operation. However, if this condition is realised during operation, then the voltage precipitously drops to near zero. This cathode-limiting current density, i_{cs}, has the following form [6]

$$i_{cs} = \frac{4F p_{O_2}^c D_{c(eff)}}{\left(\frac{p - p_{O_2}^c}{p}\right) RT l_c} \tag{14}$$

The effective cathode diffusivity contains the binary diffusivity of the relevant species, $D_{O_2-N_2}$, the volume fraction of porosity in the cathode, $V_{v(c)}$, and tortuosity, τ_c. In terms of the current density, i, and the cathode limiting current density, i_{cs}, cathodic concentration polarisation can be given by [6]

$$\eta_{conc}^c = -\frac{RT}{4F} \ln\left(1 - \frac{i}{i_{cs}}\right) \tag{15}$$

For comparable cathode and anode thicknesses and microstructures, the anodic concentration polarisation is usually much lower than cathodic concentration polarisation for two reasons: (1) The binary diffusivity of H_2-H_2O, $D_{H_2-H_2O}$ [2], is about four to five times greater than the binary diffusivity of O_2-N_2, $D_{O_2-N_2}$, due to the lower molecular weight of H_2 compared to the other species; (2) Typical partial pressure of hydrogen in the fuel, $p_{H_2}^a$, is much larger than the typical partial pressure of oxygen in the oxidant, $p_{O_2}^c$. Thus, for comparable anode and cathode thicknesses and microstructures, the anode-limiting current density is much greater than the cathode-limiting current density, i.e., $i_{as} \gg i_{cs}$. In practice, one of the electrodes is thicker than the other in an electrode-supported design. In anode-supported design, the anode thickness is much greater than the cathode thickness, i.e., $l_a \gg l_c$, and in such a case, often $i_{cs} > i_{as}$. However, even in anode-supported design, often cathode concentration polarisation can be comparable to anode concentration polarisation. Figure 9.2 shows the estimated cathodic concentration polarisation as a function of current density for a 50 micron thick cathode with different amounts of carbon added to generate various amounts of porosities [8]. The relevant effective diffusivities through porous cathodes required for the estimation of concentration polarisation were experimentally measured.

Similar anode concentration polarisation curves can be generated using equation (10) for various anode effective diffusivities. In practice, the fuel almost always is a reformed (at least partially) hydrocarbon. In such a case, internal

---
[2] Or for that matter that of H_2-CO (D_{H_2-CO}) and H_2-CO_2 ($D_{H_2-CO_2}$)

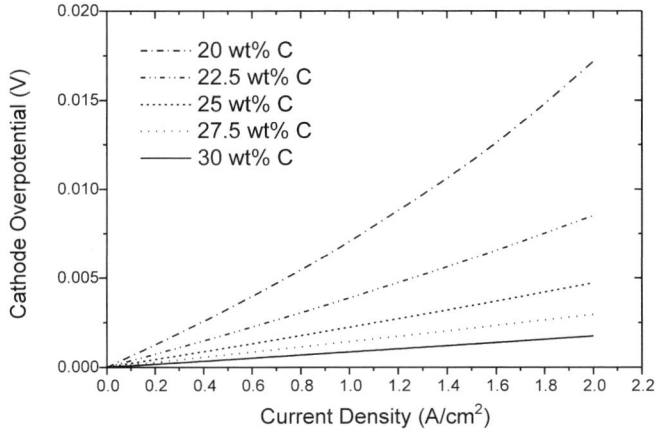

Figure 9.2 Estimated cathodic concentration polarisation vs. current density for a 50 micron thick cathode with different amounts of porosities [8]. The open porosity ranged between ~15% and ~43%.

reformation and shift reactions, as well as multi-component transport must be addressed.

The presence of gaseous hydrogen in the fuel makes gaseous transport easier, thus lowering anode concentration polarisation, even when CO and CO_2 are present. With pure hydrogen as the fuel, for an anode thickness on the order of ~1 mm, with fresh fuel the i_{as} can be as large as 5 A/cm² at 800°C or even larger. This allows for the fabrication of relatively thick anode-supported cells, without unduly increasing concentration polarisation. This is one of the principal advantages of an anode-supported design over other designs. Greater care, however, should be exercised when working with cathode-supported designs to ensure that cathodic concentration polarisation does not limit cell performance.

9.4 Activation Polarisation

Electrode reactions involve charge transfer as a fundamental step, wherein a neutral species is converted into an ion, or an ion is converted into a neutral species. Both reactions thus involve electron transfer. At the cathode, the charge transfer reaction involves the conversion of an oxygen molecule into oxide ions. The electrodes in solid state electrochemical devices may either be purely electronic conductors, or may exhibit both ionic and electronic conductivity (the so-called mixed ionic electronic conduction, MIEC). In addition, the electrodes may be either single phase or composite, two-phase. For the purposes of illustration, in what follows we will examine the overall cathode reaction in a system with a single phase, purely electronically conducting electrode.

The oxygen reduction reaction is a multi-step process, usually comprising several parallel reaction pathways. A thorough understanding of the elementary processes in SOFC cathodes under realistic operating conditions has eluded researchers because of such multiplicity of pathways. Thus, despite numerous proposed reaction mechanisms for oxygen reduction, it is virtually impossible to select any one over another as definitive [9-20]. This is also due to the fact that the reaction mechanisms are surely material and microstructure-dependent. Although it is generally difficult to isolate a single rate-determining step by virtue of the presence of a number of series steps, it is usually possible to describe the overall process in a phenomenological framework. The following describes the plausible reaction steps for one of the possible reaction schemes; it is to be emphasised that the following is only a generic sequence of steps and by no means one that has been demonstrated with any degree of certainty.

1) Surface adsorption[3] of oxygen molecules on the electronic conductor, which also is the electrocatalyst,

$$\frac{1}{2}O_2(\text{gas}) \rightarrow \frac{1}{2}O_{2\,ads}\ (\text{electrocatalyst})$$

2) Dissociation of adsorbed oxygen molecules into adsorbed atoms,

$$\frac{1}{2}O_{2\,ads}(\text{electrocatalyst}) \rightarrow O_{ads}(\text{electrocatalyst})$$

3) Surface diffusion of adsorbed oxygen atoms to a three phase boundary (TPB) between the electrocatalyst (e.g. LSM) – electrolyte (e.g. YSZ) – gas phase,

$$O_{ads}(\text{electrocatalyst}) \rightarrow O_{ads}(\text{electrocalayst/electrolyte TPB})$$

4) Formation of oxide ions by electron transfer with incorporation of these ions into the electrolyte

$$O_{ads}(\text{electrocatalyst/electrolyte TPB}) + 2e'(\text{electrocatalyst})$$
$$+ V_O^{\bullet\bullet}(\text{oxygen vacancy/electrolyte}) \rightarrow O_O^x(\text{electrolyte})$$

where Kroger-Vink notation has been assumed. Central to the above scheme is the occurrence of the charge transfer reaction at or near a TPB; in-situ ^{18}O exchange experiments under cathodic polarisation and subsequent SIMS-analysis, have confirmed the occurrence of the charge transfer reaction at a TPB [19]. Several variations of the above scheme are possible; such as, for example, the occurrence of the charge transfer reaction on the electrocatalyst surface to form an oxide ion, followed by surface diffusion of the oxide ion to a TPB, and its incorporation into the solid electrolyte at the TPB, or further surface diffusion of the oxide ion on the electrolyte surface, and its incorporation into the electrolyte at a

---

[3] Surface adsorption of oxygen molecules may also occur on the surface of the electrolyte, YSZ, followed by its dissociation into oxygen atoms and their surface diffusion to a TPB.

point some distance away from the TPB. Thus, many possibilities exist wherein there are a number of possible parallel steps to the net charge transfer reaction. Experimental work combined with continuum modeling has been conducted in order to analyse various reaction steps [15,17,20,21]; experimental methods used in these studies include steady-state current-voltage characteristics and electrochemical impedance spectroscopy. The importance of the role of TPB in $La_{1-x}(Sr,Ca)_xMnO_3$ and $La_{1-x}Sr_xMnO_3$ cathodes is documented in appended references [14,20,22,23]. These studies showed that oxygen reduction predominantly occurs at TPB's when these materials are used as cathodes. However, bulk transport of oxygen through $La_{1-x}Sr_xMnO_3$ has been reported under high applied overpotentials [9]. A wide variety of reaction mechanisms have been proposed in the literature, even using nominally similar electrode materials. This lack of consistency is apparently due to the fact that electrode processes and morphology are closely interrelated, and also that the fundamental mechanisms are not fully understood at the present time. Considering that the current and potential distribution in the 3-dimensional porous system and the available reaction zone both depend on the microstructure and the presence of secondary phases and impurities at the interface, it is clear that a simple, unified reaction mechanism is unrealistic. But this in large part is also due to a lack of a quantitative characterisation of electrode microstructure, and a lack of quantitative analysis of the relationship between microstructure and cathode performance. Such a quantitative study has been reported in only a limited number of studies, such as those by Zhao et al. [24] and Weber et al. [25].

Above reactions 1) to 4) describe a number of possible, series steps, and in the simplest model, the slowest one is the rate-determining step[4]. The remaining steps then can be assumed to be close to equilibrium. Many of the above steps are generally thermally activated. The rate of cathodic reaction is directly proportional to the net current density; or more precisely, the net current density is proportional to the cathodic reaction rate. Associated with the reaction rate, or the passage of current, is a loss in voltage, which is the activation polarisation or overpotential; the terminology being derived from the thermally activated nature of the reaction. The relationship between the cathodic activation polarisation, η_{act}^c, and the current density is usually nonlinear, except at very low current densities. In general,

$$\eta_{act}^c = f(\text{material properties, microstructure, temperature,} \\ \text{atmosphere, current density}) \tag{16}$$

---

[4] It is to be emphasised that it is not necessary that there be a single rate-determining step. If two or more steps exhibit similar kinetic barriers, it is quite possible that a simple dependency of the so-called rate-determining step on a given parameter (e.g. oxygen partial pressure) will not be reflected in the overall measured effect. That is, there can be more than one step away from equilibrium. The analysis of data in such a case can be particularly difficult.

A phenomenological theory, which gives a quantitative relation between current density and η_{act}^c, is known as the Butler-Volmer equation, and is of the form [2,26]

$$i = i_o^c \left\{ \exp\left[\frac{\beta z F \eta_{act}^c}{RT}\right] - \exp\left[-\frac{(1-\beta)zF\eta_{act}^c}{RT}\right] \right\} \tag{17}$$

where β is a dimensionless, positive number, less than one (for a one-step charge transfer process), which is known as the transfer coefficient, and i_o^c is known as the exchange current density. Note that the relationship between η_{act}^c and i is nonlinear and implicit, that is, it does not allow an explicit determination of η_{act}^c as a function of current density. Rather, the equation gives net current density for a given η_{act}^c. However, limiting forms of the Butler-Volmer equation allow one to express η_{act}^c as a function of current density, i. The low current density and high current density regimes are described in what follows.

In the low current density limit, it is possible that $\left|\frac{\beta z F \eta_{act}^c}{RT}\right| << 1$ and $\left|\frac{(1-\beta)zF\eta_{act}^c}{RT}\right| << 1$. In such a case, the Butler-Volmer equation can be simplified as

$$i \approx i_o^c \left|\frac{zF\eta_{act}^c}{RT}\right| \tag{18}$$

or

$$|\eta_{act}^c| \approx \frac{RT}{zFi_o^c} i \tag{19}$$

The term $\frac{RT}{zFi_o^c}$ has the units of area specific resistance, Ωcm^2, and is referred to as the charge transfer resistance, denoted by R_{ct}^c, and is given by $R_{ct}^c = \frac{RT}{zFi_o^c}$. An important point to note here is that a linear relationship between η_{act}^c and the current density, i, in the low current density limit does not imply ohmic relationship, since the response time for the process is long, and is determined by whatever is the underlying physical process. In the simplest case, the charge transfer process is describable by a parallel $R - C$ circuit, in which case the time constant is given as RC. Thus, in DC measurements, the capacitive part is not reflected. At the same time, in the current interruption experiment, the voltage drop across the interface is usually not separable from the other time-dependent parts of the impedance. Measurement of frequency response, however, allows one to estimate both R and C. More about this is discussed later.

An experimental measurement of η_{act}^c as a function of current density (particularly in the low current density regime) allows one to estimate the R_{ct}^c or i_o^c. The i_o^c is a measure of the rate of charge transfer process, and depends upon a number of material properties, microstructure, temperature and also on the atmosphere.

In the high current density regime, $\left|\frac{\beta z F \eta_{act}^c}{RT}\right| >> 1$ and the Butler-Volmer equation can be approximated by

$$\eta_{act}^c \approx \frac{RT}{\beta zF}\ln i_o^c - \frac{RT}{\beta zF}\ln i \approx a + b\ln i \qquad (20)$$

which is the Tafel equation [2,26]. The above equations describing the activation polarisation are from the extensive work on aqueous electrochemistry. These equations are often used to describe activation polarisation in solid state electrochemistry also. In aqueous electrochemistry, the process of charge transfer occurs across the entire liquid electrolyte/solid electrode interface. In solid state electrochemistry with the presence of gaseous species, however, this charge transfer process either involves three phases: the electrolyte, the electrode (electrocatalyst), and the gas phase, in the case of purely electronic conducting electrodes or two phase MIEC electrodes; or two phases: namely a single phase MIEC electrode and the gas phase, in the case of single-phase MIEC electrodes. If the transport of ions is restricted to the electrolyte, that of electrons through the electrocatalyst, and of the gaseous species through the porous interstices, the charge transfer reaction is presumed to occur at (or near) the three-phase boundary (TPB) where the three phases meet. This TPB is characterised by a line, extending along the electrolyte surface, and has the dimensions of cm/cm^2 or cm^{-1}. This suggests that the electrode kinetics must depend upon the TPB length, in addition to fundamental physical parameters, such as electrocatalytic activity of the electrocatalyst, and the partial pressure of the reactant (i.e. oxygen). That is, the exchange current density, i_o^c, must depend upon electrode microstructure, such as the size and the number of electrocatalyst particles per unit area of the electrolyte surface. Thus, in such a case

$$i_o^c = f(\text{TPB, partial pressure of oxygen in the atmosphere, oxygen vacancy concentration in the electrolyte, oxygen vacancy mobility in the electrolyte, electron concentration in the electrocatalyst, and temperature}) \qquad (21)$$

Figure 9.3(a) shows a schematic of such a charge transfer reaction. In a single-phase MIEC electrode, on the other hand, the charge transfer reaction is not restricted to the linear feature (e.g. TPB), but can occur over the entire electrode/gas phase interface. In such a case, the exchange current density is given by

$$i_o^c = f(\text{partial pressure of oxygen, oxygen vacancy concentration in the MIEC, oxygen vacancy mobility in the MIEC, electronic defect concentration in the MIEC, and temperature}) \qquad (22)$$

Figure 9.3(b) shows a schematic of such a charge transfer reaction. In so far as single phase, essentially electronically conducting materials as cathodes are

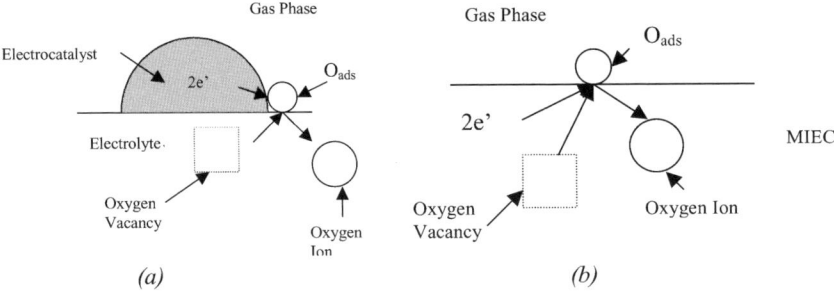

Figure 9.3 Schematic of a possible charge transfer reaction for (a) purely electronically conducting cathode material, and (b) MIEC cathode material.

concerned, most of the reported work has been on LSM. Thus, when a porous layer of LSM is applied over YSZ, the charge transfer reaction is confined to (or near) the TPB length at the LSM/YSZ interface.

No definitive relationships between R_{ct}^c or i_o^c and TPB are available for lack of definitive measurements of TPB in cases wherein a layer of porous LSM is applied over a dense YSZ surface. Nevertheless, an order of magnitude estimate can be made as follows. A typical, experimentally measured, number for R_{ct}^c, for LSM/YSZ at 800°C is on the order of ~ 2 Ωcm^2. For a LSM particle size of ~ 1 micron, and the volume fraction of porosity in LSM of $\sim 50\%$, the TPB is on the order of 2×10^4 cm^{-1}. It is convenient to define a charge transfer resistivity, ρ_{ct}^c, in terms of the charge transfer resistance and TPB length by an equation of the form [27]

$$R_{ct}^c = \frac{\rho_{ct}^c}{l_{TPB}} \qquad (23)$$

Then, approximate value of ρ_{ct}^c is $\sim 40,000$ Ωcm. An estimate of ρ_{ct}^c has been made by analysing LSM+YSZ composite electrodes by using techniques in quantitative microscopy, and comparing with the results of cell resistance [24]. The estimated ρ_{ct}^c is on the order of 50,000 to 100,000 Ωcm. Unfortunately, there are only a few measurements of this nature, and thus not much information is known on the fundamental parameter, ρ_{ct}^c, free of microstructural effects (e.g. l_{TPB}), which defines the charge transfer process for any set of materials. Nevertheless, this estimate shows that for reducing the charge transfer resistance from ~ 2 Ωcm^2 to ~ 0.2 Ωcm^2, that is by an order of magnitude, using the same set of materials, it would be necessary to decrease the particle size of LSM from ~ 1 micron to ~ 0.1 micron. This is often difficult to achieve. However, using the same particle size of LSM, for example, it is possible to substantially lower the overall charge transfer resistance by allowing the reaction of charge transfer to spread out some distance from the physically distinct electrolyte/electrode interface, well into the porous electrode. This can be achieved if the electrode exhibits MIEC characteristics.

Cathodic and anodic activation polarisations, in light of MIEC electrodes, are discussed below.

9.4.1 Cathodic Activation Polarisation

Much of the early work on SOFC cathodes has been on Sr-doped LaMnO$_3$ (LSM), which is a predominantly electronic conductor. In the case of electrolyte-supported or anode-supported cells, powder of LSM is spread (screen-printed) over the electrolyte (YSZ) surface, and fired at elevated temperatures to bond the cathode onto the electrolyte. In most practical applications, the LSM used is of a typical composition La$_{1-x}$Sr$_x$MnO$_3$, with x = 0.15 ... 0.25. These materials exhibit a diffusion coefficient of oxygen D on the order of 10^{-12} cm^2/s and a surface exchange coefficient k_{exc} of about 10^{-7} cm/s at 1000°C in air [28]. In such a case, the overall cathodic reaction occurs at the LSM-YSZ-gas phase TPB. The effective TPB length that can be realised is generally < 20,000 cm^{-1}, (equivalent to 50% surface coverage of 1 micron LSM particle size), and typically < 5,000 cm^{-1}, with the result that at temperatures below about 900°C, the cathodic activation polarisation is usually large, which limits cell performance. This limitation has now been well recognised, and the SOFC research and development community has all but abandoned this approach and has shifted focus to porous, MIEC electrodes, either single phase or composite.

The concept of porous, effectively MIEC electrodes, is not new [29]. It has been extensively studied in aqueous electrochemistry. In aqueous electrochemistry, if a porous electrode is used, the electrolyte fills the pores of the electrode. If the rate-limiting step in the electrode reaction is the overall rate of charge transfer, then increasing the electrolyte/electrode surface area should improve the rate. Over the thickness of the porous electrode, transport of electrical charges occurs in two phases – electronic through the matrix phase, and ionic through the solution (electrolyte) phase. That is, over the porous electrode, transport is by mixed ionic and electronic conduction (MIEC). In addition, convective or diffusive transport of neutral reactive species occurs through the liquid electrolyte in the pores. In this manner, the electrode reaction is spread out into the porous part of the electrode. The theory of porous electrodes in aqueous electrochemistry has been developed on this premise. In such a case, the transport of ionic and neutral species occurs through the electrolyte filling up the pores, and electron transport occurs through the solid part of the porous electrode. In solid state electrochemistry also, an analogous porous electrode should be capable of transporting both ions and electrons; that is, it must be a mixed ionic electronic conducting (MIEC) material. In solid state electrochemistry, in an analogous electrode, ion and electron transport occurs through the solid part of the MIEC electrode, and neutral (gaseous) species transport through the porous interstices [18,27,30].

Figure 9.4 shows a schematic of a porous MIEC electrode used in solid state electrochemistry, in which the pathways for the various species are shown. The MIEC characteristics can be realised in two ways: (1) Use of a single phase, porous MIEC material, such as Sr-doped LaCoO$_3$ (LSC), or (2) Use of a composite, two-phase porous mixture of an electronic conductor (e.g. LSM) and an ionic conductor (e.g. YSZ). In the case of a composite, two-phase mixture, the MIEC properties are realised globally (at the microstructural level, not at the atomistic

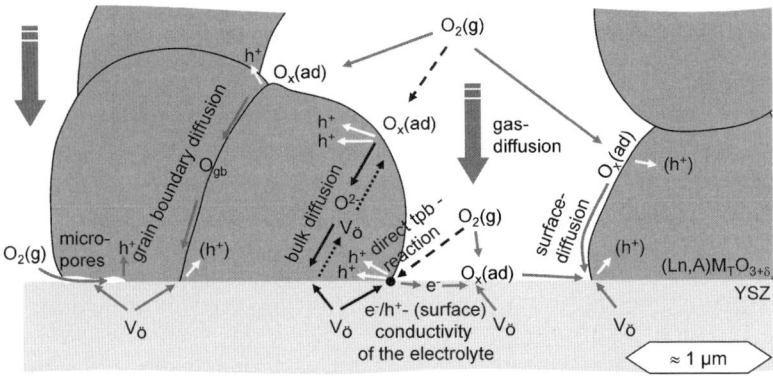

Figure 9.4 Schematic of a porous MIEC electrode with possible reaction pathways and involved species for the oxygen reduction reaction for SOFC application (adsorbed oxygen species $O_x(ad)$: $O_{2,ad}$, O_{ad}, O^-_{ad}, O^{2-}_{ad}).

level). In such composite materials used as electrodes, the TPB exists through the thickness of the electrode, and the electrochemical reaction is spread into the electrode, and not just restricted to the physically distinct electrolyte/electrode interface. In a single phase MIEC electrode, the electrochemical reaction can similarly occur over some distance into the electrode.

There are advantages and shortcomings to both approaches. If a single-phase MIEC material is used, in principle, the electrochemical reaction can occur over the entire porous surface. The potential disadvantage, however, is that careful manipulation of defect chemistry is required to ensure that both ionic and electronic conductivities are sufficiently high. This is often difficult to achieve, especially over a wide range of oxygen partial pressures and temperatures. If a two-phase MIEC material is used, it is necessary to ensure that both phases are contiguous, while at the same time exhibiting a high TPB; that is, one phase should not completely coat the other phase. This requires a careful control over the microstructure. The advantage over single-phase MIEC materials, however, is that an ability to mix two different materials allows flexibility in the choice of materials so that transport properties of the two phases can be separately optimised. In essence, by using two separate phases for the ionic and electronic transport, greater flexibility is achieved by decoupling the functions. A host of different materials for the ionic conducting part, such as YSZ, doped ceria, stabilised Bi_2O_3, LSGM, etc., can be used; and a host of electrocatalysts, such as LSM, Sr-doped $LaFeO_3$ (LSF), Sr-doped $LaCoO_3$ (LSC), etc., can be used. The use of LSC or LSF in composite electrodes is expected to be beneficial as these materials are themselves MIEC, albeit with much larger electronic conductivity compared to ionic conductivity, as they offer additional pathways for the transport of oxygen ions.

Theoretical aspects of porous MIEC electrodes, both using single-phase and two-phase materials, have been analysed by many authors [18,27,30-34]. While the particulars of the models vary from model to model, general features of the porous MIEC electrodes can be summarised as follows: (1) Gaseous species

(O_2, N_2 in the cathode; H_2, H_2O in the anode) transport through the porous interstices, which are contiguous, to or away from reaction sites. (2) In the case of two phase MIEC materials, electrons (or holes) transport through the electronically conducting (contiguous) phase, oxygen ions transport through the ionically conducting (contiguous) phase, and the charge transfer reaction occurs at or near a TPB. In single phase MIEC, both electrons (and/or holes) and oxygen ions transport through the single-phase MIEC, and the charge transfer[5] reaction occurs along the surface of the porous MIEC. In this manner, whether the electrode is two-phase or single-phase, the electrochemical reaction of charge transfer is spread from the electrolyte/electrode interface over some distance into the electrode. (3) The region over which this spreading occurs depends upon the microstructure as well as the transport properties of the electrode. Usually, the finer the microstructure, the smaller is the region over which the reaction zone is spread out. (4) Very close to the electrolyte, the current is predominantly ionic, and outside the critical thickness into the electrode, the current is predominantly electronic. Over the critical or the threshold thickness, the current varies from ionic (near the electrolyte) to electronic (towards the current collector). Thus, the electrode should exhibit MIEC characteristics at least over this critical or the threshold distance. Typically, this critical thickness is on the order of a few, to few tens of microns. This layer has been variably referred to as the electroactive layer, the electrocatalytic layer, or the interlayer. As the microstructure in this region must be fine, which enhances the rate of electrochemical reaction (lowers activation polarisation), also unfortunately impedes gas transport (increases concentration polarisation) due to the Knudsen diffusion effects, as well as due possibly to adsorption/desorption effects. The existence of a critical thickness fortunately implies that the electrode microstructure need not be fine throughout the electrode. Thus, the overall polarisation can be minimised by grading the electrode microstructure such that near the electrolyte/electrode interface, the electrode has a fine microstructure and exhibits MIEC properties; and away from the interface, the electrode has a coarse microstructure with a large pore size, and exhibits essentially electronic conduction.

While the general features of several of the models are similar, the particular analytical expressions, wherever available, vary widely depending upon the details of a given model. In what follows, some of the equations from the work of Tanner et al for composite cathodes are given to illustrate the role of various parameters [27]. In the low current density limit, over which the Butler-Volmer equation can be linearised, the effective charge transfer or the polarisation resistance (activation polarisation only) for cathode interlayer thickness greater than the critical thickness can be given by [22]

$$R_{ct(eff)}^c = R_p^c \approx \sqrt{\frac{R_{ct}^c d \rho_i^c}{(1 - V_v^c)}} \qquad (24)$$

---

[5] In the case of a single phase MIEC, the reaction may be regarded as that of oxygen incorporation (or removal) rather than that of charge transfer.

where d is the grain size of the ionic conductor (e.g. YSZ) in the composite MIEC cathode, V_v^c is the porosity in the cathode interlayer, and ρ_i^c is the ionic resistivity (inverse of ionic conductivity, $1/\sigma_i^c$) of the ionic conductor in the composite cathode. The preceding equation is illustrative from the standpoint of assessing the role of various parameters on the polarisation resistance of a composite cathode in terms of the physically measurable parameters. In this equation, R_{ct}^c has the same meaning as stated earlier. If the grain size of the ionic conductor (e.g. YSZ) in the composite cathode, d, is ~ 1 μm, the ionic resistivity, ρ_i^c is ~ 50 Ωcm, and if the cathode interlayer porosity, V_v^c is ~ 0.25, and if R_{ct}^c is ~ 2 Ωcm², then the effective charge transfer or polarisation resistance (using equation (24)) turns out to be ~ 0.12 Ωcm² — that is a reduction by a factor of 16. The critical distance, λ_c, similarly, depends upon various parameters, and has the form [27],

$$\lambda_c \sim \sqrt{\frac{R_{ct}^c d(1 - V_v^c)}{\rho_i^c}} \tag{25}$$

For the case of the LSM + YSZ composite cathode described above, and for the selected values of the parameters, the magnitude of the critical cathode interlayer thickness turns out to be ~ 17 μm. If, on the other hand, the electrode is made of porous LSM only, but of essentially the same microstructure, the value of λ_c is on the order of 1 micron, assuming ionic resistivity of LSM to be 10,000 Ωcm. This shows that if the ionic resistivity of the porous MIEC is very high, the charge transfer reaction is essentially confined to the physically distinct cathode/electrolyte interface. The corresponding polarisation resistance (as determined using the general equation given by Tanner et al [27]) or the effective charge transfer resistance is essentially the same as R_{ct}^c or ~ 2 Ωcm² in this illustration. That is, with single phase LSM, the reaction zone is confined to the physically distinct cathode/electrolyte interface, and the polarisation resistance is high since LSM is not an MIEC (or is an MIEC with very low ionic conductivity).

The preceding discussion and equations show that a fine cathode microstructure is preferred. Fabrication of such cathodes requires careful control of microstructure. It has been demonstrated that a typical high performance cathode has a particle size (of the oxide ion conductor in the composite cathode) on the order of a micron. At 800°C, the polarisation resistance less than about ~ 0.1 Ωcm² has been demonstrated with such cathodes. Figure 9.5 shows an SEM micrograph of a typical anode-supported cell. Regions adjacent to the electrolyte are the cathode and anode electrocatalytic layers of fine microstructure to facilitate electrochemical cathodic and anodic reactions, respectively. Regions next to these electrocatalytic layers have a coarse microstructure and greater porosity to facilitate easier gas transport. These regions also exhibit greater electronic conduction and serve as current collector regions.

A well-defined increase of the effective electrolyte surface area can also be achieved by a structured electrolyte surface. Sintering separate 8YSZ particles onto the electrolyte substrate and covering the increased surface area by an electrochemically active thin porous film cathode via metal-organic-deposition

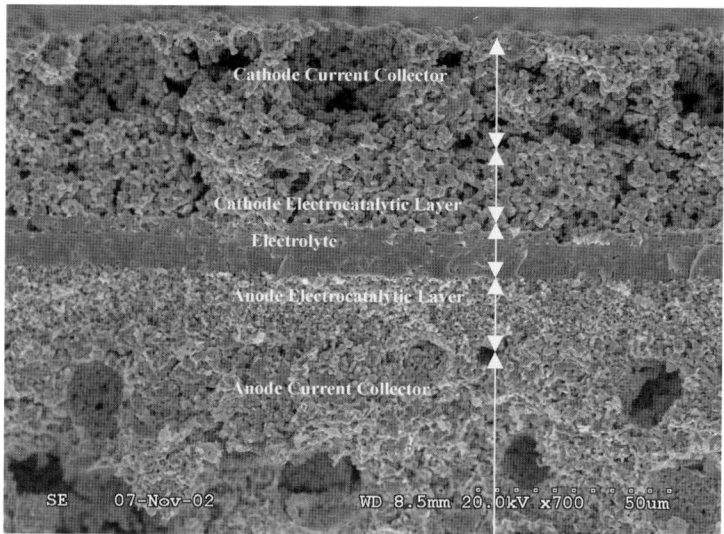

Figure 9.5 An SEM micrograph showing cross-sectional view of an anode-supported cell. Adjacent to the electrolyte are anode and cathode electrocatalytic layers of fine microstructure for enhanced electrocatalysis. Regions next to the electrocatalytic layers have higher porosity and a coarser microstructure for easier gas transport.

(MOD) is a possible approach to increasing the number of active reaction sites. An additional macroporous LSM-layer is used as a current collecting and gas distribution layer. The adhesion of the cathode is improved due to the 3-dimensional penetration structure (Figure 9.6). This approach can lead to a significant increase in power density, while ensuring long term stability against thermal cycling by structurally inhibiting delamination. By lowering the processing temperature (below about 1000°C), it is possible to use a mixed conducting LSC-thin film (LSC: (La, Sr)CoO$_3$) as a cathode, without the danger of forming unwanted secondary phases. Such cathodes showed an even higher performance (Figure 9.7), with negligible degradation over an operating period of more than 1000 h and at a current density of 0.4 A/cm^2 in air [35].

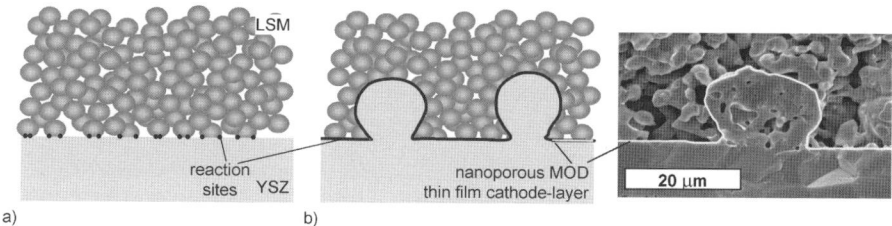

Figure 9.6 Cathode/electrolyte interface structures: (a) standard interface with smooth electrolyte surface and restricted number of active reaction sites and (b) structured electrolyte surface with nanoporous MOD thin film cathode layer leading to an enhanced reaction zone with improved performance and durability.

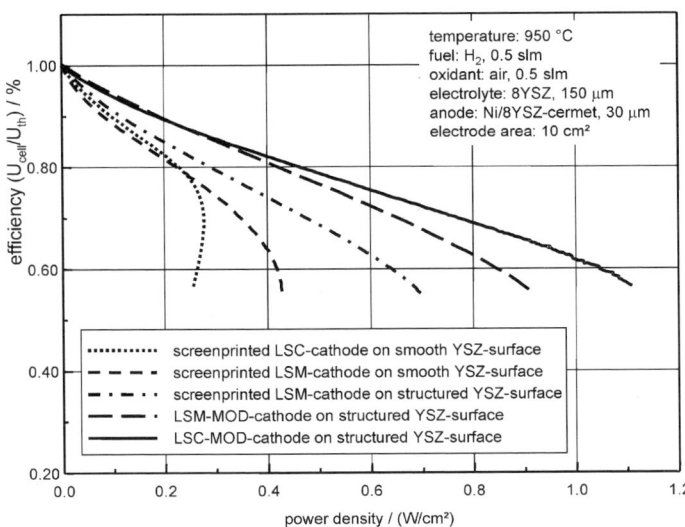

Figure 9.7   Efficiency vs. power density for electrolyte supported single cells with different types of cathode materials and cathode/electrolyte interface structures [35].

The above discussion on the role of material and microstructural parameters on the overall cathodic activation polarisation is applicable to composite MIEC cathodes, comprising a porous, two-phase, contiguous mixture of a predominantly electronic conductor and an ionic conductor. In a broader sense, the same conclusions are applicable to single phase MIEC cathodes. In the case of single-phase MIEC perovskite cathodes, the ionic conductivity is typically several orders of magnitude smaller than the electronic conductivity (albeit, still quite large in many MIEC materials) and depends on the composition, oxygen partial pressure, and temperature. Assuming that the ionic conductivity of the MIEC is much smaller than that of its electronic conductivity, the relevant bulk transport parameter of the MIEC continues to be the ionic conductivity (or ionic resistivity) of the MIEC, or the chemical diffusion coefficient of oxygen, D, in the MIEC. The relevant surface reaction parameter is the surface exchange parameter, k_{exc}, instead of $1/R_{ct}^c$ in the case of composite MIEC cathodes [36-38].

In a MIEC-cathode, at least three reaction steps have to be considered as rate determining: surface exchange at the gas phase/MIEC interface, bulk diffusion in the MIEC and incorporation of oxygen ions into the electrolyte at the MIEC/electrolyte interface. In the case that the latter is negligible, the extension of the reaction zone depends on the ratio of diffusion coefficient of oxygen, D, (or ionic conductivity of the MIEC, which could conceivably be estimated using the Hebb-Wagner polarisation technique) and surface exchange coefficient, k_{exc}, as well as the nature of porosity and microstructure. Equations similar to (24) for the effective polarisation resistance, and (25) for the extent to which reaction zone spreads, can be readily written for single phase MIEC, wherein the R_{ct}^c is replaced by $1/k_{exc}$, and an appropriate proportionality constant is introduced, which accounts for the dimensionality.

9.4.2 Anodic Activation Polarisation

The basic concepts of composite or single-phase MIEC electrodes are equally applicable to anodes. Traditionally, however, the typical anode used to date has been a composite mixture of Ni and YSZ. The presence of YSZ not only suppresses the thermally induced coarsening of Ni, but it also introduces MIEC characteristics. Other anodes currently under investigation are based on cermets of copper, which are being explored for direct oxidation of hydrocarbon fuels [39]. These types of anodes are in an early stage of development and thus their polarisation behavior is not discussed here. In so far as single-phase anodes are concerned, some work has been reported in the literature, most notably on La-SrTiO$_3$ [40, 41]. Work on this as well as other perovskite-based anodes is in its infancy, and is not elaborated upon further. The discussion in this chapter is confined to Ni + YSZ cermet anodes.

Although the basic concepts of anode reaction are similar to the cathode, the details may be different, and are not well understood at the present time. The overall anodic reaction may be given by:

$$O^{2-}(\text{electrolyte}) + H_2(\text{fuel gas}) \rightarrow H_2O(\text{fuel gas}) + 2e'(\text{anode})$$

One of the scenarios could involve the following steps:

1) Adsorption of H$_2$ on the surface of YSZ or Ni from the anode

$$H_2(\text{fuel gas}) \rightarrow H_{2_{ads}}(\text{YSZ or Ni})$$

2) Surface diffusion of adsorbed H$_2$ to TPB

$$H_{2_{ads}}(\text{YSZ or Ni}) \rightarrow H_{2_{ads}}(\text{TPB})$$

3) Anodic electrochemical reaction

$$O_O^x(\text{electrolyte}) + H_{2_{ads}}(\text{TPB}) \rightarrow H_2O(\text{fuel gas}) + 2e'(\text{anode}) + V_O^{\bullet\bullet}(\text{electrolyte})$$

In the preceding, the Kroger-Vink notation has been used. Similar to the cathodic overpotential, the anodic activation overpotential also depends upon material properties, microstructure, atmosphere, temperature and current density; that is,

$$\eta_{act}^a = f(\text{material properties, microstructure, temperature, atmosphere, current density}) \tag{26}$$

Assuming a phenomenological model, anodic polarisation can be described using the Butler-Volmer equation, and its low current density (linear) and high current density (Tafel) limits. Experimental results for some selected cases can be found in Chapter 6. The anodic exchange current density, similar to the cathodic exchange current density, depends upon a number of parameters:

$$i_o^a = f(\text{TPB, partial pressure of hydrogen in the atmosphere,} \\ \text{oxygen vacancy concentration in the electrolyte,} \\ \text{oxygen vacancy mobility, and temperature}) \quad (27)$$

Figure 9.8 shows a schematic of the anodic charge transfer – electrochemical reaction. An alternative possibility is the release of oxygen molecules into the anodic chamber, followed by reaction with hydrogen to form water vapor. That is, an alternative reaction can be of the form

$$O_O^x(\text{electrolyte}) \rightarrow \frac{1}{2}O_2(\text{fuel gas}) + 2e'(\text{anode}) + V_O^{\bullet\bullet}(\text{electrolyte})$$

followed by

$$\frac{1}{2}O_2(\text{fuel gas}) + H_2(\text{fuel gas}) \rightarrow H_2O\ (\text{fuel gas})$$

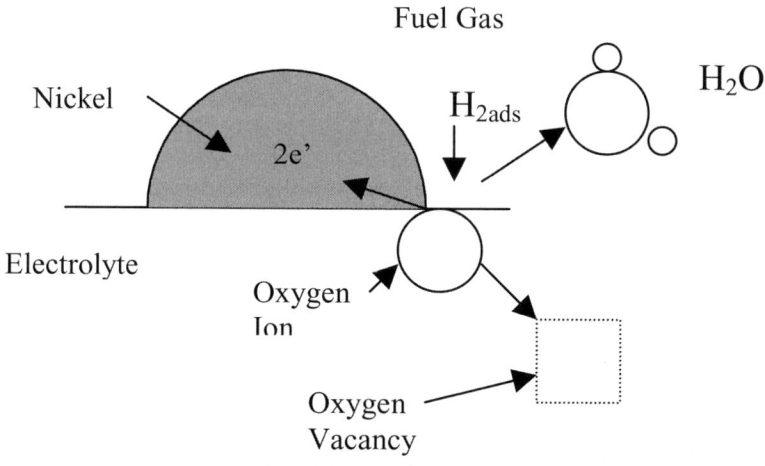

Figure 9.8   Schematic of anodic charge transfer - chemical reaction.

This latter reaction scheme does not depend upon the adsorption of fuel gas, while the former one does. The implication is that anodic activation polarisation would be independent of what the fuel is in the latter scheme, while it would be a function of the type of fuel in the former case. Recent work has shown that the total polarisation loss with CO as a fuel is much greater than that with H_2 as the fuel, and the difference cannot be attributed to differences in concentration polarisation [42]. It is possible that the differences may be due to differences in the adsorption characteristics of H_2 and CO. Thus, the preliminary conclusion is that adsorption of fuel gas must be an important step.

9.5 Measurement of Polarisation (By Electrochemical Impedance Spectroscopy)

Impedance spectroscopy has emerged over the past several years as a powerful technique for the electrical characterisation of electrochemical systems [5]. The strength of the method lies in the fact that by small-signal perturbation, it reveals both the relaxation times and relaxation amplitudes of the various processes present in a dynamic system over a wide range of frequencies.

Various polarisations exhibit different time dependence, due to different origins of the kinetic processes involved. The response time for ohmic polarisation is essentially zero, while the response time for concentration polarisation is related to the relevant gas phase transport parameters; e.g. diffusivity. In terms of an equivalent circuit, a Warburg-type element can be used to describe gas transport through porous electrodes. Similarly, the time constant for activation polarisation is related to details of the charge transfer process. In the very simplest case, it can be represented by a time constant for a parallel R-C circuit, provided the activation process can be described by a parallel R-C circuit. This, however, is an oversimplification, and an R-C element rarely describes the activation process accurately; it, nevertheless, allows some insight into the nature of time constants involved. The experimental procedure thus involves measuring impedance, $Z(\omega)$, as a function of frequency over a wide range, usually from as low as a few mHz to several hundred kHz. Often experimental difficulties in separating relevant parameters arise due to overlapping semi-circles, as well as inductive effects due to the testing setup at high frequencies.

In general, the occurrence of a multitude of chemical and physical processes in the system leads to a complicated, non-linear relationship between cell voltage and cell current. Therefore, the definition of a unique polarisation resistance is difficult, since it itself is usually a function of current density. There are two methods that can be used to measure cell polarisation; an AC method, and a DC method. The polarisation resistance determined from AC measurements can be different from that determined from DC measurements. When the system is perturbed by an AC input current signal, the AC voltage signal observed at the terminals of the cell is phase-shifted with respect to the perturbation input. The corresponding complex impedance can be determined from the current input signal, and phase shifted voltage signal. In the DC method, electrode potentials are measured with respect to suitably positioned reference electrodes, and the measured voltage differences are corrected for ohmic contributions. These two approaches are briefly described in what follows.

In the AC method, the cell is subjected to an AC source of variable frequency, and the cell response is measured as a function of frequency. Graphical representation involves a plot of negative of the imaginary part of the impedance, $-\text{Im}Z(\omega)$, on the y-axis and real part of impedance, $\text{Re}Z(\omega)$, on the x-axis; or alternatively a plot of the imaginary part of the admittance, $B(\omega)$, on the y-axis, and the real part, $G(\omega)$, on the x-axis. The plots in the ideal case are a series of semi-circles, quarter-circles, or distorted semi-circles and quarter-circles. The intercepts with the x-axis are measures of resistive losses due to various physical processes, and positions on the arcs provide information on non-ohmic terms.

The DC method is usually based on a combination of two types of measurements: ohmic contribution by current interruption, and the measurement of electrode overpotentials using reference electrodes. The placement of reference electrodes on solid-state electrochemical devices such as SOFCs presents substantial difficulties, since, unlike liquid-phase electrochemistry, they cannot be readily inserted into the electrolyte. In principle, a detailed analysis of the mixed boundary value problem for complicated specimen geometries and boundary conditions is required. These difficulties become even more serious when dealing with electrode (anode or cathode)-supported cells with thin electrolyte film [43-45]. On one hand, such cells are preferred as they exhibit considerably higher power densities; on the other hand, extracting accurate information on separate electrode polarisations becomes difficult. Detailed discussion on measurement techniques and difficulties associated with the use of reference electrodes is given in Chapter 10. For electrolyte-supported cells, measurements are often done with reference electrodes suitably placed on both the cathodic and the anodic sides. The electrical equivalent of this arrangement is shown in Figure 9.9.

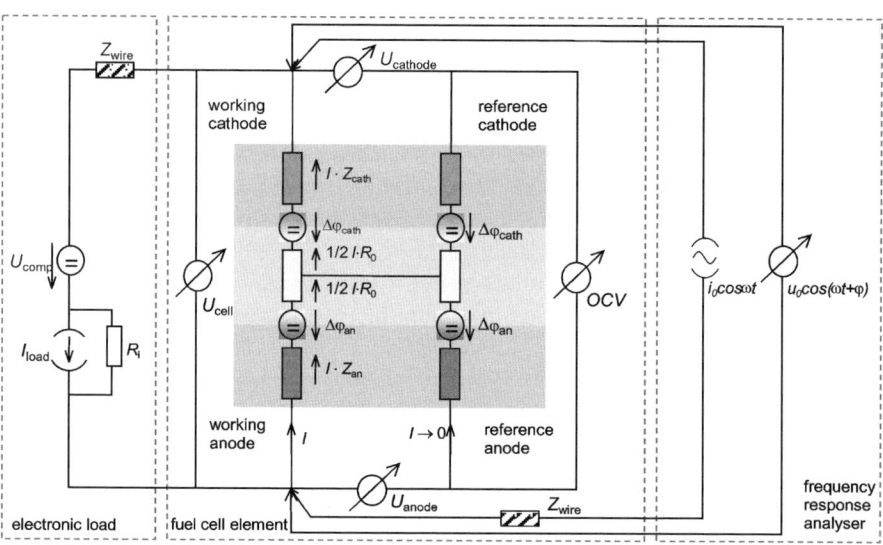

Figure 9.9 Equivalent electrical circuit description of impedance measurement arrangement for planar fuel cell elements with reference electrodes.

Here, the single cell element is modeled by a number of voltage sources representing the half-cell potentials. It is to be emphasised that the electrical equivalent shown in Figure 9.9 is only approximate, and errors in accurate determination of overpotentials cannot be entirely eliminated. The only way an accurate estimate of overpotential can be obtained is by solving the appropriate transport equations for the appropriate boundary conditions – coupled with experimental measurements.

Although measurements of separate cathode and anode overpotentials contain errors, it is possible to obtain the total cathode + anode overpotential with reasonable accuracy, by subtracting the ohmic contribution from the total voltage, at a given current density.

In a typical AC impedance measurement, the losses occurring in cell operation are represented by the ohmic resistance R_0 and the polarisation resistances $Z_{cath} + Z_{anod} = Z_{pol}$. The terminals of the measurement device can be connected either to the working electrodes or to a working electrode and a reference electrode on either side, in order to measure Z_{pol}, Z_{cath} and Z_{anod}, respectively. Due to ohmic losses in the cell a part of the potential difference across the cell is included in each of the measured I-V-characteristics or impedance curves. The distribution of the electrolyte resistance to the measured electrode impedances significantly depends on the arrangement of the electrodes. In the case of an ideal electrode alignment (i.e. electrode misalignment $<<$ electrolyte thickness) this electrode arrangement can provide useful, although not completely free of errors, information about the polarisation processes of the individual electrodes [46]. Otherwise significant errors may occur due to the inhomogeneous current density distribution [20,47].

In addition to providing information on polarisations, impedance spectroscopy is also useful in simulating reaction mechanisms; this is illustrated here for the cathodic reaction. In the reaction model presented here, only the following steps are considered: (1) Dissociative adsorption of oxygen at the cathode surface; (2) Surface diffusion of adsorbed oxygen along the cathode surface through the pores; and (3) Reduction of adsorbed oxygen at the TPB and subsequent vacancy exchange and oxygen incorporation into the electrolyte. The electrolyte surface is assumed to remain inactive because of its low electronic conductivity. For further simplification, it is assumed that the oxygen surface diffusion proceeds sufficiently fast and therefore can be neglected.

The dissociative adsorption of oxygen is assumed to proceed via the reaction

$$O_2(g) + 2s \underset{k_{des}}{\overset{k_{ads}}{\rightleftarrows}} 2O_{ad} \qquad (27)$$

where k_{ads} and k_{des} are the rate constants for adsorption and desorption of molecular oxygen, 's' is a vacant active surface site for oxygen and O_{ad} is an oxygen atom adsorbed on an active site. The adsorbed oxygen then diffuses along the pore walls of the cathodes and enters the TPB-region where it reacts according to the following equation

$$O_{ad} + V_O^{\bullet\bullet} \underset{k_{ox}}{\overset{k_{red}}{\rightleftarrows}} O_O^x + s + 2h^{\bullet} \qquad (28)$$

and transfers into the electrolyte bulk. Here, k_{red} and k_{ox} are the rate constants for the oxygen exchange reaction, in forward and reverse directions, respectively. The law of mass action applied to reactions (27) and (28) yields

$$k_{ads}p(O_2)[s]^2 = k_{des}[O_{ad}]^2 \qquad (29)$$

$$k_{red}[O_{ad}][V_o^{\bullet\bullet}] = k_{ox}[O_o^x][s] \tag{30}$$

The concentration of adsorbed oxygen $[O_{ad}]$ and the corresponding surface vacancy concentration $[s]$ depend on the oxygen partial pressure in the gas phase $p(O_2)$ for given rate constants k_{ads}, k_{ox}, k_{des} and k_{red}. Their actual values depend on the microstructural properties of the cathode surface and the interface and on temperature while the oxygen concentration $[O_o^x]$ and the oxygen vacancy concentration $[V_o^{\bullet\bullet}]$ in the electrolyte are given by the compositions of the materials. The reaction rate constants k_{red} and k_{ox} consist of a potential-dependent and a part given by the activation potential ΔG_c^0 of the form

$$k_{red} = k_{chem,red} e^{\frac{(1-\beta 2)2F}{RT}(\Delta\Phi_e+\eta)}; \quad k_{chem,red} \sim e^{\frac{-\Delta G_c^0}{RT}} \tag{31}$$

where β is the symmetry factor of the interface and the other constants have their usual meanings. Therefore, the net Faradaic current i_F that passes through the interface depends exponentially on the overvoltage η (Butler Volmer-type behavior [26]).

In the model, the quantity $x = [O_{ad}]/N_0$ denotes the oxygen surface coverage. $N_0 = [s] + [O_{ad}]$ is the concentration of active oxygen sites on the LSM surface. Its value is given by the rates of surface exchange and charge transfer at the interface and depends on the operating conditions and on the materials parameters. By combining equations (29) and (30), the mass and charge balances for oxygen in the TPB-region can be expressed as

$$\frac{dx}{dt} = 2k_{ads}p(O_2)N_0(1-x)^2 - 2k_{des}N_0x^2 \\ + k_{ox}[O_o^x](1-x) - k_{red}([V_o^{\bullet\bullet}]x \tag{32}$$

$$i_F = 2FN_0 l_{tpb} w \cdot \{k_{ox}[O_o^x]1-x) - k_{red}[V_o^{\bullet\bullet}]x\} \tag{33}$$

where F is the Faraday constant, l_{TPB} is the TPB-length of the electrode/electrolyte/gas phase and w is the lateral extension of the TPB-region. Usually, the magnitude of w is not known, but can still be included as a parameter. For a given equilibrium potential difference, $\Delta\Phi_e$, and known rate constants, the equilibrium surface coverage x is determined using the above relations. According to this model, the static characteristic $i_F(\eta)$ obeys Butler-Volmer behavior not too far from equilibrium. Under high polarisation, the current is limited by the surface adsorption process. Models of this type are discussed in greater detail in references [16,48] and literature referenced there in.

The evaluation of the non-linear $i_F(\eta)$-characteristics given by equation (33) is mathematically difficult. Therefore, the static polarisation curves are usually linearised and subjected to Fourier transformation that yields an expression of the Faradaic impedance Z_F of the interface for the given operating point.

$$Z_F(\omega) = \frac{F\{\eta(t)\}}{F\{i_F(t)\}} \approx R_{ct} + \frac{R_{ads}}{1 + j\omega R_{ads} C_{ads}} \tag{34}$$

where R_{ct} is the charge transfer resistance for oxygen incorporation at the interface and R_{ads} and C_{ads} contain the influence of the adsorption process on the impedance for a pure ac input signal[6]. The values of R_{ct}, R_{ads} and C_{ads} are functions of the reaction rates, the oxygen partial pressure, operating conditions and materials parameters that are part of the model [16]. The ratio R_{ct}/R_{ads} indicates if the reaction is controlled by adsorption or by charge transfer. The corresponding equivalent circuit is given in the inset in Figure 9.10.

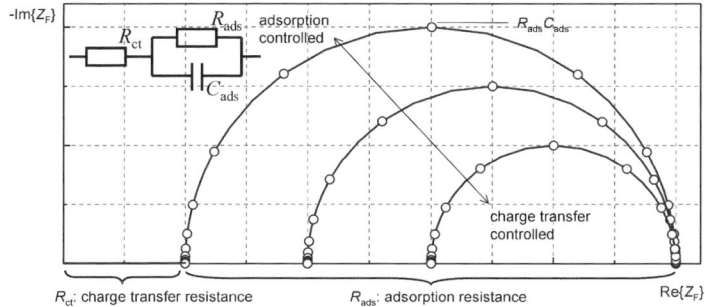

Figure 9.10 *Nyquist plot of the Faradaic impedance Z_F of the oxygen reduction reaction simulated from the above reaction model and corresponding equivalent circuit. The model describes a competition between surface adsorption and incorporation, depending on the ratio of the rate constants for the respective processes.*

Because the interface is polarised, an additional double layer capacitance occurs at the interface and is assumed to act in parallel to the Faradaic impedance. The double layer conceals the Faradaic impedance and has to be taken into account for simulation purposes.

Distributions of relaxation times can be simulated using equation (34) for different $p(O_2)$. The series of distribution functions is then compared to distributions obtained from electrochemical impedance measurements carried out under the same variation of experimental conditions as shown in Figure 9.11. The peaks in the distribution function are characterised by their frequency, shape and area. By comparing dependencies of these peak parameters on the experimental variables, that were varied in the measurement series, with the same parameter variation from simulation, physical processes described by the model may be attributed to relaxation peaks in the distribution function calculated from the impedance response of the system.

Electrochemical impedance spectroscopy is especially useful if the system performance is governed by a number of coupled processes each proceeding at a different rate. The physical and chemical processes contributing to the

---

[6] It is assumed that capacitance associated with the charge transfer process is small enough so that it is not reflected over the frequency range considered.

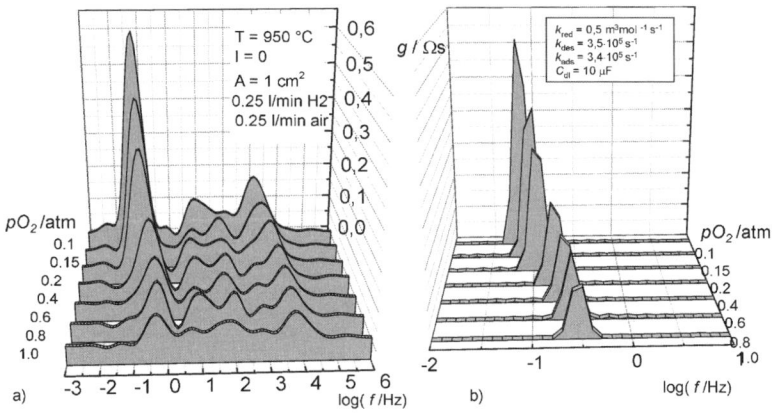

Figure 9.11 Variation of the cathodic oxygen partial pressure ($pO_2 = 0.1$ atm ... 1.0 atm): distribution functions of relaxation times (a) calculated from an impedance measurement series and (b) simulated from the physical sub-model. The simulation was performed assuming the same variation in the oxygen partial pressure as in the actual measurement series. Parameters in the simulation series show a behavior similar to the dominating peak in the measurement series.

internal resistance of the cell determine their dynamic behavior over a wide range of frequencies. The relaxation times could span more than fifteen orders of magnitude, assuming the time dependence can be described in terms of relaxation times (or time constants), reaching from fast processes that sustain cell operation, e.g. gas flow and charge transfer, to long-term degradation processes limiting the life time of the cell (Figure 9.12). Because of practical factors limiting the frequency range of impedance measurements on fuel cells, the method is useful only for processes with relaxation times ranging from µs up to tens of seconds. Slower processes exhibiting time constants from several minutes to hundreds of hours are favorably observed in the time domain, e.g. by analysing the response of the cell on a step function of the current response.

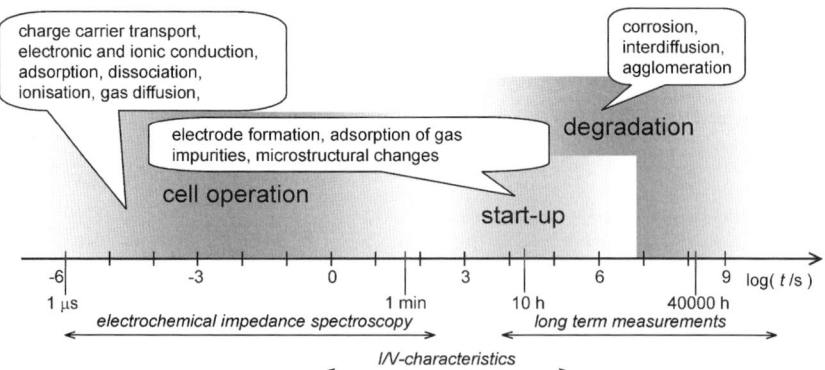

Figure 9.12 Relaxation times of physical processes present in fuel cell operation and corresponding electrical measurement techniques. The dynamic range spans over 15 orders of magnitude. Fast processes are covered by electrochemical impedance spectroscopy [46].

9.6 Summary

Electrode polarisations cause large voltage losses in SOFCs and need to be reduced to low levels for increased efficiency. The three polarisations described in this chapter are ohmic polarisation, concentration polarisation, and activation polarisation. The ohmic contribution stems from resistance to electron and ion flows in the materials, and is generally dominated by the electrolyte resistance, with the consequence that SOFCs employing thick (>100 micron) YSZ as electrolyte have high ohmic losses at temperatures below about 900°C. Now that thinner electrolytes are being used in electrode-supported cells, this resistance has dropped and it is possible to use YSZ down to about 700°C.

Concentration polarisation is caused by the resistance to mass transport through the electrodes and interfaces and is generally largest at the cathode, particularly when thick, cathode-supported cells are employed. The voltage drop is large at high current densities under conditions such that the electrolyte/electrode interface is starved of fuel (anode) or oxidant (cathode) when gaseous species cannot diffuse fast enough through the porous interstices of electrodes. In the case of anode, diffusion is generally rapid due to the presence of low molecular weight hydrogen. This means anode-supported cells usually exhibit low concentration polarisation, even with relatively thick anodes.

Activation polarisation is the voltage drop due to the sluggishness of reactions occurring at the electrode-electrolyte interfaces. Several processes are necessary for electron transfer to take place, especially at the cathode. Because LSM has little ionic conductivity, these processes are localised at the TPBs. Recently, it has become common to use MIEC (composite or single-phase) cathodes to spread the TPB and extend the reaction zones; this has had a beneficial effect on reducing the activation polarisation and allowed better SOFC performance at lower temperatures.

Bulk of the studies reported to date show that with the materials that have been researched to date, the largest contribution to polarisation is from the cathodic reaction. The kinetics of the reduction process is governed by the composition of materials as well as by the microstructure of the cathode. Minimisation of the electrode polarisations is possible by choosing appropriate materials, their compositions and morphology. From a microstructural standpoint, activation polarisation is lower if the electrode structure is fine in the immediate vicinity of the electrolyte. For minimising concentration polarisation, by contrast, electrode structure should be coarse with large amount of porosity. For this reason, an ideal electrode structure is graded, fine near the electrolyte to minimise activation polarisation, and coarse in regions away from the electrolyte to minimise concentration polarisation.

References

[1]   D. R. Gaskell, *Introduction to Thermodynamics of Materials*, Taylor and Francis, 1995.

[2] C. H. Hamann, A. Hamnett and W. Vielstich, *Electrochemistry*, J. Wiley & Sons, New York, 1998.
[3] E. L. Cussler, *Diffusion: Mass Transfer in Fluid Systems*, Cambridge University Press, Cambridge, 1984.
[4] E. A. Mason and A. P. Malinauskas, *Gas Transport in Porous Media: The Dusty-Gas Model*, Elsevier, Amsterdam, 1983.
[5] J. R. Macdonald, *Impedance Spectroscopy*, John Wiley-Interscience, New York, 1987.
[6] J. W. Kim, A. V. Virkar, K. Z. Fung, K. Mehta and S. C. Singhal, *J. Electrochem. Soc.*, **146** [1] (1999) 69-78.
[7] Peter Grathwohl, *Diffusion in Natural Porous Media*, Kluwer Academic, Boston, 1997.
[8] F. Zhao, T. J. Armstrong and A. V. Virkar, *J. Electrochem. Soc.*, **150** [3] (2003) A249-A256.
[9] E. Siebert, A. Hammouche, and M. Kleitz, *Electrochim. Acta*, **40** (1995) 1741.
[10] E. Ivers-Tiffée, M. Schießl, H.J. Oel and W. Wersing, in *Proceedings of the 14th Risø International Symposium on Materials Science: High Temperature Electrochemical Behaviour of Fast Ion and Mixed Conductors*, eds. F.W. Poulsen, J. J. Bentzen, T. Jacobsen, E. Skou and M. J. L. Østergård, Risø National Laboratory, Roskilde, Denmark, 1993, pp. 69.
[11] M. Kuznecov, M. P. Otschik, K. Eichler and W. Schaffrath, *Ber. Bunsen.*, **102** (1998) 1419.
[12] T. Inoue, K. Eguchi, T. Setoguchi and H. Arai, *Solid State Ionics*, **40/41** (1990) 407.
[13] A. Hahn and H. Landes, in *Solid Oxide Fuel Cells V*, eds. U. Stimming, S.C. Singhal, H. Tagawa and W. Lehnert, The Electrochemical Society Proceedings, Pennington, NJ, PV 97-40, 1997, pp. 595.
[14] H. Kamata, A. Hosaka, J. Mizusaki, and H. Tagawa, *Solid State Ionics*, **106** (1998) 237.
[15] J. van Herle, A.J. McEvoy and K. Ravindranathan Thampi, *Electrochim. Acta*, **41** (1996) 1447.
[16] A. Mitterdorfer and L.J. Gauckler, *Solid State Ionics*, **117** (1999) 203.
[17] A. Hammouche, E. Siebert and M. Kleitz, *J. Electrochem. Soc.*, **138** (1991) 1212.
[18] S.B. Adler, J.A. Lane, and B.C.H. Steele, *J. Electrochem. Soc.*, **143** (1996) 3554.
[19] T. Horita, K. Yamaji, M. Ishikawa, N. Sakai, H. Yokokawa, T. Kawada and T. Kato, *J. Electrochem. Soc.*, **145** (1998) 3196.
[20] M.J.L. Østergård and M.M. Mogensen, *Electrochim. Acta*, **38** (1993) 2015.
[21] F.P.F. van Berkel, F.H. van Heuveln and J.P.P. Huijsmans, *Solid State Ionics*, **72** (1994) 240.
[22] J. Mizusaki and H. Tagawa, *J. Electrochem. Soc.*, **138** (1991) 1867.
[23] J. Mizusaki, H. Tagawa, T. Saito and H. Narita, in *Proceedings of the 14th Risø International Symposium on Materials Science: High Temperature Electrochemical Behaviour of Fast Ion and Mixed Conductors*,

eds. F.W. Poulsen, J. J. Bentzen, T. Jacobsen, E. Skou and M. J. L. Østergård, Risø National Laboratory, Roskilde, Denmark, 1993, pp. 343.
[24] F. Zhao, Y. Jiang, G. Y. Lin and A. V. Virkar, in *Solid Oxide Fuel Cells VII*, eds. H. Yokokawa and S. C. Singhal, The Electrochemical Society Proceedings, Pennington, NJ, PV2001-16, 2001, pp. 501.
[25] A. Weber, R. Manner, R. Waser and E. Ivers-Tiffée, *Denki Kagaku*, **64** [4] (1996) 582-589.
[26] J.O'M. Bockris and A.K.N. Reddy, *Modern Electrochemistry*, Kluwer Academic-Plenum Press, New York, 1998.
[27] C. W. Tanner, K. Z. Fung and A. V. Virkar, *J. Electrochem. Soc.*, **144** [1] (1997) 21-30.
[28] R.A. De Souza, Ionic Transport in Acceptor-Doped Perovskites, Ph.D. Dissertation, Imperial College, London, 1996.
[29] G. Prentice, *Electrochemical Engineering Principles*, John Wiley, New York, 1991.
[30] T. Kenjo, S. Osawa and K. Fujikawa, *J. Electrochem. Soc.*, **138** (1991) 349.
[31] A. V. Virkar, K-Z. Fung and C. W. Tanner, Electrode Design for Solid State Devices, Fuel Cells and Sensors, U. S. Patent No. 5,543,239, August 6, 1996.
[32] R. M. Thorogood, R. Srinivasan, T. F. Yee, and M. P. Drake, Composite Mixed Conductor Membranes for Producing Oxygen, U. S. Patent No. 5,240,480, August 31, 1993.
[33] H. Deng, M. Zhou and B. A. Abeles, *Solid State Ionics*, **74** (1994) 75.
[34] I. V. Murygin, *Elektrokhimiya*, **23** [6] (1987) 740.
[35] E. Ivers-Tiffée, A. Weber and D. Herbstritt, *J. Europ. Cer. Soc.*, **21** (2001) 1805.
[36] H. J. M. Bouwmeester, H. Kruidhof and A. J. Burggraaf, *Solid State Ionics*, **72** (1994) 185.
[37] J. A. Kilner, R. A. De Souza and I. C. Fullarton, *Solid State Ionics*, **86-88** (1996) 703.
[38] K. Huang and J. B. Goodenough, *J. Electrochem. Soc.*, **148** [5] (2001) E203.
[39] S. Park, J. M. Vohs, and R. J. Gorte, *Nature*, **404** (2000) 265-266.
[40] O. A. Marina and L. R. Pederson, in *Proceedings of the Fifth European SOFC Forum*, ed. J. Huijsmans, Switzerland, 2002, pp. 481-489.
[41] J. Canales-Vazquez, S. W. Tao, and J. T. S. Irvine, *Solid State Ionics*, (2003) in press.
[42] Yi Jiang and A. V. Virkar, *J. Electrochem Soc.*, **150** [7] (2003) A942.
[43] M. Nagata, Y. Ito, and H. Iwahara, *Solid State Ionics*, **67** (1994) 215.
[44] T. Jacobsen and E. Skou, in *Materials and Processes: Proceedings of the IEA Workshop*, Les Diablerets, Switzerland, 1997.
[45] S. B. Adler, *J. Electrochem. Soc.*, **149** (2002) E166.
[46] A. Weber, A.C. Müller, D. Herbstritt and E. Ivers-Tiffée, in *Solid Oxide Fuel Cells VII*, eds. H. Yokokawa and S. C. Singhal, The Electrochemical Society Proceedings, Pennington, NJ, PV2001-16, 2001, pp. 952.

[47] J. Winkler, P.V. Hendriksen, N. Bonanos, and M. Mogensen, *J. Electrochem. Soc.*, **145** (1998) 1184.

[48] H. Schichlein, A.C. Müller, M. Voigts, A. Krügel and E. Ivers-Tiffée, *J. Appl. Electrochem.*, **32** (2002) 6.

Chapter 10

Testing of Electrodes, Cells and Short Stacks

Mogens Mogensen and Peter Vang Hendriksen

10.1 Introduction

This chapter describes electrochemical testing of individual electrodes and cells and gives a brief discussion of stack testing. It is extremely important in testing solid oxide fuel cells (SOFCs) to have a good understanding of the fundamental principles of electrochemistry; to this end, classical textbooks [1–3], which are based on liquid electrochemistry, are certainly helpful, but it is also extremely important to be familiar with different aspects of solid-state electrochemistry, especially those involving solid oxide electrolytes [4–12].

Testing of cells and stacks is conducted for two reasons; to assess their commercial viability and for continued cell development. The information required for these two purposes is usually different, and an interpretation of the data, taking into account the effects of the test conditions, has to be done for comparison of SOFCs from different sources. Despite the fact that many cell test data have been reported [13–19], no general agreement on test procedures exists and the actual test equipment is often not described in detail. Also the concept of area-specific resistance (ASR) is not standardised when reporting the test results. Although parameters such as temperature, inlet gas composition, fuel utilisation and current density are normally given, substantial additional information is required for complete evaluation of test data and detailed analysis of cell behaviour. Four particular issues that are considered in this chapter are: definition of area specific resistance (ASR), accuracy of temperature measurement at the cell, the effects of gas leakage through the seals, and the use of reference electrodes.

If a stack is to be tested as a commercial product, the test is usually performed in a complete system with balance-of-the-plant components added and the stack integrated into the system to the maximum possible extent. Such tests of complete SOFC systems are both expensive and complex, and it is difficult to interpret the results. Therefore, extensive testing is usually conducted only on cells and the various cell components, primarily electrodes. Nevertheless, tests on stacks and complete systems have been performed, see e.g. [20–26], but no general test methods can be derived from these reports. If an interconnect material is used for current collection from the electrodes in a single cell test, then the test is sometimes referred to as a 'short stack', a 'stack element' or a 'stack unit' test [27]. It should be noted that the contact resistance between the electrodes and the interconnect is usually significant [25], especially in cases where the interconnect is made of stainless steel or other chromia forming alloys, which have a tendency to form poorly conducting surface layers [28,29].

There are special problems in SOFC electrode testing. An electrode can only be characterised electrochemically if it is part of a cell with at least one reference electrode, that is, a cell with at least three electrodes in contact with the same electrolyte. However, it is difficult to place a usable reference electrode on a thin electrolyte film in an electrode-supported cell; this is discussed in the next section.

10.2 Testing Electrodes

The main problem with characterising the individual electrodes in single cells and short stacks is the insertion of the reference electrodes, which are used to judge the performance of the individual components and interfaces in the cell. Since a single cell stack is made up of five components, an electrolyte, two electrodes and two interconnects, there are four interfaces at which reference electrodes can be inserted. Unfortunately, such reference electrodes will not work in the geometries that are normally employed.

Results of tests on cells with thin electrolyte layers using one or more reference electrodes have been reported on many occasions [30–33] but the electrodes under investigation appeared not to be polarised; the short explanation is that the reference electrodes were not working. The geometric requirements for the position of electrodes in three-electrode set-ups have been treated in detail by several researchers [34–39], and there is general agreement among these researchers. In spite of this, there still seems to be a great need in the SOFC community for basic information on how to measure electrode potentials properly; some of these details are given below.

Figures 10.1 and 10.2 illustrate the problem. An electrode-supported cell with a 'reference' electrode is often sketched as shown in Figure 10.1a. However, such a sketch is very deceiving when it is used for an assessment of the current distribution. For this purpose, the sketch should be drawn to scale, i.e. the electrolyte thickness should be the relevant unit of length. When the correct length scale is used, as in Figure 10.1b, it is evident that the gap between the upper working electrode and the 'reference' electrode is huge. This means that the current distribution around the right-hand edge of the working electrode becomes very different from the even current distribution in the main part of the cell. Furthermore, the current in the vicinity of the 'reference' electrode becomes

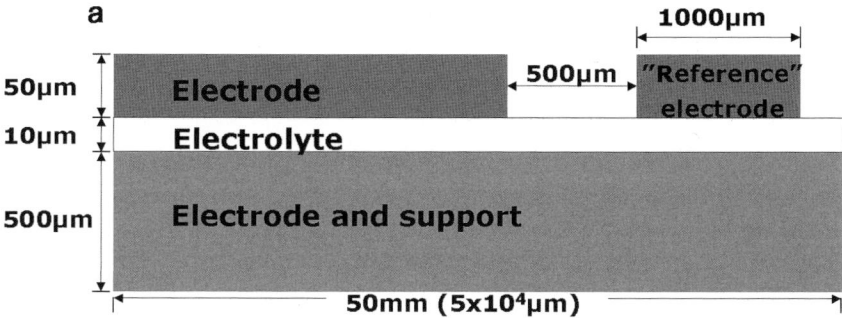

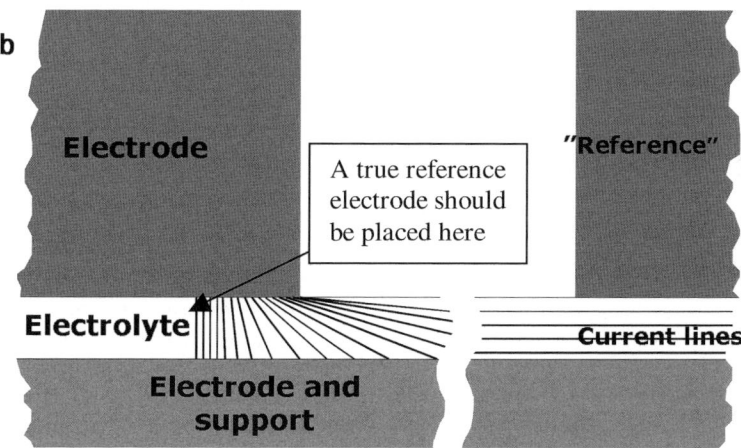

Figure 10.1 (a) A typical sketch of an electrode-supported cell. Note that such an illustration is out of scale because the gap between the top electrodes should be 50 times greater than the electrolyte thickness. (b) Expanded view of electrode corners showing the current distribution indicated by schematic current lines. The current density, apart from being approximately parallel to the electrolyte plane, is very small at the position of the 'reference' electrode, at least 50 (500 μm/10 μm) times smaller than the current density of the cell. The correct position of the reference electrode would be inside the electrolyte of the cell and some distance away from the corner, but this is difficult in a 10 μm thick electrolyte.

parallel to the electrolyte plane, i.e. there is only a minute voltage difference across the electrolyte at the 'reference' electrode position.

Figure 10.2a illustrates the potential across a cell in the open circuit voltage (OCV) condition. Note, that for a good electrolyte (ionic conduction only), there will be no potential gradient inside the electrolyte at zero current. The whole potential change across the cell is localised at the interfaces between the electrolyte and the electrodes. These regions are the so-called electrochemical double layers with thickness in the nanometre range and with high space charge concentrations as a result of the very high potential gradients. Figure 10.2b gives the potential across a cell when it is loaded, i.e. a current flows through it.

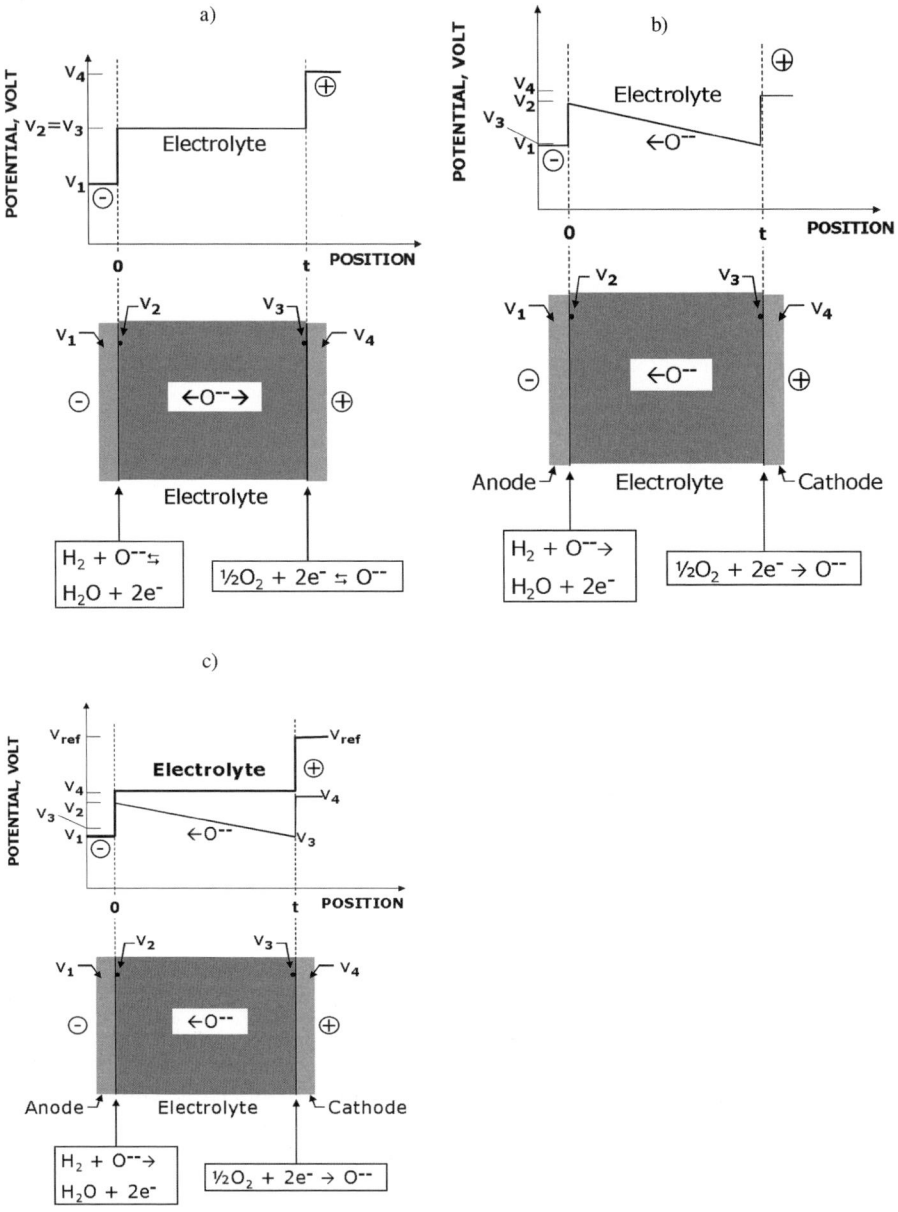

Figure 10.2 (a) Potential through the electrode-supported cell with no current and (b) through a cell with a current load. (c) The potential across the electrolyte at the 'reference' electrode position (thick line) and through the cell part with the current load (thin line). It is seen that: $(V_{ref} - V_4)/i = R_{p,anode} + R_{p,cathode} + R_{elyt}$ = the total polarisation of the cell apart from concentration polarisation.

Note that there is a potential loss across the electrolyte due to the electrolyte resistance. The potential steps at the interfaces are now smaller compared to the OCV condition due to the losses originating from the polarisation resistance of the electrode processes. Figure 10.2c illustrates the case of an anode-supported cell. The potential is given for both the position of the 'reference' electrode where no current flows across the electrolyte (i.e. the electrical potential is constant across the electrolyte), and for a position far away from the electrode edges as in Figure 10.2b.

As the potential along the anode, which is a very good electronic conductor, is the same everywhere (the anode constitutes an isopotential plane), the potential of the anode at the 'reference' electrode must be equal to the potential in the middle of the current-bearing part of the anode. Thus the two potential curves in Figure 10.2c must start at the same point. Therefore, as seen in Figure 10.2c, the potential difference between the 'reference' electrode and the upper electrode in Figure 10.1 is simply the total polarisation of the full cell. Thus it is clear that the 'reference' electrode measures only the emf of the cell with the actual gas compositions at the 'reference' and inside the support at the lateral position opposite to the reference while the current is flowing. Thus, no information on what happens on any of the working electrodes can be derived from such measurements. If the concentrations of the reactants and products at the lateral position of the reference electrode were the same as in the active electrode/electrolyte interfaces, then it would be possible to deduce the total concentration overpotential. This is, however, in general not the case, and this means that the voltage difference actually measured between the working and the 'reference' electrodes cannot be assigned any clear meaning. In the present context, it is also helpful to remember that a single-electrode potential cannot be measured directly; it is only possible to measure a potential difference between two electrodes.

To avoid such problems, it is necessary to test the electrodes using a suitable three-electrode set-up or a symmetrical two-electrode cell [40], even though both have their shortcomings. Some examples of useful set-ups for studying electrode performance are briefly presented here. The set-ups are based on zirconia pellets with an electrode arrangement suitable for three-electrode studies. Specific material choices and mounting details are given in the caption to Figure 10.3. Such pellet-like geometries, where the reference electrode can be suitably placed (in a bore as in the figure or as a ring around the pellet) are suitable for fundamental studies of electrode kinetics. The pellet-like test cell geometries depicted in Figure 10.3 suffer from two disadvantages: it is difficult to ensure that the fabrication process for the electrodes used is identical to the one used for the actual cells, and the ohmic resistance between the working and the reference electrodes is quite substantial which may result in a 'signal to noise' problem when very good electrodes are studied. An improved three-electrode geometry using a ring-shaped working electrode is currently being investigated [39].

To measure a particular electrode performance in detail, a symmetrical cell with identical electrodes on each side can be used as shown in Figure 10.4. This has a platinum mesh to make good contact with the electrodes and two platinum

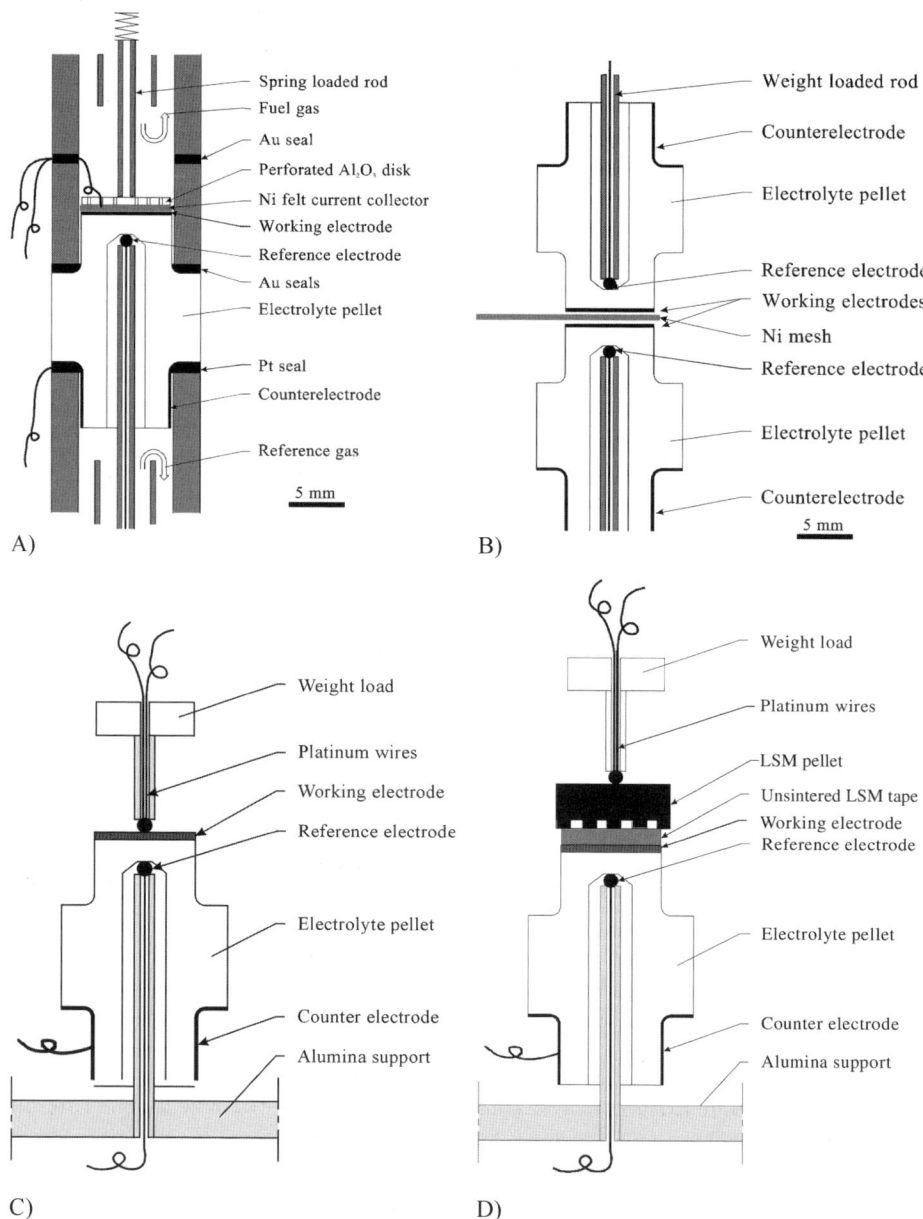

Figure 10.3 Possible three-electrode set-ups based on a cylindrical YSZ electrolyte pellet. The pellet is mounted on an alumina tube. (A) and (B) illustrate material choices and details suitable for anode studies and (C) and (D) likewise for cathode studies. Using two three-electrode pellets pressed against a common Ni mesh as depicted in (B), one may eliminate the impedance due to gas conversion. (C) In the case of a very well conducting electrode material, the current is picked up directly from the working electrode using a Pt bead. (D) In the case of a not-so-well conducting electrode, an improved current pick-up may be obtained by an unsintered LSM foil in contact with a dense, channeled LSM pellet [36,40].

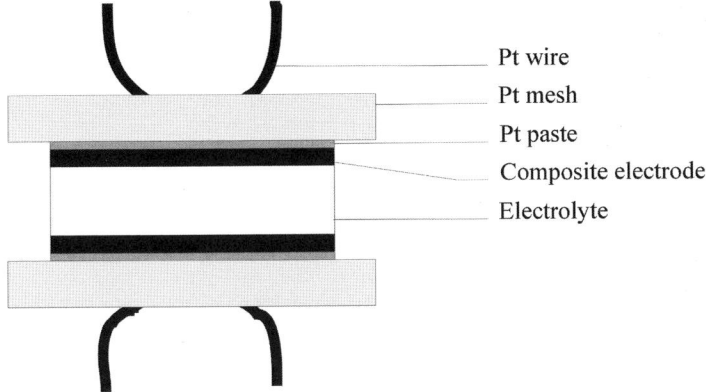

Figure 10.4. A symmetrical two-electrode cell arrangement for measurements near the OCV.

wires coming out from each side of the cell, one to measure current and the other for potential determination. Such a test cell is well suited for electrode development work because there is no ambiguity about the source of electrode properties in this case. However, its use is limited to investigations close to open circuit voltage (OCV), where the electrode loss does not depend on whether it is anodically or cathodically polarised.

10.3 Testing Cells and 'Short' Stacks

For testing planar SOFCs, a test house such as the one shown in Figure 10.5 may be used. Figure 10.5a illustrates how the cell is sandwiched between fuel and air distributor plates, which are contacted with gold or platinum foils to pick up the electrode current. This is then sealed into the test assembly shown in Figure 10.5b. The cell is sealed at its edges between two alumina blocks, which hold everything in place. The sealing is obtained using glass bars, which softens on heating. In order to prevent small leaks affecting the measurements, another seal is made several mm outside the first, and the gap between the two seals is swept with nitrogen containing 3% hydrogen. The alumina blocks have built-in gas channels for air inlet, air outlet, fuel inlet, fuel exit, and sweep gas. Current pick-up is also achieved through these alumina blocks, which also contain several voltage probes to indicate the voltage gradient along the electrodes and thermocouples to measure the temperature at the cell in a number of points.

A detailed drawing of the alumina blocks is given in Figure 10.5c to indicate the number of probes. Not shown are two oxygen sensors, one at the fuel inlet and one at the outlet to measure the amount of oxygen entering the fuel compartment through the cell electrolyte. This can be compared with the oxygen transport calculated from Faraday's law in order to check whether any leakage has occurred.

The whole assembly described in Figure 10.5 is enclosed in a furnace within a ventilated hood into which gases are fed from the manifold system shown in

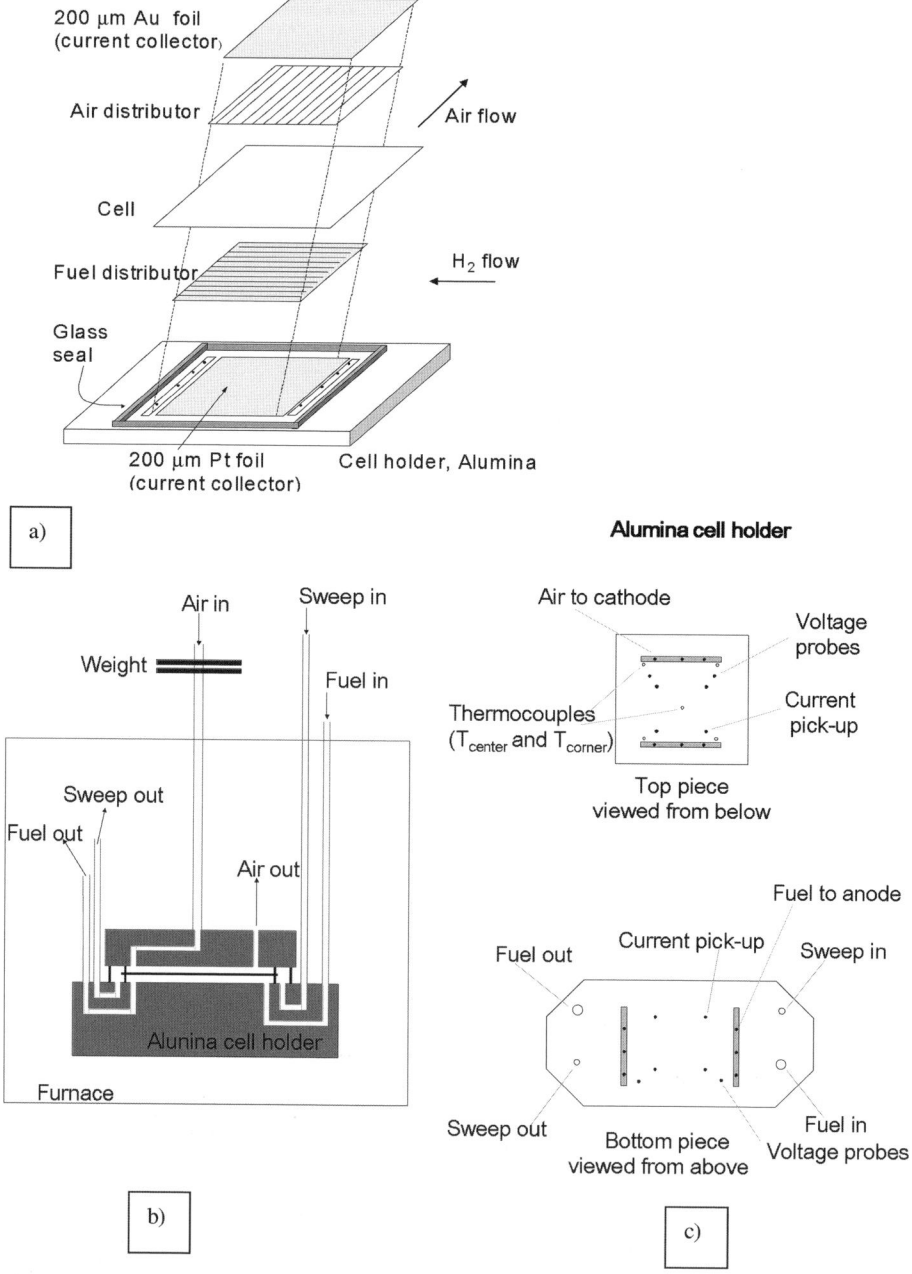

Figure 10.5 (a) The assembly of cell and distributor plates in a cross-flow pattern; (b) arrangement of the cell in the alumina test blocks; (c) the detailed design of the alumina blocks.

Figure 10.6. This allows a range of gases to be used on the fuel side, while air or oxygen can be used on the oxidant side. Typically, hydrogen is used as fuel, metered through flow controllers, and then humidified with a water bubbler before entering the cell compartment.

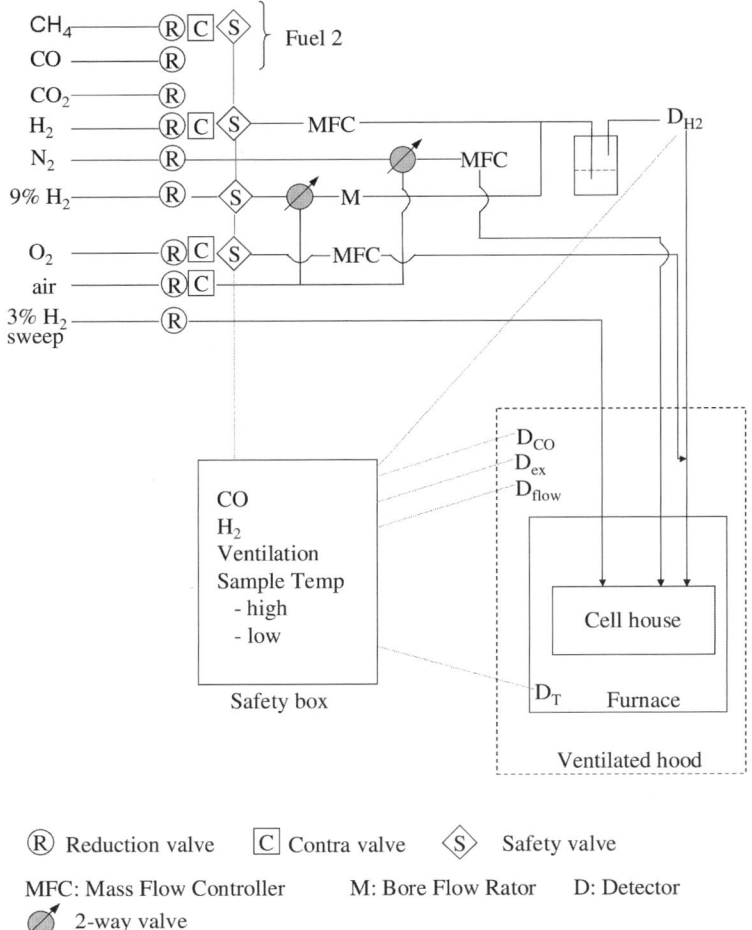

Figure 10.6 Schematic of the gas manifolding arrangement for single-cell tests.

Results of a recent cell test at Risø of an anode-supported cell are shown in Figure 10.7. The fuel utilisation was calculated from the current flowing in the cell and also from the oxygen analysis at the fuel inlet and outlet. It is seen that there is an excellent agreement between the fuel utilisations derived from the two independent methods. Furthermore, the OCV and the oxygen potential of the inlet fuel gas agreed within 1 mV. When obtaining the data in Figure 10.7, the cell was operated at constant gas flows, which may be convenient for cell characterisation purposes, but is obviously not the way the cells will eventually

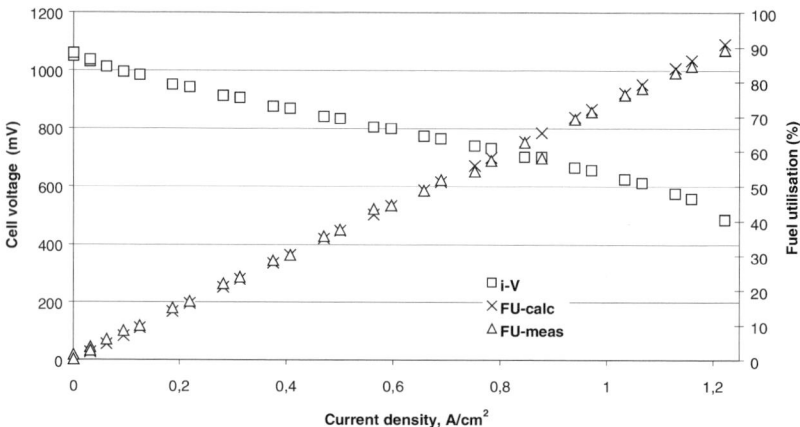

Figure 10.7 A cell test with varying fuel utilisation (FU). The test gases were air (170 l/h) and hydrogen with 5% water (9 l/h). The cell temperature was 845°C. The agreement between the fuel utilisation calculated from the oxygen potentials of the fuel in and out, and the FU derived from the measured current using Faraday's law, is good over the entire range from 0 to 90 % FU. A maximum power density of 0.65 W/cm² (0.61 V × 1.07 A/cm²) was obtained at 80% FU. Cell area: 16 cm².

operate in a practical power system; there, gas feed will be reduced at part-load in order to maintain high fuel utilisation. Data on the same cell but operated at 'constant' fuel utilisation are depicted in Figure 10.8. Note that a straight line very well describes the i–V curve.

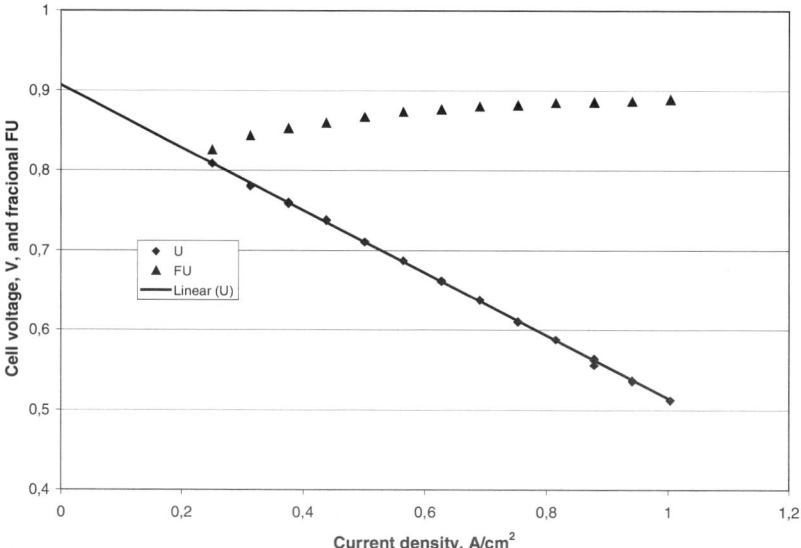

Figure 10.8 Cell voltage, U, as a function of current density with the fuel utilisation (FU) kept within 80–90%. The cell was operated on hydrogen containing ca. 5% water vapour at the inlet and the air flow was kept constant at 170 l/hr. The cell temperature was 835–840°C. The solid line is the 'best fit' straight line.

Such single-cell tests are useful to determine the performance of SOFCs under ideal conditions where gold or platinum is used to make excellent current contact with the electrodes. However, there can still be contact problems, as shown in Figure 10.9 which shows how cell voltage falls with time as a result of an interface problem between the platinum foil and the gas distributor plate. Figure 10.9 gives the variation in cell voltage over time of a cell tested at a constant current density. Also shown is the voltage loss over the interface between the anode gas distributor plate (Ni/YSZ-cermet) and the Pt current collection foil, as measured by potential probes connected to the two components. Evidently, the degradation of the cell voltage is not due to a degradation of the cell – the increasing loss occurs at the interface between Pt foil and the gas distributor plate. Though the degradation mechanism is not fully understood, it appears to be related to Ni loss from the interface. Much better long-term stability can be obtained using Ni rather than Pt foils for current collection on the anode side and this is preferred for long-term testing.

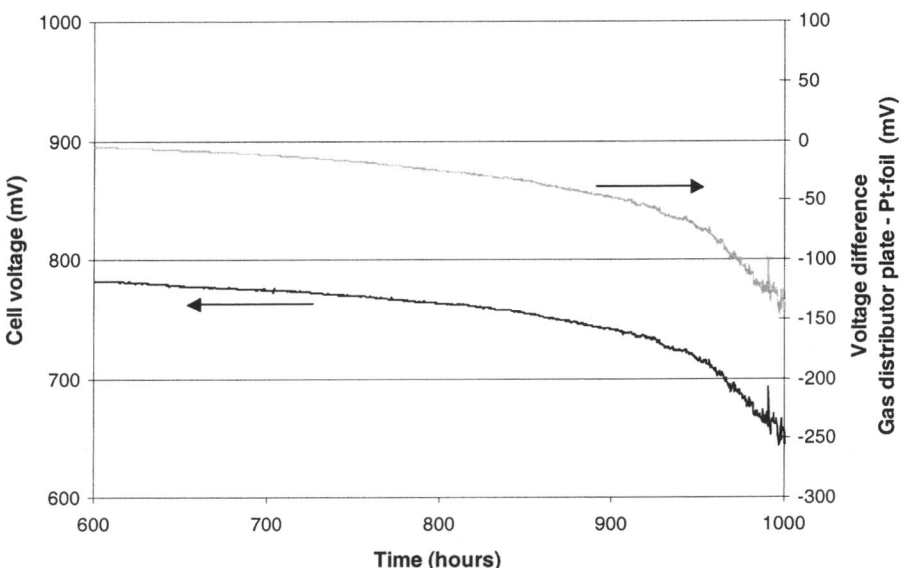

Figure 10.9 Voltage degradation with time of a cell tested at constant current density (0.5 A/cm^2) in a hydrogen/water vapour gas mixture (48% hydrogen, 52% water vapour). The cell temperature was $850°C$. Also shown (right-hand scale) is the voltage loss over the interface between the Pt current collection foil and the Ni/YSZ gas distributor plate, which evidently accounts for the observed loss in 'cell voltage'.

The detailed and precise SOFC tests described above are ideal for research purposes. For industrial screening of SOFCs, it may be, under certain circumstances, preferable to use a simpler test equipment [19] with easily assembled seals, for example gold rings, and safer fuels at a single concentration, say nitrogen containing typically < 9% hydrogen. The usual result from such a test is a current density–voltage curve (i–V curve), which can indicate the effects of the cell design, component materials, and fabrication processes, performance stability with time, and performance variation with temperature. Often the results of screening tests are reported as power density (W/cm^2) figures, and record values of maximum power density are extensively quoted. However, this measure can be confusing because power density varies greatly with fuel composition and with electrode polarisation. The i–V curves for SOFCs are often linear and therefore allow an interpretation in terms of area-specific resistance (ASR). Even though ASR has no generally accepted definition, it is much less dependent on test conditions than power density, and it is preferable to use it to compare screening test results. The concept of ASR is discussed in the next section.

10.4 Area-Specific Resistance (ASR)

A fuel cell stack can be regarded as a 'black box' into which hydrogen (gas) and oxygen (air) are inputs, and electricity and exhaust gases are outputs. For such a stack, ASR is defined as:

$$ASR = \frac{Emf - U}{i} \tag{1}$$

where Emf is the electromotive force with the inlet fuel and air, and U is the cell voltage at the current density, i, at the design point. A possible design point, for example, might be 0.6 V at 1000°C, a fuel utilisation of 85% and an air flow of 4 times the stoichiometric amount ('4 stoichs').

The cell voltage, U, should be measured independently of the leads carrying the current, i.e. separate potential probes should be used. This ASR is in most cases not very sensitive to small variations in cell voltage and fuel utilisation. By determining the ASR at a few different temperatures, an apparent activation energy, E_A, may be derived. Thus, in a voltage interval (from say 0.5 to 0.7 V) and a temperature interval (say from 650 to 1050°C), the cell may be characterised, with fair approximation, by only two characteristic numbers, namely ASR at one temperature and E_A.

In case the i–V curve is concave, it may be tempting to use a differential ASR (i.e. the tangent) at high current densities as this gives a nice low value. Such a number has the drawback that it does not reflect the cell performance over the full polarisation range as does the quantity defined by Eq. (1).

Often, cell tests are conducted with very low fuel and air utilisations because these are easier to perform than the 'realistic' tests with high fuel utilisation. In case of insignificant fuel and oxygen utilisations, the relevant definition of ASR is again that of Eq. (1), but the insignificant utilisation makes this value incomparable to ASR derived from experiments with high fuel utilisation, because the concentration polarisation resistance due to the fuel and air conversion can be a considerable fraction of the total cell resistance. Thus, it is, in general, necessary to specify also the fuel and oxygen utilisation together with temperature and apparent activation energy. Furthermore, using almost dry hydrogen, as is the common practice, it is not easy to conduct experiments with a real negligible fuel utilisation, since even small current densities will create enough water to change significantly the Emf of the hydrogen/water fuel gas versus air, e.g. if the inlet gas contains 0.1% H_2O and the fuel utilisation is 0.1%, this changes the H_2O/H_2 ratio by a factor of 2, which in turn changes the Emf by 34 mV at 850°C. Therefore, and in order to be able to compare results for different fuel utilisations, the ASR value should be corrected for the effect of fuel utilisation. Before describing how this may be done, various contributions to the total ASR are examined below.

ASR may be divided into ohmic resistance, R_s, and electrode polarisation resistance, R_p. The ohmic resistance originates from the electrolyte, the electrodes materials and the current collection arrangement. This is very much dependent on geometric factors such as thickness of the cell components and the detailed geometry of the contact between current collection and electrodes, and between electrodes and electrolyte as current constrictions may be important [41]. The electrode polarisation resistance is further divided into contributions from the various rate-limiting steps. Thus, ASR can be broken down in five terms:

$$ASR = R_{elyt} + R_{connect} + R_{p,elchem} + R_{p,diff} + R_{p,conver} \quad (2)$$

where R_{elyt} is the electrolyte resistance calculated from the measured specific conductivity and the thickness; $R_{connect} = R_s - R_{elyt}$ is the resistance due to non-optimised contact and current collection; $R_{p,elchem}$ is the electrode polarisation originating from all the limiting chemical and electrochemical processes on the electrode surfaces, in the bulk electrode material and on the electrolyte/electrode interfaces; $R_{p,diff}$ is the contribution from the gas phase diffusion; and $R_{p,conver}$ is the contribution due to gas conversion, i.e. fuel oxidation and oxygen reduction. This division of ASR is based on what is possible to measure and calculate reliably rather than on any physical or electrochemical basis. Some terms in Eq. (2) can therefore be thought of as 'equivalent resistances', e.g. the Emf drop due to changes in gas composition resulting from the fuel utilisation is translated to an equivalent resistance. Depending on the exact type of electrode, different types of contributions are possible as derived from more basic electrochemical point of view. For example, current constriction may be important if the electrode has coarse porous structure but of less or no importance in case of a fine structured electrode. In one type of cathode, the surface diffusion may be important, but in another the diffusion of oxide ions (and electrons) through the electrode particles may cause the main polarisation loss. Values for 'the Eq. (2)' ASR contributions for an anode-supported cell (short stack) with a 1 mm thick support fed with hydrogen (with 3% steam) are given in Table 10.1 for 5 and 85% fuel utilisations (FU).

It is seen that the contribution from the concentration polarisation, $R_{p,diff} + R_{p,conver}$ is dominating. In an electrode-supported cell, the limitation of gas diffusion through the support is a cell-relevant resistance, whereas $R_{p,conver}$

Table 10.1 Contributions to ASR for a Risø-type anode-supported cell (Ni-YSZ/YSZ/LSM-YSZ) at 850°C tested in a plug flow-type configuration at 5 and 85% fuel utilisation (FU). R_{elyt} is calculated using a specific conductivity of YSZ of 0.045 S/cm, $R_{connect}$ is an estimation, $R_{p,elchem}$ is the sum of typical anode and cathode polarisation resistances measured in separate electrode experiments, $R_{p,diff}$ is calculated using a diffusion coefficient of 10 cm²/s, 30% porosity, a tortuosity factor of 3 and a thickness of 0.1cm, and $R_{p,conver}$ is calculated using Eq. (10) with $i = 0.5$ A/cm²

Resistance type	Contribution to ASR (Ω cm²)	
	5% FU	85% FU
R_{elyt}	0.06	0.06
$R_{connect}$	~0.1	~0.1
$R_{p,elchem}$	0.15	0.15
$R_{p,diff}$	0.06	0.02
$R_{p,conver}$	0.06	0.31

results from operation demands, and it is thus of special interest to be able to correct ASR for the effect of fuel conversion.

If a significant amount of fuel is consumed in the cell (or stack) under test, a resistance derived on the basis of Emf of the inlet gas (cf. Eq. (1)) will be an overestimation of the 'true' cell resistance. The larger the fuel utilisation the larger will be the overestimation and results of cell tests performed with different fuel (and air) utilisations are thus not directly comparable. A comparison between cell test results obtained under different and non-negligible fuel utilisations must thus, to be meaningful, be based on a resistance measure, ASR_{cor}, where the effects of changes in gas composition over the cell area have been taken into account. How the correction is applied depends on how the gases are fed to the cell. Here, two idealised cases are considered, namely the case where the fuel compartment may be considered a continuously stirred tank reactor (CSTR) or a plug flow reactor.

If the fuel compartment can be considered CSTR-like due to effective mixing because of a turbulent gas stream and fast gas diffusion, ASR_{cor} can be calculated from the expression

$$ASR_{cor} = \frac{Emf_{avg} - U}{i} \quad (3)$$

where Emf_{avg} signifies the average Emf, which in this case is the same as the Emf of the outlet gas. An example of such conditions is reported in reference [42].

The plug flow case is slightly more complex. Under the assumptions that the local area-specific resistance is independent of the position along the fuel and air flow channels, and the flow pattern is co-flow, ASR_{cor} may be calculated from the expression [43]

$$ASR_{cor} = \left\{ \frac{i}{X_{H_2}^i - X_{H_2}^o} \int_{X_{H_2}^o}^{X_{H_2}^i} \frac{dX_{H_2}}{Emf(X_{H_2}) - U} \right\}^{-1} \quad (4)$$

where

$$Emf(X_{H_2}) = E^0 - \frac{RT}{2F} \ln \left\{ \frac{X^i_{H_2O} + X^i_{H_2} - X_{H_2}}{X_{H_2} \sqrt{X^i_{O_2} - \frac{N_f}{2N_a}(X^i_{H_2} - X_{H_2})} \sqrt{P_a/atm}} \right\} \quad (5)$$

Here X is the molar fraction, P_a is the air pressure and superscripts i, o signify inlet and outlet, respectively. N_f and N_a are the molar flows of fuel and air.

The assumption of a position-independent local area specific resistance is an approximation, which is not always justifiable. Part of the anode polarisation resistance is dependent on fuel composition. However, often this part is small. If the cell is not isothermal, the local resistance will vary with position due to its temperature dependence. Also the actual flow pattern may be much more complex than just co-flow. Even so, if the fuel utilisation is large, ASR_{cor} derived from Eqs. (4) and (5) will always be a better characteristic of a cell than a value derived neglecting the fuel utilisation (Eq. (1)). More precise evaluation of ASR_{cor} requires a rigorous 3-D modelling of the cell test.

For purposes of evaluating Eq. (4), the integral may be approximated by a sum [44]:

$$ASR_{cor} = \left\{ \frac{i}{N} \sum_{j=0}^{N-1} \frac{1}{Emf(X_{H_2}(j)) - U} \right\}^{-1} \quad (6)$$

where $Emf(X_{H_2}(j))$ is given by Eq. (4) with

$$X_{H_2}(j) = X^i_{H_2} + \frac{j + \frac{1}{2}}{N} \left(X^o_{H_2} - X^i_{H_2} \right) \quad (7)$$

The more terms are included in the sum, the better the approximation.

A 'first-order' correction for the effects of finite fuel and air utilisations may be obtained taking only one term in the sum in which case ASR_{cor} should be evaluated from Eq. (3) with

$$Emf_{avg} = E^0 - \frac{RT}{2F} \ln \left\{ \frac{\bar{P}_{H_2O}}{\bar{P}_{H_2} \sqrt{\bar{P}_{O_2}/atm}} \right\}, \quad (8)$$

where the bar indicates 'average', i.e.

$$\bar{P}_{H_2O} = \frac{P^i_{H_2O} + P^o_{H_2O}}{2}, \quad \bar{P}_{H_2} = \frac{P^i_{H_2} + P^o_{H_2}}{2} \quad \text{and} \quad \bar{P}_{O_2} = \frac{P^i_{O_2} + P^o_{O_2}}{2} \quad (9)$$

If there are no significant leaks in the cell or the test equipment, then both fuel and air utilisation, and from this the compositions, may be calculated from the flow rates and the current using Faraday's law. Alternatively, the composition of the outlet fuel and air may be obtained by gas analysis. The conversion resistance, $R_{p,conver}$ may be calculated using the concept of Emf_{avg}:

$$R_{p,conver} = \frac{\text{Emf}_{inlet} - \text{Emf}_{avg}}{i} \tag{10}$$

However, the anisotropic nature (temperature, gas composition, current flow) of a real cell under current flow often invalidates this simple approach. Gas conversion under electric load causes an uneven distribution of the current density with decreasing current density in the downstream direction. Fuel composition gradients over the cell originating from leaks cause different driving potentials at different points. In extreme cases, this may result in high Emf areas driving low Emf areas in electrolyser mode. Thus, internal currents may flow in the cell even at open circuit voltage (OCV). Gas leaks in the cell also affect the current density distribution under *load* and cause localised heating by combustion. For tubular cell designs with high in-plane resistance, the current density distribution may be affected; furthermore, the temperature may not be constant over the whole cell length with flowing gases adding to the inhomogeneity of the current density.

Modelling can simplify or reduce the extent of experimental task and predict likely behaviour under a broad range of test conditions. However, subsequent validation by comparison with relevant cell and stack data is always important.

An example of the magnitude of $R_{p,conver}$ under different conditions can be deduced from Figure 10.10, which shows i–V characteristics obtained in

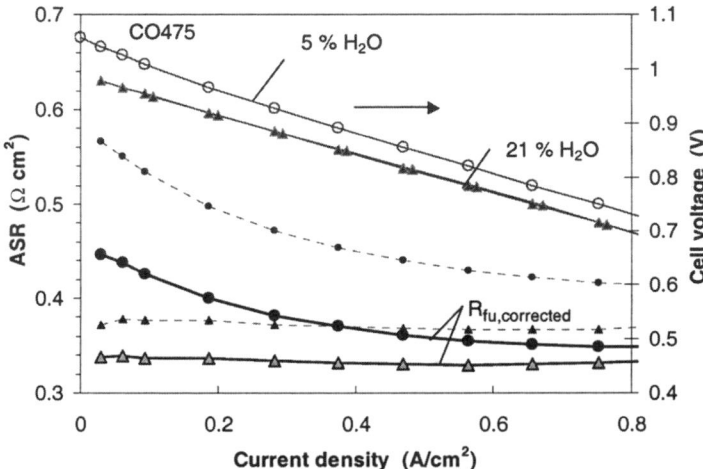

Figure 10.10   i–V characteristics (right-hand scale) of an anode-supported cell tested in hydrogen with different amounts of water vapour in the fuel. Dashed lines are the ASRs as deduced from Eq. (1), and the solid lines are the cell resistances after correction for conversion of the fuel, ASR_{cor} (Eq. (3)). The cell temperature was 850°C and the fuel flow was 24 l/h in the case with 5% H_2O and 20 l/h for the 21% H_2O case. The air flow was 170 l/h. Cell area: 16 cm^2.

hydrogen with either 5 or 20% water vapour for an anode-supported cell. Also shown are the area specific resistance deduced directly from the curves (OCV-U)/i curves and corrected for the fuel conversion (Eqs. (6) and (7)). The correction is largest for the dry gas, where $R_{p,conver}$ is ca. 0.12 Ω cm^2, reflecting the gas composition dependence of the Nernst voltage.

After correcting for the effect of non-negligible fuel utilisation, the cell resistance is still significantly smaller when measured with 20% water in the feed than with 5%. This reflects a gas composition dependence of some of the loss terms in Eq. (2). In reference [45], it is argued that the observed composition dependence is primarily due to the composition dependence of the diffusive losses on the anode side (diffusion overvoltage), and it is shown how one may utilise characteristics obtained with different water vapour/hydrogen ratios to assess the magnitude of the diffusion loss [45].

10.5 Comparison of Test Results on Electrodes and on Cells

As mentioned earlier, cell performance within a certain operational envelope can be fairly well described by just two parameters, namely a resistance (ASR) and an overall activation energy (E_A). Selected ASR values from tests on cells and stacks from various sources, with apparent activation energies (E_A) derived by linear approximations of i–V curves for both low current density (<100 mA/cm^2) and mid current density (100<i<1000 mA/cm^2) over the stated temperature intervals, are summarised in Table 10.2 [46]. The listed data are not strictly comparable because gas composition, flow rate, fuel and air utilisations, etc., are not known in all cases.

Values for R_{Anode}, $R_{Cathode}$ ($R_{p,elchem} = R_{Anode} + R_{Cathode}$) and R_{Elyte} derived from tests of single electrodes and electrolytes are given in Table 10.3, selected on the basis of being comparable with cell results in Table 10.2. The typical E_A for ASR

Table 10.2 Apparent thermal activation of cells and stacks as reported in the literature [46]

	E_A (eV)	$T_{interval}$ (°C)	ASR (Ω cm^2) at T (°C)
Risø (thin electrolyte)	0.6–0.8 (mid i)[a]	650–850	0.30 at 850
Risø (thick electrolyte)	0.6–0.9 (mid i)	800–1050	1.1 at 850
Allied Signal [47]	0.50 (low i)	700–1100	
	0.55 (mid i)	700–900	0.5 at 800
Northwestern University [48]	0.77 (low i)	550–800	2.0 at 700
Forschungszentrum Jülich [49]	0.45 (low i)	800–950	
	0.45 (mid i)	800–950	1.2 at 800
Lawrence Berkeley Laboratory [50]	0.80 (low i)	650–800	
	1.10 (mid i)	650–800	0.20 at 800
Westinghouse [51]	0.45 (mid i)	900–1000	1.0 at 900

Low i: i <100 mA/cm^2; mid i: linear i–v in the range 100 < i < 1000 mA/cm^2.
[a] Often lower E_A at higher temperatures.

Table 10.3 Resistances and apparent activation energy E_A for selected single electrodes, that is Ni–YSZ cermet anodes, LSM–YSZ composite cathodes, and for selected electrolytes from the literature [46]

Electrode type	R_{Anode} ($\Omega\,cm^2$)	$E_{A,Anode}$ (eV)	$R_{Cathode}$ ($\Omega\,cm^2$)	$E_{A,Cathode}$ (eV)	R_{Elyt} ($\Omega\,cm^2$)	$E_{A,Elyt}$ (eV)
Risø anode-supported cell technology	0.06 (850°C) [52]	0.8–0.9 [53]	0.08 (1000°C) [54]	1.4 [54]	0.05 (850°C)[a]	0.8[a]
Risø electrolyte-supported cell technology	0.11 (850°C) [52]	0.8–0.9 [53]	0.12 (1000°C) [54]	1.8 [55]	0.33 (850°C)[a]	0.8[a]
Northwestern University	?	?	1.25 (700°C) [56]	1.5–1.6 [56]	0.24 (640°C) [48]	0.93 (>640°C) [48]

[a] Measured on symmetric Ni-YSZ/YSZ/Ni-YSZ cells, 650–800°C.

values obtained in tests of Risø anode-supported thin-electrolyte cells [57] is about 0.6–0.8 eV, often with a tendency towards lower values at increasing temperatures as mentioned under Table 10.2. The same observation applies for Risø thick electrolyte-supported cells with bi-layer cathodes [58]. This is comparable to the E_A of the electrolyte and the anode, whereas the cathode has a substantially higher E_A.

In Figure 10.11, ASR values are calculated from the electrode data of Table 10.3 and plotted, together with the ASR values measured on single cells, against temperature within the relevant temperature intervals. In the calculation of ASR values, minimum values are used, e.g. the lowest E_A, if a range is listed. Despite a well-documented E_A for LSM-based cathodes of 1.4–2 eV [54,59], the resulting high calculated ASR value at lower temperatures is not reflected in the measured ASR as obtained from full cell tests. Possible causes for this discrepancy are examined in the following.

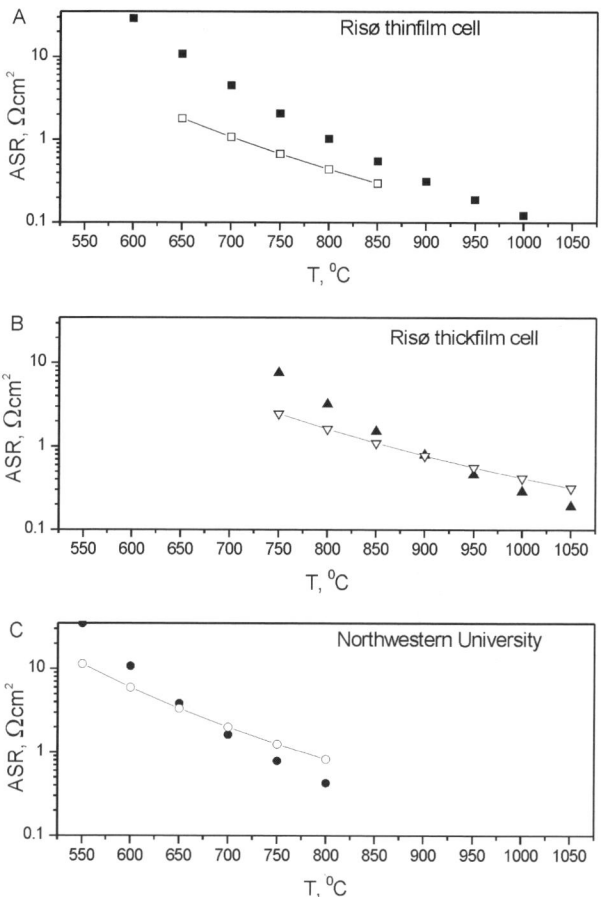

Figure 10.11 Measured ASR values based on data in Table 10.2 (open symbols, lines) compared to ASR values calculated from the electrode data in Table 10.3 (filled symbols).

10.5.1 Non-activated Contributions to the Total Loss

Evident causes of the big difference between $E_{A,cell}$ and $E_{A,cathode}$ are contributions to ASR which are fairly constant with temperature such as $R_{p,conver}$ and contributions with a metallic-type dependence on temperature such as from Ni/YSZ or metallic connections. These would be most significant at higher temperatures where the thermally activated contributions are smallest. This is part of the explanation for the declining E_A observed for Risø cells (as stated under Table 10.2). A similar tendency is seen in the analysis of Allied-Signal data [38]. Conservatively, such non-activated contributions can be estimated as the difference between measured and calculated ASR at high temperatures where the non-activated contributions prevail.

The data in Table 10.2 also indicate a slight increase in E_A with increasing current i. Part of the explanation is the relative importance of $R_{p,conver}$ plus $R_{p,diff}$. Correction for this effect would increase E_A at low i. From Tables 10.2 and 10.3, the difference between E_A for the calculated ASR (the sum of the individual contributions, Table 10.3) and ASR as measured on practical cells (Table 10.2), is about a factor of 2 for the three cells plotted in Figure 10.11; it is not likely that this effect could be due to uncertainties and non-linearities.

10.5.2 Inaccurate Temperature Measurements

Another reason for the cell E_A values being much smaller than E_A of the electrodes and the electrolyte could be the differences between actual and assumed cell temperatures. Measurement of cell temperature during testing is not trivial. A cell larger than a few cm^2 in area is likely to experience temperature gradients due to cooling by feed gases and heating by current. For this reason, measurements of temperature at multiple points on the cell itself are recommended; measurements of furnace temperature or gas stream temperature will typically be misleading. Even sheathed thermocouples in contact with the cell surface or contacting structure may influence the actual cell temperature by heat conduction. Using such thermocouples (diameter = 1.5 mm), a temperature rise of 20°C has been observed at 1 A/cm^2 at 860°C [46]. Within the first 3 s, a rise of < 1°C was observed, and within 20 s the temperature was up by 4°C. The total rise occurs over 2–3 min. This means that in testing, significant heating of the cell is avoidable for only a few seconds, and that i–V curves should be recorded with this in mind, or corrected for the observed T(i) effect. The difference in measured and calculated ASR, at, for example, 750°C, for Risø thick-electrolyte cells, if assumed to be (Figure 10.11) entirely due to an inaccurate temperature measurement, would correspond to an over-temperature of about 70°C, which is highly unlikely given the measured temperature increases under load. Hence, the difference between the measured E_A for cells and the value expected from the known activation energies of the electrode processes and the electrolyte does not seem to be due to erroneous temperature measurement.

10.5.3 Cathode Performance

Due to rather low E_A of cell ASR compared to E_A of cathodes, a tempting explanation appears to be that cathodes just operate better on cells than in single electrode tests. This might be due to water being present on the cathode side in full cell tests, affecting the exchange properties of the electrolyte [60]; however, any possible effect of water on the cathode performance was tested for Risø cathodes and no effect was found. Also having hydrogen as the reference gas in tests of cathodes on pellets showed no effect. Finally, an effect of small amounts of Ni, which could diffuse to the cathode side during sintering was investigated but only a very minor effect was observed [60]. However, differences in fabrication methods of various types of full cells and cathodes on electrolytes may cause unintended contaminations and segregations as well as differences in microstructure, which might affect the cathode performance.

Figure 10.12 shows a number of ASR values obtained at different temperatures for an anode-supported cell together with values modelled from the available knowledge of the cell components [39]. The measured ASRs have been corrected for fuel utilisation. The electrolyte resistance and the electrode polarisations only approximately follow Arrhenius expressions in reality. The assumed values of activation energies and 'pre-exponentials' are given in the figure caption. For the diffusion resistance, which is the only non-temperature activated term of the considered losses, a conservative estimate is used.

To account for the observed temperature dependence, it is necessary to assume that the cathode performance of full cells is much better than measured on separate cathodes on thicker electrolytes, prepared by a very similar procedure. The main point is that the activation energy of the cathode reaction must be

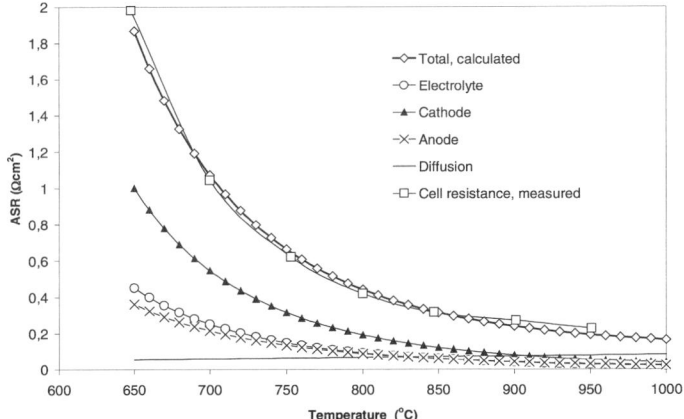

Figure 10.12. Temperature dependence of the ASR of an anode-supported thin-film cell and an estimated breakdown of ASR into individual components. The fuel flow was 24 l/h (94% hydrogen, 6% water vapour) and the air flow was 170 l/h. The following values were assumed: $E_{a,cat} = 0.94\ eV$, $E_{a,an} = 0.8\ eV$, $E_{a,electrolyte} = 0.9\ eV$ and $R_{cat,850C} = 0.12\ \Omega\ cm^2$, $R_{an,850C} = 0.06\ \Omega\ cm^2$, $R_{electrolyte,850C} = 0.06\ \Omega\ cm^2$ and $R_{diffusion,850C} = 0.07\ \Omega\ cm^2$.

below 1 eV in order to obtain a reasonable good modelling of the cell performance, whereas there seems to be a general agreement in the literature of an activation energy in the vicinity of 1.4–2 eV [42].

The resistance measure used in Figure 10.12 is the minimum resistance measure, that is the resistance at high polarisation. Better agreement between cell performance and the performance expected from single electrode studies is achieved if the comparison is based on impedance data, where the polarisation is very small [45]. This is due to the non-linearity of the cell response at low temperatures. As pointed out in Figures 10.8 and 10.10, the cell characteristic is quite linear at least at a high temperature (850°C) and when measured in high water vapour content. However, at lower temperatures the i–V curves are non-linear even when taken in moist hydrogen. An example is shown in Figure 10.13 [61]. Obviously from Figure 10.13, describing the resistance at small current load rather than at high current would result in a larger $E_{A,cell}$.

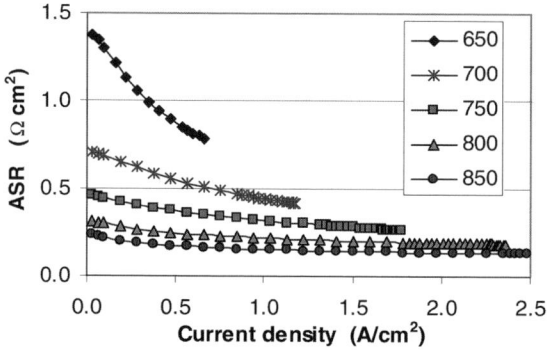

Figure 10.13 Area-specific cell resistances corrected for fuel utilisation (ASR_{cor}) measured at various temperatures of an anode-supported cell. The fuel was hydrogen with ca. 5% water vapour at a flowrate of 30 l/h and the air flow was 140 l/h. Cell area: 16 cm².

10.5.4 Impedance Analysis of Cells

As realised from the above issues in the comparison of test results on the electrodes and on the cells, it is a non-trivial task to break down the total loss measured on a single cell into its components using the results from the electrode studies. Impedance spectroscopy on practical cells is, however, a technique by which a partial break down can be made. Though the impedance spectra obtained in general are difficult to interpret due to the many processes involved, the spectra can at least provide a break down of the total loss into an ohmic resistance ($R_s = R_{elyt} + R_{connect}$) and a polarisation resistance reflecting losses due to chemical, electrochemical, and transport processes, as described in more detail in Chapter 9.

Examples of impedance spectra obtained on a 4 cm × 4 cm cell in various gas atmospheres are illustrated in Figure 10.14 [45]. Clearly, R_s is independent of the gas composition, and by fitting the impedance curve to an equivalent circuit, which takes into account the inductance in the measuring loop, a precise

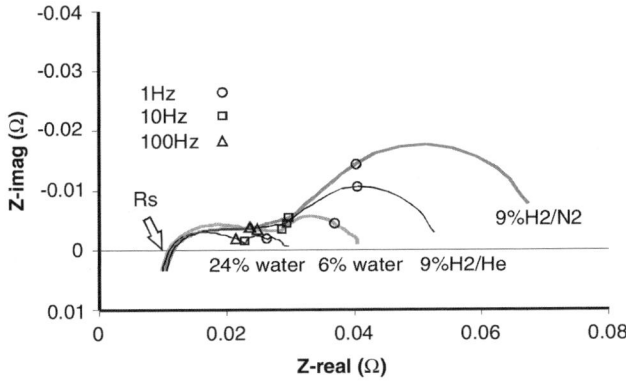

Figure 10.14 Impedance spectra of an anode-supported cell tested at 850°C with different fuel gases. Two measurements were performed in hydrogen/water vapour gas mixtures with either 24 or 6% water vapour and two measurements were performed in diluted hydrogen, with either nitrogen or helium as the diluent. The water vapour content in the diluted gas experiments was ca. 6%. R_s is invariant to changes in the fuel but both the water vapour content and the type of diluent strongly affect the low frequency semicircle.

measure of R_s can be deduced from the spectra [62]. The low-frequency semicircle is seen to be strongly dependent on water content as well as on whether He or N_2 is used as the diluent, showing that it is related to gas conversion and gas diffusion.

10.6 The Problem of Gas Leakage in Cell Testing

A major source of error in cell testing is due to gas leakage, either of air from the outside into the anode compartment or from air crossing over from the cathode through cracks or holes. Figure 10.15 shows an i–V curve obtained at 1000°C on a thick-electrolyte cell, which at a first glance looks normal; only the open circuit voltage (OCV) is much lower than the calculated Emf, and the curve bends slightly downwards [44]. A low OCV usually means leakage of air into the anode compartment because normally the air flow is much higher than the fuel flow, and thus, the pressure is slightly higher in the cathode compartment than in the anode compartment. In the case of Figure 10.15 about 50% of the H_2 is converted to H_2O. If the OCV value is used in Eq. (1) instead of the Emf of the inlet gas, an ASR of 0.16 Ω cm^2 at 1000°C is calculated. However, an ASR between 0.3 and 0.4 Ω cm^2 is well established for this type of cell at 1000°C. Using the Emf instead of the OCV and the current density at the cell voltage of 0.75 V, an ASR of ~0.35 Ω cm^2 is in fact obtained. The good numerical agreement with the expected value may be somewhat fortuitous, but it is a general experience with cells leaking during test that their power producing capability is similar to or lower than that of comparable cells despite the low resistance values calculated based on measured OCV values.

A large localised leak of air into the anode compartment (for instance at the rim of the cell) causes an increase in temperature due to hydrogen combustion;

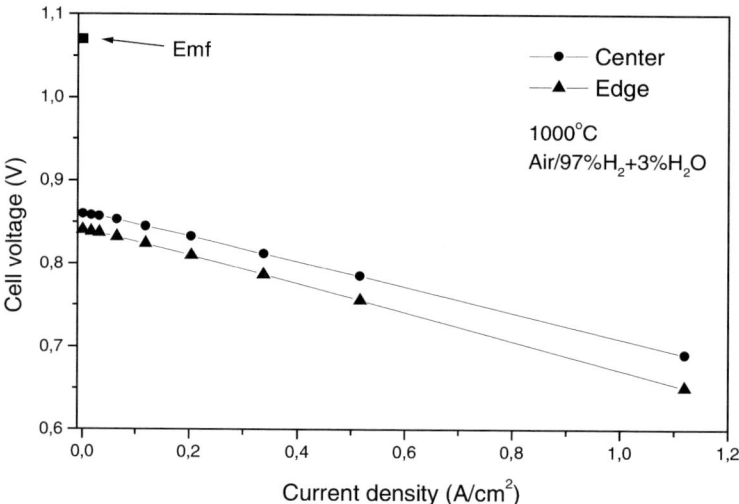

Figure 10.15   i–V curve for a 'thick electrolyte' Ni-cermet/YSZ/LSM cell in a test with a significant leak of air into the anode gas stream [44]. The cell was tested in the set-up described in [19], where gold wires are used for sealing.

this temperature increase may not be registered if the thermocouple is positioned at the cell centre. The increased temperature decreases the cell resistance locally.

However, the temperature increase cannot explain more than half of the deviation of the apparent ASR from the true value. Significant variations in the fuel composition due to localised leaks over the cell area can induce internal currents in the cell, i.e. the parts of the cell in areas with high local Emf are loaded in the fuel cell mode whereas other parts with Emf below the OCV are loaded in the electrolyser mode. When the cell is externally loaded, internal currents decrease with decreasing cell voltage. Temperature and fuel compositions vary accordingly and thus the local internal resistance also varies.

10.6.1 Assessment of the Size of the Gas Leak

The effect of gas leaks on cell tests may be separated into two components: a loss of driving potential (Emf) caused by O_2 entering to convert H_2 to H_2O, and a volumetric loss of fuel affecting the fuel utilisation, caused by a pressure gradient versus the ambient. A gas leak may be before, in, or after the cell in a test set-up fuel line. Leaks before the cell affect the intended fuel composition and quantity. However, a true average fuel composition may be obtained from the cell OCV. Leaks after the cell are of no importance to the cell test, unless off-gas analysis is carried out. Leaks in the cell or in seals cause inhomogeneous gas composition over the cell, even at OCV.

Assuming oxygen (air) leakage into the fuel compartment, an estimate of the loss of fuel can be based on observed OCV deviation from the expected Emf as calculated by the Nernst equation. A number of such curves at 850°C are given in Figure 10.16 [46]. From a given feed percentage of H_2O, the loss of H_2 can be

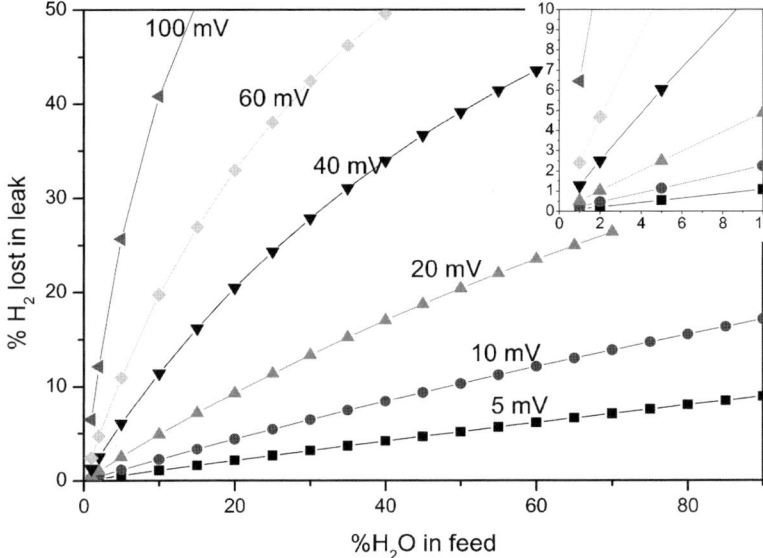

Figure 10.16 Observable deviation (OCV – Emf in mV) translated into percent loss of feed hydrogen as a function of water concentration in the fuel gas, H_2/H_2O mixtures at 850°C [46].

estimated from the OCV–Emf deviation. Reading the curve in another way, a given loss of H_2 (e.g. 5% of the feed) causes a loss of driving potential of 60 mV if the feed contains only 3% water, whereas for a feed with 50% water the loss is only 5 mV.

One way to exploit this knowledge is to assess the leak-rate of a test system while running water-lean (sensitive) gas mixtures, and transfer the observed leak to tests carried out in more water-rich feed gases. A practical approach to minimising the seal-leak problem in testing could include establishing a reducing sweep atmosphere, e.g. 3% H_2 in N_2 outside the test compartment as illustrated in Fig. 10.5a. This would reduce the pH_2 gradient to well below one order of magnitude, and prevent incoming leakage of O_2. At the same time the potential of the cathode gas is far less sensitive to reaction of O_2 with H_2 and production of H_2O, for example, a 5% injection of H_2 in air reduces the potential by only 7 mV at 850°C.

An electronic leak through the electrolyte is in principle another explanation for the differences in Emf and OCV. If the electronic conductivity in such a leak decreases with cell voltage, a very flat i–V curve may be obtained.

Great care should always be taken when the OCV is far from the theoretical Emf. When the measured OCV lies close (~10 mV) to the calculated Emf, it is justified to calculate resistances based on OCV rather than Emf when the purpose is a detailed cell performance evaluation, thereby avoiding lumping problems of gas-tightness of the electrolyte or seals into the cell resistance. (If the purpose of the calculation is economic assessment or system studies, the calculations should obviously be based on Emf.) However, if large differences between Emf and OCV are observed, the OCV value may be used instead of the Emf of the inlet gas only in case where it is in some way verified that the leak is evenly distributed over the cell area.

10.7 Summary

This chapter has considered the main types of electrochemical tests which have been applied to SOFCs and has outlined the main issues which require detailed attention for obtaining meaningful test results.

One important aspect in electrode testing is to assure a correct geometry in three-electrode set-ups. This is very difficult in practice in case of electrode-supported cells with thin electrolytes. Unfortunately, even testing the individual electrodes in sound geometry set-ups is not a perfect procedure either, because the sum of the contributions from individual cell components to the cell resistance does not add up to the actual measured total cell resistance. This is probably due to differences in the fabrication of the special cells for electrode characterisation and the practical cells.

Another important issue is that of gas leakage, which can cause significant errors in performance data. Especially in the case of gas leakage, cells can easily be at a higher temperature than their surrounding environment, causing cells to give better apparent performance than individual electrodes tested under better controlled conditions. Also, the gas composition at the electrodes (usually at the anode) may be different from the intended composition in case of gas leakage. A method for estimating the size of the gas leakage has been presented here.

It is recommended that cell test results be reported in a way that makes it easy to derive area specific resistance (ASR) from the i–V curves. Sufficient information should be provided so that the ASR values can be corrected for effects of finite fuel utilisation. Also, the choice of fuel composition should preferably reflect real cell operation conditions. The ASR should be derived using the Emf and a cell voltage in the range of 0.5–0.7 V and its corresponding current density. In case of a grossly non-linear I–V curve, a differential ASR value is of little practical use.

References

[1] K. J. Vetter, *Electrochemical Kinetics – Theoretical and Experimental Aspects*, Academic Press, New York, 1967.
[2] J. O'M. Bockris and A. K. N. Reddy, *Modern Electrochemistry*, Plenum Press, New York, 1970.
[3] Southampton Electrochemistry Group, *Instrumental Methods in Electrochemistry*, Ellis Horwood, Chichester, 1985.
[4] R. A. Rapp and D. A. Shores, in *Physicochemical Measurements in Metal Research, Part 2*, ed. R. A. Rapp, Wiley Interscience, New York, 1970, pp. 123–192.

[5] M. Kleitz, P. Fabry and E. Schouler, in *Fast Ion Transport in Solids*, ed. W. van Gool, North-Holland, Amsterdam, 1973, pp. 439–451.
[6] P. H. Bottelberghs, in *Solid Electrolytes*, eds. P. Hagenmuller and W. van Gool, Academic Press, New York, 1978, pp. 145–172.
[7] J. A. Kilner and B. C. H. Steele, in *Non-stoichiometric Oxides*, ed. O. T. Sørensen, Academic Press, 1981, pp. 233–269.
[8] H. L. Tuller, *ibid.*, pp. 271–335.
[9] H. Rickert, *Electrochemistry of Solids*, Springer-Verlag, Berlin, 1982.
[10] I. Riess, in *Solid State Electrochemistry*, eds. P. J. Gellings and H. J. M. Bouwmeester, CRC Press, New York, 1997, pp. 233–268.
[11] I. Riess and J. Schoonman, *ibid.*, pp. 269–294.
[12] W. Weppner, *ibid.*, pp. 295–327.
[13] S. C. Singhal, in *Solid Oxide Fuel Cells V*, eds. U. Stimming, S. C. Singhal, H. Tagawa and W. Lehnert, The Electrochemical Society Proceedings, Pennington, NJ, PV97-40, 1997, p. 37.
[14] H. J. Beie, L. Blum, W. Drenckhahn, H. Greiner, B. Rudolf and H. Schichl, *ibid.*, p. 51.
[15] K. Honneger, E. Batawi, Ch. Sprecher and R. Diethelm, *ibid.*, p. 321.
[16] K. Kendall, R. C. Copcutt and G.Sales, *ibid.*, p. 283.
[17] K. Eguchi, H. Mitsuyasu, Y. Mishima, M. Ohtaki and H. Arai, *ibid.*, p. 358.
[18] J.-W. Kim, A. V. Virkar, K.-Z. Fung, K. Mehta and S. C. Singhal, *J. Electrochem. Soc.*, **146** (1999) 69.
[19] C. Bagger, E. Hennesø and M. Mogensen, in *Solid Oxide Fuel Cells III*, eds. S. C. Singhal and H. Iwahara, The Electrochemical Society Proceedings, Pennington, NJ, PV93-4, 1993, p. 756.
[20] S. C. Singhal, *Solid State Ionics*, **135** (2000) 305.
[21] J. Sukkel, in *Proceedings of the 4th European SOFC Forum*, ed. A. J. McEvoy, Oberrohrdorf, Switzerland, 2000, p. 159.
[22] R. Diethelm, E. Batawi and H. Honneger, *ibid.*, p. 183.
[23] K. Kendall, C. Finnerty and G. A. Tompsett, *ibid.*, p. 115.
[24] K. Ahmed, J. Gamman and K. Foger, *Solid State Ionics*, **152–153** (2002) 485.
[25] C. Bagger, M. Juhl, P. Vang Hendriksen, P. Halvor Larsen and M. Mogensen, in *Proceedings of the 2nd European SOFC Forum*, ed. B. Thorstensen, Oberrohrdorf, Switzerland, 1996, pp. 175–184.
[26] L. G. J. de Haart, I. C. Vinke, A. Janke, H. Ringel and F. Tietz, in *Solid Oxide Fuel Cells VII*, eds. H. Yokokawa and S. C. Singhal, The Electrochemical Society Proceedings, Pennington, NJ, PV2001-16, 2001, p. 111.
[27] K. A. Nielsen, S. Linderoth, B. Kindl, J. B. Bilde-Sørensen and P. H. Larsen, in *Proceedings of the 5th European SOFC Forum*, ed. J. Huijsmans, Oberrohrdorf, Switzerland, 2002, p. 729.
[28] W. J. Quadakkers, H. Greiner, W. Kock, H. P. Buchkremer, H. Hilpert and D. Stover, in *Proceedings of the 2nd European SOFC Forum*, ed. B. Thorstensen, Oberrohrdorf, Switzerland, 1996, p. 297.
[29] Y. Larring and T. Norby, *J. Electrochem. Soc.*, **147** (2000) 3251.

[30] M. Gödickemeier, K. Sasaki and L. J. Gauckler, in *Solid Oxide Fuel Cells IV*, eds. M. Dokiya, O. Yamamoto, H. Tagawa and S. C. Singhal, The Electrochemical Society Proceedings, Pennington, NJ, PV95-1, 1995, p. 1072.

[31] J. W. Erning, W. Schafffrath, U. Stimming, E. Syskakis and K. Wipperman, *ibid.*, p. 492.

[32] F. H. van Heuveln, F. P. F. van Berkel and J. P. P. Huijsmans, in *14th Risø International Symposium on Materials Science*, eds. F. W. Poulsen, J. J. Bentzen, T. Jacobsen, E. Skou and M. J. L. Østergård, Risø National Laboratory, Roskilde, Denmark, 1993, p. 53.

[33] C. Kleinlogel, M. Gödickemeier, K. Honneger and L. J. Gauckler, in *Proceedings 3rd Third International Symposium on Ionic and Mixed Conducting Ceramics*, eds. T. A. Ramanarayanan, W. L. Worrell, H. L. Tuller, A. C. Khandkar, M. Mogensen and W. Göpel, The Electrochemical Society Proceedings, Pennington, NJ, PV97-24, 1998, p. 1072.

[34] M. Nagata, Y. Ito and H. Iwahara, *Solid State Ionics*, **67** (1994) 215.

[35] T. Jacobsen and E. Skou, in *Materials and Processes*, Proceedings IEA Workshop, 28–31 Jan. 1997, Les Diablerets, Switzerland.

[36] J. Winkler, P. V. Hendriksen, N. Bonanos and M. Mogensen, *J. Electrochem. Soc.*, **145** (1998) 1184.

[37] G. Reinhardt and W. Göpel, in *Proceedings Third International Symposium on Ionic and Mixed Conducting Ceramics*, eds. T. A. Ramanarayanan, W. L. Worrell, H. L. Tuller, A. C. Khandkar, M. Mogensen and W. Göpel, The Electrochemical Society Proceedings, Pennington, NJ, PV97-24, 1998, p. 610.

[38] S. B. Adler, *J. Electrochem. Soc.*, **149**, E166 (2002).

[39] M. Mogensen, P. V. Hendriksen and K. Kammer, in *Proceedings of the 5th European SOFC Forum*, ed. J. Huijsmans, Oberrohrdorf, Switzerland, 2002, p. 893.

[40] M. J. Jørgensen, S. Primdahl and M. Mogensen, *Electrochem. Acta*, **44** (1999) 4195.

[41] J. Fleig and J. Maier, *J. Electrochem. Soc.*, **144** (1997) L302.

[42] S. Primdahl and M. Mogensen, *J. Electrochem. Soc.*, **145** (1998) 2431.

[43] A. Solheim, R. Tunold and R. Ødegård, in *Solid Oxide fuel Cells II*, eds. F. Gross, P. Zegers, S. C. Singhal and O. Yamamoto, Commission of the European Communities, Luxembourg, EUR 13564EN, 1991, p. 297.

[44] M. Mogensen, P. H. Larsen, P. V. Hendriksen, B. Kindl, C. Bagger and S. Linderoth, in *Solid Oxide fuel Cells VI*, eds. S. C. Singhal and M. Dokiya, The Electrochemical Society Proceedings, Pennington, NJ, PV99-1, 1999, p. 904.

[45] P. V. Hendriksen, S. Koch, M. Mogensen, Y. L. Liu and P. H. Larsen, in *Solid Oxide Fuel Cells VIII*, eds. S. C. Singhal and M. Dokiya, Electrochemical Society Proceedings, Pennington, NJ, PV2003-07, 2003, p. 1147.

[46] S. Primdahl, P. V. Hendriksen, P. H. Larsen, B. Kindl and M. Mogensen, in *Solid Oxide Fuel Cells VII*, eds. H. Yokokawa and S. C. Singhal, The

Electrochemical Society Proceedings, Pennington, NJ, PV2001-16, 2001, p. 932.
[47] N. Q. Minh, in *1994 Fuel Cell Seminar Abstracts*, Courtesy Associates, Washington, DC, 1994, p. 577.
[48] T. Tsai and S. A. Barnett, in *Solid Oxide Fuel Cells V*, eds. U. Stimming, S. C. Singhal, H. Tagawa and W. Lehnert, The Electrochemical Society Proceedings, Pennington, NJ, PV97-40, 1997, p. 274.
[49] L. G. J. de Haart, T. Hauber, K. Mayer and U. Stimming, in *Proceedings of the Second European SOFC Forum*, ed. B. Thorstensen, Oberrohrdorf, Switzerland, 1996, p. 229.
[50] S. de Souza, S. J. Visco and L. C. de Jonghe, in *Proceedings of the Second European SOFC Forum*, ed. B. Thorstensen, Oberrohrdorf, Switzerland, 1996, p. 677.
[51] S. C. Singhal, in *Proceedings of the 17th Risø International Symposium on Materials Science: High Temperature Electrochemistry: Ceramics and Metals*, eds. F. W. Poulsen, N. Bonanos, S. Linderoth, M. Mogensen and B. Zachau-Christiansen, Risø National Laboratory, Roskilde, Denmark, 1996, p. 123.
[52] S. Primdahl and M. Mogensen, in *Solid Oxide Fuel Cells VI*, eds. S. C. Singhal and M. Dokiya, The Electrochemical Society Proceedings, Pennington, NJ, PV99-19, 1999, p. 530.
[53] M. Brown, S. Primdahl and M. Mogensen, *J. Electrochem. Soc.*, **147** (2000) 475.
[54] M. J. Jørgensen and M. Mogensen, *J. Electrochem. Soc.*, **148** (2001) A433.
[55] M. J. Jørgensen, Risø National Laboratory, personal communication.
[56] E. P. Murray, T. Tsai and S. A. Barnett, *Solid State Ionics*, **110** (1998) 235.
[57] S. Primdahl, M. J. Jørgensen, C. Bagger and B. Kindl, in *Solid Oxide Fuel Cells VI*, eds. S. C. Singhal and M. Dokiya, The Electrochemical Society Proceedings, Pennington, NJ, PV99-19, 1999, p. 793.
[58] Unpublished data, Risø National Laboratory.
[59] M. Kleitz, T. Kloidt and L. Dessemond, in *Proceedings of the 14th Risø International Symposium on Materials Science: High Temperature Electrochemical Behaviour of Fast Ion and Mixed Conductors*, eds. F. W. Poulsen, J. J. Bentzen, T. Jacobsen, E. Skou and M. J. L. Østergård, Risø National Laboratory, Roskilde, Denmark, 1993, p. 89.
[60] N. Sakai, K. Yamaji, H. Negishi, T. Horita, H. Yokokawa, Y. P. Xiong and M. B. Phillipps, *Electrochemistry*, **68** (2000) 499.
[61] W. G. Wang, R. Barfod, P. H. Larsen, K. Kammer, J. J. Bentzen, P. V. Hendriksen and M. Mogensen, in *Solid Oxide Fuel Cells VIII*, eds. S. C. Singhal and M. Dokiya, The Electrochemical Society Proceedings, Pennington, NJ, PV2003-07, 2003, p. 400.
[62] N. Bonanos, P. Holtappels and M. Juhl Jørgensen, in *Proceedings of the 5th European SOFC Forum*, ed. J. Huijsmans, Oberrohrdorf, Switzerland, 2002, p. 578.

Chapter 11

Cell, Stack and System Modelling

Mohammad A. Khaleel and J. Robert Selman

List of terms

c_i	mass fraction of species i
c_p	specific heat
$c_{r,electrode}$	reactant concentration at electrode
$c_{r,bulk}$	reactant concentration at gas channel
$c_{p,electrode}$	product concentration at electrode
$c_{p,bulk}$	product concentration at gas channel
d	half-thickness of the plate
D_h	hydraulic diameter
$D_{ij}, D_{O2\text{-}N2}, D_{H2\text{-}H2O}$	binary diffusion coefficients
D_{im}	multicomponent diffusion coefficient of species i in a mixture
$D_{eff(a)}$	anode effective diffusion coefficients
$D_{eff(c)}$	cathode effective diffusion coefficients
dr_y/dt	rate of mole change of species y (y = H_2, O_2, H_2O, CO, CO_2, CH_4)
dr_f/dt	rate of change due to forward reaction
dr_b/dt	rate of change due to backward reaction
E	Young's modulus
E_{eq} or E^o	the equilibrium (open circuit) voltage
E_{H2}	activation energy
E_{total}	total energy (delivered per unit time)
E^k	turbulent kinetic energy
F	Faraday constant
g	acceleration due to gravity
ΔG^o	standard free-energy change of a reaction (eq.8a)
ΔG^0_{shift}	standard Gibbs free-energy change of the shift reaction
h	heat transfer coefficient

h_1 and h_2	thickness of layer 1 and 2, respectively
ΔH	enthalpy of formation
I, i	current density
i_0	the exchange current density
i_{H2}	limiting current density due to H_2 transport
i_{O2}	limiting current density due to O_2 transport
i_c	concentration limiting current
k_1	rate constant for shift reaction
k_{H2}	rate constant for H_2 oxidation
k_{CO}	rate constant for CO oxidation
K_{shift}	equilibrium constant for the shift reaction
l	the length of the flow path
l_a, l_c	thickness of anode and cathode, respectively
m_1, m_2, m_3, m_4	reaction order parameter
m	Weibull parameter
$\mathbf{n}$	unit vector normal to the boundary
p_{H2}	H_2 partial pressure in the anode fuel channel
p_{H2O}	H_2O partial pressure in the anode fuel channel
$p_{O2}, p_{O2,c}$	oxygen partial pressure in the cathode air channel
$p_{O2,a}$	oxygen partial pressure in the anode fuel channel
P	flow pressure
P_{cell}	cell power density
P_{stack}	stack power density
P_{ex}	electric power
p	stands for 'product'
Q	nonviscous volumetric heat generation term
Q_{gen}	heat generation
Q_{ohm}	ohmic heat
$Q_{irr.}$	heat generation due to irreversible process
$Q_{rev,a}$	reversible heat generation at the anode
$Q_{rev,c}$	reversible heat generation at the cathode
$Q_{rev,total}$	total reversible heat generation
Q_{vis}	viscous heat generation term
R	gas constant
R_i	ohmic resistance
Re	Reynolds number
r	stands for 'reactant'
S_i	entropy of species i ($i = O_2, O^{2-}$, el)
T	temperature
T_s and T_f	solid and fluid temperatures
ΔT	change in temperature
t	time
$\mathbf{U}_i$	diffusion velocity of species i
U_f	fuel utilisation
V	voltage

V_c	cathode porosities
V_a	anode porosities
V^{thn}	'thermoneutral voltage'
v	fluid velocity
W	Weibull function, the probability of failure
W^v	viscous work
z	number of electrons participating in the electrode reaction

Greek letters

α	anodic transfer coefficient
β	thermal expansion coefficient
γ_i	resistivity
ε_{el}	fuel cell electrical efficiency
η	overpotential or polarisation
η_A	anode polarisation
η_a	activation polarisation
η_C	cathode polarisation
η_c	concentration polarisation
η_{Ca}, η_{Cc}, η_{Aa} and η_{Ac}	cathode activation, cathode concentration, anode activation, and anode concentration polarisations, respectively
λ	thermal conductivity
μ_e	effective viscosity
ν	Poisson's ratio
ρ_i	density of the species i
σ_e	electronic conductivity
σ_{eff}	effective ionic conductivity
σ_{ion}	ionic conductivity
σ, σ_1, σ_x, σ_y	thermal stress
σ_0	material-specific characteristic stress of the Weibull function
τ_a and τ_c	anode and cathode tortuosities
τ_i	non-Newtonian viscous losses
Φ	electrical potential
ω_i	rate of production of species i
Ω_k	kth direction momentum source term

11.1 Introduction

Mathematical models that predict performance can aid in understanding and development of solid oxide fuel cells (SOFCs). A mathematical simulation of a SOFC is helpful in examining issues such as temperatures, materials, geometries, dimensions, fuels, and fuel reformation and in determining their associated performance characteristics. When physical properties or reaction kinetics are not known reliably, they can be estimated by fitting performance data on small-size, laboratory-scale cells to a mathematical model. The performance of a small-size, laboratory-scale cell, by fitting an appropriate model, can yield input parameters for the performance of a larger cell or stack. This cell or stack simulation can be used to determine the effects of various design and operating parameters on the power generated, fuel conversion efficiency, maximum cell temperature reached, stresses caused by temperature gradients, and the effects of thermal expansion for electrolytes, electrodes, and interconnects.

Thus, modelling is an important tool in design optimisation, helping to answer important practical questions such as what air and fuel flow rates must be used to avoid excessive temperature or pressure drop. On the other hand, by providing answers to questions such as how much the electrical properties of the cell materials must be improved, simulations at the cell and electrode level can guide the development of new and improved materials. Mathematical simulation, therefore, has the potential to guide technology development, test the significance of various design features, assess the effectiveness of developments in materials or fabrication procedures, and select optimum operating conditions from a set of feasible parameters.

Various modelling approaches exist. The modelling may focus on individual thermal-mechanical, flow, chemical, and electrochemical subsystems or on coupled integrated systems. Because the subsystems are typically characterised by different length scales, modelling may also take place on different levels, ranging from the atomistic/molecular-level via the cell component-level, the cell-level to the stack-level, and finally to the system-level performance simulations.

This chapter discusses SOFC modelling primarily from the viewpoint of cells and stacks, although some information on system modelling and more extensive information on electrode modelling are also presented. After an introductory discussion of modelling levels, the SOFC cell and stack are first examined from the viewpoint of fluid dynamics and transport phenomena (SOFC as a heat and mass exchanger). This is followed in Section 11.3 by an exposition of electrochemical modelling at the 'continuum level', suitable for integration into modelling of full-scale stacks (SOFC as an electrochemical generator). In Section 11.4, the chemical reactions depending on fuel composition and the heat effects associated with their electrochemical conversion are discussed in detail (SOFC as a chemical reactor). Section 11.5 discusses cell- and stack-level modelling; and Section 11.6 briefly describes major approaches in SOFC system modelling (SOFC as a system component); Section 11.7 links the thermal analysis of the SOFC cell and stack with the modelling of thermal stresses; and Section 11.8 discusses in more detail the electrochemical modelling at the μm level suitable for electrode design and microstructure. Finally, Section 11.9 sketches possible approaches of molecular modelling suitable for elucidating kinetic and mechanistic issues relevant to SOFC performance.

11.2 Flow and Thermal Models

In a fuel cell operation, the flow, thermal, chemical, and electrochemical systems are intrinsically coupled. Heat generation and absorption affect the temperature distribution and gas flow rate. Undesirable or even dangerous operating conditions may arise from the flow distribution [1]. Due to differences in coefficients of thermal expansion, temperature gradients during transient or stationary operation cause stresses that may lead to failure. Interdiffusion of materials used for the anode, the electrolyte, and the cathode may lead to gradual performance degradation. In order to calculate flow and temperature, the conservation laws in fluid mechanics are used [2].

11.2.1 Mass Balance

A species' mass in a reacting mixture of gases is determined by solving the species continuity equations:

$$\partial \rho_i/\partial t + \nabla \cdot [\rho_i(\mathbf{v} + \mathbf{U}_i)] = \omega_i \tag{1}$$

where ρ_i is the species density, $\mathbf{v}$ is the fluid velocity, $\mathbf{U}_i$ is the species diffusion velocity, t is time, and ω_i is the rate of production of species *i* due to chemical (or electrochemical) reactions. The mass flux of species *i* ($\rho_i \mathbf{U}_i$) due to diffusion can be approximated for most applications using Fick's law:

$$\rho_i \mathbf{U}_i = -\rho D_{im} \nabla c_i \tag{2}$$

where c_i is the species mass fraction (ρ_i/ρ), and D_{im} is the multicomponent diffusion coefficient of species *i* in the mixture. D_{im} is a weighted average of binary diffusion coefficients D_{ij}, that is, of the diffusion coefficients of species *i* with respect to each of the other species, *j*. Depending on the composition of the gas mixture, D_{im} can often be assumed to remain fairly constant. If there is one dominant species, *k*, in the mixture, the multicomponent diffusion coefficient D_{im} may often be approximated by the binary diffusion coefficient D_{ik}.

11.2.2 Conservation of Momentum

Mass balances must be used with the flow pattern (known *a priori* from theory or experimental measurements) to establish species concentrations and fluxes at any point in the fuel cell. When the flow pattern is *a priori* unknown, conservation-of-momentum equations (also called equations of motion) must be used with mass balance equations to establish the velocity and concentration profiles. Conservation of momentum for gases leads to the following equations (Navier–Stokes equations), in which *k* represents one of the three orthogonal directions in the coordinate system (x, y, and z):

$$\partial(\rho v_k)/\partial t + \nabla \cdot (\rho v_k \mathbf{v}) = \rho g_k - \partial P/\partial x_k + \nabla \cdot (\mu_e \nabla v_k) + \Omega_k + \tau_k \tag{3}$$

Here P is the pressure, g is the acceleration due to gravity, and μ_e is the effective viscosity. The term τ_k represents other than Newtonian viscous losses and may be neglected for SOFC flows. A user-defined source term, Ω_k, can be used to represent in- and outflows due to electrode reactions at the boundary of the flow field.

Equation (3) provides details of gas flow movements. The full treatment requires a rigorous computational fluid dynamics (CFD) tool. Startup and transient processes as well as variations in certain operating parameters may have a sizeable effect on flow and concentration profiles, but the effect on overall electrochemical performance of the cell is not necessarily of the same order. Sometimes it is desirable to make a simplification such as assuming laminar flow to reduce the computation cost and allow quick estimates of certain flow properties. For example, the pressure drop of a laminar flow through a channel can be estimated as

$$\Delta P = (1/2)\rho v^2 fl/(ReD_h) \qquad (4)$$

where Re is the D_h-based Reynolds number, D_h is the hydraulic diameter, l is the length of the flow path, and f depends on the shape of the cross section of the channel, e.g., $f = 56.8$ and 64 for a square and a round channel, respectively [3]. Such simplification can reduce the computation cost significantly [4].

11.2.3 Energy Balance

The temperature field and local heat fluxes in the gas phase are governed by the energy balance:

$$\partial(\rho c_p T)/\partial t + \nabla \cdot (\rho c_p T \mathbf{v}) = \nabla \cdot (\lambda \nabla T) + Q + \partial P/\partial t + Q_{vis} + W^v + E^k \qquad (5)$$

Here c_p is specific heat, λ is thermal conductivity, Q is the nonviscous volumetric heat generation term, Q_{vis} is the viscous heat-generation term, W^v is viscous work, and E^k is turbulent kinetic energy. The volumetric heat source Q represents heat generation by the electrochemical reactions (planar heat sources being expressed on volumetric basis), chemical reactions (e.g., hydrocarbon reforming and CO water shift reaction), and Joule heating (due to ohmic resistance of electrolyte and electrodes). Without the last four terms, Eq. (5) also applies to the solid components of the fuel cell. These components consist of the positive electrode, the electrolyte, and the negative electrode (PEN) elements and the interconnect (IC) or bipolar plate. The PEN is sometimes assigned lumped properties for heat transfer modelling.

Heat transfer between cell components must also be accounted for, either as boundary conditions of Eq. (5) (boundary heat flows) or as a volumetric heat source (contributing to Q in Eq. (5)). These heat source terms due to interfacial heat transfer occur mainly in two ways [5]:

- Between cell component layers and flowing gas streams, e.g., between the anode or anode side of the PEN and the fuel gas stream or between the interconnect and the oxidant gas stream. This type of heat transfer is best described in terms of convective heat transfer coefficient h.

- Between adjacent solid layers with different thermal conductivities, λ_i (where i = cathode, anode, electrolyte, or interconnect). This type of heat transfer may be folded into a lumped effective conductivity, for example, for the PEN.

Alternatively, the heat transfer from the fuel gas stream to the oxidant gas stream via a solid layer such as the PEN element or the interconnect may be described in terms of an overall heat transfer coefficient.

For convective heat transfer at the boundary between a solid layer and a fluid, the following continuity condition may be imposed [6]:

$$\lambda(x)\nabla T_s(x)\cdot \mathbf{n} = h[T_f(x) - T_s(x)] \tag{6}$$

where $\mathbf{n}$ is the unit vector normal to the boundary, h is the heat transfer coefficient, and $T_s(x)$ and $T_f(x)$ are the temperatures of the solid and fluid, respectively, at location x on the boundary. Heat transfer may also take place by radiation from solid to gas phase or from solid to solid across a gas phase. This can usually be represented by variants of Eq. (6). Radiative heat transfer is especially important in higher temperature (900–1000°C) SOFC systems, for example, the tubular design SOFC generator [7,8].

For steady-state simulation, the equations above are simplified by deleting the time-dependent terms. However, the general forms are necessary for simulating transient operating conditions such as startup and 'load' variation, i.e., change in electrical output.

The combined flow and thermal models can be a powerful tool for addressing various SOFC design issues. For example, during fast startup or fast cool-down, which may be needed in automotive applications, thermal stresses that develop within the fuel-cell stack must not exceed acceptable levels. It is therefore necessary to model in detail the gas flows as well as heat and mass transfer throughout the fuel-cell stack to analyse the transient temperature distribution. The latter, in turn, may be used to predict the thermal stresses.

As an example, Figure 11.1 shows a typical planar cell stack model geometry [9]. The upper-left portion of the figure shows the full stack geometry. Preheated air is introduced at the bottom left side of the stack. The air travels across the interconnect channels, is further heated in contact with the PEN, and exits downward at right. In the fuel electrode (anode) side manifolds, as in the air electrode manifold, the outlet manifold is wider than that at the inlet. The 'zoom' view of the stack at the upper right in Figure 11.1 shows more detail of the grid. Details of the individual flow channels are simulated using a porous media model in the active area.

Obtained using the commercial computational fluid dynamics (CFD) software, STAR-CD, Figure 11.2 shows the temperature distribution within the interconnect which is subject to the largest temperature gradient, 5 minutes after startup.

Predictions of the stress created by thermal gradients within the stack can be used to establish control parameters for transient operations and to minimise

298   High Temperature Solid Oxide Fuel Cells: Fundamentals, Design and Applications

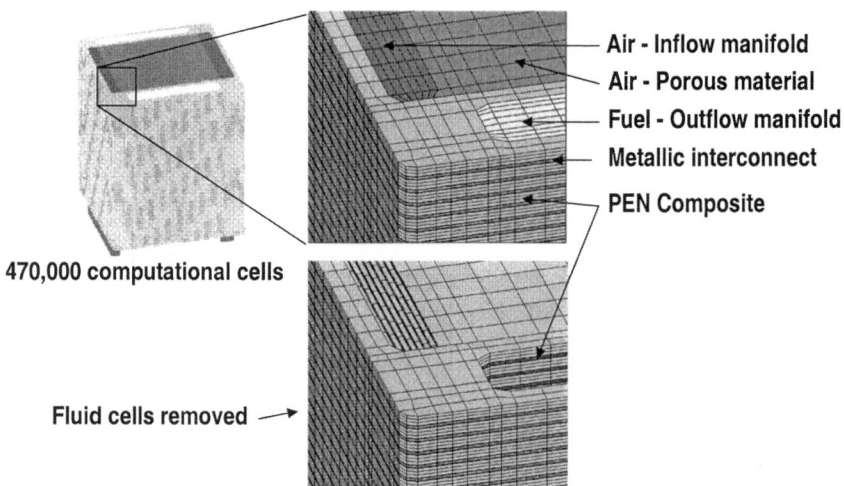

Figure 11.1.   Stack model geometry for combined flow and thermal calculations.

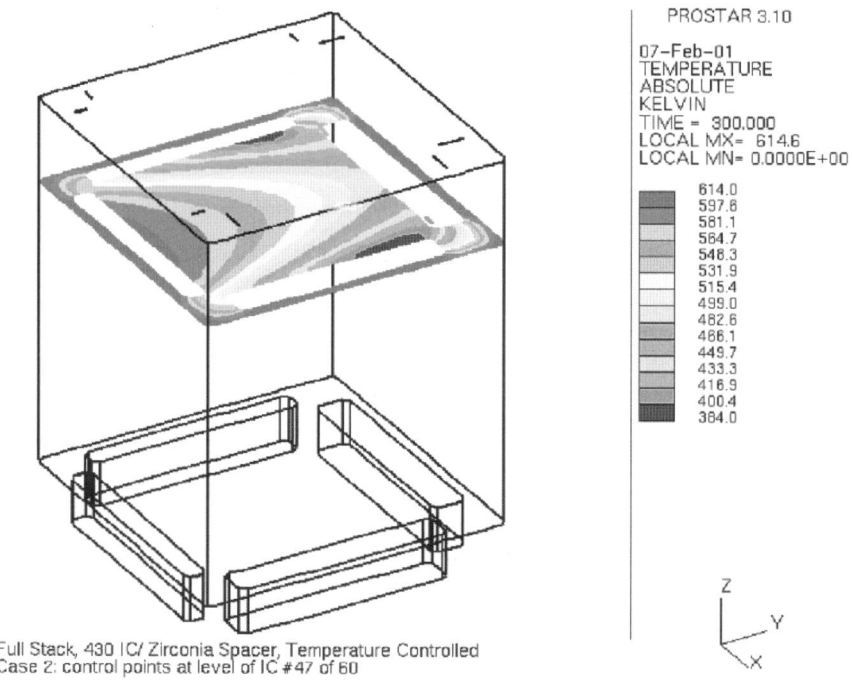

Figure 11.2   Temperature over interconnect 5 minutes after startup.

thermal stresses. Knowledge of thermal stress patterns can also provide guidance for geometry changes that may help to reduce stresses. The dimensioning of the individual PEN and IC elements is dictated in large part by the thermal stresses that develop during startup, caused by the thermal gradients within each element and depending on the mismatch in coefficients of thermal expansion between cell components. The configuration of the manifolds and overall cell dimensions can be modified based on the thermal stress results. The simulation can also be used to optimise the stack geometry for flow uniformity.

11.3 Continuum-Level Electrochemistry Model

One of the most important aspects of SOFC design is the voltage and current distribution in the PEN. This couples with the temperature distribution from the flow model and also with the electrochemical reactions at the electrodes. The electrochemical process generates electrical power and heat, but excessive heat generation must be avoided since it may cause thermal stresses affecting the structural stability of the SOFC.

At the effective property or continuum level, the simulation of electrode and cell performance basically requires only a parameterised electrochemical model. Such an electrochemical model is usually described as a current–voltage relation, or I–V curve, for a single cell, in terms of parameters that are effective cell properties and operational parameters. The I–V relation describes the voltage (potential) loss at a specified current with respect to the ideal thermodynamic performance, which is called overpotential or polarisation (η). This cell I–V curve is specific for the materials, structural characteristics, and operational parameters (gas compositions, pressure, temperature) of a given PEN element.

As an analogy to mass and energy balances, one can write a potential balance of the fuel cell as follows [10]:

$$V(i) = E_{eq} - iR_i - \eta_C - \eta_A = E_{eq} - iR_i - \eta_{Ca} - \eta_{Cc} - \eta_{Aa} - \eta_{Ac} \qquad (7)$$

Here E_{eq} is the equilibrium (open circuit) voltage, or emf (electromotive force) of the cell, i is the current density, iR_i is the ohmic potential drop, and η_C and η_A are the polarisation of the cathode and the anode, respectively. As shown in Eq. (7) each of the polarisation may be further split in an activation-related contribution (subscript a) and a concentration (i.e., diffusion) related contribution (subscript c).

The thermodynamic cell potential, E_{eq}, depends on reactant and product partial pressures as well as temperature. For example, for the hydrogen/oxygen fuel cell

$$E_{eq} = -\frac{\Delta G^0}{2F} + \frac{RT}{4F} \ln \frac{P_{O_2} P_{H_2}^2}{P_{H_2O}^2} = E^0 + \frac{RT}{4F} \ln \frac{P_{O_2} P_{H_2}^2}{P_{H_2O}^2} \qquad (8a)$$

where R is the gas constant, T is the temperature, and F is the Faraday constant. ΔG^0 is the standard free-energy change of the reaction $H_2 + 1/2 O_2 \rightarrow H_2O$; i.e., the free-energy change when reacting species and products are all at the standard pressure of 1 atm. The first term on the right-hand side of Eq. (8a), denoted by $E°$, is therefore called the standard cell potential or standard emf. It depends only on temperature. In the second term on the right-hand side of Eq. (8a), p_{O2} is an abbreviation of $p_{O2,c}$ for notational simplicity. The quantities p_{H2} and p_{H2O} are, respectively, the H_2 and H_2O partial pressures in the anode fuel channel.

In the most general way, one can express the thermodynamic cell potential of an SOFC as the cell potential of an oxygen concentration cell using the Nernst equation:

$$E_{eq} = (RT/4F)\ln(p_{O2,c}/p_{O2,a}) \tag{8b}$$

In Eq (8b), $p_{O2,c}$ and $p_{O2,a}$ represent the oxygen partial pressures in the cathode air channel and anode fuel channel, respectively. Notice from Eq. (8b) that air leakage reduces the open-circuit voltage and is detrimental to cell operation. Good sealing technology is required to minimise the leakage.

In Eq. (7), R_i represents the total area specific ohmic resistance. R_i is the sum of the cathode, electrolyte, anode, interconnect, and contact ohmic resistances expressed in $\Omega\ m^2$. Typically, R_i is dominated by the electrolyte resistance and decreases with increasing operating temperature. To account for any electronic conductivity in the electrolyte, the effective ohmic resistance should be used in Eq. (7). The effective conductivity depends on the applied voltage and can be expressed as a correction to the ionic conductivity, σ_{ion}, by a term involving the electronic conductivity, σ_e as follows [11,12]:

$$\sigma_{eff} = \sigma_{ion} - \sigma_e/[\exp(2eV/kT) - 1]/[1-\exp(-2e(E_0 - V)/kT)]\} \tag{9}$$

The final terms in Eq. (7), η_{Ca}, η_{Cc}, η_{Aa} and η_{Ac}, are the cathode activation, cathode concentration, anode activation, and anode concentration polarisations, respectively. In general, their dependence on the current density is nonlinear, although at low polarisation they may be approximated by linear relationships.

The activation polarisation terms are controlled by the electrode reaction kinetics of the respective electrodes. They represent the voltage loss incurred due to the activation necessary for charge transfer. The activation polarisation, η_a, is usually related to the current density by the phenomenological Butler–Volmer equation [13]:

$$i = i_0\{\exp[-\alpha zF\eta_a/RT] - \exp[(1-\alpha)zF\eta_a/RT]\} \tag{10a}$$

In this equation, i_0 is the exchange current density, α is the anodic transfer coefficient ($0 < \alpha < 1$), and z is the number of electrons participating in the electrode reaction. The exchange current density corresponds to the dynamic electron transfer rate at equilibrium, which is thermally activated. Therefore, the exchange current density can be expressed as $i_0 = P_x \exp(-E_{act}/RT)$, where the prefactor, P_x, and the activation energy, E_{act}, are properties specific for the electrode-electrolyte interface in question. The kinetic properties (i_0, α, z, P_x, E_{act})

depend not only on the materials forming the reaction interface but also on its microstructure[13,14].

The reasons for the importance of microstructure are twofold. Electrode reactions are interfacial reactions, i.e., surface bound, and therefore intrinsically slow compared with the reactions of gases. Moreover, in an SOFC they must take place at particular locations on the electrode–electrolyte interface, namely at or near a triple-phase boundary (TPB), where solid electrocatalyst, electrolyte, and gaseous reactants or products meet. In a typical SOFC porous electrode, the TPB is geometrically a serpentine line. The TPB by itself would form an extremely limited 'area' available for electron transfer. However, in SOFC electrodes the electron transfer step is only one of several steps in a reaction mechanism that may be quite complicated (as discussed in Chapter 9 and further referred to in Section 11.8). At the microscopic level, the active area of SOFC electrodes appears to be a nm-to-μm-wide zone bordering the TPB where surface diffusion of intermediate reactants or product species occurs. Nevertheless, the entire internal area of a porous electrode usually is not active. Typically, SOFC electrodes must have a reaction surface large enough to generate an internal current density (usually called transfer current density and denoted j) that is two to four orders of magnitude smaller than the projected (or external) current density at the electrode. Thus, to reduce the activation losses at an SOFC electrode, a large internal surface area is needed.

Therefore, in simplified continuum treatment of the electrochemical performance, e.g., in the potential balance (Eq. (7)), the activation polarisation is frequently calculated from Eq. (10a) by dividing the projected current density, i, by a dimensionless quantity, a, which represents the ratio of active internal area to external area:

$$j = i/a = i_0\{\exp[-\alpha z F \eta_a / RT] - \exp[(1-\alpha)z F \eta_a / RT]\} \qquad (10b)$$

The parameter a is specific for a given electrode microstructure and may be estimated from known or estimated microstructural parameters, physicochemical surface area measurements (e.g., by the BET technique, yielding total pore volume), or special-purpose electrochemical measurements (which yield the product $i_0 a$). However, a is rarely known accurately and may vary significantly with current load. Electrode-level models may be used to determine this variation and calculate polarisation without recourse to Eq. (10b) (see Section 11.8).

In practice, the activation loss of the SOFC cathode is larger than that of the anode; that is, the cathode reaction has a smaller $i_0 a$ than the anode reaction. In fact, greater kinetic limitation of oxygen reduction than of hydrogen oxidation is common to all types of fuel cells at high (e.g. 1000°C) as well as low (ambient) temperatures.

The concentration polarisation is the voltage loss associated with the resistance to transport of reactant species to and product species from the reaction sites. This transport occurs by diffusion because convection is negligible in the pores of SOFC electrodes. The concentration difference between the bulk gas and the gas contacting the reaction site forms a concentration cell whose cell voltage (emf), opposing the overall SOFC voltage, is observed as a voltage loss contribution, i.e., concentration polarisation. Thus, oxygen partial pressure in the cathode pores near the cathode/electrolyte is lower than that in the air channel. The more difficult the transport of oxygen through the porous medium, the greater the concentration polarisation at the cathode. Thus, a thick cathode in cathode-supported cells gives rise to high concentration polarisation even at moderate current densities. To lower concentration polarisation at high current densities to acceptable levels, the cathode should be as thin as practically feasible and the porosity and pore size as large as possible.

Excessive mass transfer resistance may cause a current limitation if the reactant concentration at the reaction site becomes small. In the extreme, that concentration may become zero (or rather, negligibly small). The current, in that case, reaches a plateau called the limiting current for the reactant species in question. With a number of simplifying assumptions, the limiting current concept can be used to derive a simple one-parameter expression for the concentration polarisation:

$$\eta_c = (RT/nF))\ln(1 - i/i_c) \tag{11}$$

Here i_c is the limiting current for the reacting species, i.e., O_2 for the cathode, with n = 4, and H_2 or CO for the anode, with n = 2 [9]. The limiting current of a species depends on its diffusivity in the surrounding gas mixture, its partial pressure, and the porosity, tortuosity, and thickness of the electrode. For H_2 fuel, the limiting current density can be calculated [15] as:

$$i_{H2} = 2p_{H2}D_{eff(a)}/(Cl_a) \tag{12}$$

while for air as oxidant, the O_2 limiting current density is

$$i_{O2} = (p_{O2}D_{eff(c)}/Cl_c)[P/(P - p^0_{O2})] \tag{13}$$

where P is the air pressure and l_a and l_c are respectively the anode and cathode thickness. The effective diffusion coefficients are given in terms of binary diffusion coefficients, porosities ($V_{c/a}$), and tortuosities ($\tau_{c/a}$):

$$D_{eff(c)} = V_c D_{O2-N2}/\tau_c \tag{14a}$$

$$D_{eff(a)} = V_a D_{H2-H2O}/\tau_a \tag{14b}$$

Analogous results can be obtained with CO as fuel. Because the anode binary diffusion coefficient, D_{H2-H2O}, is about four times that of the cathode counterpart, D_{O2-N2}, the cathode would have a much larger concentration polarisation than that of the anode for similar thickness, porosity, and tortuosity. Fairly thick anodes may be used without incurring excessive voltage loss. This is one of the reasons why anode-supported designs are preferred over cathode-supported designs in the thin-electrolyte intermediate temperature SOFCs.

As discussed above, the I–V relation of a PEN element depends on material properties and electrode structures as well as on operating parameters such as gas composition, pressure, and temperature. Using a simple first-order electrochemical model and the potential balance, Eq. (7), combined with simplified expressions for the various polarisation contributions such as Eqs. (8a), (10b), and (11)–(14b), I–V curves can be predicted. These predicted curves may be used to fit experimental I–V data and deduce, from an optimal fit, certain material and structure properties such as R_i and $i_o a$, which cannot be measured directly. In cells with sizable electrode area, which tend to have appreciable fuel and oxidant utilisation, temperature and gas partial pressures are local quantities dependent on the extent of the electrochemical and chemical conversion (i.e., the fuel and oxidant utilisations). The electrochemical model predicting the I–V curve simultaneously yields the current distribution, temperature distribution, and other quantities of interest.

Figure 11.3 shows the theoretical and experimental I–V relations of a small-size cell (considered isothermal) for different fuel compositions at a set of temperatures [14]. The material properties were obtained by fitting the theory to the experimental data for 97% H_2 + 3% H_2O fuel. As shown, the simplified theory can predict variation in the I–V curve with fuel composition reasonably well.

To obtain accurate information about microstructural characteristics of SOFC electrodes, a set of experimental i–η curves for a given electrode may be fitted, similar to Figure 11.3, against predictions of a more complex porous electrode model, as discussed in Section 11.8.

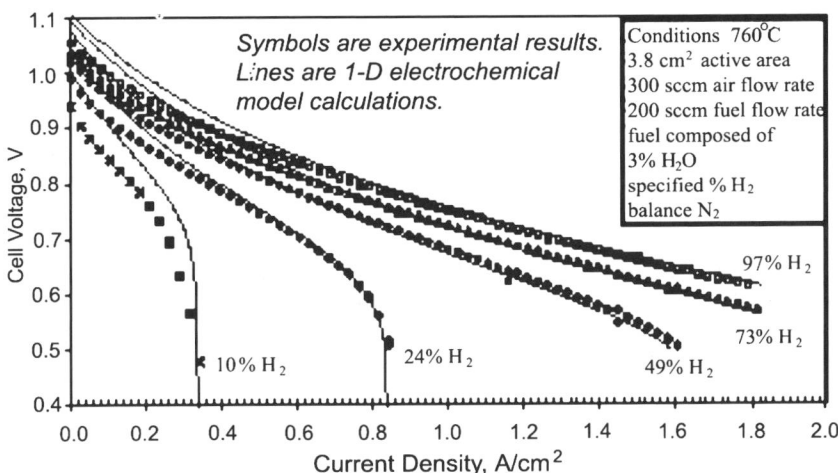

Figure 11.3. Predicted and measured cell I–V curves [14].

11.4 Chemical Reactions and Rate Equations

When fuel cells are operating, the heat generation rate (the source term needed in the thermal-fluid model) depends on the rates of the various chemical and electrochemical reactions. These reactions are not always simple; methane fuel, for example, in the presence of steam may undergo steam reforming upstream of the electrode reaction sites, so the overall heat generation may be due to a multitude of anode reactions such as

$$H_2 + O^{2-} \rightarrow H_2O + 2e^- \tag{15a}$$

$$CO + O^{2-} \rightarrow CO_2 + 2e^- \tag{15b}$$

$$CH_4 + H_2O \rightarrow 3H_2 + CO \tag{15c}$$

$$CO + H_2O \rightarrow CO_2 + H_2 \tag{15d}$$

The cathode reaction, on the other hand, has a single stoichiometry:

$$1/2 O_2 + 2e^- \rightarrow O^{2-} \tag{15e}$$

The heat source is related to the enthalpy change of the reactions, and the free-energy change of reactions (15a) and (15b) combined with (15e) determines the fuel cell Nernst potential. If chemical equilibrium is achieved in the system, the fuel composition, heat generation, and Nernst potential can be determined from thermodynamic theory. However, chemical equilibrium is usually not attained. In such cases, fuel composition and other information cannot be rigorously determined and must be approximated. The details of the reaction mechanism are complicated and usually not well understood, both for electrochemical and chemical reactions.

For example, anodic hydrogen oxidation probably involves dissociative hydrogen adsorption on the electrocatalyst (e.g., nickel) surface, surface diffusion of hydrogen adatoms, electron transfer under oxidation of H adatoms by an oxide ion at an adjacent active reaction site, and desorption of the H_2O molecule formed. Unstable and bulk unknown species such as OH may function as reaction intermediates. In chemical reactions, too, intermediates may play a role. Partial oxidation pathways for CH_4 may exist: e.g., $CH_4 + O \rightarrow CH_2O + H_2$ and $CH_4 + O \rightarrow CH_3OH$, resulting in the formation of chemicals such as CH_3OH or CH_2O with concomitant energy loss for the SOFC. Similarly, reaction intermediates such as O adatoms or adions may play a role.

Therefore, when equilibrium cannot be plausibly assumed, apparent kinetic parameters (effective rate constants) must be used to express the reaction rate. The parameters that describe the electrochemical reaction rate include the above-mentioned exchange current density, the transfer coefficient, the activation enthalpy, and the pre-exponential factor as well as the reaction order of the species involved. These parameters are not necessarily related to a single rate-determining step, as is often assumed in electrochemical theory. By investigating i–η curves as functions of electrode potential, temperature, and concentration of the reacting species, insight may be gained into the reaction mechanism and microscopic transport processes (such as surface diffusion) that determine the kinetic rate. This requires specialised electrochemical models, as discussed in Section 11.8.

Chemical reactions, too, may be characterised in a similar manner, following a strategy of effective kinetic parameters. When detailed knowledge of the reaction mechanism is lacking, the effective rate constants and other reaction-kinetic parameters can be determined by fitting a simplified kinetic model to the experimental data. For example, the steam-reforming conversion rate of CH_4 (reaction (15c)) may be expressed by the following empirical equation [6]:

$$dr_{CH_4}/dt = k_{CH_4} p_{CH_4}^{m_1} p_{H_2O}^{m_2} \exp(-E_{CH_4}/RT) \tag{16}$$

Different materials and designs will have different values for the parameters in Eq. (16). Depending on cell materials, the manufacturing process, and even the operating temperature, different values of m_1 and m_2 are possible [16–18]. Other rate expressions based on various kinetic models have also been proposed [19]. For realistic modelling, parameters that fit well with the desired systems should be used. Equation (16) represents the mass sink term for CH_4 in the CH_4 mass balance as given, in general form, by Eq. (1).

As Eq. (16) suggests, the effective rate constant strategy is particularly useful in the case of the anode fuelled by CH_4 because five or more gaseous species must be accounted for. Hydrogen is generated by steam reforming of CH_4 and by the forward process of the shift reaction $CO + H_2O \rightarrow CO_2 + H_2$. The backward process of the shift reaction and anodic hydrogen oxidation both consume hydrogen. The total rate of the mole change of H_2 is then

$$dr_{H2}/dt = 3dr_{CH4}/dt + dr_f/dt - dr_b/dt - dr_{H2}/dt(\text{oxidation}) \tag{17a}$$

Here,

$$dr_f/dt = k_1 p_{CO} p_{H2O} \tag{17b}$$

and

$$dr_b/dt = k_1 K_{shift} p_{CO2} p_{H2} \tag{17c}$$

where k_1 is a constant with dimensions of [kmol H_2 m^{-3} s^{-1} bar^{-2}]. From thermodynamics, the equilibrium constant of the shift reaction is given by

$$K_{shift} = \exp(-\Delta G_{shift}^0/RT) \tag{18}$$

where ΔG_{shift}^0 is the standard Gibbs free-energy change of the shift reaction at temperature T and can be calculated from the standard Gibbs free-energy change at 298 K (ΔG_{298K}^0) and the standard enthalpy change at 298 K (ΔH_{298K}^0) with the help of heat capacities of reactant and product expressed as functions of temperature.

The rate of anodic hydrogen oxidation is proportional to the current density of hydrogen oxidation (i_{H2}). As discussed in Section 11.3, this current density according to the electrochemical model follows a Butler–Volmer-type equation (Eqs. (10a) and (10b)) with a concentration-dependent exchange current density, i_o, which is specific for hydrogen oxidation. For gas-phase reaction calculations it is convenient to simplify Eq. (10b) in the form of an effective rate expression having an empirical rate constant, k_{H2}, reaction order, m_3, and activation energy, E_{H2}:

$$dr_{H_2}/dt(oxidation) = k_{H_2} p_{H_2}^{m_3} \exp(-E_{H_2}/RT) \qquad (19)$$

Similarly, the rate of anodic oxidation of CO may be expressed by means of effective rate parameters:

$$dr_{CO}/dt(oxidation) = k_{CO} p_{CO}^{m_4} \exp(-E_{CO}/RT) \qquad (20a)$$

The change in the CO concentration can then be written as

$$dr_{CO}/dt = dr_{CH4}/dt + dr_b/dt - dr_f/dt - dr_{CO}/dt(oxidation) \qquad (20b)$$

Consequently, the rate of change of H_2O concentration is

$$dr_{H2O}/dt = dr_{H2}/dt(oxidation) + dr_b/dt - dr_f/dt - dr_{CH4}/dt \qquad (20c)$$

At the cathode, too, effective rate expressions for oxygen reduction, if available, may be convenient, but the gas-phase mass balances are basically simpler than those at the anode. When current density is known, the rate of O_2 loss is

$$dr_{O2}/dt = i/4F \qquad (21)$$

Assuming the current vector is everywhere transverse to gas flow direction, that is, perpendicular to the PEN element, Eq. (21) can be used to determine the variation of O_2 in an air channel. Similarly, if there is no current component parallel to the PEN element

$$dr_{H2}/dt(oxidation) + dr_{CO}/dt(oxidation) = i/2F \qquad (22)$$

Because numerical error is inherent in modelling software, for exact mass balance, $i/2F - dr_{H2(CO)}/dt$ (oxidation) can determine $dr_{CO(H2)}/dt$ (oxidation), or dr_{H2}/dt (oxidation) + dr_{CO}/dt (oxidation) can determine $i/2F$. However, when a shift equilibrium reaction (15d) is achieved or assumed, the chemical equilibrium condition and Eq. (22) uniquely determine the sum of the rates of hydrogen and CO consumption. Therefore, separate rate parameter measurements for CO and hydrogen are not necessary. In fact, experience suggests that in most SOFCs the dominant anodic process is hydrogen oxidation, while CO is consumed by the shift equilibrium [10].

Similarly, if CH_4 is present and steam reforming is at equilibrium, chemical equilibrium conditions determine the rates of consumption of each fuel component uniquely for a given current production.

Combining the above rate equations and mass balances with the flow and thermal model equations presented in Section 11.2, detailed information about the variation of gas composition, fuel or oxidant utilisation, etc., in the flow channels may be generated.

The reactions presented thus far assume that the fuel mixture is CH_4/H_2O. The same approach would apply to other fuels such as H_2, CH_3OH, or dry CH_4 with corresponding changes in the reaction paths. If pure H_2 is used with a small amount of H_2O, the fuel composition and reaction mechanism are simplified. The number of experimental parameters and mathematical equations needed is reduced and the simulation is easier.

The equilibrium theory is very useful in addressing fuel processing issues whether or not equilibrium is attained. For example, with an internal reforming fuel cell the carbon forming reactions, decomposition of methane according to $CH_4 \rightarrow 2H_2 + C$, and Boudouard reaction, $2CO \rightarrow C + CO_2$, can be suppressed by providing a proper molar ratio of water to methane. For the external reforming subsystem, the theory can determine the optimal fuel-to-air ratio and operating temperature to maximise stack fuel (H_2 and CO) production while minimising equilibrium-predicted carbon formation. The equilibrium theory can also guide some cell design issues. Because steam reforming is an endothermic process, excessive cooling of the stack at the fuel inlet can occur with internal reforming. Nickel as an anode is known to be a good catalyst to promote cracking. A possible improvement is kinetic suppression of the cracking reaction using catalysts that are not as effective at promoting the cracking reaction. An alternative approach would use catalysts that promote electrochemical oxidation of hydrocarbons at lower operating temperatures. Figure 11.4 shows the equilibrium constant of the CH_4 steam reforming reaction as a function of temperature. Clearly, temperature has a significant effect on the resulting CH_4 content. CH_4 is stable against reforming at lower operating temperatures. Because suppression of steam reforming is also beneficial in full utilisation of the chemical energy of hydrocarbons, resulting in higher energy efficiency, considerable interest exists in direct electrochemical oxidation of natural gas and other hydrocarbons [20–22] in SOFCs.

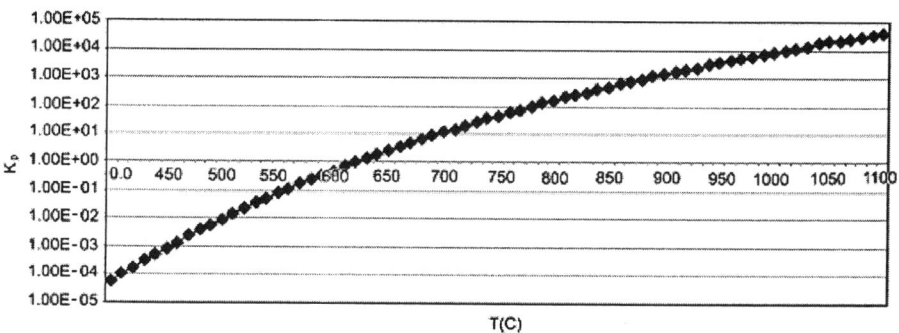

Figure 11.4   Equilibrium constant of methane steam reforming reaction as a function of temperature.

11.5 Cell- and Stack-Level Modelling

The coupled continuum-level electrochemical, flow, and thermal models are usually discretised in a finite element mesh [23,24]. When the necessary material properties, geometrical parameters, operating parameters, and boundary conditions are supplied, cell- and stack-level performance can be analysed. The combined models can determine the cell/stack voltage, the total current output, temperature distribution, species concentration, etc.

An important outcome of the combined models at the cell level is the cell efficiency, and at the stack level, the stack efficiency as well. The electrical efficiency of a cell and of the stack is defined as

$$\varepsilon_{el} = \text{electrical power output/chemical energy input per unit time} \quad (23)$$

for the cell in question, and for the stack, respectively.

The electrical efficiency therefore depends to some extent on the definition of the fuel energy input and on whether power for gas pumping and the like is subtracted from the generated power. ε_{el} may be expressed in terms of current, voltage, etc., as follows:

$$\varepsilon_{el} = VI/(\Delta H dn/dt) \quad (24a)$$

This can be considered the product of three additional fundamental efficiencies, namely the ideal or thermodynamic efficiency $\{\Delta G/\Delta H\}$, the voltage efficiency $\{V/E_0\}$, and the fuel utilisation $\{U_f\}$:

$$\varepsilon_{el} = VI/(\Delta H dn/dt) = \{\Delta G/\Delta H\}\{V/E_0\}\{U_f\} \quad (24b)$$

The fuel utilisation, U_f, is the ratio of the delivered current to the stoichiometric current equivalent to the fuel flow rate:

$$U_f = I/(2Fdn/dt) \quad (24c)$$

The ideal efficiency $\Delta G/\Delta H$ may be considered a measure of the thermodynamic reversibility of the reaction and depends only on the operating temperature and fuel used. It is typically between 80 and 100%. The voltage efficiency and fuel utilisation, as well as the electrical energy efficiency, are useful measures for the success of the cell and stack design.

Another important figure of merit predicted by the combined model is the power density of a cell, or of the stack as a whole. The power density of a cell is usually defined on the basis of the cell or electrode area, that is

$$P_{cell} = I(\text{current}).V(\text{cell voltage})/A(\text{area}) \quad (25a)$$

The power density of a stack is conveniently defined on the basis of stack volume:

$$P_{stack} = I(\text{current}).V(\text{stack voltage})/V(\text{stack volume}) \quad (25b)$$

The cell- and stack-level models can improve understanding of the complex interactions between fluid dynamic, thermal, chemical, and electrochemical phenomena. The combined models can therefore help maximise efficiency or power density by optimising PEN element design, cell configuration, and stack architecture for a given set of operating conditions. Most SOFC modelling focuses on cell- and stack-level performance for exactly this purpose.

Cell-level models simulate the performance of a single cell, also called a unit cell. This is the repeating unit of a stack and consists basically of a PEN element, an interconnect layer, and a gas channel/current collector structure. The desired output is the current–voltage relationship, the temperature distribution, and the heat production in the cell. Cell-level models frequently treat the major cell components, the PEN element, the interconnect layer, and the gas flow, as two-dimensional (2-D) – having negligible thickness compared with the dimensions of the flow direction [12,25–27]. When the continuum flow/thermal/electrochemical model is treated as a 2-D model, calculations of heat production at any point of the cell plane (node of the 2-D model) can be simplified substantially. In such a 2-D cell model (or quasi-2-D stack model), heat generation may be determined by evaluating the change in gas composition between the inlet and exit conditions. The total energy (heat and work) delivered per unit time at each control volume (node) is simply

$$E_{total} = (\Delta H/nF)I \qquad (26a)$$

where ΔH is the enthalpy of formation, representing the maximum chemical energy for the simplest $H_2 + 1/2O_2 \rightarrow H_2O$ reaction. Note that E_{total} is negative, in agreement with thermodynamic sign conventions for heat and work (heat positive when absorbed by, and work positive when performed on, the system). The heat generation, Q_{gen}, is then determined by subtracting from the total energy generation the electrical work delivered per unit time externally, i.e., the electric power:

$$P_{ex} = -I.V \qquad (26b)$$

Therefore, the heat generated per unit time at each node is

$$Q_{gen} = (\Delta H/nF).I + V.I \qquad (26c)$$

which is sometimes expressed in terms of the 'thermoneutral voltage' $V^{thn} = -(\Delta H/nF)$ as

$$Q_{gen} = (V - V^{thn}).I \qquad (26d)$$

The expressions (26c) and (26d) for Q_{gen}, because they are derived from the overall change in thermodynamic state, account for all the various kinds of heat generation, including Joule (or ohmic) heating, heating due to polarisation, and entropic ($T\Delta S$) heat effects. The heat development at each node, $Q_{gen(i,j)}$, is balanced by heat convection and conduction to and from adjacent nodes. Thus, all nodes in the 2-D model are thermally coupled to one another. Note, however, that in the 2-D model, the principal variable that determines the nonuniformity of heat generation, and therefore temperature, over the 2-D surface is the current I through each node (perpendicular to the 2-D plane). Applying the 2-D model to a planar stack with reasonably low in-plane electrical resistance (compared with the impedance of the electrochemical reaction), the variable V, the voltage at a given node, is relatively constant from node to node. Because the term $(V - V^{thn})$ is then also relatively uniform, the nonuniformity of I is the key to the temperature distribution. The principal cause of the nonuniformity of I, in turn, is the asymmetry of utilisation (hence of the local driving force, the local E_{eq}) imposed by the flow configuration of the planar cell.

In the 2-D cell simulation, as well as in simplified (quasi-2-D) stack-level simulations, it is usually assumed that each side of the electrode/interconnect is at equal potential over the whole 2-D plane of the cell. As mentioned above, this is justified because the ohmic voltage drop in the plane of the electrodes and interconnect layer is usually much smaller than the ohmic voltage drop across the electrolyte and the combined polarisation of the two electrodes. Nevertheless, in such a quasi-2-D stack model, individual fuel cells in the stack may have different cell voltages due to different temperature, fuel distribution, and other factors. However, the total current flow through each cell (integrated over the plane of the PEN elements and the gas flows) must be the same. The total stack output voltage is the sum of each individual cell voltage.

A true three-dimensional model [28–32] is necessary for a more accurate thermal analysis of a stack or for a detailed analysis of the temperature profile at the contact regions between PEN element, current collector/gas channel profiles, and the interconnect layer. In those cases, a more detailed heat source calculation is also needed. It is necessary to distinguish three different types of heat effects acting in specific components of the fuel cell: chemical, electrical, and electrochemical.

Chemical reactions (reforming and shift reactions) take place at the anode side, and chemical heat effects represent an important heat source (sink) term for the anode and the fuel channel.

The electrical heat effects are caused by resistance to current flow, which yields ohmic heating (also called Joule heating). Ohmic heating takes place throughout the solid structure wherever electrical current flows, for instance, from PEN element to interconnect layer. The total ohmic resistance can be decomposed into contributions from various cell components. If the component material has an ohmic resistivity γ_i (expressed in Ω m), the ohmic heat generated per unit volume of that computational region can be calculated from

$$Q_{ohm} = i^2 \gamma_i \tag{27}$$

where i is the local current density.

The electrochemical heat effect has two components: reversible or entropic heat effect (positive or negative, endothermic or exothermic), and irreversible heat generation (always exothermic). Reversible heat generation is associated with the change of entropy occurring as a result of the electrochemical reaction. It is generated at the two electrodes in unequal amounts. In the case of hydrogen oxidation, the reversible heat generation at the anode per unit of projected area of the anode is

$$Q_{rev,a} = T(S_{H_2O} - S_{H_2} - S^*_{O^=} + 2S^*_{el})i/2F \tag{28a}$$

Here S_i is the entropy of the species i; that is, S_{O_2} is the entropy of O_2, $S^*_{O^=}$ is the transported entropy of the oxygen ion, and S^*_{el} is the transported entropy of the electrons. The effect represented by Eq. (28a) is positive but relatively small [33]. The reversible heat generated at the cathode per unit projected area

$$Q_{rev,c} = T(S_{O^=} - \tfrac{1}{2}S_{O_2} - 2S^*_{el})i/2F \tag{28b}$$

is much larger and negative. Because the sum of the effects is equal to

$$Q_{rev,total} = T(S_{H2O} - S_{H2} - 1/2 S_{O2})i/2F \tag{28c}$$

it follows that almost the entire entropic heat effect of the hydrogen oxidation reaction is released at the cathode. In some designs (with relatively thick electrolyte or thick anode) this may lead to significant temperature gradients, especially because cathode polarisation is usually dominant over anode polarisation, which further contributes to local heating at the cathode.

Conceptually, one can split the overall entropic effect, Eq. (28c), in two equal but opposite heat effects occurring at the two electrode-electrolyte interfaces. For example, the heat effect at the anode is

$$Q_{rev,a} = T(\tfrac{1}{2}S_{O_2} - S^*_{O^=} - 2S^*_{el})i/2F \tag{28d}$$

while that at the cathode is given by Eq. (28b). In that case, the overall reversible heat effect due to hydrogen oxidation (Eq. 28c) must be accounted for separately in the anode fuel gas channel. The advantage of introducing such a symmetric expression for the reversible heat effect is that in principle it allows taking into account heat development due to diffusion effects in the solid electrolyte upon current passage. However, in an SOFC these effects are minor compared with the Joule heating due to the ohmic resistance of the electrolyte included in Q_{ohm} (Eq. 27).

Irreversible heat generation due to the electrochemical reaction can be concisely represented by the local planar heat source for a two-electron reaction:

$$Q_{irr} = -(\eta_a + \eta_c)i/2F \tag{29}$$

Using an approximate electrochemical performance model, as discussed in Section 11.2, or a more detailed electrode-level model, as will be discussed in Section 11.8, the polarisation components can be estimated and the heat generation calculated. In an approximate 3-D cell or stack model, it is customary to use a lumped electrochemical heat generation; that is, one assumes a uniform distribution of the reversible and irreversible heat generation over the respective electrode. Using a more advanced electrode-level model, the distribution of heat generation in the electrodes can be analysed, which may yield important information for structural stability of the electrodes and their interfaces with the electrolyte and the current collector/gas channel profiles.

The heat source terms discussed above are used in the energy balance equations of the flow and thermal models for both cell and stack simulations. To manage the computation cost, most simulations avoid the full CFD treatment of the flow. However, full CFD treatments provide more accurate and detailed information on the flow in an SOFC system. With increased computing power, more and more full CFD simulations are expected to appear.

Figure 11.5 shows sample cell modelling results: the steady-state temperature, current density, and species concentration distributions. As can be seen by comparison with Figure 11.2, the steady-state temperature distribution is quite different from that during the rapid startup.

Figure 11.6 shows the steady-state modelling results for a tubular SOFC design, as reported by Ferguson *et al.* [29]. Clearly, the characteristic operating

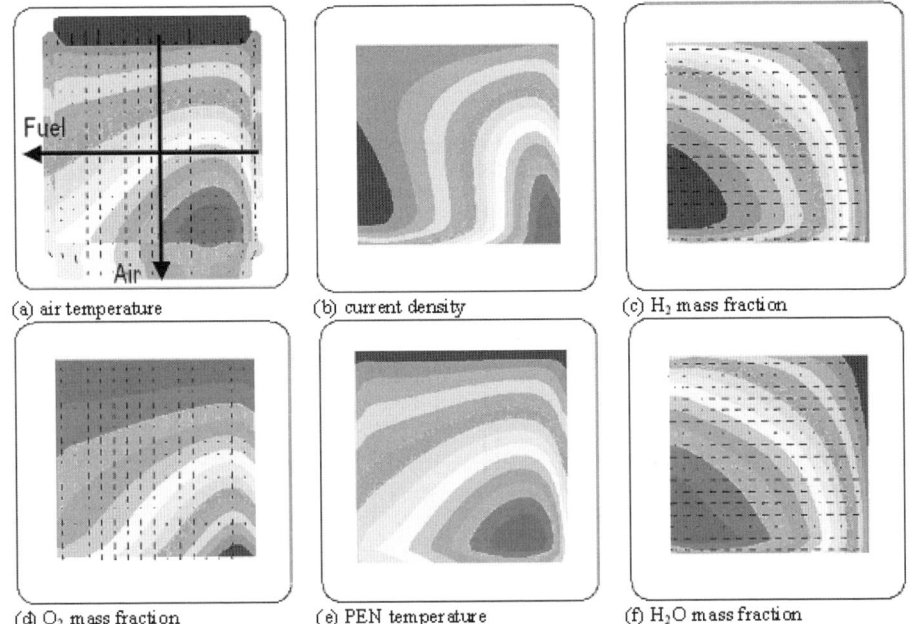

Figure 11.5 Sample results for planar SOFC in cross-flow configuration: a, air temperature, max. 899°C, min. 625°C; b, current, A/cm^2: max. = 1.46, min. = 0.300; c, H$_2$ mass fraction, kg/kg: max. 0.0385, min. 0.00472 (CO distribution similar); d, O$_2$ mass fraction, kg/kg: max. 0.231, min. 0.182; e, PEN temperature, °C: max. 911, min. 643; f, H$_2$O mass fraction, kg/kg: max. 0.224, min. 0.0387 (CO$_2$ distribution similar).

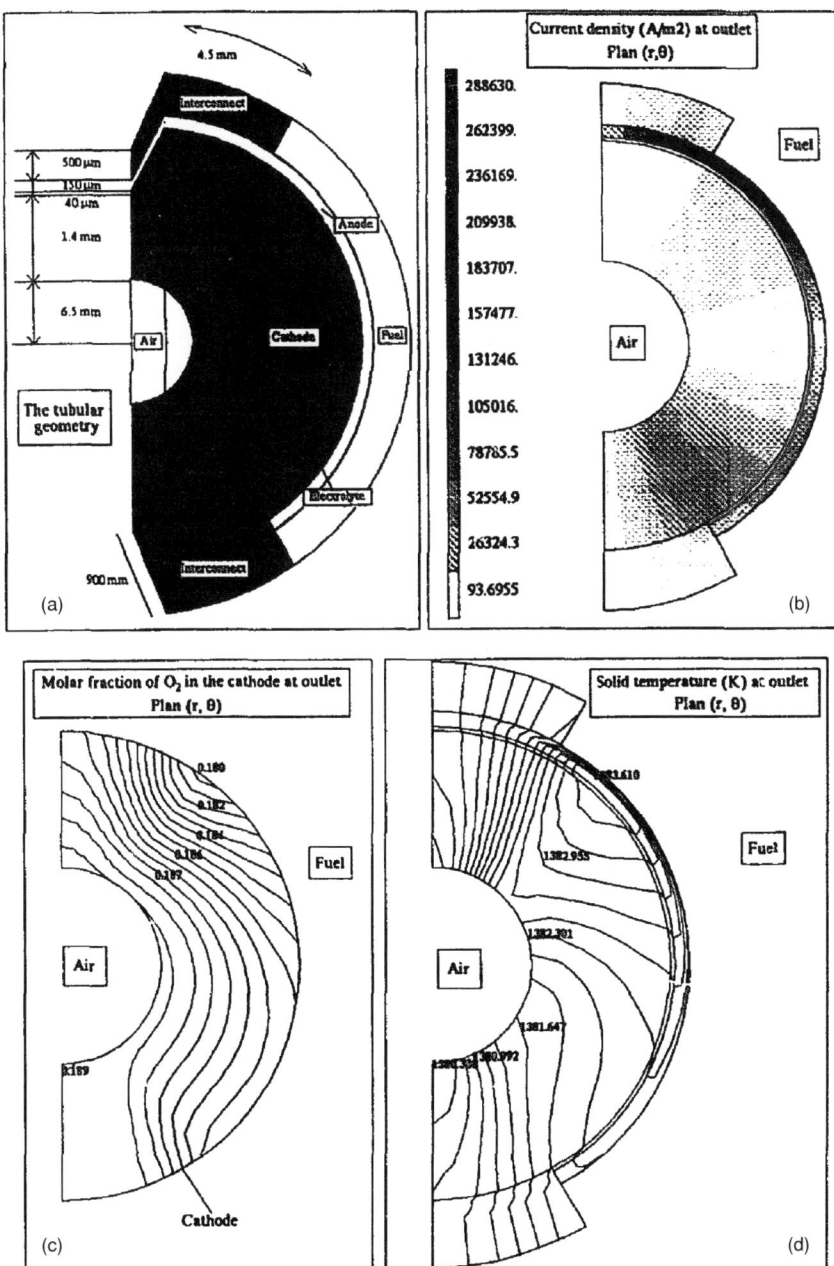

Figure 11.6 (a) Three-dimensional section of a tubular design; (b) current density; (c) molar fraction of O_2; (d) temperature at the outlet section.

properties for tubular designs are very different from that of planar SOFCs. Dedicated efforts are required for each individual design.

11.6 System-Level Modelling

In a system-level model, stack models are combined with models of system components including reformer, contaminant removal unit, compressors, topping or bottoming turbines, inverters, etc. The principal objective of system modelling is to determine the energy efficiency and heat/power ratio of the system. Such a model is also an excellent tool for making initial sketches of the system design and for initial sizing of components [14,34–38]. One example is the Excel spreadsheet model developed by Keegan et al. [14]. It combines models for gas preheater, reformer with recycle, stack, interface between stack and inlet gas, external reformer and exit gas, and combustor and uses heat balances and variable recycle ratio (with reformate composition as a function of recycle ratio) to evaluate (i) the overall system configuration and connectivity options, including various other recycle options, heat exchanger types, locations, and sizes; (ii) required energy transfer, resulting temperatures, and overall system efficiency, including pumping power and other parasitics, for the different proposed system designs; (iii) subsystem requirements associated with specific selected configurations, including required stream mass flows and allowable branch pressure drops; (iv) system performance at various load conditions; and (v) dynamic system performance during startup to determine additional constraining requirements, including allowable thermal mass and required heat transfer. The model also evaluates the system cycle efficiency, that is, the overall system performance as a function of the system start-up and shut-down cycling.

Another example is the systematic analysis undertaken by Palsson et al. on combined SOFC and gas turbine cycles [36]. In combination with a robust and accurate 2-D SOFC model, the system-level model attempts to provide an unbiased evaluation of performance prospects and operational behaviours of such systems. The 2-D SOFC model was integrated into a process simulation tool, Aspen Plus™, as a user-defined model, whereas other components constituting the system are modelled as standard unit operation models. Parametric studies can be carried out to gain knowledge of stack and system behaviour such as the influence of fuel and air flow rate on the stack performance and the mean temperature and the effects of cell voltage and compressor pressure on the system efficiency. The pressure ratio is shown to have a large impact on performance and electrical efficiencies of higher than 65% are possible at low-pressure ratios.

Extensive system modelling for SOFC systems has been carried out and published by Winkler et al. [37,38]. Their publications cover the methodology of system modelling as well as the effect of hardware design variations on efficiency and cost of integrated SOFC-GT hybrid systems. The market acceptance ultimately depends on the system cost, which is influenced by the process design, hardware design, production (materials and handling), and market (production quantity). Based mainly on the thermodynamics, the process design examines issues such as the choice of heat engine and the integration of heat engine and heat exchangers. The hardware design examines the geometric effects on the system cost through its effects on the power density, thermal insulation, and the wall of the pressure vessel. Such analysis provides valuable information for design optimisation.

These and other system modelling analyses show how strongly system characteristics such as efficiency depend on accurate input data for the electrochemical model used in simulating stack performance. On the other hand, these studies also show the strong effect of turbine operating parameters (e.g., pressure ratios and maximum allowable temperature) on the system performance. Such studies clarify that the ultimate design of the stack and the required accuracy of stack modelling are best determined after preliminary system design studies have been performed using rough stack, reformer, and turbine models.

11.7 Thermomechanical Model

Avoiding thermomechanical failure is critical to the applications of the SOFC technology. SOFCs are produced by processing at elevated temperatures. As the cells are cooled to room temperature, stresses due to mismatch in coefficients of thermal expansion (CTEs) are developed. Additional residual stresses develop in the stack during the assembly and sealing process. The factors that affect the magnitude of the stresses include (i) differences in CTEs of the material parts, (ii) the differential between stress-free (processing) temperature and operation temperature, (iii) elastic constants of the components, and (iv) the thickness of the cell components. Because the cell thickness is much less than the lateral dimensions, the elasticity problem may be approximated as 2-D and the state of stress is thus biaxial. For the state-of-the-art PEN materials, cathode (LSM) and electrolyte (YSZ) have similar CTEs, while the CTE of the anode (Ni + YSZ) is higher. Thus, when cooled from a high temperature, stresses in the electrolyte and the cathode would tend to be compressive, while stresses in the anode would be tensile. In an anode-supported cell, the tensile stress can cause a delamination crack between the anode and the electrolyte.

The residual stress in a cell when cooled from stress-free temperature to room temperature can be calculated [39]:

$$\sigma_1 = (\beta_2 - \beta_1) E_1 \Delta T / [1 + h_1 E_1 / h_2 E_2] \tag{30}$$

where β is the thermal expansion coefficient, ΔT is the change in temperature, h is the layer thickness, and E is the biaxial modulus. The subscripts '1' and '2' denote two neighbouring layers of the cell. From Eq. (30) it can be seen that thin layers suffer higher residual stresses than thick layers. In the anode-supported cell, the electrolyte has a much higher residual stress than the anode. Fortunately, the electrolyte is strong against compressive stresses. For the anode, the tensile stress is a concern. Assuming $\Delta T = 1000$ K, $\beta_2 - \beta_1 = 1.7 \times 10^{-6}/°C$, $E_2 = 200$ GPa, and $h_1/h_2 = 60$, the residual stress is about 6 MPa. For cathode-supported cells, however, the anode residual stress could be one order of magnitude higher and is very undesirable. The anode residual stress is the highest in the electrolyte-supported cells.

When temperature distribution in the cell structure is known, the finite-element structure model can analyse the thermal stresses. The thermal stresses dictate the process of heating and cooling required in SOFC applications. Thermal stresses also dictate how uniform the temperature should be in steady-state operations. Modelling results indicate that temperature gradient makes the largest contribution to overall stress. Moreover, it is important to maintain a uniform temperature gradient along the cell in minimising the anode/electrolyte/cathode stress. Figure 11.7 demonstrates the dramatic reduction in stress going from a parabolic temperature distribution to uniform temperature gradient.

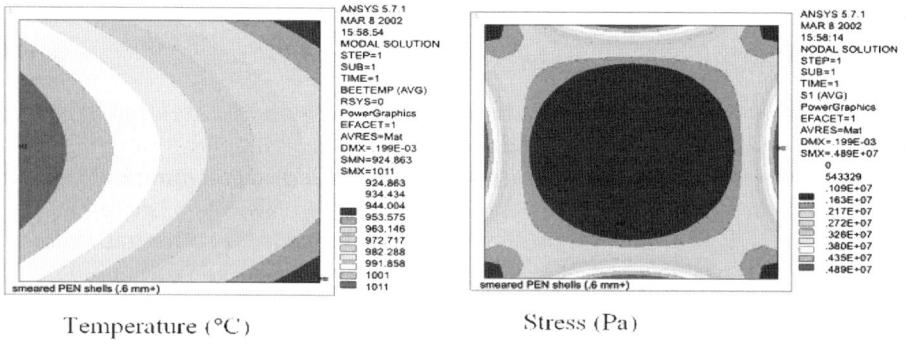

(a) Parabolic Temperature Distribution

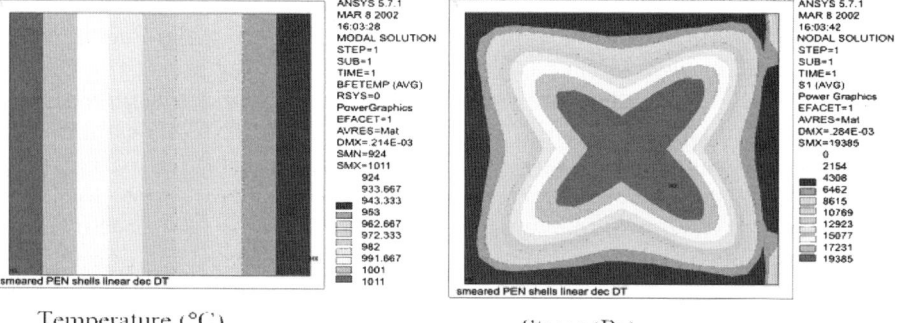

(b) Uniform Temperature Gradient

Figure 11.7. Effect of temperature profile on SOFC stress.

The probability of failure under stress σ can be calculated from the Weibull function:

$$W = 1 - \exp[-(\sigma/\sigma_0)^m] \qquad (31)$$

The Weibull parameter m and the characteristic stress σ_0 are material-specific parameters.

Mechanical failure can be caused by various mechanisms. As discussed, the electrolyte and the cathode are subjected to compressive stresses due to thermal mismatch. For thin-film coating under compressive stress, a common failure mode is buckle-driven delamination, or blistering [40]. The failure entails the film first buckling away from the substrate in some small region where adhesion is poor or nonexistent. Buckling then loads the interface crack between the film and the substrate, causing it to spread. Another failure mode of the fuel cell structure is thermal shock spalling. During thermal cycling, biaxial tensile residual stress can develop in and spall the surface layer. The spall depth and the time elapse can also be analysed with the finite-element method once the temperature gradient is known.

The mechanical strength of a metallic interconnect such as stainless steel 430 decreases significantly at elevated temperatures. Modelling results indicate that the portion of the interconnect near the fuel cell edge often suffers from high tensile stress. Therefore, optimising designs and operating conditions to reduce the interconnect stress is also a focus of the modelling activity.

The thermal stress consideration also limits the design and material choice for the seal. The seal is responsible for the gas-tight separation of the air and the fuel gas chambers and air manifold from the fuel electrode and fuel manifold from the air electrode porosities. In addition, to prevent gas crossover the sealant should be strong and stiff so that stacks are mechanically stable, can be handled, and can withstand pressure differences during operation. On the other hand, the sealant must be soft enough to reduce mechanical stresses during fabrication and operation. Moreover, the requirement of chemical compatibility with other cell components (electrolyte, electrode, interconnect) as well as stability in both oxidising and reducing gas atmospheres should also be satisfied. These considerations affect whether the design should be rigid glass seal; flexible, glass-free, compression seal; or a combination of the two [41,42].

Dimensional changes of components may arise due to a change in temperature. Nonstoichiometric oxides exhibit an expansion behaviour depending on oxygen stoichiometry due to reduction or oxidation upon changes in oxygen partial pressures. Interconnect and electrolyte are exposed to different oxygen partial pressures at the anode and cathode side, respectively. An expansion behaviour depending on oxygen nonstoichiometry can therefore lead to different expansions on each side of the interconnect. Bending and mechanical failure may result. As a simple one-dimensional example, the steady-state thermal stress, σ, in an infinitely wide, free plate subject to a temperature distribution, $T(z)$, which varies only in the direction of the thickness, z, can be expressed as [43]

$$\sigma_x = \sigma_y = -\frac{\beta E}{1-\nu}\left\{T + \frac{1}{2d}\int_{-d}^{d} Tdz + \frac{3z}{2d^3}\int_{-d}^{d} Tzdz\right\} \tag{32}$$

where β is the CTE, E is the Young's modulus, v is the Poisson's ratio, and d is the half-thickness of the plate. Assuming that the lattice expansion-induced strain can be treated the same way as the thermal strain, βT, Eq. (32) can be recast into a stress–isothermal strain relation. The experimental data for the isothermal expansion characteristics can then be used to obtain the nonstoichiometry-induced strain. Consequently, the isothermal stress due to nonstoichiometry can be determined. Delamination will occur if the elastic energy release rate by crack formation exceeds a critical value where the critical value is an interface property and must be determined experimentally. Alternatively, mechanical failure occurs if the thermal stress exceeds a certain value.

11.8 Electrochemical Models at the Electrode Level

Electrode-level models describe the performance of SOFC electrodes in detail. They take into account the distribution of species concentrations, electric potential, current, and even temperature in the electrode. Their purpose is to (i) interpret the performance (polarisation curve) of electrodes in terms of rate-limiting resistances such as kinetic (activation), mass transfer, and ohmic resistance; and (ii) predict the local polarisation in full-scale cell and stack models.

To predict the local polarisation in a full-scale cell or stack at any point, its dependence on composition, pressure, and temperature of the gas flowing in the gas channel contacting the electrode must be known. In a large cell, these bulk gas properties vary from one point to the next. Electrode polarisation or overpotential – the difference between the local potential of the electrode under load and the potential at open circuit (equilibrium potential) – is also a local quantity because it depends not only on the bulk gas composition but also on the current density. In a large cell the current is usually distributed nonuniformly, as discussed in Sections 11.2–11.5. Similar to Eq. 7, one can express the local cell voltage under load, i.e., when current is passed, as the thermodynamic cell potential minus three loss terms: the ohmic loss, the cathode polarisation, and the anode polarisation:

$$V(i) = E_{eq} - iR_i - \eta_C - \eta_A \tag{33}$$

As discussed in Section 11.2, the total polarisation of each electrode consists of two contributions, activation polarisation (due to electrode kinetic resistance) and concentration polarisation (due to mass transfer resistance), so

$$\eta_C = \eta_{Ca} + \eta_{Cc} \quad \text{and} \quad \eta_A = \eta_{Aa} + \eta_{Ac} \tag{34}$$

For cell- and stack-level modelling it is necessary to have reliable values of the total polarisation of cathode, η_C, and anode, η_A, as a function of local bulk gas composition, pressure, and temperature, as well as the local current density.

In principle, one could make a large set of measurements of cathode and anode polarisations in a small-size cell with a reference electrode (three-electrode cell) and express the total polarisation as a function of local bulk gas composition, pressure, temperature, and current density. The essential condition is that the small-size cells ('button cells') use very little fuel gas and oxidant gas, so that the measured polarisation is representative for the bulk gas composition, pressure, and temperature at a given current level. The cell then functions as a differential reactor that provides data for the cell- and stack-level (integral reactor) modelling. Although small-size cell data are obviously useful and many such measurements are made, the effort implied in a full 'polarisation mapping' of this kind for each electrode is usually prohibitive. Moreover, the results are valid only for the range over which the operating parameters are varied and for the electrode–electrolyte assembly microstructure and configuration used in the small-size cell.

In lieu of an experimental 'map' of polarisation, it is often desirable to have an electrode model that provides reliable predictions of polarisation of either electrode over a wide range of operating and structural variables. This is the first purpose of the electrode model. But, conversely, to be a good predictor the model should be capable of interpreting available polarisation data for well-defined conditions, that is, for small cells at low utilisation of fuel or oxidant. Thus, the second purpose of an electrode model is to enable a more efficient process of collecting, correlating, and interpreting polarisation data. The electrode model is capable of extracting the kinetic and mass transfer (diffusion) resistance information by fitting small-size cell polarisation data. It provides these resistance characteristics in a form suitable as input to full-scale cell and stack models.

An electrode model is especially advantageous if it can be used to relate the kinetic and mass transfer resistance to electrode geometry and microstructure; for instance, to thickness, porosity, pore or particle size, contact areas of phases, and/or grain size of electrode and electrolyte materials. A well-tested and validated electrode model, therefore, may serve to assist in the design of optimised electrode structures or electrode/electrolyte interfaces to minimise polarisation loss.

11.8.1 Fundamentals and Strategy of Electrode-Level Models

The objective of an electrode model is to analyse the point-to-point distribution of the reaction in an SOFC electrode, leading to current, potential, and species concentration distributions. The result of the analysis is a prediction of the polarisation of the electrode due to (i) kinetic resistance, (ii) mass transfer resistance, and (iii) ohmic resistance.

The analysis includes a whole set of material properties and structural parameters. In principle it is based on the same fundamental laws used in full-scale cell analysis. Thus, mass transfer is subject to mass balances (Eqs. (1), (2)), heat flow to energy balances (Eqs. (5), (6)), and fluid flow to Eqs. (3), (4), but it is usually negligible in the pores of the electrodes. In addition, current flow is subject to the electrical conservation equation (under assumption of electroneutrality):

$$(\nabla \cdot i) = 0 \tag{35}$$

where i is the current density vector. This yields for the potential distribution Laplace's equation:

$$(\nabla^2 \Phi) = 0 \tag{36}$$

with the appropriate boundary conditions for conducting and nonconducting boundaries. Equations (35) and (36) may be applied separately to the ionic current and its associated electric potential, respectively, and to the electronic current and its associated electric potential. This is very helpful in formulating the electrochemical rate at each point of the electrode/electrolyte interface by equating the local potential difference $\Phi_{\text{electronic}} - \Phi_{\text{ionic}}$ with the total overpotential $\eta = V - E_{\text{eq}}$ at that point of the interface. This coupling of the potential distribution, which obeys Eq. (36), with the electrochemistry and thermodynamics of the electrode reaction leads to a generalised potential balance equation, of which Eq. (7) is a specific form valid for thin planar cells.

In a similar manner, the species mass balance equations, Eqs. (1), (2), may be coupled with the electrochemical rate at each point of the reaction zone (at or near the TPB). In the continuum-level modelling discussed in Section 11.2, the concentration polarisation of the electrode, η_{conc}, was related to a limiting current of the reactant, e.g., Eq. (9). A more fundamental and general expression for the concentration overpotential (the term 'overpotential' denotes exclusively the local polarisation) at any point of the electrode reaction zone is the so-called Nernst equation; for example

$$\eta_c = RT/nF \ \ln[(c_{r,\text{electrode}}/c_{r,\text{bulk}})/(c_{p,\text{electrode}}/c_{p,\text{bulk}})] \tag{37}$$

This is valid for a simple electron transfer reaction $r + ne^- = p$ but may easily be generalised. Because the fundamental mass balance equations, Eqs. (1), (2), in the absence of convection become diffusion equations, the solution of the diffusion equations for species r and p yields the concentration overpotential, η_c, at any point of the reaction zone.

Once the local concentration overpotential is known, the activation overpotential, η_a, is obtained by subtracting η_c from total η. The local activation overpotential is the actual driving force of the electrochemical reaction. It is related to the local current density at any point of the reaction zone by an electrochemical rate equation such as the Butler–Volmer equation (Eq. (10a)). Therefore, the rate equation, the Nernst equation (Eq. (37)), and the potential balance in combination couple the electric field with the species diffusion field. In addition, the energy balance applies also at the electrode level. Although this introduces another complication, a model including a temperature profile in the electrode is very useful because heat generation occurs mainly by electrochemical reaction and is localised at the reaction zone, while the strong temperature sensitivity of properties like electrolyte conductivity and electrode kinetics may skew the reaction distribution from that expected in isothermal operation.

Thus, even at the electrode level the interactions between electrochemical reaction, mass transfer, ionic conduction, and heat transfer yield a very complicated set of equations. Of course, this has given rise to many attempts to use simplified models wherever possible. Some of these are summarised below. Several are based on an assumption of one or more dominant rate-controlling resistances, for instance, mass transfer or ohmic resistance or neglect of coupling conditions that complicate the reaction distribution. Others introduce linearised electrode kinetics and neglect mass transfer resistance. Of course, these simplifying assumptions must be validated.

Validation by fitting empirical polarisation curves is helpful, especially if the objective is input for full-scale performance models. But it is of limited value if the parameter space of the fitted curves is restricted, especially when the objective is optimisation of electrode design. Benchmarking results of simplified models against a set of more complete model equations is also helpful but limited by uncertainty about some important coupling conditions. The important role of the reaction mechanism in determining kinetic rates has been recognised early and has led to specialised electrode modelling focused on this aspect of the electrode process. In addition, it has recently been realised that computational studies of the electrochemical reaction steps may contribute to greater insight in those aspects of the electrochemical rate process that are specific for the SOFC.

The following summarises a few types of simplified electrode models proposed in the literature.

11.8.2 Electrode Models Based on a Mass Transfer Analysis

If the reaction kinetics of the electrode is assumed to be very rapid, mass transfer and ohmic resistance are the dominant resistances. Assuming a reaction zone that coincides with the electrode-electrolyte interface, the diffusion fluxes in stationary operation can be expressed simply in terms of bulk gas partial pressures and gas-phase diffusivities. This is illustrated schematically in Figure 11.8, which compares anode- and cathode-supported cell designs for the simple case of a H_2/O_2 fuel cell. The decrease in concentration polarisation at the cathode, η_{Cc}, is obvious in the case of an anode-supported cell, while the model shows that concentration polarisation at the anode, η_{Ac}, is relatively insensitive to anode thickness. The advantage of the mass transfer-based approach is that analytical expressions are obtained for the polarisation behaviour. These are rather simple if activation overpotential is excluded but may still become elaborate in the case of an internally reforming anode where a number of reactions (discussed in Section 11.3) may occur simultaneously within the pores of the anode.

In further development of this model, a finite reaction zone may be introduced and activation overpotential added to the polarisation [44–47]. Kinetic resistance is believed to be particularly important for the cathode (η_{Ca} is

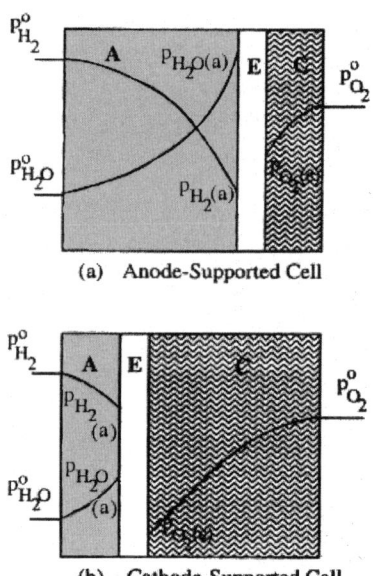

Figure 11.8. Schematic diagram of (a) anode- and (b) cathode-supported SOFCs [47].

not negligible); however, this adds further parameters of uncertain magnitude to the polarisation expressions and detracts from their original conciseness. The simplicity of the mass-transfer-based approach makes it attractive as a first-order approach in scale-up modelling [5,48,49] because it is axiomatic in practice that a good fuel cell electrode should have rapid kinetics. However, for analysis of transients other simplified models may be preferable, and for optimisation of electrode microstructure other specialised models may be equally suitable.

11.8.3 One-Dimensional Porous Electrode Models Based on Complete Concentration, Potential, and Current Distributions

Simplified models that do not make *a priori* assumptions about one or more dominant resistances are often of the 1-D macrohomogeneous type. The 1-D assumption is similar to that in mass transfer-based models. The assumption of macrohomogeneity, based on work by Newman and Tobias [50], has proven useful in battery and fuel cell electrode modelling. It implies that the microstructure of the electrode is homogeneous at the level of the continuum equations governing mass transfer, heat transfer, and current conduction in the electrode (Eqs. (1)–(7) and (33)–(37)). This type of model can exploit solutions available in chemical reaction engineering practice and has been elaborated by several researchers in that field [51–55].

Mathematically, the three phases, the solid electrocatalyst, the solid electrolyte, and the gas, are assumed to be present simultaneously at each point. The microstructure of the electrode, which produces the interface between the three phases, is represented by a volumetric interfacial reaction area (designated by a_m) with a meaning similar to that of the dimensionless interfacial area factor a in Eq. (10b). Across the interface represented by a_m, the electrochemical reaction takes place, generating electronic and ionic current fluxes and their associated potentials, as discussed under Eqs. (35) and (36). The species concentration and ionic or electronic current fluxes are projected with respect to the macrohomogeneous electrode cross section. This implies that the volume fractions of electrocatalyst, electrolyte, and gas-filled pores are necessary structural parameters, in addition to a_m.

The simplification inherent in the 1-D macrohomogeneous model is that of the microstructure. For the model to be useful in optimising electrode microstructure, the parameter a_m must be related to microstructural characteristics such as pore size and porosity. There are various techniques available from percolation theory to accomplish this and relate a_m and other model parameters to empirical pore-size distribution and total pore volume.

One of the advantages of the 1-D macrohomogeneous approach is that complete diffusion, reaction, and potential profiles are obtained, which is advantageous when the relative rates of competing reactions (for example, anodic oxidation of H_2 compared with direct anodic oxidation of CO or even direct anodic oxidation of CH_4) are compared. Another advantage is that no a priori assumption is made about the location of the reaction zone. The zones of maximum reaction are identified from the current and potential profiles and can be correlated with structural characteristics and operating variables. Finally, the very general formulation of the fundamental equations makes it possible to use dimensional analysis as a guide in correlating results and fitting against experimental data [55].

To illustrate the detailed nature of results from such a model, Figure 11.9 shows the distribution of local overpotential in the pore of an internally

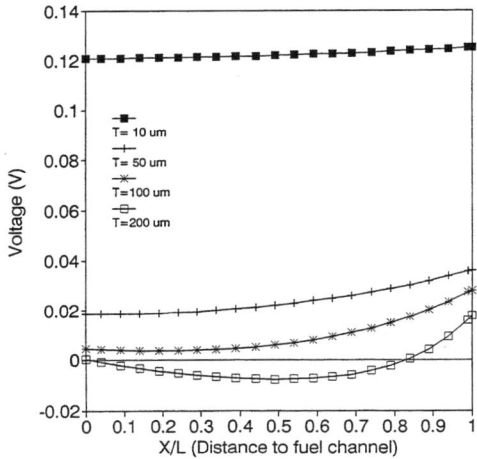

Figure 11.9 *Distribution of local overpotential at the pore wall of an internally reforming anode, with fuel gas containing 33% CH_4, 66% H_2O, balance CO and H_2, at 0.2 A/cm^2 [51].*

reforming anode as a function of pore size. In extremely wide pores, the overpotential is negative over a large part of the pore wall (away from the interface with the electrolyte), indicating that the electrochemical process may be reversed away from the interface and hydrogen generation occurs not only by reforming but electrochemically if pore dimensions are not optimised.

11.8.4 Monte Carlo or Stochastic Electrode Structure Model

A third distinct type of electrode model developed in response to the need for modelling the composite structure of SOFC electrodes more accurately is the Monte Carlo, or stochastic structure, model. This model is based on a random number-generated 2- or 3-D structure of electrode particles, electrolyte particles, and holes (for gas pores). It has been shown to represent the composite conductivity quite well and may be able to model polarisation behaviour adequately [56–58]. This is of interest because microstructure, and in particular hard-to-control variations in local microstructure, may have an important effect on overall polarisation, perhaps more so than the intrinsic kinetic characteristics measured at an 'ideal' interface.

The Monte Carlo-type electrode model is also called the particle connectivity model because its physics is straightforwardly based on Kirchhoff's law for an electrical network, with particle resistance and interconnection resistances defined by a set of rules to mimic the current flow and electrochemical current generation within the microstructure. The electrochemical process is considered to take place with a constant resistance in agreement with intuitive notions about the mechanism. Variants of this concept attach correlated values to the resistances in the network to model polarisation more closely according to a percolation concept of active sites and passive connections [59]. Other specialised types of electrode models are mentioned briefly below.

11.8.4.1 Electrode or Cell Models Applied to Ohmic Resistance-Dominated Cells
These models start from solving Laplace's equation (Eq. (36)) with appropriate boundary conditions, sometimes including polarisation. The most important application is the correct design of test cells with reference electrodes because small deviations in reference electrode placement may cause appreciable deviations in polarisation readings [60–63].

11.8.4.2 Diagnostic Modelling of Electrodes to Elucidate Reaction Mechanisms
Because the electrode kinetics of both anode and cathode and their dependence on microstructure are so important for performance, much attention has been given to elucidating reaction mechanisms based on independent electrochemical measurements (usually with respect to a reference electrode). AC impedance measurements are particularly favoured. The interpretation of these measurements requires specialised models that reflect in part the hypothesised kinetics and in part the electrode structure. It seems certain that

eventually the results will be integrated with both macro- and molecular modelling [64–69].

11.8.4.3 Models of Mixed Ionic and Electronic Conducting (MIEC) Electrodes
These specialised electrode models usually consider the MIEC electrode in combination with the electrolyte and focus on correlating performance with the semiconductor characteristics of the electrode (and sometimes electrolyte) [70–72]. Recent modelling of oxygen reduction and oxygen permeation at perovskite electrodes includes both MIEC effects and classical diffusion-type analysis [73–75].

11.9 Molecular-Level Models

Molecular-level SOFC models aim to understand (i) the kinetics of the reaction at the interface between electrode and electrolyte, (ii) the conduction process in the electrolyte, and (iii) the conduction process in the electrodes. Catalytic activity at TPB, activation energy for oxygen ion transport, and surface exchange current are application examples for such models.

Within the last two decades, enormous progress has been achieved in the ability to calculate the structures, the properties (e.g., thermodynamic, mechanical, transportation properties), and the reactivity of solids starting from atomistic approaches. The molecular-level models can be classified into three categories.

- *Empirical interatomic potential models.* Such simulations start from a given effective potential that describes the interatomic forces in a system of atoms using essentially classical techniques. The simulation algorithms are based on static minimisation methods to calculate the structural configuration of the lowest potential energy. One popular approach is the molecular dynamics method. Classical molecular dynamics can use the simple interatomic potential as well as kinetic energy to simulate fast diffusion and high-temperature properties as well as other material properties. Molecular dynamics has been performed to investigate the grain boundary phenomena in cubic zirconia at constant temperatures up to 2673 K with a system of 1920 atoms [76]. Simulations indicate that the interfaces between perfect zirconia crystals are sources of resistance in these ionic conducting systems. Another approach is Monte Carlo methods, computing random changes in the structure with results accepted or rejected on the energy criterion. Monte Carlo methods are suitable to treat disordered systems and, for example, the vacancy distribution and ion motions in heavily doped, fast ionic conducting fluorite oxides such as CeO_2 [77].
- *Quantum mechanical electronic structure calculations, or the ab initio methods.* Ab initio methods are based, at some level of simplification, on the

self-consistent solution of the Schrödinger equation for a cluster (up to about 100) of atoms and a set of (usually periodic) boundary conditions [78–80]. *Ab initio* calculations can be useful in furthering the understanding of the electrode process. For example, the computed barrier to desorption of H_2O_{ad} on Ni may be used to see whether it is the rate-determining step. Calculations can also be used to understand the reaction of H_{ad} and O_{ad}, H_{ad} and OH_{ad} as well as adsorption and diffusion of H, OH, and H_2O on the YSZ. This could be very useful in finding new catalysts capable of providing a current orders of magnitude larger than Ni as only a very small fraction of H_{ad} on Ni participates in the oxidation reaction [79,80]. By effectively eliminating the need for the self-consistent iterative process, accurate and robust *ab initio* molecular dynamics are now available, and simulations for systems of more than 100 atoms can be performed in a single CPU [81,82].

- *Hybrid techniques.* These use, for example, quantum mechanical techniques or their simplified variants to provide the effective potentials needed for the simulation of interatomic forces [83]. The computational efficiency and accuracy of the hybrid methods fall between the *ab initio* methods and the empirical methods.

With the improvement in hardware and software tools, the *ab initio* electronic structure calculations will gain importance because they can deal with increasingly complex systems and yield higher precision in the result. Along with this trend, the hybrid techniques will grow in relevance. It is expected that the hybrid methods will play an important role in the molecular-level modelling of SOFCs in the near future.

11.10 Summary

Modelling of SOFCs is advancing at a rapid rate, facilitating quick predictions of SOFC performance at a number of levels, and aiding the design of SOFC systems. Macroscopic flow and thermal models are the best known and have followed from straightforward chemical engineering principles of mass and energy balance. At the nanoscale of atoms and molecules, predictions of material behaviour and of interface interactions are also becoming possible. Most significant advances are now taking place in the understanding of complex composite structures of electrodes and three phase boundaries. Ultimately these should lead to predictions of cell behaviour which at present are measured empirically and inserted into stack models. Stack modelling has advanced to the point where acceptable start-up rates can be predicted and where overall performance can be optimised. The integration of these stacks into complete systems can also be predicted with some precision, leading to new design possibilities for hybrid SOFCs. In the immediate future, it is anticipated that models which combine the macroscopic and atomistic approaches will develop rapidly.

References

[1] E. Arato, P. Costamagna and P. Costa, *Chem. Biochem. Eng. Q.*, **8**(20) (1994) 85.
[2] J. C. Tannehill, D. A. Anderson and R. H. Pletcher, *Computational Fluid Mechanics and Heat Transfer*, 2nd edition, Taylor and Francis, Washington, DC, 1996, Ch. 5.
[3] R. J. Boersma and N. M. Sammes, *J. Power Sources*, **63** (1996) 215–219.
[4] E. Achenbach and U. Reus, in *Solid Oxide Fuel Cells VI*, eds. S. C. Singhal and M. Dokiya, The Electrochemical Society Proceedings, Pennington, NJ, PV99-19, 1999, pp. 1125–1134.
[5] J. Yuan, M. Rokni and B. Sunden, in *Solid Oxide Fuel Cells VI*, eds. S. C. Singhal and M. Dokiya, The Electrochemical Society Proceedings, Pennington, NJ, PV99-19, 1999, pp. 1099–1108.
[6] H. Yakabe, T. Ogiwara, I. Yasuda and M. Hishinuma, in *Solid Oxide Fuel Cells VI*, eds. S. C. Singhal and M. Dokiya, The Electrochemical Society Proceedings, Pennington, NJ, PV99-19, 1999, pp. 1087–1098.
[7] M. Suzuki, A. Hirano, T. Ioroi, Z. Ogumi and Z. Takehara, in *Solid Oxide Fuel Cells V*, eds. U. Stimming, S. C. Singhal, H. Tagawa and W. Lehnert, The Electrochemical Society Proceedings, Pennington, NJ, PV97-40, 1997, pp. 1359–1367.
[8] P. Costamagna and E. Arato, in *Solid Oxide Fuel Cells V*, eds. U. Stimming, S. C. Singhal, H. Tagawa and W. Lehnert, The Electrochemical Society Proceedings, Pennington, NJ, PV97-40, 1997, pp. 1339–1348.
[9] M. A. Khaleel, K. P. Recknagle, Z. Lin, J. E. Deibler, L. A. Chick and J. W. Stevenson, in *Solid Oxide Fuel Cells VII*, eds. H. Yokokawa and S. C. Singhal, The Electrochemical Society Proceedings, Pennington, NJ, PV2001-16, 2001, p. 1032.
[10] EG&G Services, Parsons, Inc., and Science Applications International Corporation, in *Fuel Cell Handbook*, 5th edition, U.S. Department of Energy, National Energy Technology Laboratory, Morgantown, WV, 2000.
[11] I. Riess, *J. Electrochem. Soc.*, **128** (1981) 2077–2081.
[12] C. Milliken, S. Guruswamy and A. Khandkar, *J. Electrochem. Soc.*, **146** (1999) 872–82.
[13] H. L. Tuller, J. Schoonman and I. Riess, *Oxygen Ion and Mixed Conductors and their Technological Applications*, Kluwer Academic, Dordrecht, 2000.
[14] K. Keegan, M. Khaleel, L. Chick, K. Recknagle, S. Simner and J. Deibler, in *Proceedings of the 2000 Society of Automotive Engineers Congress*, 2002-02-0413, 2002.
[15] J. W. Kim, A. V. Virkar, K. Z. Fung, K. Metha and S. C. Singhal, *J. Electrochem. Soc.*, **146** (1999) 69–78.
[16] T. Tojo, T. Atake, T. Mori and H. Yamamura, *J. Chem. Thermodynamics*, **31** (1999) 831–845.

[17] S. Srinivasan, R. Mosdale, P. Stevens and C. Yang, *Annu. Rev. Energy Environ.*, **24** (1999) 281–328.

[18] S. Nagata, A. Momma, T. Kato and T. Kasuga, *J. Power Sources*, **101** (2001) 60–71.

[19] K. Ahmed and K. Foger, *Catalysis Today*, **63** (2000) 479–487.

[20] E. P. Murray, T. Tsai and S. A. Barnett, *Nature*, **400** (1999) 649–651.

[21] S. Park, J. M. Vohs and R. J. Gorte, *Nature*, **404** (2000) 265–267.

[22] T. Hibino, A. Hashimoto, T. Inoue, J. Tokuno, S. Yoshida and M. Sano, *Science*, **288** (2000) 2031–2033.

[23] R. D. Cook, D. S. Malkus and M. E. Plesha, *Concept and Applications of Finite Element Analysis*, 3rd edition, John Wiley, New York, 1989.

[24] J. M. Fiard and R. Herbin, *Comput. Methods Appl. Mech. Eng.*, **115** (1994) 315–338.

[25] C. G. Vayenas, P. G. Debenedetti, I. Yentekakis and L. L. Hegedus, *Ind. Eng. Chem. Fundam.*, **24** (1985) 316–324.

[26] S. Ahmed, C. McPheeters and R. Kumar, *J. Electrochem. Soc.*, **138** (1991) 2712–2718.

[27] P. Costamagna and K. Honegger, *J. Electrochem. Soc.*, **145** (1998) 3995–4007.

[28] E. Achenbach, *J. Power Sources*, **49** (1994) 333–348.

[29] J. R. Ferguson, J. M. Fiard and R. Herbin, *J. Power Sources*, **58** (1996) 109–122.

[30] J. Yuan, M. Rokni and B. Sunden, in *Solid Oxide Fuel Cells VI*, eds. S. C. Singhal and M. Dokiya, The Electrochemical Society Proceedings, Pennington, NJ, PV99-19, 1999, pp. 1099–1108.

[31] H. Yakabe, M. Hishinuma, M. Uratani, Y. Matsuzaki and I. Yasuda, *J. Power Sources*, **86** (2000) 423–431.

[32] M. Iwata, T. Hikosaka, M. Morita, T. Iwanari, K. Ito, K. Onda, Y. Esaki, Y. Sakaki and S. Nagata, *Solid State Ionics*, **132** (2000) 297–308.

[33] S. Kjelstrup Ratkje and K. S. Forland, *J. Electrochem. Soc.*, **138** (1991) 2374–2376.

[34] C. Haynes and W. J. Wepfer, *Energy Conversion & Management*, **41** (2000) 1123–1139.

[35] K. W. Bedringas, I. S. Ertesvag, S. Byggstoyl and B. F. Magnussen, *Energy*, **22** (1997) 403–412.

[36] J. Pålsson, A. Selimovic and L. Sjunnesson, *J. Power Sources*, **86** (2000) 442–448.

[37] W. G. Winkler, in *Solid Oxide Fuel Cells VI*, eds. S. C. Singhal and M. Dokiya, The Electrochemical Society Proceedings, Pennington, NJ, PV99-19, 1999, pp. 1150–1159.

[38] W. Winkler and H. Lorenz, in *Solid Oxide Fuel Cells VII*, eds. H. Yokokawa and S. C. Singhal, The Electrochemical Society Proceedings, Pennington, NJ, PV-2001-16, 2001, pp.196–204.

[39] A. Atkinson and A. Selcuk, *Acta Mater.*, **47** (1999) 867–874.

[40] J. W. Hutchinson and A. G. Evans, *Surface and Coatings Technology*, **149** (2002) 179–184.

[41] S. P. Simner and J. W. Stevenson, *J. Power Sources*, **102** (2001) 310–316.

[42] S. Taniguchi, M. Kadowaki, T. Yasuo, Y. Akiyama, Y. Miyake and K. Nishio, *J. Power Sources*, **90** (2000) 163–169.

[43] I. Yasuda and M. Hishinuma, in *Solid Oxide Fuel Cells IV*, eds. M. Dokiya, O. Yamamoto, H. Tagawa and S. C. Singhal, The Electrochemical Society Proceedings, Pennington, NJ, PV95-1, 1995, 924–933.

[44] K. Z. Fung and A. V. Virkar, in *Solid Oxide Fuel Cells IV*, eds. M. Dokiya, O. Yamamoto, H. Tagawa and S. C. Singhal, The Electrochemical Society Proceedings, Pennington, NJ, PV95-1, 1995, pp.1105–1114.

[45] C. W. Tanner, K. Z. Fung and A. V. Virkar, *J. Electrochem. Soc.*, **144** (1997) 21–30.

[46] F. Zhao, Y. Jiang, G. Y. Lin and A. V. Virkar, in *Solid Oxide Fuel Cells VII*, eds. H. Yokokawa and S. C. Singhal, The Electrochemical Society Proceedings, Pennington, NJ, PV2001-16, 2001, pp. 501–510.

[47] J. W. Kim, A. V. Virkar, K. Z. Fung, K. Mehta and S. C. Singhal, *J. Electrochem. Soc.*, **146** (1999) 69–78.

[48] T. Ackmann, L. G. J. de Haart, W. Lehnert and F. Thom, in *Proceedings of the 4th European SOFC Forum*, ed. A. J. McEvoy, Switzerland, 2000, pp. 431–438.

[49] J. Yuan, M. Rokni and B. Sunden, in *Proceedings of the Fifth European SOFC Forum*, ed. J. Huijsmans, Switzerland, 2002, pp. 921–928.

[50] J. Newman and C. W. Tobias, *J. Electrochem. Soc.*, **109** (1962) 1183–1191.

[51] Y. C. Hsiao, Porous anode with internal reforming in a SOFC: Modeling analysis, MS thesis, IIT, Chicago (1992).

[52] S. Al-Hallaj and J. R. Selman, Porous-electrode model for SOFC electrodes, in *Proceedings NETL Workshop on Fuel Cell Modeling*, U.S. Department of Energy, National Energy Technology Laboratory, Morgantown, PA., 2000.

[53] M. L. Perry, J. Newman and E. J. Cairns, *J. Electrochem. Soc.*, **145** (1998) 5–15.

[54] P. Costamagna, P. Costa and V. Antonucci, *Electrochim. Acta*, **43** (1998) 375–394.

[55] S. H. Chao and Z. T. Xia, *J. Electrochem. Soc.*, **148** (2001) A388–A394.

[56] S. Sunde, *J. Electrochem. Soc.*, **143** (1996) 1123–1132.

[57] S. Sunde, *J. Electrochem. Soc.*, **143** (1996) 1930–1939.

[58] S. Sunde, *J. Electroceramics*, **5** (2000) 153–182.

[59] J. Abel, A. A. Kornyshev and W. Lehnerts, *J. Electrochem Soc.*, **144** (1997) 4253–4259.

[60] F. H. van Heuveln, Characterisation of porous cathodes for application in solid oxide fuel cells, Doctoral dissertation, University of Twente,

Netherlands, Appendix: The electrode configuration of a three-electrode cell (1995), pp. 167–183.
[61] J. Winkler, P. V. Hendriksen, N. Bonanos and M. Mogensen, *J. Electrochem. Soc.*, **145** (1998) 1184–1192.
[62] J. Fleig and J. Maier, in *Solid Oxide Fuel Cells V*, eds. U. Stimming, S. C. Singhal, H. Tagawa and W. Lehnert, The Electrochemical Society Proceedings, Pennington, NJ, PV97-40, 1997, pp. 1374–1383.
[63] S. Primdahl, P. Hendriksen, P. Larsen, B. Kindl and M. Mogensen, in *Solid Oxide Fuel Cells VII*, eds. H. Yokokawa and S. C. Singhal, The Electrochemical Society Proceedings, Pennington, NJ, PV2001-16, 2001 pp. 932–941.
[64] H. Schichlein, M. Feuerstein, A. Müller, A. Weber, A. Krügel and E. Ivers-Tiffée, in *Solid Oxide Fuel Cells VI*, eds. S. C. Singhal and M. Dokiya, The Electrochemical Society, Pennington, NJ, PV99-19, 1999, pp. 1069–1077.
[65] A. Weber, A. Müller, D. Herbstritt and E. Ivers-Tiffée, in *Solid Oxide Fuel Cells VII*, eds. H. Yokokawa and S. C. Singhal, The Electrochemical Society Proceedings, Pennington NJ, PV2001-16, 2001, pp. 952–962.
[66] A. Bieberle, The electrochemistry of solid oxide fuel cell anodes: experiments, modeling, and simulation, Doctoral dissertation, ETH Zürich, Ch. 2, The state-space modeling approach, 2000, pp. 28–37.
[67] A. Mitterdorfer and L. J. Gauckler, *Solid State Ionics*, **117** (1999) 187–202.
[68] A. Mitterdorfer and L. J. Gauckler, *Solid State Ionics*, **117** (1999) 203–218.
[69] M. Prestat and L. J. Gauckler, in *Solid Oxide Fuel Cells VII*, eds. H. Yokokawa and S. C. Singhal, The Electrochemical Society Proceedings, Pennington, NJ, PV2001-16, 2001, pp. 574–582.
[70] I. Reiss, in *CRC Handbook of Solid State Electrochemistry*, eds. P. J. Gellings and H. J. M. Bouwmeester, Ch. VII, 1997, pp. 223–268.
[71] S. Yuan and U. Pal, *J. Electrochem. Soc.*, **143** (1996) 3214–3222.
[72] P. Soral, U. Pal and W. Worrell, *J. Electrochem. Soc.*, **145** (1998) 99–106.
[73] S. Diethelm, A. Closset, J. van Herle and K. Nisancogliu, *Electrochemistry Japan (Denki Kagaku)*, **68** (2000) 444.
[74] S. Sunde, K. Nisancogliu and T. Gür, *J. Electrochem. Soc.*, **143** (1996) 3497–3504.
[75] S. Diethelm, Electrochemical characterization of oxygen transport and nonstoichiometry in mixed conducting perovskite-type oxides, Doctoral dissertation, EPFL Lausanne (2001), Ch. 5, 6.
[76] C. Fisher, H. Matsubara, *Computational Mater. Sci.*, **14** (1999) 177–184.
[77] G. E. Murch, C. R. Catlow and A. D. Murray, *Solid State Ionics*, **18/19** (1986) 196.
[78] Z. Lin, J. Jaffe and A. Hess, *J. Phys. Chem. A*, **103** (1999) 2117–2127.

[79]   J. Snyder, J. Jaffe, Z. Lin, A. Hess and M. Gutowski, *Surf. Sci.*, **445** (2000) 495–505.
[80]   T. Kato, S. Y. Kang, X. Xu and T. Yamabe, *J. Phys. Chem. B*, **105** (2001) 10340–10347.
[81]   R. Car and M. Parrinello, *Phys. Rev. Lett.*, **55** (1985) 2471–2474.
[82]   Z. Lin and J. Harris, *J. Phys. Condens. Matter*, **5** (1993) 1055–1080.
[83]   W. M. C. Foulkes, *Phys. Rev. B*, **48** (1993) 14216–14225.

Chapter 12

Fuels and Fuel Processing

R. Mark Ormerod

12.1 Introduction

The high operating temperature (700–1000°C) of solid oxide fuel cells (SOFCs) has a number of consequences, the most important of which is the possibility of running the cells directly on practical hydrocarbon fuels without the need for a complex and expensive external fuel reformer that is necessary for low-temperature fuel cells. Low-temperature proton exchange membrane (PEM) fuel cells are poisoned by even a small quantity of carbon monoxide and require very pure hydrogen as the fuel, therefore placing significant demands, and hence cost, on a complex external fuel processor. This is shown schematically in Figure 12.1. By contrast, in an SOFC, the hydrocarbon fuel is catalytically converted (internally reformed), generally to hydrogen and carbon monoxide (synthesis gas) together with some carbon dioxide, within the cell stack, and the carbon monoxide and hydrogen are then electrochemically oxidised to carbon dioxide and water at the anode, with production of electrical power and high-grade heat.

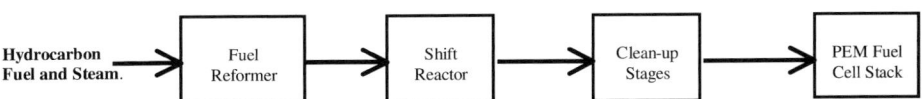

Figure 12.1 Schematic of fuel processing for a PEM fuel cell system showing the external fuel processing and the several reactor stages.

Internal reforming of the fuel can either be achieved indirectly using a separate fuel reforming catalyst within the SOFC stack, or directly on the anode. This is shown schematically in Figure 12.2. Internal reforming of the fuel within the SOFC stack is preferred, since this both increases the operational efficiency to well above that of low-temperature fuel cells and reduces the complexity of the system, hence reducing cost [1–3]. Thus fuel processing has been and continues

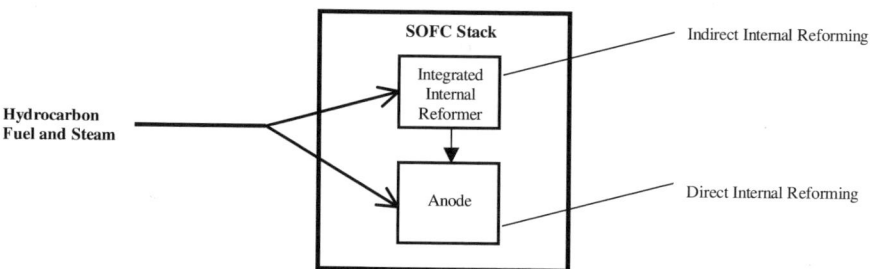

Figure 12.2 Schematic of internal reforming in an SOFC stack showing both direct and indirect reforming approaches.

to be extensively studied, and the development of a successful, cost-effective fuel processing strategy is vital for economic production of any SOFC system.

The elevated operating temperature of an SOFC also leads to production of high-temperature heat as a by-product in addition to the electrical power. This high-quality heat is not wasted, but can be used in various ways, for example in combined heat and power (CHP) systems, or to drive a gas turbine to generate more electricity, in addition to providing the heat for the endothermic internal fuel reforming process. This significantly increases the overall efficiency of an SOFC compared to lower temperature fuel cells. Figure 12.3 shows the heat flows in an SOFC system, showing the utilisation of heat by the internal reformer.

Another key advantage of SOFCs over other types of fuel cells is the flexibility

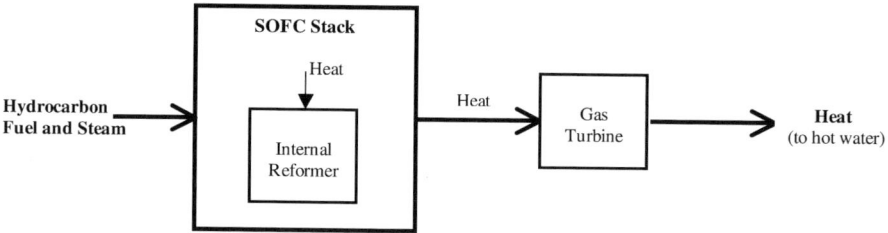

Figure 12.3 Heat flows in an advanced internally reforming SOFC system showing the absorption of heat by the internal reformer.

in the choice of fuel, which derives from the elevated operating temperature and the tolerance to carbon monoxide, and to some extent, other impurities in the fuel. SOFCs also show greater tolerance to other impurities in the fuel, which poison PEM fuel cells, and to variations in the fuel composition. This opens up opportunities for the possible use of renewable fuel sources such as biogas, vegetable matter and landfill gas in SOFC systems which are precluded for other types of fuel cells. The choice of fuel, in particular, and the operating temperature, which in turn can influence the choice of fuel, are to a large extent dictated by the intended application. The flexibility in the choice of fuel, and in particular the ability to operate SOFCs directly on practical hydrocarbon fuels, in addition to the modular nature of fuel cells, makes SOFCs ideally suited for small-scale, stand-alone and remote applications, for example on gas pipelines and farms. Fuels such as coal or biogas are ideally suited for SOFC applications in larger combined heat and power (CHP) and multi-MW gas turbine hybrid systems.

This chapter starts by discussing the range of possible fuels for SOFCs; a brief discussion on the possibility of using renewable fuels in SOFCs is also included. The remainder of the chapter is devoted to approaches in fuel processing in SOFCs, and some of the issues and problems inherent in such fuel processing. The possibility of direct electrocatalytic oxidation of the hydrocarbon fuels at the anode is also discussed. The chapter concludes with a brief consideration of future prospects.

12.2 Range of Fuels

The most common fuel (Figure 12.4), especially for stationary applications, is natural gas, which is low-cost, clean, abundant and readily available, with a supply infrastructure already in existence in many places. Natural gas varies in composition, being predominantly methane, but also containing significant quantities of higher hydrocarbons, the amount of which decreases logarithmically with increasing carbon chain length. Natural gas can be internally reformed by co-fed steam or oxygen within the SOFC stack at temperatures as low as 600°C, which means that even SOFCs operating at lower temperature can be run on natural gas without the need for a complex external fuel reformer. The presence of higher hydrocarbons and sulphur-containing compounds is significant in terms of deleterious carbon deposition and sulphur

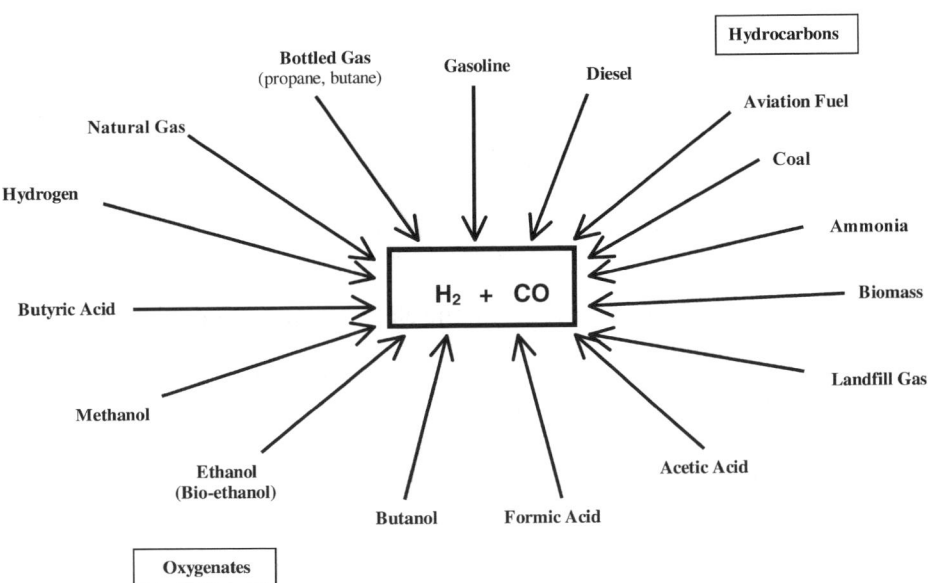

Figure 12.4 Range of potential practical fuels for SOFCs.

poisoning, respectively. An excess of steam or oxygen is generally required to prevent carbon deposition and coking, which results in deactivation and poor cell durability. Generally sulphur compounds which are added to the natural gas as odorants are removed upstream of the SOFC stack to prevent poisoning.

In remote, small-scale, stand-alone applications where there is no existing natural gas supply, bottled gas (consisting of propane/butane) offers significant practical advantages over natural gas. Such bottled gas can be internally reformed within the SOFC stack just like natural gas, and a small tubular SOFC device powered by butane has recently been demonstrated (Figure 12.5) [4]. However, the problems of carbon deposition and coking are considerably more severe with higher hydrocarbons than with methane. Thus SOFCs operating on bottled gas are more susceptible to deactivation and poor durability due to carbon deposition, and thus the challenge to avoid carbon deposition is considerably greater. The SOFC power systems represent a viable alternative to conventional power generation methods in remote areas with no natural gas supply, where diesel is generally used, which is both inefficient and highly polluting. It is also likely that bottled gas will be used as an emergency stand-by fuel source for larger-scale SOFCs operating on natural gas, in the event of an interruption to the natural gas supply.

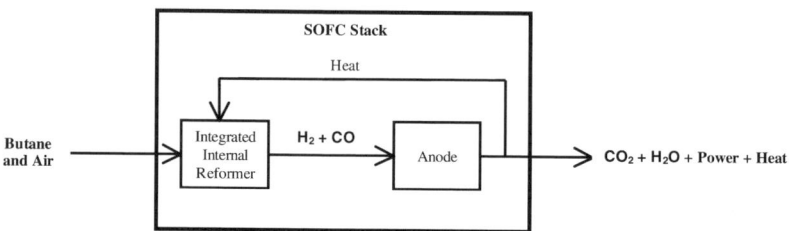

Figure 12.5 Schematic of a small-scale SOFC device powered by butane.

Recently, dimethyl ether (DME) has also been tested as a fuel for SOFCs [5]. DME is an attractive fuel because, though it is a gas at room temperature, it is easily liquefied under practical conditions (it is liquid above 3 atm) making it easier to store and handle than hydrogen.

In addition to being dictated by the specific application, the choice of fuel is also partly governed by the operating temperature of the cell. For intermediate temperature SOFCs operating at temperatures as low as 500°C, methanol is considered the most likely fuel, since these operating temperatures are below those at which natural gas and higher hydrocarbons can be effectively reformed to synthesis gas, whereas methanol can be efficiently reformed at 300–600°C [6]. Intermediate temperature SOFCs operating directly on methanol offer some potential for transportation and small portable applications. At such low operating temperatures, it is generally the cathode which limits the performance of the SOFC rather than issues of fuel processing if methanol is used as the fuel, though the development of composite cathodes is leading to cathode materials with improved performance at lower temperatures [7,8].

There is also the possibility of utilising higher hydrocarbons such as gasoline and diesel, for internally reforming SOFCs. However, internally reforming such liquid fuels within an SOFC system represents a major challenge in terms of avoiding coking on any of the active components of the cell, though autothermal reformers have been developed based on platinum group metal catalysts, which can efficiently convert diesel into a hydrogen-rich fuel [3].

SOFCs can also be operated on the output from coal gasification systems [1]. The sulphur content of these gases poisons the anode, and can also poison any internal reforming catalyst, causing loss of performance and eventual cell deactivation. Thus the sulphur has to be at least partially removed from the gas prior to entering the SOFC stack.

Recently, the possibility of using waste biogas generated from vegetable matter, and landfill gas, in SOFC systems has been demonstrated (Figure 12.6) [9,10]. At present the difficulty in utilising these gases, because of their variable composition and often low level of methane [11], means that large quantities are simply vented wastefully to the atmosphere, making a significant contribution to greenhouse gas emissions, whilst at the same time wasting a potentially clean, renewable energy resource, which is currently under-exploited. As with utilising the output from coal gasification systems, the high sulphur content of these gases is a problem in terms of poisoning and poor cell durability, and again the sulphur has to be at least partially removed from the gas prior to entering the SOFC stack [12]. Although the efficiencies of SOFCs operating on biogas, landfill gas or the output from coal gasification systems are lower than for SOFCs operating on natural gas, these provide the potential of significantly cleaner and more efficient power generation compared to combustion engine generators, as well as offer a valuable use for generally poor-quality biogas that is currently wasted.

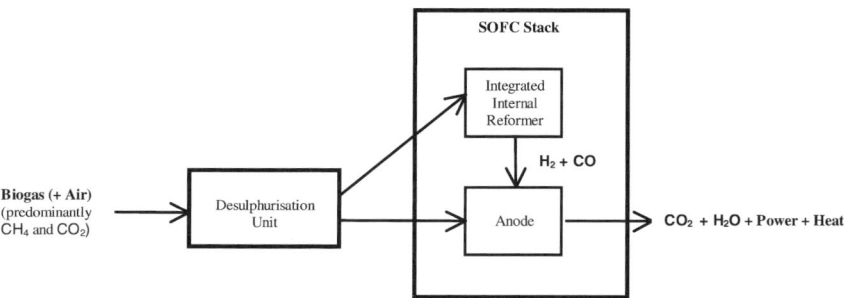

Figure 12.6   Schematic of small-scale SOFC powered by waste biogas.

The elevated operating temperature of SOFCs makes them particularly suitable for combined heat and power (CHP) applications, ranging from less than 1 kW to several MW, which covers individual households, larger residential units and industrial premises. For such applications, long-term durability of the fuel processing system is a key requirement. Figure 12.7 shows schematically the fuel processing scheme for the Sulzer CHP system running on

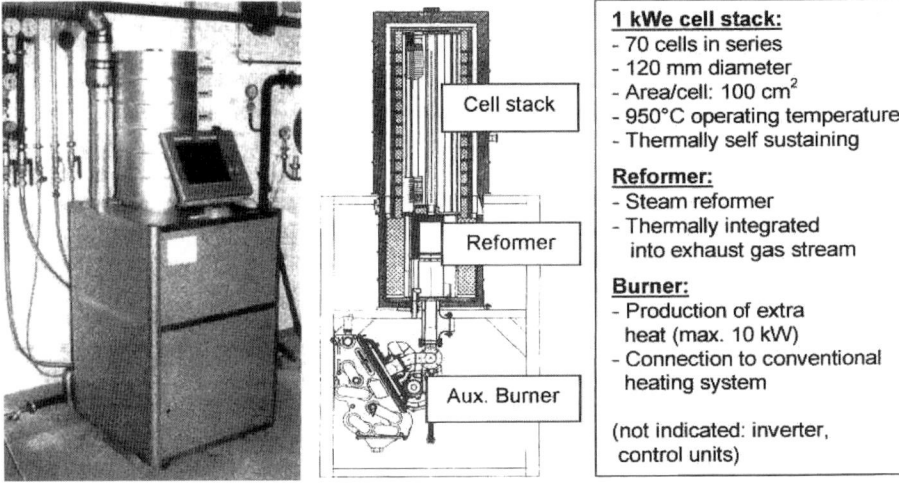

Figure 12.7   Schematic of the fuel processing scheme for the Sulzer SOFC CHP system running on natural gas.

natural gas [12]. Another advantage of such CHP units for both residential and commercial use is the reliability in the fuel supply, which is becoming increasingly important, especially in premium power applications like computing.

12.3 Direct and Indirect Internal Reforming

Although solid oxide fuel cells can be coupled to a separate fuel processor, external reforming negates the major efficiency advantage and cost benefits of SOFC systems over low-temperature fuel cells. The ability to internally reform a range of practical hydrocarbon fuels within the SOFC, together with the ability to electrochemically oxidise CO, and their increased tolerance to other impurities in the fuel, including sulphur, represent significant advantages of SOFCs over low-temperature fuel cells [1,3,6,13]. Internal reforming of the hydrocarbon fuels in SOFC systems significantly increases the system efficiency by recuperating waste heat from the stack into the fuel supply, whilst at the same time substantially reducing the complexity and cost of the system by elimination of the external reformer and associated heating arrangements and by reduction in the stack cooling air requirements and associated equipment. Thus internally reforming SOFCs offer significantly higher system efficiencies and reduced complexity compared to lower temperature fuel cells, as well as offer flexibility in the choice of fuel.

In principle, therefore, SOFC technology is both simpler, more flexible and more efficient than other fuel cell types, with potentially significant cost benefits. However, there are several major problems associated with internal reforming in SOFCs, which can lead to deactivation and a loss in cell performance, and in some cases materials failure, and hence result in poor cell durability. A particular problem with internal reforming in SOFCs is that of carbon deposition resulting from hydrocarbon pyrolysis (Eqs. (1) and (2)), especially on the nickel cermet anode [3,14–17], as well as on other active components within the SOFC system. This is shown in Eq. (1) for methane ($\Delta H = +76$ kJ mol^{-1}) and in Eq. (2) for general higher hydrocarbons:

$$CH_4 \rightarrow C + 2H_2 \tag{1}$$

$$C_nH_{2n+2} \rightarrow nC + (n+1)H_2 \tag{2}$$

These reactions are inhibited by adding steam to the fuel. Hydrocarbon steam reforming (Eqs. (3) and (4) for methane and general higher hydrocarbons, respectively) is a strongly endothermic reaction; in the case of methane steam reforming (Eq. (3)), the heat of reaction, ΔH, is $+206$ kJ mol^{-1}. This can give rise to potential instabilities in the coupling between the slow exothermic fuel cell reactions (Eqs. (5) and (6)) and the fast strongly endothermic reforming reactions (Eqs. (3) and (4)):

$$CH_4 + H_2O \rightarrow CO + 3H_2 \tag{3}$$

$$C_nH_{2n+2} + nH_2O \rightarrow nCO + (2n+1)H_2 \tag{4}$$

$$H_2 + O^{2-} \rightarrow H_2O + 2e^- \tag{5}$$

$$CO + O^{2-} \rightarrow CO_2 + 2e^- \tag{6}$$

In addition, self-sustaining internal reforming is precluded during start-up from cold and operation at low power levels, where the heat available from the stack is insufficient to meet the endothermic requirements of hydrocarbon reforming [13]. There is therefore considerable effort being devoted to developing stable internal reforming approaches for the full range of possible SOFC operating conditions, namely from start-up and zero power, through operation at low power loads, to operation at full load.

As water is formed as the product of the electrochemical oxidation of hydrogen (Eq. (5)) at the anode, this water can be recirculated and reintroduced into the hydrocarbon fuel feed, rather than continuously adding water to the system. This is shown schematically in Figure 12.8. In an SOFC where the exit gas from the anode is recirculated, the steam (and CO_2)/hydrocarbon ratio is governed by the fraction of the exit gas that is recirculated.

Internal reforming of the fuel can be achieved either indirectly using a separate reforming catalyst within the SOFC stack, or directly on the nickel anode, or by a combination of indirect and direct approaches using a separate catalyst within the SOFC system to convert a significant proportion of the hydrocarbon fuel to synthesis gas with the balance of the fuel reforming occurring directly on the nickel anode.

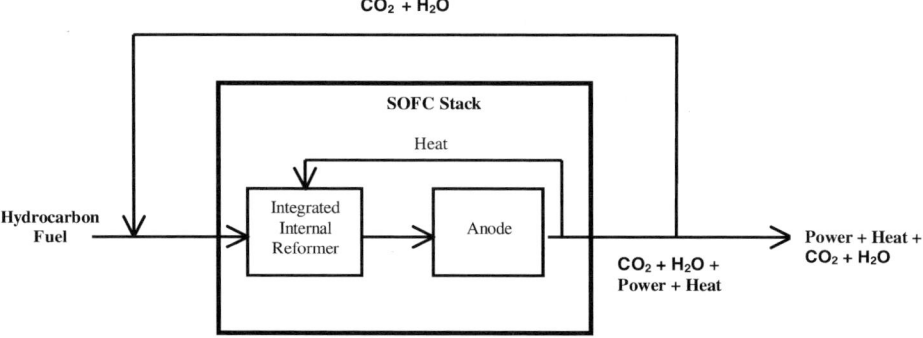

Figure 12.8 Schematic of an internally reforming SOFC with recirculation of the anode exit gas.

12.3.1 Direct Internal Reforming

Direct reforming of the fuel on the anode offers the simplest and most cost-effective design for an SOFC system, and in principle provides the greatest system efficiency with least loss of energy [1,3,13]. In direct reforming, the anode must fulfil three roles; firstly as a hydrocarbon reforming catalyst, catalysing the conversion of hydrocarbons to hydrogen and carbon monoxide; secondly as an electrocatalyst responsible for the electrochemical oxidation of H_2 and CO to water and CO_2, respectively; and finally as an electrically conducting electrode. Fulfilling all these criteria places severe demands on the anode, and hence the development of suitable anode materials for directly internally reforming SOFCs remains a major challenge.

High-efficiency results from utilising the heat from the exothermic electrochemical oxidation reaction to reform the hydrocarbon fuel, which is a strongly endothermic reaction. A further advantage of direct internal reforming is that by consuming the hydrogen to form steam, which can then reform more of the hydrocarbon fuel, the electrochemical reactions help drive the reforming reaction to completion. However, one of the major problems with direct reforming is that it gives rise to a sharp endothermic cooling effect at the cell inlet, generating inhomogeneous temperature distributions and a steep temperature gradient along the length of the anode, which is very difficult to control and can result in cracking of the anode and electrolyte materials. Significant reductions in operating temperature of the cell due to the endothermic reforming reaction have been reported [3,18]. The kinetics of methane reforming over nickel/zirconia cermet anodes have been fairly well established, at least for high-temperature operation [3,18–21]. There is some evidence of mass transfer influences on the reforming kinetics, and this has been used as an explanation for a reported dependency of the rate of reforming on the steam/methane ratio.

Another problem in direct reforming is the susceptibility of the nickel anode to catalyse the pyrolysis of methane and higher hydrocarbons (Eqs. (1) and (2)), which results in deleterious carbon deposition and subsequent build-up of

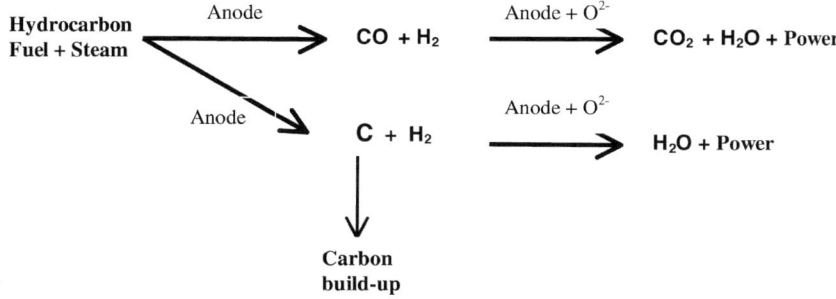

Figure 12.9 Possible reaction pathways in a directly reforming SOFC.

deactivating carbon, and leads to rapid deactivation of the cell [3,14–17, 22–24]. The possible reaction pathways which can occur on the anode in a directly reforming SOFC are shown in Figure 12.9. The high metal content of the anode precludes the use of expensive precious metals such as rhodium or platinum which are more resistant to carbon deposition than nickel [25]. Much research is currently being carried out to develop nickel-based anodes which are sufficiently active for hydrocarbon reforming under fuel cell operating conditions but are more resistant to carbon deposition, especially at lower steam/carbon ratios in the fuel feed [16,24,26–28]. Approaches include the incorporation of small amounts of dopants such as gold, molybdenum and copper into the nickel anode [16,26,28,29], and the addition of ceria to nickel/zirconia cermets [30–32]. Another problem with hydrocarbon reforming directly on the anode is that high steam partial pressures can cause sintering of the nickel anode particles, resulting in a significant reduction in the catalytic activity of the anode and a loss in cell performance [3,33–36]. This problem provides an additional incentive for developing anodes which are resistant to carbon deposition at lower steam/carbon ratios.

12.3.2 Indirect Internal Reforming

In indirect internal reforming, a separate catalyst, which reforms the hydrocarbon fuel to synthesis gas, is integrated within the SOFC stack upstream of the anode. The heat from the exothermic fuel cell reaction is still utilised. Figure 12.10 shows schematically the reaction pathways in an SOFC with

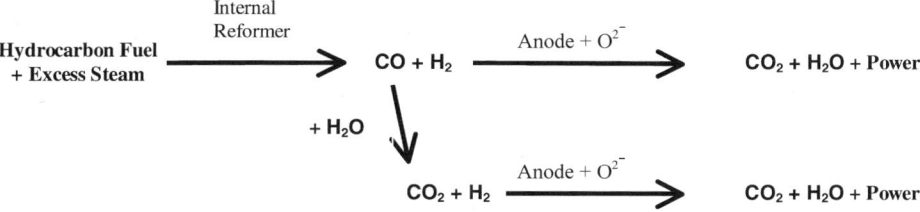

Figure 12.10 Schematic of reaction processes in an SOFC with indirect internal reforming.

indirect internal reforming. Although indirect internal reforming is less efficient and less straightforward than direct reforming, it still represents a significantly more efficient, simpler and more cost-effective approach than using an external reformer.

The major advantage of indirect over direct reforming is that it is much easier to control from a thermodynamic standpoint. One approach involves the development of mass transfer controlled steam reforming catalysts with reduced activity [37]. It is also easier to develop dispersed catalysts which do not promote carbon deposition to the same extent as the nickel anode. As the indirect reforming catalyst does not need to be electrically conducting, unlike the anode, more highly dispersed catalysts may be used, and so although nickel remains the metal of choice for cost reasons, the use of highly dispersed rhodium and platinum catalysts, which show greater resistance to carbon deposition and sulphur poisoning, can potentially be considered.

Consequently, the majority of designs currently being developed employ a separate catalyst within the SOFC stack, upstream of the anode, to indirectly reform the majority of the hydrocarbon fuel, with some residual hydrocarbon reforming occurring directly on the anode.

12.4 Reformation of Hydrocarbons by Steam, CO_2 and Partial Oxidation

Depending on the temperature and the steam to methane ratio, the water gas shift reaction (Eq. (7)) ($\Delta H = -42$ kJ mol^{-1}) can occur, whereby some of the CO is converted to CO_2, with production of one mole of hydrogen for every mole of CO converted:

$$CO + H_2O \rightarrow CO_2 + H_2 \tag{7}$$

The H_2 and CO are then electrochemically oxidised to H_2O and CO_2 (Eqs. (5) and (6)) at the anode by oxide ions electrochemically pumped through the solid electrolyte.

Conventional steam reforming catalysis has been extensively studied and reviewed over the last two decades [38–41]. Generally, in steam reforming catalysis, steam to carbon ratios of around 2.5–3 are used, i.e. well in excess of the stoichiometric requirement of Eq. (3), such that the equilibrium of the water gas shift reaction (Eq. (7)) lies to the right to maximise H_2 production, and minimise carbon deposition through hydrocarbon pyrolysis (Eqs. (1) and (2)) and the Boudouard reaction (Eq. (8)):

$$2CO \rightarrow C + CO_2 \tag{8}$$

Figure 12.11 shows the possible reaction pathways in an internally reforming SOFC running on natural gas and steam. In some SOFC applications, especially small-scale and directly reforming systems, a lower steam to carbon ratio is

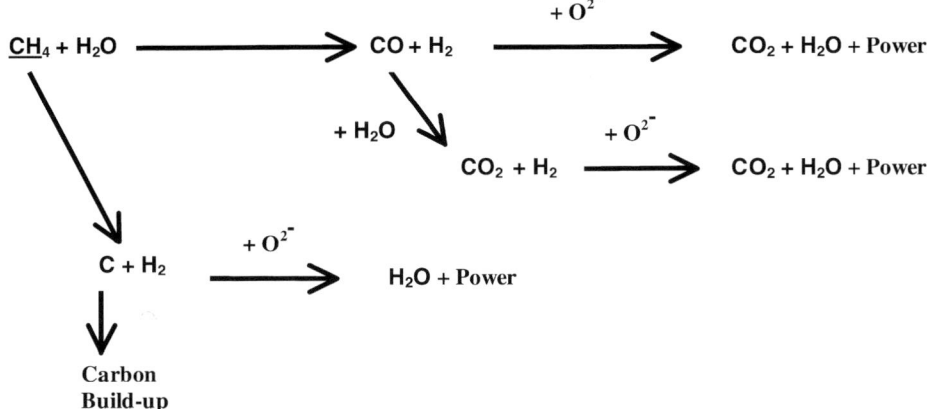

Figure 12.11 Possible reaction pathways in an internally reforming SOFC running on natural gas and steam.

desirable because of the costs and complexity associated with using large quantities of steam, and the problem of sintering of the nickel anode particles at high steam levels [33–36]. Consequently, there is considerable interest in developing catalyst and anode formulations which are resistant to carbon deposition at lower steam contents.

Carbon dioxide, formed by the water gas shift reaction (Eq. (7)) and by electrochemical oxidation of carbon monoxide (Eq. (6)), present in the exit gas leaving the anode, can be recirculated in the fuel supply at the cell inlet. It is well known that CO_2 can act as an oxidant for hydrocarbons in the presence of a suitable catalyst, so called dry reforming (Eqs. (9) and (10) for methane and a general higher hydrocarbon, respectively) [38,39,42,43]:

$$CH_4 + CO_2 \rightarrow 2CO + 2H_2 \tag{9}$$

$$C_nH_{2n+2} + nCO_2 \rightarrow 2nCO + (n+1)H_2 \tag{10}$$

Carbon deposition is a particular problem with dry reforming, especially with nickel-based catalysts [44–46]. Platinum and rhodium based catalysts show greater tolerance to carbon deposition [47,48]. Therefore, in addition to steam, CO_2 can also reform the methane, though it also represents a possible source of carbon deposition. Like steam reforming, dry reforming is also a strongly endothermic reaction. In the case of the dry reforming of methane (Eq. (9)), the heat of reaction ΔH is $+248$ kJ mol^{-1}.

In certain applications, especially in small-scale devices being developed for stand-alone or remote applications, oxygen, or simply air in many cases, is used as the oxidant rather than steam, because of the cost and complexity associated with using large quantities of steam. Recent work has demonstrated the feasibility of using partial oxidation in SOFCs running on natural gas and bottled gas (propane/butane) [4,49–51]. Using oxygen, or air, is much simpler, more convenient and less costly in terms of system configuration. However, it does lead to a lower electrical efficiency due to the energy loss in oxidising the hydrocarbon, partial oxidation being an exothermic reaction, in contrast to the strongly endothermic steam reforming reaction, which can utilise the heat from the exothermic fuel cell reactions. Further, in order to maximise the power output from the SOFC, it is necessary for the internal reforming catalyst or the anode to be selective for the partial oxidation of the hydrocarbon to synthesis gas. This is shown in Eqs. (11) and (12) for methane and a general higher hydrocarbon, respectively:

$$CH_4 + 1/2O_2 \rightarrow CO + 2H_2 \qquad (11)$$
$$C_nH_{2n+2} + n/2O_2 \rightarrow nCO + (n+1)H_2 \qquad (12)$$

The catalytic partial oxidation of hydrocarbons has been studied over many years [52–62]. A particular problem is to develop a catalyst and operating regime where high selectivity to the partial oxidation products is obtained at sufficiently high activity, whilst avoiding carbon deposition on the catalyst via Eqs. (1) and (2) [53,61,63,64]. For cost reasons, supported nickel catalysts are generally preferred, although highly dispersed platinum and rhodium catalysts are also commonly used because of their greater activity and resistance to carbon deposition [55,59,62]. Clearly carbon deposition is very undesirable both in heterogeneously catalysed hydrocarbon conversion processes and in internally reforming solid oxide fuel cells. However, if an excess of oxygen is used, then there is a marked tendency for complete oxidation (combustion) to CO_2 and H_2O to occur (Eqs. (13) and (14)):

$$CH_4 + 2O_2 \rightarrow CO_2 + 2H_2O \qquad (13)$$
$$C_nH_{2n+2} + (3n+1)/2O_2 \rightarrow nCO_2 + (n+1)H_2O \qquad (14)$$

CO_2 and H_2O cannot be electrochemically oxidised, so there is a further loss of efficiency compared to using steam as the oxidant, if any complete oxidation products are formed in the catalytic oxidation process. Figure 12.12 shows

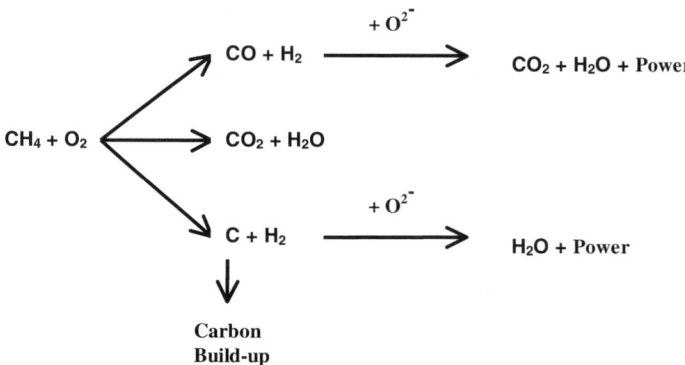

Figure 12.12 Possible reaction pathways in an internally reforming SOFC running on natural gas and air.

possible reaction pathways in an internally reforming SOFC running on natural gas and air. Recent work has shown that nickel-based anodes can have excellent selectivity for the partial oxidation of methane with minimal carbon deposition [29,49]. Figure 12.13 shows the exit gas composition from a nickel/zirconia anode exposed to a methane/oxygen gas feed at 900°C. The high selectivity towards partial oxidation, and hence synthesis gas formation, can clearly be seen.

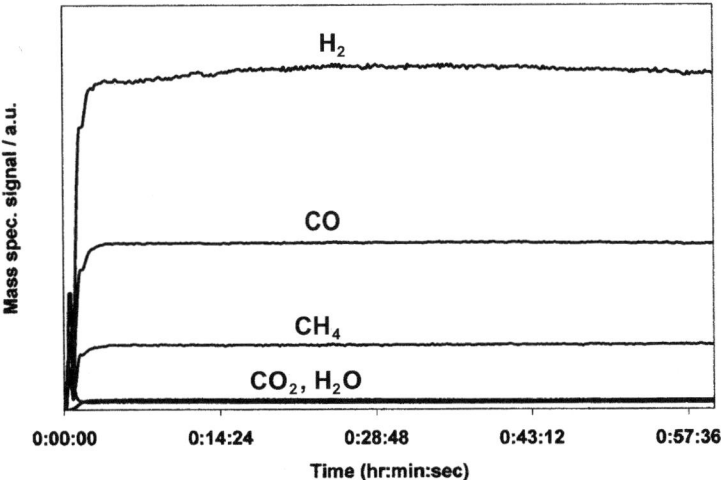

Figure 12.13 Exit gas composition resulting from a nickel/zirconia anode exposed to a methane/oxygen gas feed at 900°C, showing high selectivity towards partial oxidation and synthesis gas formation.

Although for most SOFCs under normal operation, steam (and CO_2) will be used to internally reform the natural gas, self-sustaining internal reforming is precluded during start-up from cold and operation at low power levels because of the strongly endothermic nature of steam (and CO_2) reforming. Partial oxidation of hydrocarbons is an exothermic process; in the case of the partial oxidation of methane (Eq. (10)), the heat of reaction, ΔH, is -37 kJ mol^{-1}. The complete oxidation of hydrocarbons to CO_2 and water is considerably more exothermic, with the heat of reaction for the total oxidation of methane (Eq. (12)) being -193 kJ mol^{-1}. Partial oxidation, and the use of oxygen as the oxidant, therefore offers the potential for start-up and self-sustaining operation of internally reforming SOFCs running on natural gas and other hydrocarbons at low power. It is likely that a combination of partial oxidation and steam reforming will be used as the basis for operation from zero power through low power loads to operation at full load; at zero and self-sustaining low power operation, partial oxidation will be used, and at high load, i.e. normal operation, exclusively steam (and CO_2) reforming will be used, whilst in between there will be a balance between the exothermic partial oxidation and endothermic steam reforming, so called autothermal reforming. Using partial oxidation in this way removes the need for any external source of heat, and provides the basis for much faster start-up times than steam reformers. In addition the system can be started without the addition of steam, which means that the fuel can be supplied to the cell before steam is generated, and thus if the anode exit gas is recirculated, there is no need for any external steam. Autothermal reformers which utilise precious metal-based catalysts have been developed which can convert diesel into a hydrogen-rich fuel whilst avoiding carbon deposition [3].

12.5 Direct Electrocatalytic Oxidation of Hydrocarbons

In principle, SOFCs can operate on natural gas or other hydrocarbon fuels without the addition of any oxidant to the fuel, instead directly oxidising the hydrocarbon fuel on the anode using the oxide ions which have passed through the solid electrolyte from the cathode [65–71]. The hydrocarbon can either be partially oxidised by the oxide ions to carbon monoxide and hydrogen (Eqs. (15) and (16)), or fully oxidised to CO_2 and water (Eqs. (17) and (18)), or undergo a mixture of partial and total oxidation:

$$CH_4 + O^{2-} \rightarrow CO + 2H_2 + 2e^- \tag{15}$$

$$C_nH_{2n+2} + nO^{2-} \rightarrow nCO + (n+1)H_2 + 2ne^- \tag{16}$$

$$CH_4 + 4O^{2-} \rightarrow CO_2 + 2H_2O + 8e^- \tag{17}$$

$$C_nH_{2n+2} + (3n+1)O^{2-} \rightarrow nCO_2 + (n+1)H_2O + 2(3n+1)e^- \tag{18}$$

The possibility of directly oxidising hydrocarbon fuels on the SOFC anode without any added oxidant is an extremely attractive one [65–71]. This is especially true if the hydrocarbon is partially oxidised to CO and H_2 (Eqs. (15) and (16)), rather than fully oxidised to CO_2 and H_2O (Eqs. (17) and (18)), since this results in synthesis gas from which other useful chemicals can be produced, in addition to electricity and heat. In this case the SOFC can be viewed as an electrocatalytic reactor [72,73].

The major problem with direct electrocatalytic oxidation of the hydrocarbon fuel at the anode is the marked tendency towards carbon deposition via hydrocarbon decomposition (Eqs. (1) and (2)). It is extremely difficult to avoid carbon deposition in the absence of a co-fed oxidant. However, some recent studies have reported anodes which show considerable promise for the direct electrocatalytic oxidation of hydrocarbons [29,68,69]. The conditions under which these anodes can be used may present problems for their widespread application, whilst their long-term durability with respect to carbon deposition must be established. Electrically conducting oxides have also been proposed in recent years as having potential for use as anodes for the direct electrocatalytic oxidation of hydrocarbons [67,70,74].

The concept of using an SOFC as an electrocatalytic reactor for chemical cogeneration has attracted much interest [72,73], offering the possibility of achieving higher product selectivity using electrochemically pumped oxide ions compared to gas-phase oxygen, whilst at the same time using air rather pure oxygen as the oxidant, which would bring significant cost benefits, since generation of pure oxygen from air constitutes a significant cost. In addition to synthesis gas, it has also been shown that oxidative coupling of methane to ethene and ethane can be carried out in an SOFC electrocatalytic reactor [73] (Eqs. (19)–(21)):

$$2CH_4 + O^{2-} \rightarrow C_2H_6 + H_2O + 2e^- \tag{19}$$

$$2CH_4 + 2O^{2-} \rightarrow C_2H_4 + 2H_2O + 4e^- \tag{20}$$

$$C_2H_6 \rightarrow C_2H_4 + H_2 \tag{21}$$

12.6 Carbon Deposition

Nickel in particular is well known for its propensity to promote hydrocarbon pyrolysis and the build-up of carbon [14–17,22,38,39,75]. Figure 12.14 shows an electron micrograph showing the formation of carbon filaments on a supported nickel catalyst following exposure to a hydrocarbon feed [39].

Carbon deposition can also occur on the reforming catalyst and anode in the SOFC via the disproportionation of CO (the Boudouard reaction) (Eq. (8)), and by reduction of CO by H_2 (Eq. (22)):

$$CO + H_2 \rightarrow C + H_2O \tag{22}$$

The build-up of carbon (coking) on either the internal reforming catalyst or on the anode, or indeed anywhere else in the fuel supply inlet manifold, is a critical problem to be avoided, or at least minimised, since over time this can lead to a loss of reforming activity and blocking of active sites on the reforming catalyst and the anode, and a loss of cell performance and poor durability. In extreme cases, growth of carbon filaments or whiskers can restrict the gas flow in the fuel supply system and result in actual physical blockages [3,75]. The phenomenon of carbon deposition on steam reforming catalysts has been extensively studied [38–40,75]. Carbon deposition through hydrocarbon decomposition is most prevalent at the inlet of the reforming catalyst or directly reforming anode where there is almost no hydrogen present, and the carbon-forming reactions proceed at a faster rate than the carbon-removal reactions.

It is well known that higher hydrocarbons are more reactive and show a much greater propensity towards carbon deposition than methane [3,39,75]. Coke formation occurs by cracking of the hydrocarbon to the corresponding alkene, followed by subsequent formation of a carbonaceous overlayer, which undergoes further dehydrogenation to form coke. In reality, it is the presence of these higher hydrocarbons in natural gas, rather than the methane itself, that represents the most likely source of deleterious carbon build-up in SOFCs operating on natural gas.

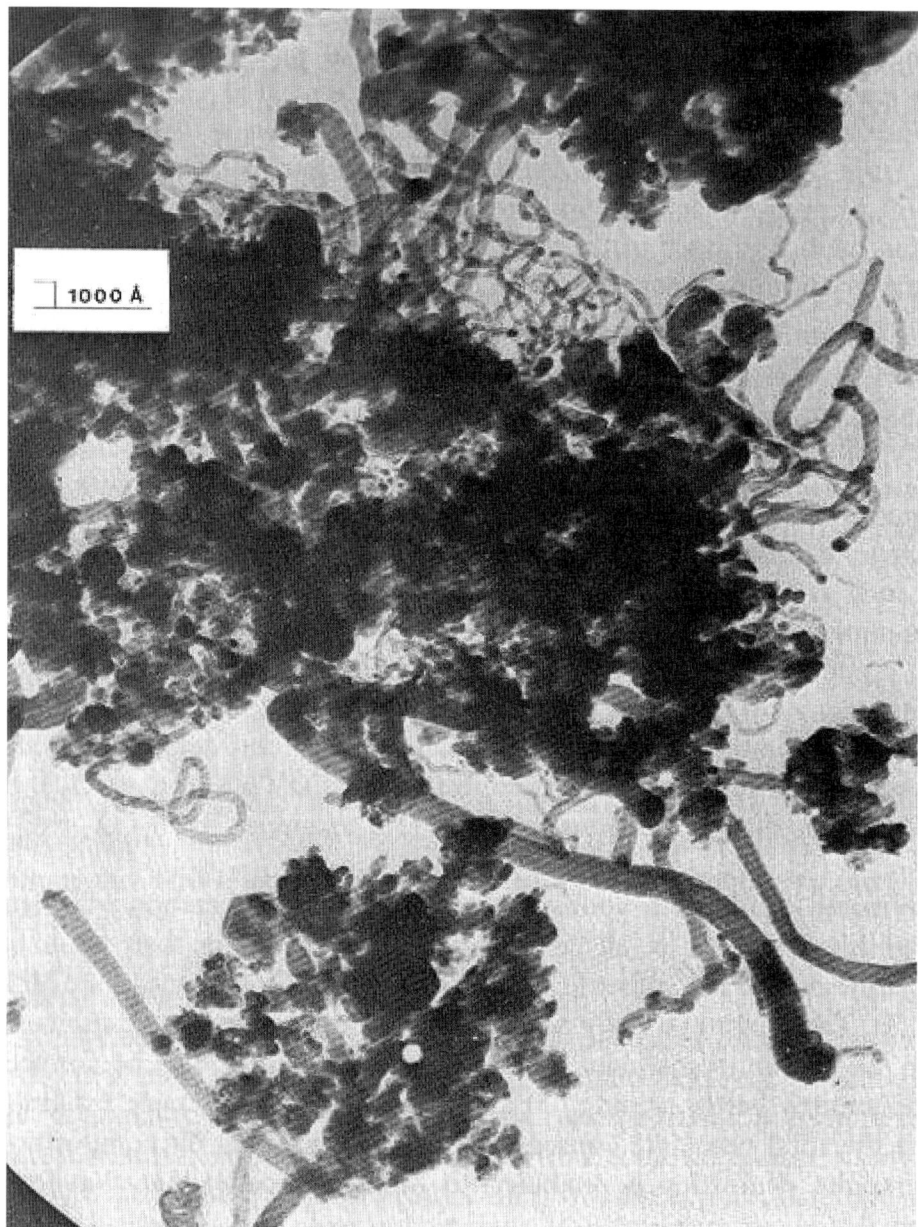

Figure 12.14   Electron micrograph showing filamentous carbon growth on a supported nickel catalyst.

It is clear, therefore, that the problem of carbon deposition and deactivation becomes more severe if bottled gas (propane/butane) is used as the fuel, and more acute still if gasoline or diesel is being contemplated as fuel. To counter the problem of carbon deposition from higher hydrocarbons, a low temperature pre-reforming stage, at 250–500°C, is often employed to remove the higher hydrocarbons. At these temperatures, pre-reforming of higher hydrocarbons can be carried out with no resultant carbon deposition [3].

Much research has been devoted to studying and characterising carbon deposition and carbon build-up on reforming catalysts and SOFC anodes [14–16,24,26,38,39,75,76], and in establishing oxidant/hydrocarbon operating envelopes which do not give rise to carbon formation on the reforming catalyst or on the anode surface at a given operating temperature. Activity has focused on developing catalyst and anode formulations which show greater resistance to carbon deposition [15,16,23,24,26,29,76]. Resistance to carbon deposition becomes especially important if the aim is to run at lower oxidant/carbon ratios to lower operating costs and prevent sintering of the anode, whilst carbon deposition is the principal barrier which must be overcome in the development of SOFCs operating on pure hydrocarbon fuels with direct electrocatalytic oxidation of the hydrocarbon at the anode. For nickel-based catalysts, this is generally achieved by the addition of small quantities of dopants. Dopants that have been added to nickel anodes include gold, copper, ruthenium and molybdenum [16,26–28]. It has been demonstrated that the incorporation of small amounts of gold or molybdenum into nickel/zirconia and nickel/ceria anodes can result in significantly improved resistance to carbon deposition when operating at low oxidant/carbon ratios [16,24,26,28,29,76]. Table 12.1 shows the amount of carbon deposited on gold-doped nickel/zirconia anodes following reforming at temperatures in the range 700–900°C. The effect of gold in reducing carbon deposition can clearly be seen. Incorporation of ceria into nickel/zirconia anodes has also been shown to give significantly improved resistance to carbon deposition [30–32,77]. Potassium is generally added to commercial nickel steam reforming catalysts to minimise carbon deposition [3,40], and the addition of small quantities of gold to nickel catalysts has been shown to be particularly effective in preventing carbon deposition when using fuel mixtures with a low oxidant/carbon ratio [28,78].

Table 12.1 Influence of gold doping of nickel/zirconia anodes on the amount of carbon deposited during reaction of a 2:1 methane/O_2 gas mixture at different reaction temperatures

Reaction temperature (°C)	Carbon deposition (a.u.)			
	Undoped Ni anode	2 mol% Au-doped Ni anode	5 mol% Au-doped Ni anode	20 mol% Au-doped Ni anode
700	4.3	0.7	0.3	0.0
800	4360.0	90.0	5.9	0.4
850	3500.0	175.0	15.0	1.4
900	440.0	364.0	150.0	46.0

Electrically conducting oxide materials have been investigated as potential anode materials. One of the main attractions of such materials is their resistance to carbon deposition compared to nickel cermet anodes under direct reforming conditions [67,70,74,79,80]. Rhodium, platinum and ruthenium have also been shown to exhibit very good activity for steam reforming and partial oxidation of hydrocarbons but, unlike nickel, carbon filaments are not formed [3,39,75]. They are, however, still susceptible to deactivation by carbon deposition on the metal surface, which blocks active sites, under conditions where hydrocarbon decomposition occurs. The major disadvantage of rhodium and ruthenium compared to nickel is their high cost which precludes their use as anodes, and essentially requires them to be present in a highly dispersed form. Despite this cost disadvantage, highly dispersed rhodium, platinum and ruthenium catalysts are used in some catalytic reformers, especially smaller scale autothermal reformers [3].

Carbon formation on catalysts and anodes can be determined using the technique of temperature-programmed oxidation [15,16,24,26,76,81,82]. This method can be used to determine safe operating envelopes for avoiding carbon deposition in terms of oxidant/carbon ratio, oxidant and operating temperature for each reforming catalyst or anode [15,16,24,26,76]. Furthermore the technique is quantitative, enabling the amount of carbon deposited on the catalyst or anode to be precisely determined, and carbon build-up to be followed as a function of operating time. In addition, the temperature at which the carbon is removed from the catalyst gives an indication of the strength of interaction of the carbon with the material, and hence how easily it may be removed during fuel cell operation [83]. Figure 12.15 shows the results of a typical temperature-programmed oxidation (TPO) experiment carried out following reforming over an anode. The figure shows the carbon removed, as CO_2,

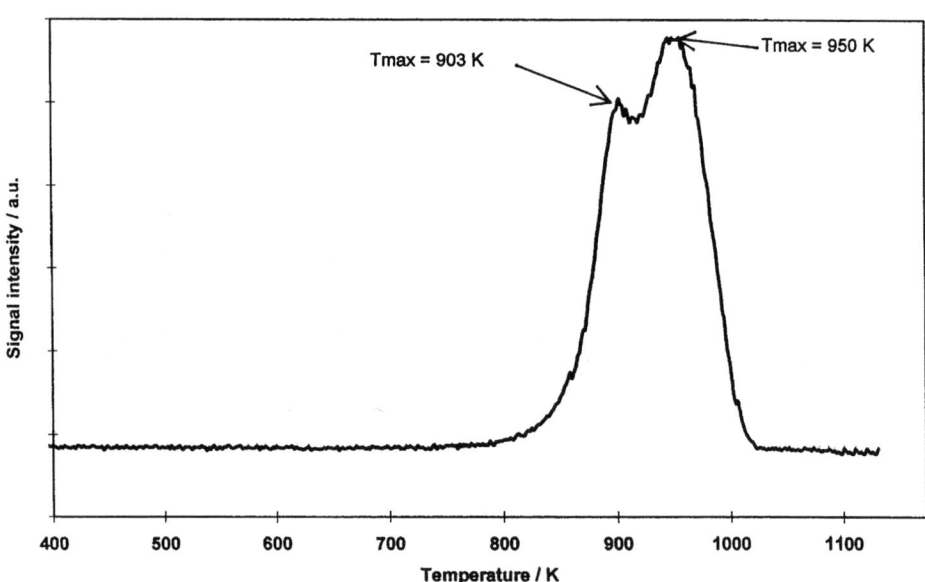

Figure 12.15 Typical temperature programmed oxidation profile obtained following reforming of a methane/steam mixture over a nickel/zirconia anode at 900°C. The figure shows the carbon removed from the anode as CO_2.

from a nickel/zirconia anode, following reforming of a methane/steam mixture over the anode at 900°C. Figure 12.16 shows the TPO spectrum following reaction of a 2:1 methane/O_2 gas mixture over a 2 mol% gold-doped nickel/zirconia anode and the corresponding undoped nickel/zirconia anode at 800°C. The effect of gold in dramatically increasing the resistance to carbon deposition can clearly be seen [29].

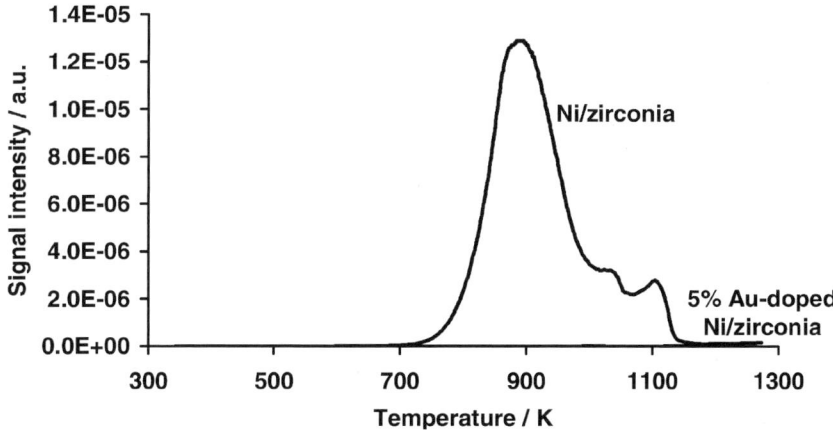

Figure 12.16 Post-reaction temperature-programmed oxidation following reaction of a 2:1 methane/O_2 gas mixture over a 5 mol% gold-doped nickel/zirconia anode and the corresponding undoped nickel/zirconia anode at 800°C.

12.7 Sulphur Tolerance and Removal

Sulphur-containing compounds such as dimethyl sulphide ((CH_3)$_2$S), diethyl sulphide ((C_2H_5)$_2$S), ethyl mercaptan ((C_2H_5)HS), tertiary butyl mercaptan ((CH_3)$_3$CHS) and tetrahydrothiophene (C_4H_8S), are added to natural gas as odorants at the level of ~5 ppm of sulphur, whilst hydrogen sulphide (H_2S) and carbonyl sulphide (COS) are also frequently present in natural gas at low levels. Although at the elevated operating temperature of SOFCs, the nickel anode, and any internal reforming catalyst, show some tolerance to sulphur [3,84,85], generally the majority of the sulphur is removed from the natural gas prior to entering the SOFC to prevent poisoning of the anode and the reforming catalyst.

At low concentrations of sulphur-containing compounds, the adsorption of sulphur on nickel is reversible, and thus low concentrations of sulphur in the feed gas can be tolerated, especially at higher operating temperatures, since the tolerance of the anode and reforming catalyst to sulphur progressively increases with temperature [3,84,85]. Any adsorbed sulphur can be removed, and the activity restored to the original activity, by switching to a sulphur-free fuel feed or by a short exposure to steam. However, at higher sulphur concentrations, irreversible sulphidation of the catalyst or anode can occur [3].

Sulphur can be removed by various methods. These include high-temperature desulphurisation (hydrodesulphurisation) [3,86–88], whereby sulphur-containing organic compounds are reduced to hydrogen sulphide and the corresponding hydrocarbon by adding a small quantity of hydrogen to the fuel feed and passing this over a molybdenum or cobalt catalyst supported on alumina [3]. This is shown in Eq. (23) for diethyl sulphide. The hydrogen sulphide is then removed by adsorption on zinc oxide (Eq. (24)). Hydrodesulphurisation is generally carried out in the temperature range 350–400°C.

$$(C_2H_5)_2S + 2H_2 \rightarrow 2C_2H_6 + H_2S \qquad (23)$$

$$H_2S + ZnO \rightarrow ZnS + H_2O \qquad (24)$$

Low-temperature desulphurisation catalysts have also been developed which remove organic sulphur compounds and hydrogen sulphide at room temperature [3].

In smaller SOFC systems, sulphur can be removed by adsorption methods, most frequently using activated carbon or molecular sieves [3]. The ability of activated carbon to adsorb H_2S can be enhanced by chemically impregnating it. Such adsorbent systems can be reactivated by thermal treatment. However, their use becomes impractical for larger scale SOFC systems because of the large quantities of adsorbent required and the subsequent problems associated with adsorbent reactivation and disposal of the desorbed sulphur [3].

12.8 Anode Materials in the Context of Fuel Processing

The zirconia component in a Ni cermet anode can be considered as essentially analogous to the support in a supported metal catalyst, and hence may influence the catalytic behaviour of the anode. There is currently considerable interest in studying the catalytic properties of nickel-based anodes, and modifying and optimising their composition [15,16,23,24,26,29,89,90], particularly for directly internally reforming SOFCs that avoid carbon deposition during operation, whilst having sufficiently high hydrocarbon reforming activity. There is also an incentive to develop anodes which are more resistant to carbon deposition at lower oxidant/carbon ratios, since higher steam levels have been shown to cause increased sintering of the nickel anode particles [3,33–36], in addition to adding to the cost and complexity of the system. Similarly, the development of anodes which show a greater tolerance to sulphur is highly desirable since this places less severe demands on the sulphur removal system, both in terms of the sulphur level to which the fuel has to be desulphurised and the frequency at which the desulphurisation unit has to be regenerated or replaced.

The kinetics of methane reforming over nickel cermet anodes have been fairly well established for high-temperature operation [3,20,21]. It has been found that different nickel cermet anodes can have markedly different reforming activities [3,24,26,76,91,92].

A number of researchers have studied nickel/ceria cermet anodes for both zirconia and ceria–gadolinia electrolyte-based SOFCs [30,93–96]. Nickel/ceria cermet anodes have been shown to still give sufficiently good performance in ceria electrolyte-based SOFCs operating at temperatures as low as 500°C. Ceria has also been added to nickel/YSZ anodes to improve both the electrical performance and also the resistance to carbon deposition [30–32,77,93,97]. Although cobalt and ruthenium offer potential advantages over nickel, including high sulphur tolerance, and in the case of ruthenium, higher reforming activity and greater resistance to sintering, and cobalt/YSZ and ruthenium/YSZ anodes have been developed, the cost of these materials effectively precludes their use in practical SOFCs [13,98].

Various dopants have been incorporated into nickel/zirconia and nickel/ceria anodes, in an attempt to modify their behaviour. Dopants that have been studied include molybdenum, gold, ruthenium and lithium [16,26–29]. The effect of gold in dramatically increasing resistance to carbon formation and build-up was previously shown in Table 12.1 and Figure 12.16.

Electrically conducting oxides, which are stable under both oxidising and reducing conditions have been actively studied in recent years as potential alternative anode materials to nickel [30,67,70,74,79,80]. In principle, the use of such oxides as anodes would overcome many of the problems associated with nickel cermet anodes for use in direct reforming SOFCs, in particular those of carbon deposition, sulphur poisoning and sintering and the possible undesirable formation of nickel oxide under oxidising conditions. Such oxides also offer potential for direct electrocatalytic hydrocarbon oxidation. Oxides that have been investigated include materials based on $LaMnO_3$ [67], $LaCrO_3$ [99–102], and $SrTiO_3$ [102–104]. In the case of $LaCrO_3$, calcium, strontium and titanium substituents have been used [101,102], e.g. $La_{0.7}Sr_{0.3}Cr_{0.8}Ti_{0.2}O_3$, whilst for $SrTiO_3$, niobium and lanthanum substitution has been investigated, e.g. $Sr_{0.6}Ti_{0.2}Nb_{0.8}O_3$ and $La_{0.4}Sr_{0.6}TiO_3$ [102,104]. Doped titanium oxides have also been studied in this context. Mixed conducting oxides such as terbia- and titania-doped YSZ [105,106] and yttria- and gadolinia-doped ceria [101,107–111] have also attracted interest as potential anode materials for direct methane oxidation, with laboratory tests showing some promise in the temperature range 800–1000°C. Ruthenium-doped ceria anodes have also been studied [95,112,113], with the ruthenium dopant enhancing the catalytic activity.

Recent studies have also identified some alternative anodes, one being copper-based and incorporating significant quantities of ceria in addition to YSZ [68,114] and the other adding yttria-doped ceria to nickel and YSZ [69], both of which have been reported to show considerable promise for the direct electrocatalytic oxidation of the hydrocarbon fuels, without the need for any co-fed oxidant. However, the conditions under which such anodes can be used for direct hydrocarbon oxidation may be a problem for their widespread application, whilst their long-term performance in terms of deactivation resulting from carbon deposition remains to be investigated.

Another approach that is currently being investigated to give greater control of the reforming reactions in the SOFC is the development of mass transfer controlled steam reforming catalysts/anodes with reduced activity [37]. The development of such materials would minimise the temperature gradients within an SOFC and hence reduce problems due to cracking of the anode and electrolyte materials. This is particularly true for planar designs of SOFCs where even small temperature gradient across the plane can induce mechanical cracking of the cell components.

12.9 Using Renewable Fuels in SOFCs

SOFCs can be run directly on biogas and landfill gas [9,10]. Biogas is predominantly a mixture of methane and carbon dioxide, the composition of which varies both with location and over time, which presents major difficulties in its use, especially at low methane levels. At CO_2 levels which are too high for conventional power generation systems, SOFCs could, in theory, still extract the power available from the methane content of biogas, however low the methane content. Furthermore, as CO_2 is inherently present in biogas in addition to methane, in principle biogas may be used directly in an SOFC without the addition of either steam or oxygen. It has recently been shown that SOFCs can be run on biogas over a wide compositional range of methane and CO_2, with internal dry reforming of the methane by the CO_2 in the biogas (Eq. (9)) [9,10]. Figure 12.17 shows the power output from a small tubular SOFC running on biogas at 850°C as a function of the methane content of the biogas, whilst Figure 12.18 shows the corresponding exit gas compositions from an identical but unloaded SOFC, which shows clearly that internal dry reforming of the methane by the CO_2 in the biogas is occurring. Power can still be drawn at even the lowest methane contents. Thus at methane contents below which

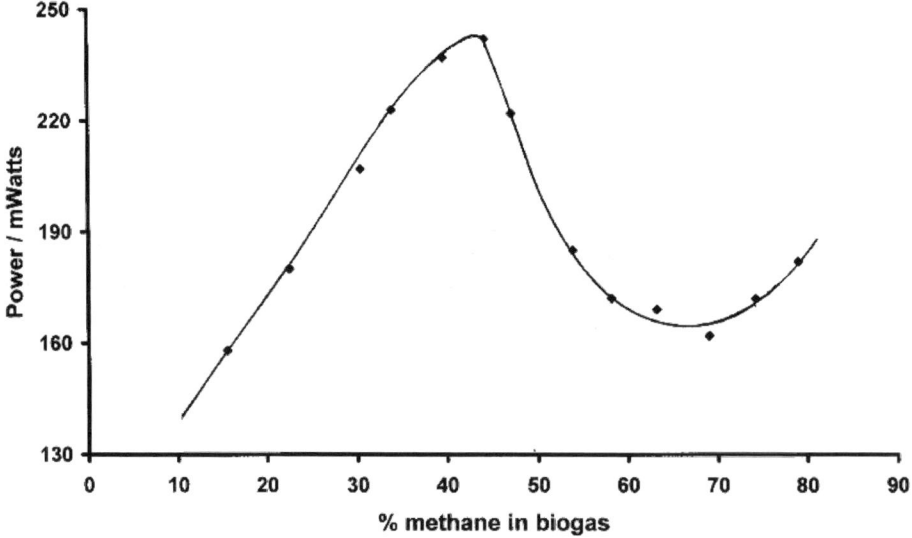

Figure 12.17 Power output from a small tubular SOFC running on biogas at 850°C as a function of the methane content in the biogas.

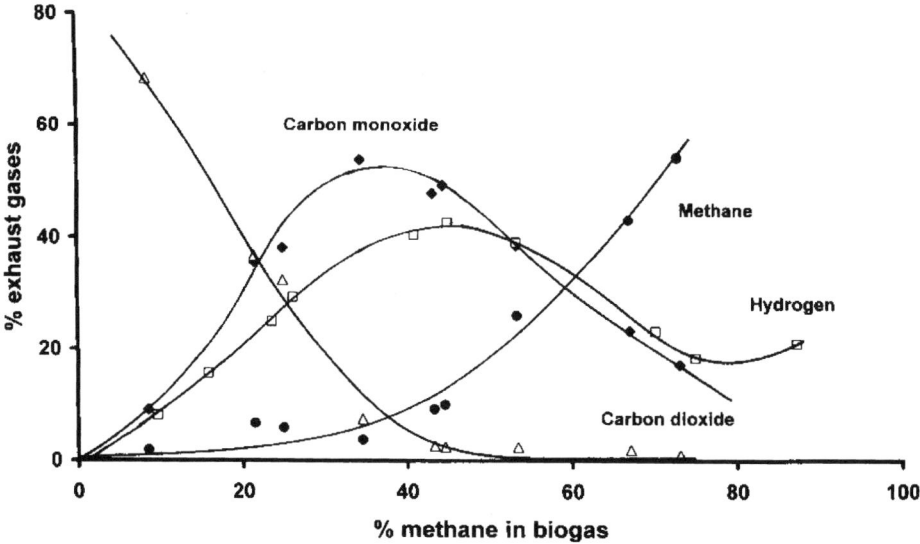

Figure 12.18   Exit gas compositions from an unloaded small tubular SOFC running on biogas at 850°C, as a function of the methane content in the biogas.

conventional heat engines would have long since stopped working (~45%), the SOFC is able to still produce useful power from poor quality biogas, which is presently disposed of by simply venting wastefully and detrimentally to the atmosphere. For any practical application using biogas, its high sulphur content requires an efficient means of at least partial sulphur removal prior to it entering the SOFC. Biogas can also contain other impurities, such as halides, which can poison the anode and any reforming catalyst, causing deactivation of the SOFC.

12.10 Summary

The elevated operating temperature of SOFCs, combined with the ability to utilise carbon monoxide, means it is possible to operate SOFCs directly on a full range of practical hydrocarbon fuels without the need for a separate external fuel processor, such as that required for low-temperature PEM fuel cells. The fuel of choice for SOFC systems for most applications is natural gas due to its low cost, abundance, and the existing supply infrastructure, though bottled gas (propane/butane) offers significant practical advantages in certain small-scale applications where there is no natural gas supply infrastructure. There is currently interest in developing methanol powered SOFCs for operation at temperatures as low as 500°C, with possible uses including transport and portable applications. In addition, there is considerable interest in developing internally reforming SOFCs, running on fuels such as diesel, gasoline and kerosene. Recently, the possibility of using waste biogas in SOFCs has emerged;

biogas is cheap, readily available and totally renewable and is at present an under-exploited energy resource.

Direct reforming of the fuel on an SOFC anode, although offering the simplest and most cost-effective design for an SOFC system, remains a major challenge. Significant problems need to be overcome including the susceptibility of the nickel anode to carbon build-up and subsequent deactivation, and sintering of the nickel anode particles, together with the problem of steep temperature gradients across the cell caused by the strongly endothermic nature of the reforming reaction.

Some small-scale SOFC devices are currently being developed which use oxygen rather than steam as the oxidant, using partial oxidation to convert the hydrocarbon fuel to syngas. Whilst this confers significant advantages in terms of system cost and convenience for small-scale stand-alone and remote applications, catalytic partial oxidation is a much more difficult process than steam reforming in terms of achieving high selectivity towards syngas, rather than combustion products, whilst avoiding any carbon deposition. Autothermal reforming approaches to methane conversion in internally reforming SOFCs which use a combination of exothermic partial oxidation and endothermic steam reforming are currently being developed, since partial oxidation removes the need for any external source of heat and provides the basis for much faster start-up times than any approach based solely on steam reforming.

There is much interest at present in the development of anodes which are resistant to carbon deposition at lower oxidant/carbon ratios in the fuel feed. Recent studies using doped nickel cermet anodes and certain electrically conducting oxide materials show great promise. The development of such anodes is particularly important for SOFCs using co-fed oxygen as the oxidant or small quantities of steam added to the fuel. Furthermore, the possibility of developing SOFCs with such anodes which can run directly on practical hydrocarbon fuels, without any co-fed oxidant, something which has long been regarded as an ultimate goal for SOFCs, is now receiving renewed and vigorous effort, with some encouraging results.

References

[1]   D. Hart, *Chem. Ind.* (1998) 344.
[2]   A. C. Lloyd, *Scientific American*, **281** (1999) 64.
[3]   A. L. Dicks, *J. Power Sources*, **61** (1996) 113.
[4]   C. M. Finnerty, J. Staniforth and K. Kendall, in *Proceedings of the 4th European SOFC Forum*, ed. A. J. McEvoy, Switzerland, 2000, pp. 151.
[5]   A. Tatemi, S. Wang, T. Ishihara, H. Nishiguchi, and Y. Takita, in *Solid Oxide Fuel Cells VIII*, eds. S. C. Singhal and M. Dokiya, The Electrochemical Society Proceedings, Pennington, NJ, PV 2003-7, 2003, pp. 1260.
[6]   B. C. H. Steele and A. Heinzel, *Nature*, **414** (2001) 345.
[7]   J. M. Ralph, A. C. Schoeler and M. Krumpelt, *J. Materials Sci.*, **36** (2001) 1161.

[8]   R. Doshi, V. L. Richards, J. D. Carter, X. P. Wang and M. Krumpelt, *J. Electrochem. Soc.*, **146** (1999) 1273.
[9]   J. Staniforth and R. M. Ormerod, *Green Chemistry*, **3** (2001) G61.
[10] J. Staniforth and R.M Ormerod, *Catalysis Letters*, **81** (2002) 19.
[11] J. Huang and R. J. Crookes, *Fuel*, **77** (1998) 1793.
[12] J. Staniforth and K. Kendall, *J. Power Sources*, **86** (2000) 401.
[13] R. M. Ormerod, *Chem. Soc. Reviews*, **32** (2003) 17.
[14] R. H. Cunningham, C. M. Finnerty and R. M. Ormerod, in *Solid Oxide Fuel Cells V*, eds. U. Stimming, S. C. Singhal, H. Tagawa and W. Lehnert, The Electrochemical Society Proceedings, Pennington, NJ, PV97-40, 1997, pp. 973.
[15] C. M. Finnerty, N. J. Coe, R. H. Cunningham and R. M. Ormerod, *Catalysis Today*, **46** (1998) 137.
[16] R. M. Ormerod, *Stud. Surf. Sci. Catal.*, **122** (1999) 35, and refs. therein.
[17] J. R. Rostrup-Nielson and L. J. Christiansen, *Applied Catalysis A: General*, **126** (1995) 381.
[18] W. Lehnert, J. Meusinger, E. Riensche and U. Stimming, in *Proceedings of the 2nd European SOFC Forum*, ed. B. Thorstensen, Switzerland, 1996, pp. 143.
[19] E. Achenbach, *J. Power Sources*, **49** (1994) 333.
[20] A. L. Lee, R. F. Zabransky and W. J. Huber, *Ind. Eng. Chem. Res.*, **29** (1990) 766.
[21] E. Achenbach and E. Riensche, *J. Power Sources*, **52** (1994) 283.
[22] S. H. Clarke, A. L. Dicks, K. Pointon, T. A. Smith and A. Swann, *Catalysis Today*, **38** (1997) 411.
[23] C. M. Finnerty, R. H. Cunningham and R. M. Ormerod, in *Proceedings of the 3rd European SOFC Forum*, ed. P. Stevens, Switzerland, 1998, pp. 217.
[24] C. M. Finnerty, R. H. Cunningham and R. M. Ormerod, in *Solid Oxide Fuel Cells VI*, eds. S. C. Singhal and M. Dokiya, The Electrochemical Society Proceedings, Pennington, NJ, PV99-19, 1999, pp. 568.
[25] M. A. Pena, J. P. Gomez and J. L. G. Fierro, *Applied Catalysis A: General*, **144** (1996) 7.
[26] C. M. Finnerty and R. M. Ormerod, in *Solid Oxide Fuel Cells VI*, eds. S. C. Singhal and M. Dokiya, The Electrochemical Society Proceedings, Pennington, NJ, PV99-19, 1999, pp. 583.
[27] N. J. Coe, R. H. Cunningham and R. M. Ormerod, in *Proceedings of the 3rd European SOFC Forum*, ed. P. Stevens, Switzerland, 1998, pp. 39.
[28] A. L. Hopkin and R. M. Ormerod, in preparation.
[29] I. A. Proctor, A. L. Hopkin and R. M. Ormerod, *Ionics*, (2003), in press.
[30] M. Mogensen, N. M. Sammes and G. A. Tompsett, *Solid State Ionics*, **129** (2000) 63.
[31] T. Tsai and S. A. Barnett, *J. Electrochem. Soc.*, **145** (1998) 1696.
[32] V. D. Belyaev, T. I. Politova, O. A. Marina and V. A. Sobyanin, *Appl. Catal. A: General*, **133** (1995) 47.
[33] D. Simwonis, F. Tietz and D. Stöver, *Solid State Ionics*, **132** (2000) 241.

[34] T. Kawada, N. Sakai, H. Yokokawa, M. Dokiya, M. Mori and T. Iwata, *Solid State Ionics*, **40/41** (1990) 402.
[35] B. de Boer, M. Gonzales, H. J. M. Bouwmeester and H. Verweij, *Solid State Ionics*, **127** (2000) 269.
[36] G. J. M. Janssen, J. P. de Jong and J. P. P. Huijsmans, in *Proceedings of the 2nd European SOFC Forum*, ed. B. Thorstensen, Switzerland, 1996, pp. 163.
[37] P. Aguiar, E. Ramirez-Cabrera, A. Atkinson, L. S. Kershenbaum and D. Chadwick, in *Solid Oxide Fuel Cells VII*, eds. H. Yokokawa and S. C. Singhal, The Electrochemical Society Proceedings, Pennington, NJ, PV2001-16, 2001, pp. 703.
[38] C. N. Satterfield, *Heterogeneous Catalysis in Industrial Practice*, McGraw-Hill, New York, 1991.
[39] J. R. Rostrup-Nielson, *Catalysis. Today*, **18** (1993) 305.
[40] D. E. Ridler and M. V. Twigg, in *Catalyst Handbook*, ed. M. V. Twigg, Manson, London, Ch. 5, 1996.
[41] J. R. Rostrup-Nielson, in *Catalytic Steam Reforming*, Catalysis, Science and Engineering, ed. J. R. Anderson and M. Boudart, Springer, Berlin, Vol. 5, 1984.
[42] H. M. Swann, V. C. H. Kroll, G. A. Martin and C. Mirodatos, *Catalysis Today*, **21** (1994) 571.
[43] P. D. F. Vernon, M. L. H. Green, A. K. Cheetham and A. T. Ashcroft, *Catalysis Today*, **13** (1992) 417.
[44] V. R. Choudhary and A. M. Rajput, *Ind. Eng. Chem. Res.*, **35** (1996) 3934.
[45] G. Xu, K. Shi, Y. Gao, H. Xu and Y. Wei, *J. Mol. Catalysis A*, **147** (1999) 47.
[46] D. Duprez, M. C. Demichali, P. Marecot, J. Barbier, O. A. Ferretti and E. N. Ponzi, *J. Catalysis*, **124** (1990) 324.
[47] A. M. Gadalla and M. E. Sommer, *Chem. Eng. Sci.*, **44** (1989) 2825.
[48] A. T. Ashcroft, A. K. Cheetham, M. L. H. Green and P. D. F. Vernon, *Nature*, **352** (1991) 225.
[49] R. H. Cunningham and R. M. Ormerod, in *Proceedings of the 4th European SOFC Forum*, ed. A. J. McEvoy, Switzerland, 2000, pp. 507.
[50] M. Pastula, J. Devitt, R. Boersma and D. Ghosh, in *Solid Oxide Fuel Cells VII*, eds. H. Yokokawa and S. C. Singhal, The Electrochemical Society Proceedings, Pennington, NJ, PV2001-16, 2001, pp. 180.
[51] E. Batawi, U. Weissen, A. Schuler, M. Keller and C. Voisard, in *Solid Oxide Fuel Cells VII*, eds. H. Yokokawa and S. C. Singhal, The Electrochemical Society Proceedings, Pennington, NJ, PV2001-16, 2001, pp. 140.
[52] M. Prettre, C. Eichner and M. Perrin, *Trans. Faraday Soc.*, **43** (1946) 335.
[53] A. T. Ashcroft, A. K. Cheetham, J. S. Foord, M. L. H. Green, C. P. Grey, A. J. Murrell and P. D. F. Vernon, *Nature*, **344** (1990) 319.
[54] D. Dissanayake, M. P. Rosynek, K. C. C. Kharas and J. H. Lunsford, *J. Catalysis*, **132** (1991) 117.
[55] Y. Boucouvalas, Z. Zhang and X. E. Verykios, *Catalysis Letts.*, **40** (1996) 189.
[56] D. A. Hickman and L. D. Schmidt, *Science*, **259** (1993) 343.

[57] Y. H. Hu and E. Ruckenstein, *J. Catalysis*, **158** (1996) 260.
[58] S. C. Tsang, J. B. Claridge and M. L. H. Green, *Catalysis Today*, **23** (1995) 3.
[59] S. S. Bharadwaj and L. D. Schmidt, *J. Catalysis*, **146** (1994) 11.
[60] C. T. Au and H. Y. Wang, *Catalysis Letts.*, **41** (1996) 159.
[61] V. A. Tsipouriari, Z. Zhang and X. E. Verykios, *J. Catalysis*, **179** (1998) 283.
[62] K. H. Hofstad, O. A. Rokstad and A. Holmen, *Catalysis Letts.*, **36** (1996) 25.
[63] S. C. Tsang, J. B. Claridge and M. L. H. Green, *Catalysis Today*, **23** (1995) 3.
[64] A. M. Diskin, R. H. Cunningham and R. M. Ormerod, *Stud. Surf. Sci. Catal.*, **122** (1999) 393.
[65] Y. L. Sandler, *J. Electrochem. Soc.*, **118** (1971) 1378.
[66] K. Otsuka, S. Yokoyama and A. Morikawa, *Bull. Chem. Soc. Japan*, **57** (1984) 3286.
[67] B. C. H. Steele, I. Kelly, H. Middleton and E. Rudkin, *Solid State Ionics*, **28–30** (1988) 1547.
[68] E. P. Murray and T. Tsai, *Nature*, **400** (1999) 649.
[69] S. Park, R. Craciun, V. Radu and R. J. Gorte, *J. Electrochem. Soc.*, **146** (1999) 3603.
[70] E. S. Putna, J. Stubenrauch, J. M. Vohs and R. J. Gorte, *Langmuir*, **11** (1995) 4832.
[71] J. T. S. Irvine, D. P. Fagg, J. Labrincha and F. M. B. Marques, *Catalysis Today*, **38** (1997) 467.
[72] V. V. Galvita, V. D. Belyaev, V. N. Parmon and V. A. Sobyanin, *Catalysis Letters*, **39** (1996) 209.
[73] V. A. Sobyanin, V. D. Belyaev and V. V. Galvita, *Catalysis Today*, **42** (1998) 337.
[74] P. Han and W. L. Worrell, *J. Electrochem. Soc.*, **142** (1995) 4235.
[75] C. H. Bartholomew, *Catal. Rev. Sci. Eng.*, **24** (1982) 67.
[76] C. M. Finnerty and R. M. Ormerod, *J. Power Sources*, **86** (2000) 390.
[77] J. Staniforth and K. Kendall, in *Solid Oxide Fuel Cells VI*, eds. S. C. Singhal and M. Dokiya, The Electrochemical Society Proceedings, Pennington, NJ, PV99-19, 1999, pp. 603.
[78] F. Besenbacher, I. Chorkendorf, B. S. Clausen, B. Hammer, A. M. Molenbrook, J. K. Norskov and I. Stensgaard, *Science*, **279** (1998) 193.
[79] O. A. Marina, C. Bagger, S. Primdahl and M. Mogensen, *Solid State Ionics*, **123** (1999) 199.
[80] P. Tsiakaras, P. C. Athanasiou, G. Marnellos, M. Stoukides, J. E. ten Elshof and H. J. M. Bouwmeester, *Applied Catalysis A: General*, **169** (1998) 249.
[81] C. M. Finnerty, R. H. Cunningham, K. Kendall and R. M. Ormerod, *J. Chem. Soc. Chem. Commun.*, (1998) 915.
[82] C. M. Finnerty, R. H. Cunningham and R. M. Ormerod, *Catalysis Letters*, **66** (2000) 221.
[83] N. J. Coe, R. H. Cunningham and R. M. Ormerod, *Catalysis Letters*, **49** (1997) 189.
[84] K. Kinoshita, F. R. McLarnon and E. J. Cairns, *Fuel Cells, A Handbook*, DOE/METC-88/6096, (1998).

[85] L. A. Shockling and P. Reichner, in *Fuel Cells*, eds. R. E. White and A. J. Appleby, The Electrochemical Society Proceedings, Pennington, NJ, PV89-14, 1989, pp. 106.
[86] H. Topsoe, B. S. Clausen and F. E. Massoth, in *Hydrotreating Catalysis Science and Technology*, Springer Verlag, Berlin, 1996.
[87] V. H. J. de Beer, T. H. M. van Sint Fiet, G. H. A. M. van der Steen, A. C. Zwaga and G. C. A. Schuit, *J. Catalysis*, **35** (1974) 297.
[88] K. Foger and B. Godfrey, in *Proceedings of the 4th European SOFC Forum*, ed. A. J. McEvoy, Switzerland, 2000, pp. 167.
[89] A. S. Carrillo, T. Tagawa and S. Goto, *Mat. Res. Bull.*, **36** (2001) 1017.
[90] A. L. Sauvet and J. Fouletier, *J. Power Sources*, **101** (2001) 259.
[91] C. A.-H. Chung, N. J. E. Adkins and R. M. Ormerod, in *Solid Oxide Fuel Cells VII*, eds. H. Yokokawa and S. C. Singhal, The Electrochemical Society Proceedings, Pennington, NJ, PV2001-16, 2001, 693.
[92] S. J. A. Livermore, J. W. Cotton and R. M. Ormerod, *J. Power Sources*, **86** (2000) 411.
[93] H. Uchida, H. Suzuki and M. Watanabe, *J. Electrochem. Soc.*, **145** (1998) 615.
[94] B. C. H. Steele, P. H. Middleton and R. A. Rudkin, *Solid State Ionics*, **40–41** (1990) 388.
[95] M. Watanabe, H. Uchida, M. Shibata, N. Mochizuki and K. Amikura, *J. Electrochem. Soc.*, **141** (1994) 342.
[96] M. B. Joerger and L. J. Gauckler, in *Solid Oxide Fuel Cells VII*, eds. H. Yokokawa and S. C. Singhal, The Electrochemical Society Proceedings, Pennington, NJ, PV2001-16, 2001, pp. 662.
[97] M. B. Joerger, C. M. Kleinlogel, D. Perednis and L. J. Gauckler, in *Proceedings of the 4th European SOFC Forum*, ed. A. J. McEvoy, Switzerland, 2000, pp. 489.
[98] N. Q. Minh, *J. Am. Ceram. Soc.*, **76** (1993) 563.
[99] P. Vernoux, J. Guindet and M. Kleitz, *J. Electrochem. Soc.*, **145** (1998) 3487.
[100] P. Vernoux, *Ionics*, **3** (1997) 270.
[101] A. L. Sauvet, J. Guindet and J. Fouletier, in *Proceedings of the 4th European SOFC Forum*, ed. A. J. McEvoy, Switzerland, 2000, pp. 567.
[102] G. Pudmich, W. Jungen and F. Tietz, in *Solid Oxide Fuel Cells VI*, eds. S. C. Singhal and M. Dokiya, The Electrochemical Society Proceedings, Pennington, NJ, PV99-19, 1999, pp. 577.
[103] J. T. S. Irvine, P. R. Slater, A. Kaiser, J. L. Bradley, P. Holtappels and M. Mogensen, in *Proceedings of the 4th European SOFC Forum*, ed. A. J. McEvoy, Switzerland, 2000, pp. 471.
[104] C. M. Reich, A. Kaiser and J. T. S. Irvine, in *Proceedings of the 4th European SOFC Forum*, ed. A. J. McEvoy, Switzerland, 2000, pp. 517.
[105] A. Kelaidopoulou, A. Siddle, A. L. Dicks, A. Kaiser and J. T. S. Irvine, in *Proceedings of the 4th European SOFC Forum*, ed. A. J. McEvoy, Switzerland, 2000, pp. 537.

[106] A. Kaiser, A. J. Feighery and J. T. S. Irvine, in *Solid Oxide Fuel Cells VI*, eds. S. C. Singhal and M. Dokiya, The Electrochemical Society Proceedings, Pennington, NJ, PV99-19, 1999, pp. 541.

[107] M. Mogensen, T. Lindegaard, U. R. Hansen and G. Mogensen, *J. Electrochem. Soc.*, **141** (1994) 2122.

[108] O. A. Marina and M. Mogensen, *Applied Catalysis A: General*, **189** (1999) 117.

[109] E. Ramirez-Cabrera, A. Atkinson and D. Chadwick, in *Proceedings of the 4th European SOFC Forum*, ed. A. J. McEvoy, Switzerland, 2000, pp. 49.

[110] E. Ramirez-Cabrera, A. Atkinson and D. Chadwick, *Solid State Ionics*, **136–137** (2000) 825.

[111] O. A. Marina, C. Bagger, S. Primdahl and M. Mogensen, *Solid State Ionics*, **123** (1999) 199.

[112] M. Watanabe, H. Uchida, M. Shibata, N. Mochizuki and K. Amikura, *J. Electrochem. Soc.*, **141** (1994) 342.

[113] H. Uchida, M. Sugimoto and M. Watanabe, in *Solid Oxide Fuel Cells VII*, eds. H. Yokokawa and S. C. Singhal, The Electrochemical Society Proceedings, Pennington, NJ, PV2001-16, 2001, pp. 653.

[114] S. D. Park, J. M. Vohs and R. J. Gorte, *Nature*, **404** (2000) 265.

Chapter 13

Systems and Applications

Rob J. F. van Gerwen

13.1 Introduction

There has not yet been a commercial, cost-effective application of the solid oxide fuel cell (SOFC) technology for power generation. The closest commercial application of the solid oxide electrolyte technology at present is the oxygen sensor, which is used to control the operation of the exhaust catalyst in an automobile; this is essentially a single fuel cell based on yttria-stabilised zirconia electrolyte, which gives a small voltage output related to the oxygen concentration in an automobile's exhaust gas stream. Similar oxygen sensors are also used in the food, metal and burner industries, in which oxygen measurement is of prime importance.

Yet numerous potential opportunities exist for using SOFC systems at various power levels, as shown in Figure 13.1. At low power levels, around 1–10 W, small SOFC devices could be used as battery replacements on remote sites, where the cost of recharging or replacement batteries is high but where fuel for fuel cells is readily available. This is the market where thermoelectric generators are now employed, e.g. on pipelines or gas rigs [1]. At a somewhat higher power level, from about 100 W to 1 kW, military applications requiring lightweight portable power sources to be carried by troops for communication and weapon power become possible; batteries are too heavy and short-lived at present. Similar products could also be used in leisure applications, such as yachting and camping, where navigation systems, computers and telephones are required. In military applications, power may cost up to \$30,000/kW. Another high price market for SOFC technology is in the production of high-purity oxygen, for laboratory and respiratory applications. On a larger scale, SOFCs can be used to produce oxygen for the chemical industry (for production of syngas), metal industry, and for coal gasification.

A major application of SOFC systems is perceived at the 1–10 kW level to supply power to residential buildings and as auxiliary power units in vehicles and trucks. A number of projects are underway at present to show the feasibility

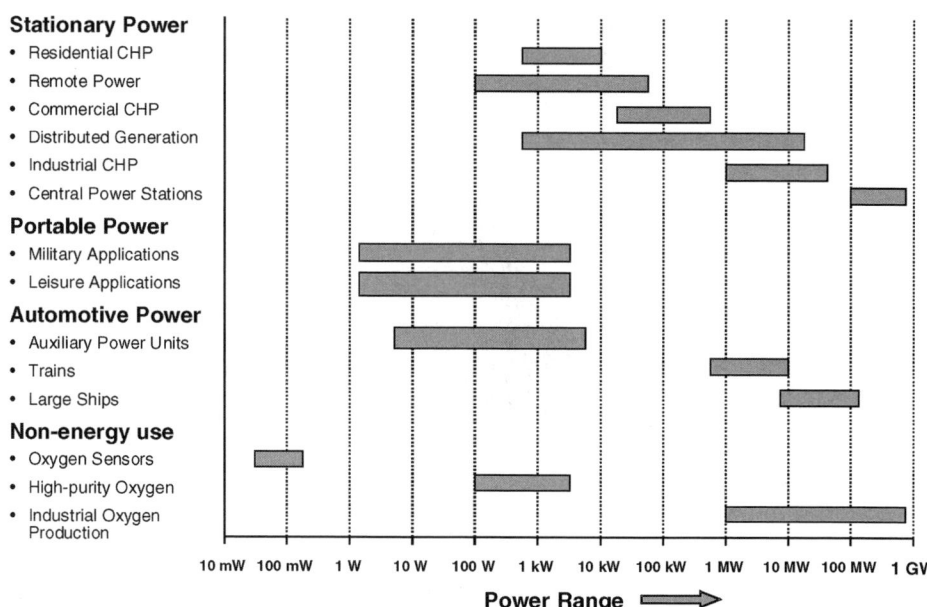

Figure 13.1 Potential applications of the SOFC technology.

of this concept, in which the capital cost of the SOFC system needs to be in the range of $400–1000/kW. The potential market for such SOFC systems worldwide is very large, approaching several hundreds of billions of dollars annually, once the technical and economic criteria are met. At the larger end of this market is the distributed power generation and the cogeneration (combined heat and power, CHP) sector, which, at present, is served by combustion engines such as diesels or small turbines, with outputs of about 10 kW to a few MW. For commercial acceptance, the SOFCs must be reliable and should have low maintenance to compete with diesel engines, which are relatively low in cost but detrimental to the environment. Above this power range, greater than 1 MW, SOFC systems can augment existing turbine power plants to improve efficiency and reduce emissions. Integrated with gas turbines, large SOFC hybrid power systems could be up to 70% efficient in converting fuel to electricity.

For SOFC commercialisation, it is important to identify suitable applications and realise successful demonstrations early in light of the concept of an early market and a mainstream market for any new technology; Moore [2] discussed the concept of a *chasm* between these two markets. The early market is formed by people who are enthusiastic about the technology and people who see possibilities for its exploitation (visionaries); this market is usually relatively small and quickly saturated. On the other side of the chasm is the mainstream market, where pragmatists expect reliable systems at a competitive cost now with only a little concern about any bright future. Therefore, identifying suitable SOFC applications, demonstrating system reliability and lifetime, and lowering system cost is vital in order to convince the pragmatists to buy and thus to enter

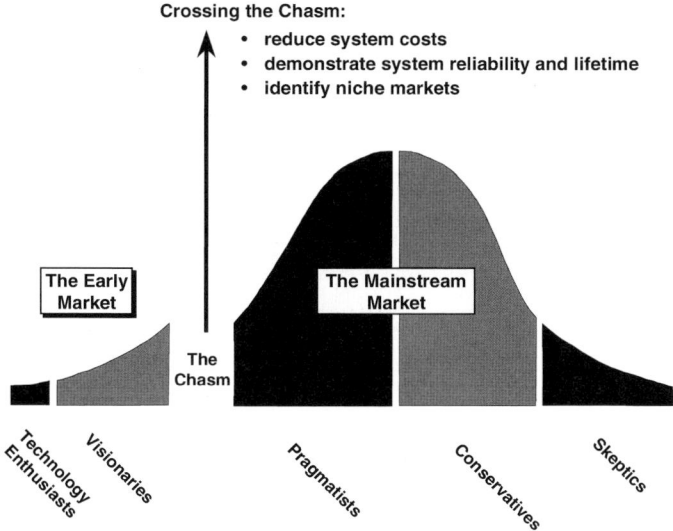

Figure 13.2 *The main challenge for the SOFC technology is to cross the* chasm *between the early market and the mainstream market [2].*

the mainstream market. Crossing the *chasm* (Figure 13.2) is, at the moment, the main challenge for solid oxide fuel cell technology.

This chapter first considers the trends in the energy markets that are driving towards utilisation of SOFC technology for electricity generation in various applications. Then SOFCs are compared with combustion engine generators and the potential application areas of SOFC systems are explored in detail. Designs and performance of SOFC systems are then analysed, followed by examples of SOFC systems and their demonstrations.

13.2 Trends in the Energy Market and SOFC Applicability

Three major trends currently dominate energy markets; liberalisation, growing environmental concerns, and a move towards distributed power generation. Liberalisation of the energy markets leads to less predictable energy prices. A power generation capacity shortage, or maybe just an apparent shortage, may lead to high electricity prices as was the case in California in 2000. A surplus of generation capacity, as existed in Europe in the same year, causes a fall in electricity prices. Industrial electricity prices in European countries dropped more than 30% in the 1995–2000 period [3]. This resulted in many combined heat and power generation (CHP or cogeneration) projects becoming economically unfeasible, especially those projects that were heat load following, meaning that the system was sized to deliver base heat load and surplus electricity was sold to the grid. During night time, the price received for electricity delivered to the grid was barely higher than the avoided fuel cost for large power plants and not nearly enough to pay the on-site fuel cost. As a result, Germany and the Netherlands now subsidise electricity delivered to the grid by certain cogeneration systems with 0.6 to 2.6 US cents per kWh.

Our environment is closely affected by power generation. Use of fossil fuels is the major source of emissions of carbon dioxide and the so-called regulated pollutants: particulate matter (PM), oxides of nitrogen (NO_x), oxides of sulphur (SO_x), and unburned CO and hydrocarbons (HC). On local, regional and global levels, these emissions are harmful to our health and our climate. Reducing global use of fossil fuels in order to reduce emissions is the most sensible option but that is not likely in the foreseeable future. The US Department of Energy [3] expects the world energy consumption to rise by 60% from 1999 to 2020, to 650 trillion MJ. Of this, electricity consumption will increase 66% to 22 trillion kWh. Use of renewable energy is another desirable option to diminish carbon dioxide emissions and, in the case of solar and wind power, emissions of regulated pollutants. Although the use of renewable energy is expected to increase more than 50%, its share in total consumption is expected to drop from currently 9% to less than 8%. As both reducing the world energy consumption and increasing the share of renewable energy seem unfeasible, other options must be envisioned. Clean, highly efficient fossil power plants, combined heat and power generation, and CO_2 abatement are considered feasible options [3,4].

Another effect of liberalisation of energy markets has been that electricity from the grid and natural gas have become commodities. Customers are free to buy their gas or electricity from anywhere they want. They pay separately for the generation and for the transmission of electricity. Before liberalisation of the markets, the cost of the transmission of electricity through the grid was more or less included in the electricity price. Now, a connection to the grid might be seen as an investment that must be weighted against the investment in an on-site generation system. Reliability of power supply is another decision factor with respect to the choice between on-site generation and a connection to the grid.

Distributed power generation, the on-site generation of electricity rather than buying it from the grid, also impacts the electricity market. Reasons for implementing distributed generation include [5]:

- **Power reliability.** Businesses such as computer centres, mail distribution facilities, credit card processors, and industries with high start-up or shut-down cost for their processes, are potential implementers of reliable distributed generation. The reliability of the grid is three to four 'nines' (99.9–99.99%, meaning 1–9 hours of outage per year). The reliability of distributed generation using SOFCs could be five to six nines, that is 30 seconds to 5 minutes of outage per year, due to trading the electricity grid for a less vulnerable gas supply that does not require immediate balancing. Reliability required for 'digital power' for computer centres, etc., is nine to ten nines, i.e. 3–30 ms outage per year [6].
- **Power quality.** Some industries, for instance chip manufacturers, are very sensitive to disturbances in the electric power quality. One way to solve this problem is to generate high quality, on-site power using fuel cells.

- **Absence of a grid** in certain areas automatically leads to a need to install on-site power generation capacity.
- **Solar and wind power** are also distributed sources of electricity which can readily be connected to the grid to minimise environmental concerns.

SOFC systems fit well with these trends in the energy markets. Although liberalisation of the markets means more uncertain electricity prices, the higher efficiency of these systems lowers the marginal cost of the generated electricity, making it easier to compete with the grid. The environmental benefits from SOFC technology are evident and fit remarkably well with the trend towards clean, highly efficient fossil power plants, combined heat and power generation, and CO_2 abatement. Reliability of SOFC systems has to be proven yet, although prospects are very favourable. The quality of the power delivered from the SOFC systems is generally high, due to the advanced power electronics used. In short, the SOFC systems, from a technological point of view, are extremely promising with respect to present energy market trends. However, their cost-competitiveness with other established power generation technologies remains a major issue; this is discussed in the next section.

13.3 Competing Power Generation Systems and SOFC Applications

Figure 13.3 shows the electrical efficiency of different power generation systems versus the system size. Advantages of fuel cell systems are a high electrical

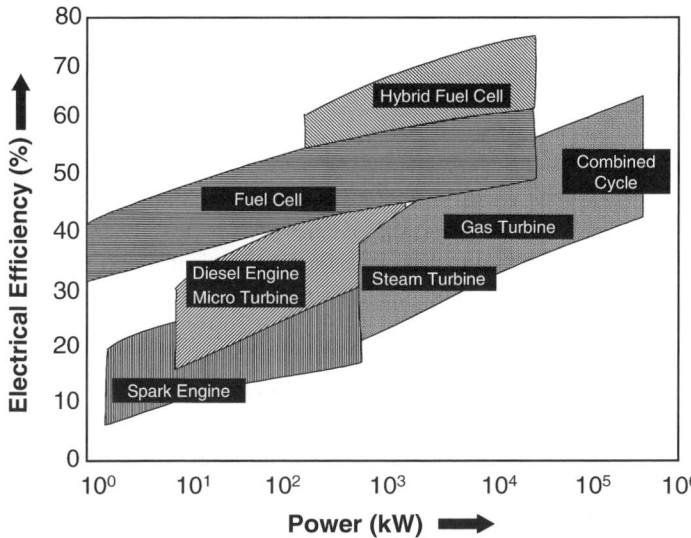

Figure 13.3   The electrical efficiency versus size for conventional and fuel cell power systems.

efficiency, even in small systems, plus the best electrical efficiency in fuel cell/turbine hybrid systems.

Table 13.1 gives a more detailed overview of the main competitors of the SOFC systems. The cost range for gas turbines and gas engines is large; the lower costs are for systems in the higher system sizes and vice versa. The same is true for the efficiency. Emissions of CO and hydrocarbons are not included in Table 13.1 but scale more or less with the NO_x emissions.

Table 13.1 Overview of competitive conventional natural gas-fired power generation systems [7–11]

	Microturbine	Gas engine	Gas turbine	Combined cycle
Power range	25–100 kW	25 kW–5 MW	3–100 MW	>10 MW
Electrical efficiency (%)	26–33	20–40	30–42	50–58
Lifetime (year)	>10	15–20	15–20	15–25
NO_x emission (g/GJ)	20–50	100–150	20–50	15–35
Investment cost ($/kW)	1000–2000	500–1200	500–1100	400–600
Maintenance and operational cost (cent/kWh)	1.0–2.0	1.0–1.5	0.3–0.6	0.2–0.3
Technology	Developing	Mature	Mature	Mature

The gas turbine/steam turbine combined cycle is the main competition to large SOFC/turbine hybrid systems from a cost-of-electricity point of view. Specific investment costs and maintenance costs are low, resulting in a very low cost of electricity (approximately 2.5 US cents per kWh). This can be regarded as a reference commodity price for electricity from the grid. Gas engines like diesels and gas turbines have a lower electrical efficiency together with higher emissions and higher maintenance cost, but the installed cost is attractive. Microturbines are still under development; their power range will probably extend to that of a conventional gas turbine, providing serious competition for the gas engine. It is expected that the cost of microturbines will eventually go down to the $500/kW range [12]. A distinct advantage of the microturbine, the gas engines and the conventional gas turbines, as compared with the SOFC systems, is the shorter start-up time. Dynamic properties (control range and control speed) might also be better, although the electrical efficiency significantly decreases at part load, whereas SOFC systems generally show an increasing efficiency at part load.

Thus, SOFC/turbine hybrid systems are better than conventional combustion systems in terms of electrical efficiency, part-load efficiency, and emissions. Reliability and maintenance cost of mature SOFC systems will probably be comparable with gas turbine systems and certainly better than gas engines. Attributes of SOFC systems in the field of control speed, control range, investment cost and lifetime have yet to be fully established. The longer start-up time of SOFC systems is a clear disadvantage for certain market applications, for instance, the uninterruptible power supply market and the auxiliary power units for the transportation market.

SOFC prototype systems of up to 1 MW size are currently planned [13], with further SOFC systems going up to 20 MW size [14,15]. In the higher size range, the competition from gas turbine/steam turbine combined cycle systems will probably be too strong, but in the lower size range, two important markets can be anticipated: distributed power generation and combined heat and power generation (CHP). Other markets like uninterruptible power and peak shaving are not as attractive for SOFC systems because they require fast-starting, low-investment systems, whilst efficiency and emissions are of less importance.

The market for distributed power can be divided into a residential market (1–10 kW), a commercial market (50–500 kW), and an industrial market (1–20 MW). Residential market includes single-family homes and apartments, etc. Commercial market includes computer centres of financial institutions, insurance companies, stock exchanges, etc., hospitals, nursing facilities, department stores, restaurant chains, airports, traffic control centres, military bases, and police stations, etc. A typical example of an industrial market is the microchip industry where distributed power is one of the top priority areas [12]. In 2010, at least 20% of the newly installed capacity in the USA is expected to be distributed generation. Given an average yearly US capacity increase of 10.5 GW from 2000 to 2010 [16], this means at least 2.1 GW of distributed capacity per year. In Europe, the average yearly capacity increase for this period is 7.5 GW, resulting in a distributed generation potential of 1.5 GW yearly, also assuming 20% distributed generation.

In the USA, CHP accounts for approximately 7% of the total generation capacity [12]. The industrial market (most importantly chemicals, paper, refining, food and metal) accounts for 90% of the current CHP installations, the commercial market for the other 10%. The CHP potential in the USA is large: 88 GW for industrial cogeneration and 75 GW for commercial cogeneration. If only a 2–3% of this potential is realised per year, it means a yearly CHP potential of 3.3–4.8 GW.

In 1997, the European Commission set a target to double CHP output as a proportion of electricity generation from 9 to 18% by 2010 [19]. A scenario study showed that in the most pessimistic case this percentage would not increase at all, while in the most optimistic scenario gas-fired CHP would increase from 35 GW in 1997 to 135 GW in 2010, an average growth rate of 4.2 GW per year. According to this optimistic prediction, approximately 22% of the total electricity output in the European Union will be CHP-based by 2010. The conclusion [19] was that this percentage could only be achieved with the creation of a significant cogeneration market in the residential sector.

Residential cogeneration is seen as a relatively low-risk market (Figure 13.4 [19]), partly because development times required for small systems are shorter and the investment level is lower. Retail prices for residential electricity are relatively high. In countries with many rural areas and a weak or unreliable grid, cogeneration will significantly increase the reliability of the electricity supply and will, in most cases, be less costly than upgrading many miles of transmission and distribution lines. In the US, the nation's 900 rural electric

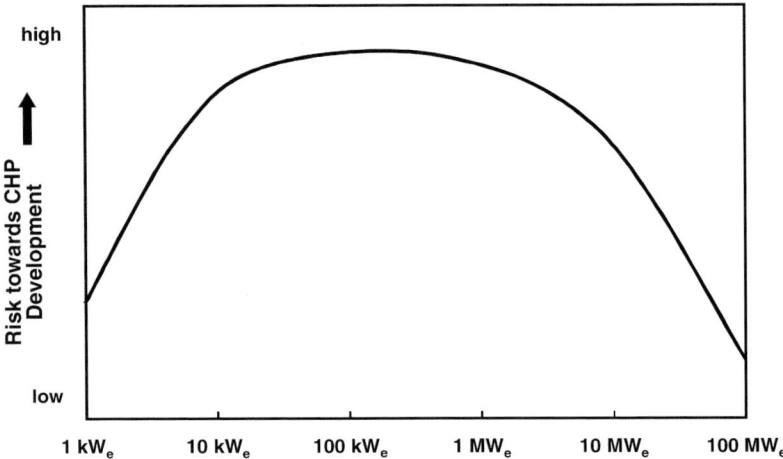

Figure 13.4   Perceived risk towards the development of cogeneration plants versus the size of the plants [19].

co-operatives sell 8% of the total electricity but own nearly 50% of the electric distribution lines [20]. The market for CHP in these areas is estimated to be several hundreds of MW [20].

Besides these stationary power markets, mobile applications can be predicted in ships, locomotives and auxiliary power units for automobiles, trucks, and recreational vehicles. One of the significant advantages of SOFC technology in these applications is the fuel flexibility.

13.4 SOFC System Designs and Performance

In the SOFC systems, fuel and air enter the SOFC stack, and electricity, exhaust gas and possibly hot water or steam exit the system. Such systems include atmospheric SOFC CHP systems; pressurised SOFC/turbine hybrid systems; atmospheric SOFC residential and auxiliary power systems; and oxygen separating systems. The difference between an SOFC stack and an SOFC system is generally referred to as the balance-of-plant (BOP). BOP equipment may differ for each application, depending on the size of the system, the operating pressure, and the fuel used.

13.4.1 Atmospheric SOFC Systems for Distributed Power Generation

Figure 13.5 shows an example of the main balance-of-plant equipment for an atmospheric SOFC system; it is a simplified scheme of a system built by the Siemens Westinghouse Power Corporation by using tubular SOFCs and demonstrated in Westervoort, the Netherlands [21]. In this example, the SOFC stack is integrated with a pre-reformer for reforming higher hydrocarbons in the natural gas and a catalytic burner into a single SOFC generator module.

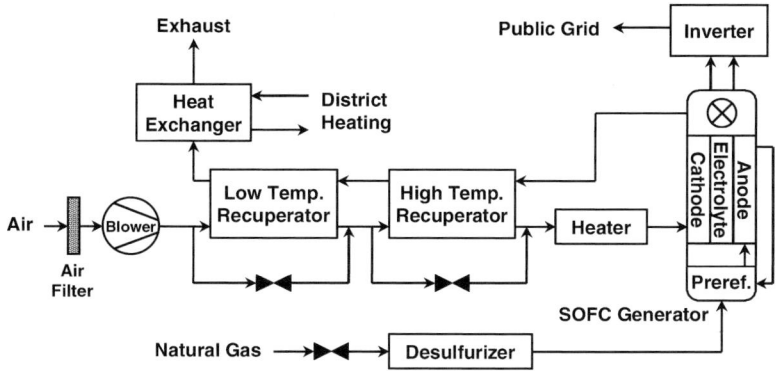

Figure 13.5 The main components of a typical atmospheric SOFC system [21].

Figure 13.5 shows that fuel is desulphurised before it enters the SOFC stack. Sulphur in any form is harmful to the SOFC stack; it poisons the nickel-containing anode and reduces the stack performance. The sulphur level must generally be below 0.1 ppm to avoid a performance loss [22]. However, any performance loss due to low concentrations of sulphur in the fuel is a reversible process [23,24]. Natural gas contains far less sulphur than oil-derived liquid fuels, but generally a sulphur-containing odorant (mercaptans or tetrahydrothiophene) is added to a level of approximately 4–5 ppm [22,25]. A common way to remove sulphur compounds is to hydrogenate them in a catalytic reactor with hydrogen to hydrogen sulphide and absorb hydrogen sulphide at an elevated temperature (up to 450°C) in a separate reactor containing zinc oxide. Absorbing sulphur compounds by means of activated carbon is easier because it works at room temperature, but it is almost ten times as costly [22,26].

The heavy hydrocarbons in the fuel must be pre-reformed to methane, carbon monoxide and hydrogen to avoid carbon formation in the SOFC stack. Generally, steam reforming is used. Recirculating part of the SOFC exhaust stream, as shown in Figure 13.5, can be used to supply steam to the reformer and this has the additional advantage that the overall fuel utilisation increases.

Due to the cell voltage drop as a result of the decreasing fuel concentration towards the exhaust side of the stack, only a certain percentage of the available fuel can be electrochemically converted to electricity and heat. An overall utilisation of 85–90% is considered a practical maximum. At higher fuel utilisations, nickel may oxidise locally. A catalytic burner is used to burn the remaining fuel from the anode side with the surplus air from the cathode side.

The exhaust gas is led through a recuperator to heat the air before it enters the stack. A temperature of at least 500°C is necessary to avoid thermal shock that may cause possible irreversible damage to the stack. An additional heat exchanger is used to produce hot water. By positioning this heat exchanger between the two recuperators, as shown in Figure 13.5, process steam can be generated instead of hot water.

An SOFC stack produces direct current (DC). An inverter is used to convert the DC to grid-quality alternating current (AC). State-of-the art inverters are based on fast switching semi-conductors. In general, a two-step converter is used. The first step is a DC–DC stabiliser to transform the stack voltage to a stable DC voltage. In the second step, DC power is converted to grid quality AC power. Most inverters are able to supply as well as to take in reactive power. Reactive power or idle power is typical for alternating current grids. It refers to power flows with voltage and current in opposite phase. Just like normal power flows, reactive power flows must be balanced in the grid. Reactive power flows are caused by capacitive and reactive loads in the grid, such as electric motors and strip lights. Fast switching semiconductor inverters can be used for locally balancing the grid and maintaining the required grid voltage. The inverter efficiency typically ranges from 94 to 98% [22,26,27].

Additional SOFC system components are:

- A blower to supply air to the system (the only source of noise, and a major parasitic electricity consumer).
- Air filters; the air does not have to be particularly clean, and therefore a common air filter suffices.
- An air heater, to start the SOFC generator module and sometimes used to run at low load, when stack temperature may be low.
- Control equipment and user interface for remote load control, reactive power control and grid-independent operation.
- Purge gas systems to avoid damage to the stack during start-up or to flush the stack during an emergency shutdown.
- Start-up steam generator to supply steam to the pre-reformer.

For such atmospheric CHP systems, an electrical efficiency of 45–50%, based on lower heating value (LHV) of the fuel, is considered the upper limit [21,26,28], while a total thermal and electrical efficiency of 85–90% is feasible.

Emissions from such natural gas-fuelled SOFC systems are negligible except for CO_2. This is partly due to the fact that a clean fuel is used, with a desulphuriser to remove sulphur from natural gas, leading to very low levels of SO_x and particulates. But the main cause of low emissions is the lower operating temperature of an SOFC stack compared with a conventional burner, preventing the formation of NO_x from oxygen/nitrogen reaction which typically starts above 1000°C. Higher hydrocarbons are converted to hydrogen and carbon monoxide at the stack inlet and catalytic burning of the spent fuel at the stack exit removes all remaining carbon monoxide.

Another concept for atmospheric SOFC systems is a multi-stage fuel cell generator. In this concept, the SOFC stack is divided into different stages, operating at different temperatures and fuel utilisation. In theory, a higher stack voltage at a higher fuel utilisation can be obtained. Some sources calculate efficiencies of 75–80% for multi-stage SOFC systems [27,29], although other sources predict only a small efficiency gain compared to a single-stage stack [22,26].

13.4.2 Residential, Auxiliary Power and Other Atmospheric SOFC Systems

Residential and auxiliary power units (APUs) are also atmospheric systems; at a power level of 1–10 kW, pressurisation is not useful. The general layout of these systems does not differ much from the atmospheric system outlined above. Figure 13.6 shows the main sub-systems of a 5 kW size auxiliary power unit fuelled by gasoline [30]. Typical of this small system is the close thermal integration as relative heat losses increase with decreasing size and the use of a partial oxidation fuel reformer to pre-process the gasoline. Partial oxidation reforming is generally less efficient than steam reforming, but needs a much smaller reactor. The APU unit in Figure 13.6 also includes a battery to provide for peak power and to level the load of the SOFC stack. An electrical efficiency of up to 30% may be expected from such gasoline-fuelled APUs. As for residential systems operating on natural gas, this efficiency is expected to be up to 40%.

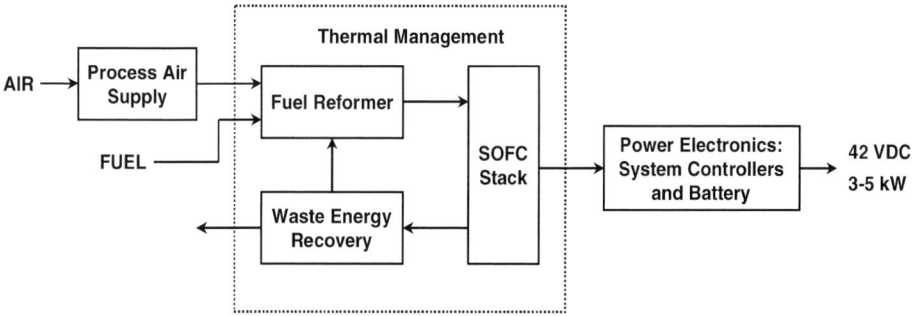

Figure 13.6 The main subsystems of an SOFC auxiliary power unit [30].

The SOFCs can also serve as an excellent generator of synthesis gas [31–33]. This 'syngas' is a mixture of carbon monoxide and hydrogen and is used as a raw material in many chemical processes, for instance in the production of hydrogen, methanol and other liquid fuels from natural gas. The traditional way of cryogenic separation of oxygen from air is an important fraction of the cost of syngas production (25–45%). So, the application of SOFC technology for the combined production of power and syngas, or the combined production of heat, power and syngas is promising.

It is relatively easy to separate carbon dioxide from water in the anode exhaust stream, thus opening the possibility for carbon sequestration. In a conventional natural gas combined cycle plant, the electrical efficiency drops from 55 to 47–48% if CO_2 is captured. The investment cost also almost doubles due to CO_2 capture measures, thus increasing the cost of electricity by about 1 US cent per kWh. The penalty for separating CO_2 from SOFC anode gas is far lower (in both investment and electrical efficiency). A CO_2 separating SOFC system based on a pressurised SOFC generator combined with a gas turbine is expected to hardly lose any efficiency when CO_2 is captured, with the additional advantage that pressurised CO_2 becomes available for other applications [34].

The key to CO_2 capture from an SOFC is omitting the conventional after-burner. Instead, the fuel side is sealed from the air side and the anode exhaust gas is electrochemically oxidised in a special after-burning section through an oxygen-selective ceramic membrane. Cooling the exhaust gas separates the water vapour from carbon dioxide (Figure 13.7). Developing an effective high-temperature seal and a suitable oxygen-selective membrane are the challenges for this design. Another challenge is to design and control for exactly 100% oxidation of the anode gas [34].

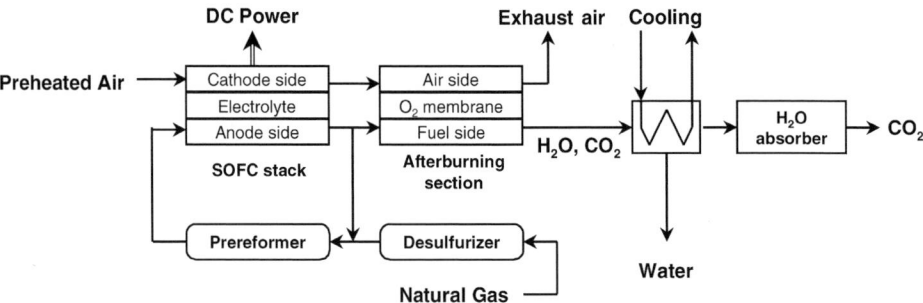

Figure 13.7 Simplified configuration for CO_2 capture from an SOFC system.

13.4.3 Pressurised SOFC/Turbine Hybrid Systems

Combining a pressurised SOFC stack with a gas turbine [18,27,29,35,36] promises very high electrical efficiencies (60–75%), even at a relatively small 1 MW scale, with negligible harmful emissions except for CO_2. Figure 13.8.1 shows the simplest design for a gas turbine (GT). A compressor compresses air to 3–30 bar; fuel is burned with the compressed air (typically 3 to 4 times the stoichiometric amount); and exhaust gas (800–1300°C) is expanded in a turbine that is mechanically coupled to a generator and to the compressor. Due to the expansion in the turbine, the exhaust gas temperature decreases to 250–600°C. The higher the pressure ratio between the turbine inlet and exhaust, the lower the exhaust temperature. The overall electrical efficiency ranges from 20% for small size turbines to 35% for large industrial turbines.

Figures 13.8(2)–13.8(6) show several configurations for conventional gas turbines [36,37] that can be used for hybrid SOFC/GT systems as well. In principle, all that has to be done is to replace the conventional combustion chamber in a turbine with a (pressurised) SOFC generator. The different configurations in Figure 13.8 are discussed below.

- Configuration 2: using a recuperator decreases the amount of natural gas necessary to heat the air, thus increasing the electrical efficiency to 40% or higher. However, it is a more expensive and complicated system, and there will be a slight loss in maximum power due to the pressure losses in the recuperator.

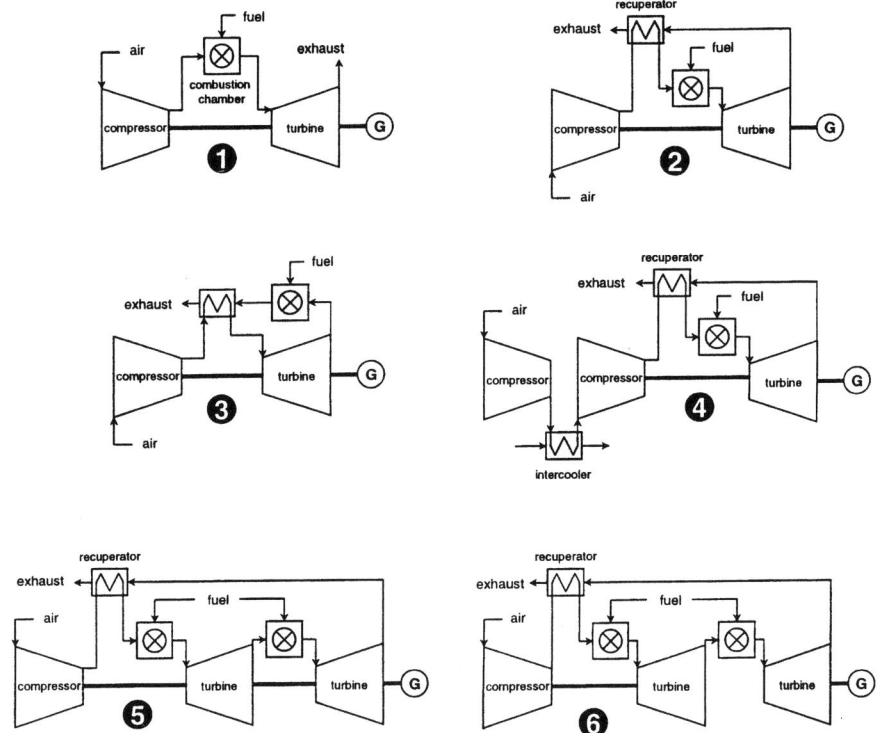

Figure 13.8 Possible gas turbine configurations [36,37]. Configurations 2 to 6 can be used for hybrid systems by replacing the conventional burner with a (pressurised) SOFC generator.

- Configuration 3: this configuration uses an indirectly fired turbine. The main advantage is that the SOFC generator (replacing the conventional burner) operates at atmospheric pressure, avoiding expensive pressurisation of the generator and a more complicated control system.
- Configuration 4: by using an intercooler between consecutive compressor stages, the compression becomes more efficient as it approaches ideal isothermal compression. It also means a trade-off between a higher initial investment and a higher electrical efficiency.
- Configuration 5: at the turbine side, an ideal isentropic expansion can be approached by reheating the exhaust gas between turbine stages. Again, it is a trade-off between a higher initial investment and a higher electrical efficiency.
- Configuration 6: to obtain a more flexible part-load behaviour, a separate power turbine can be used. Generally, the part-load efficiency is higher in the case of a double-axis gas turbine than in the case of a single-axis gas turbine.

It is important to realise that the air inlet temperature of an SOFC generator must be at least 500–650°C [22,27,36] to avoid thermal shock and possible damage to the SOFC stack. Therefore, heat recuperation is necessary. The choice of the pressure ratio between the gas turbine inlet and outlet is also limited because at high pressure ratios, the exhaust gas temperature becomes too low to heat the air to the required inlet temperature of at least 500°C. A pressure ratio of 2–4 is necessary, unless there is additional heating by conventional burners. Without additional heating by conventional burners, the SOFC stack delivers about 65–80% of the total electrical power output of the hybrid system.

The efficiency of a SOFC/GT hybrid system depends on the system size, the configuration used (Figures 13.8(2)–13.8(6)), the use of additional heating by conventional burners and the performances of the particular SOFC and gas turbine technologies used. Efficiencies may range from 55 to 60% for a small, simple SOFC/GT system in the 250 kW to 1 MW range to almost 68% for a 5–10 MW SOFC/GT system with intercooler and reheating by a separate SOFC generator. A total efficiency (thermal and electrical combined) of 85–90% may be expected. Emissions of NO_x mainly depend on whether a conventional burner is used in the hybrid system, but will, in any case, be considerably lower than that for a comparable conventional gas turbine.

Another interesting hybrid concept consists of an SOFC combined with a low-temperature (polymer electrolyte) fuel cell. The basic idea is to generate a mixture of carbon monoxide and hydrogen (syngas) in the SOFC stack. This syngas is then converted with steam to produce hydrogen by using a shift reactor and then used in the polymer electrolyte fuel cell stack to produce additional electricity. Any remaining carbon monoxide in the feed gas must be removed first, for instance by preferential oxidation. At lower temperatures less than 100°C, the conversion efficiency of hydrogen to electricity is considerably higher than at higher SOFC generator temperatures. At an operating temperature of 80°C, the intrinsic maximum efficiency for a hydrogen fuel cell is 93% whereas at 1000°C it is only 73% [23].

13.4.4 System Control and Dynamics

When controlling SOFC systems, the main control parameter is the electrical power output of the system. This AC output is determined by the inverter connected to the SOFC stack. The fuel supply to the stack must follow the DC power demand by the inverter, which is needed to deliver the required AC power. This is more complicated than controlling gas turbines, whose power output is related directly to the fuel flow. In the case of a SOFC/GT hybrid system, power control is even more complicated since both the gas turbine generator and the SOFC stack deliver power. A major concern in load following is the risk of retaining residual unburned hydrogen and carbon monoxide in the stack, due to a sudden load drop [35], which can be difficult to handle.

If grid-independent operation is required, a more complicated control system is necessary. A frequency control must be added. The inverter has to be able to supply as well as to demand reactive power. Especially when an asynchronous generator (the most common type of generator) is used in a hybrid system, the inverter must supply reactive power for grid-independent operation.

Synchronising the unit to the grid, when going from grid-independent operation to operation parallel to the grid, is necessary. If several SOFC systems are used to support an independent grid, a load sharing control is necessary.

Closely related to the power control is the fuel utilisation control. At the same power level, different levels of fuel utilisation are possible. If the fuel utilisation is too low, a relatively large amount of anode exhaust gas is burned in the catalytic after-burner. This may cause problems with overheating and thermal shock. If the fuel utilisation is too high, the oxygen partial pressure at the anode side may locally become so high that nickel might oxidise, causing irreversible damage to the nickel anode. Both effects limit the fuel utilisation within certain limits. With respect to system efficiency, there is an optimum fuel utilisation. At a lower utilisation, the efficiency decreases because fuel is burned in the after-burner instead of being electrochemically converted to DC power. At higher fuel utilisation, the cell voltage drops, also causing an efficiency drop. A generally used level of fuel utilisation is approximately 85%.

Another important control is the SOFC stack temperature control. Theoretically, there is an optimum stack temperature. Lowering the stack temperature means a more favourable electrochemical conversion of hydrogen in the stack; increasing the stack temperature means a lower electrolyte resistance. At the optimal stack temperature, these effects are balanced. A higher stack temperature also increases the stack degradation, and in practical applications there will be an economic trade-off between a higher system efficiency and replacement cost of the SOFC stack. Changing the amount of cooling air through the stack controls the stack temperature. At lower power levels, less cooling air is needed. As the stack efficiency increases at lower power levels, the required amount of cooling air decreases more than proportional to the power level. As for small SOFC systems, it is, therefore, possible that heat losses at part-load may be higher than the internal heat production and additional heating might be needed.

To avoid carbon formation in the SOFC stack, a minimum steam-to-carbon molar ratio of 2.5–3.0 is needed [7]. This ratio is maintained by controlling the amount of recirculation of H_2O-rich anode exhaust gas.

As for SOFC/GT hybrid systems, it is important to avoid surge in the compressor. There will be surge when the air flow is low in relation to the pressure ratio over the compressor. If the air flow is too low to maintain the upstage pressure level, the flow suddenly collapses, causing the output pressure to drop to a lower level. Pressure will then build up again until the air flow collapses once more. This causes repeated jolts in the compressor which can mechanically damage it.

Important dynamic features of an SOFC or SOFC/GT system are control speed, control range and start-up time. Control speed is mainly limited by the inertia of the rotating machinery (for hybrid systems), the fuel gas volume in the stack and the fuel ducts up to the fuel control valve, and the tuning of the power control loop. Load excursions can, depending on the duration, be handled by the fuel cell alone or, in case of hybrid systems, by the total system [35]. Control range is important in case of load following applications (for instance grid independent operation). The danger of surge in the compressor may limit the control range of an SOFC/GT system. Reducing the power to 60–70% of full load does not seem to pose any problems [24,38,39]. In order to extend the control range further down, additional air or exhaust bypass/bleed valves may be needed.

Due to a high SOFC stack mass, heating the stack from the ambient temperature to an operating temperature of approximately 1000°C can take several hours. Using an SOFC for stand-by power is only possible if the unit is kept at a minimal operating temperature at all times. Depending on the system size, this might cause considerable stand-by losses. It is estimated that it will still take several minutes to proceed from hot stand-by to actual power delivery.

13.4.5 SOFC System Costs

There is a large potential market for distributed power and CHP that can be fulfilled by SOFC systems. The high SOFC system cost is the main obstacle to penetration of this market. A capital cost of $1000–$1500 per kW is very often quoted for SOFC systems to be competitive. Current capital costs are considerably higher, although these costs are continually coming down as the development proceeds.

In order to accelerate the commercialisation of solid oxide fuel cells, the US Department of Energy initiated the Solid State Energy Conversion Alliance (SECA) programme in 2001. It is a collaborative effort between the funding agencies (primarily the department of energy), US industry, and universities and other research organisations. SECA programme's strategy is based on developing a common 3–10 kW size SOFC module that can be mass-customised for use in multiple products for stationary, transportation and military applications. The combined large volume in these multiple applications is expected to lead to a reduction of cost to $800 per kW in 2005 and $400 per kW in 2010 [40].

The total SOFC system cost consists of the costs of the cells, the generator (taking the cells and connecting them with interconnects, seals, etc., and building a stack with thermal insulation), and the balance-of-plant. At present, all three costs are high, although the main emphasis in cost reduction has been on reducing the cost of the cells. The cost of cells depends on the cell geometry, particular materials used (e.g. metallic or ceramic interconnect), and the particular fabrication processes used, and varies widely. The estimates for raw materials costs vary from generally $70 per kW for planar SOFC stacks with metal interconnects to $300 per kW for planar all-ceramic SOFC stacks [17,41]. However, such cost estimates are very approximate and depend on a large number of assumptions. It is fair to say that one needs to reduce the cost of every item in a SOFC system and simplify the system as much as possible to meet the commercial cost targets for the various SOFC systems.

The main operational costs for a SOFC system are desulphuriser absorbent and catalyst replacement; regular gas turbine maintenance (including air filter replacement) for hybrid systems; SOFC stack replacement; and plant operation and administration. Siemens Westinghouse expects Operation and Maintenance

(O&M) costs for hybrid systems from 0.45 to 0.56 cents per kWh [22,26]. This includes stack and gas turbine replacements. It might be expected that O&M costs of hybrid systems will be comparable to those of gas turbines (0.3–0.6 cents per kWh, Table 13.1), as the cost of desulphurising fuel is not very significant. Therefore, an O&M cost estimate of 0.45–0.56 seems realistic. O&M costs for atmospheric systems will be lower since only the air blower needs to be maintained.

13.4.6 Example of a Specific SOFC System Application

This section shows an example of the economic feasibility of using a SOFC/GT hybrid system for greenhouse cultivation of roses. Greenhouse cultivation is an interesting market for fuel cells because it needs heat, electricity and CO_2 to stimulate the growth of flowers or vegetables. All three can be supplied by a SOFC/GT hybrid system. In this specific application, the greenhouse has a fairly constant need of electricity for assimilation illumination for 4000 hours per year [42]. As assimilation illumination is mostly needed in the evening and at night time in order to extend the growth period of the roses, the system delivers electricity to the grid during daytime when the price for electricity delivered is higher than during night time. The generated heat and exhaust gas can be used to heat the greenhouse and for CO_2 fertilisation. In this calculation, the avoided CO_2 and NO_x emissions compared with a conventional gas engine CHP are capitalised. The maximum cost of avoided CO_2 is estimated to be $50 per ton, based on abatement cost for central power stations [43,44]. The cost of avoided NO_x is $1700–$2800 per ton, based on cost for NO_x removal technologies for central power stations [45].

Figure 13.9 shows a sensitivity diagram regarding the cost of electricity. The electrical efficiency of the SOFC/GT hybrid system is assumed to be 60%, and

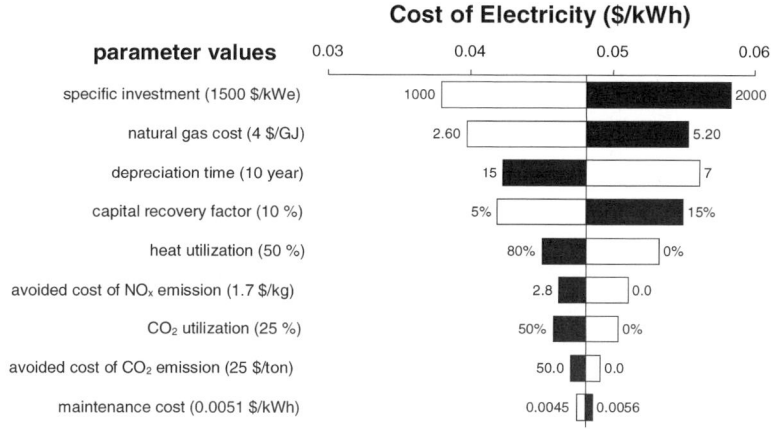

Figure 13.9 Cost of electricity (COE) for a SOFC/GT hybrid system in a greenhouse application. For every parameter, the sensitivity of the COE for this parameter is shown, including the assumed minimum and maximum parameter values.

the thermal efficiency 30%, with 8000 operating hours per year. The SOFC system cost is estimated to vary between $1000 and $2000 per kW. Other parameter values are shown in Figure 13.9, which shows that the cost of electricity (COE) is approximately $0.048 per kWh at average parameter values. As might be expected, the COE mainly depends on the specific investment cost for the SOFC/GT hybrid system and the natural gas price. Also, the cost of electricity depends more on the financial appreciation of the investment (capital recovery factor and depreciation time) than on the cost of avoided CO_2 and NO_x emissions. The avoided CO_2 and NO_x emissions account for a decrease in the electricity price of $0.004 per kWh. In other words, the capitalised avoided emissions correspond to a capital cost benefit of $200 per kW for the SOFC system. At current costs, this is insignificant, but it may tip the balance in the future market penetration.

13.5 SOFC System Demonstrations

Table 13.2 lists the major SOFC system manufacturers worldwide; this list does not include research institutes, universities, and manufacturers of solely ceramic components. Many of these manufacturers have built and tested SOFC stacks in their own facilities. Only Siemens Westinghouse, Sulzer Hexis and a few other companies have built fully integrated SOFC systems, and operated them at customer sites; these systems are described in this section.

13.5.1 Siemens Westinghouse Systems

Since 1986, Siemens Westinghouse has tested at least 12 fully integrated SOFC systems on customer sites, ranging from 0.4 kW to 220 kW [47]. The most recent demonstrations were a 100 kW atmospheric SOFC system, operated in Westervoort, the Netherlands, by a consortium of Dutch and Danish energy companies and a 220 kW SOFC/GT hybrid system, installed and tested at the National Fuel Cell Research Center on the campus of the University of California-Irvine for Southern California Edison [48–51]. These two systems are discussed below.

13.5.1.1 100 kW Atmospheric SOFC System

The 100 kW atmospheric SOFC system layout was shown in Figure 13.5; Figure 13.10 shows a photograph of the system. This system delivered electricity to the Dutch grid and hot water to the local district heating system. The SOFC stack consisted of 1152 tubular cells, in 48 cell bundles of 24 cells each, with eight cells in electrical series and three in parallel. Each cell had an active area of 834 cm^2 and produced approximately 110 W AC at nominal power conditions. System operation started in February 1998. In June 1998, the system was shut down because of observed voltage degradation from the stack. Inspection revealed a broken baffle between the depleted fuel plenum and the combustion zone, fuel leakage around the active stack area, and separation of some nickel felt

Table 13.2 SOFC manufacturers and status of their technology [13.46]

Manufacturer	Country	Achieved	Year	Attributes and status
Acumentrics Corp.	USA	2 kW	2002	Microtubular SOFCs, 2 kW systems for uninterruptible power
Adelan	UK	200 W	1997	Microtubular, rapid start-up and cyclable
Ceramic Fuel Cells Ltd	Australia	5 kW 25 kW	1998 2000	Planar SOFC, laboratory stack testing, 600 operating hours for 5 kW stack, developing 40 kW fuel cell system
Delphi/Battelle	USA	5 kW	2001	Developing 5 kW units based on planar cells
Fuel Cell Technologies (with Siemens Westinghouse Power Corporation)	Canada	5 kW 2 kW	2002 2002	5 kW prototype SOFC under test, 40% electrical efficiency. Several field trials planned in Sweden, USA, Japan, etc.
General Electric Power Systems (formerly Honeywell and Allied Signal)	USA	0.7 kW 1 kW	1999 2001	Planar SOFCs, atmospheric and pressurised operation, laboratory stack testing, developing atmospheric and hybrid systems
Global Thermoelectric	Canada	1 kW	2000	Planar SOFCs, 5000 hours fuel cell test
MHI/Chubu Electric	Japan	4 kW 15 kW	1997 2001	Planar SOFC, laboratory stack testing, 7500 operating hours
MHI/Electric Power Development Co.	Japan	10 kW	2001	Tubular SOFC, pressurised operation, 10 kW laboratory testing for 700 hours
Rolls-Royce	UK	1 kW	2000	Planar SOFC, laboratory testing, developing 20 kW stack for hybrid systems
Siemens Westinghouse Power Corporation	USA	25 kW 110 kW 220 kW	1995 1998 2000	Tubular SOFC, several units demonstrated on customer sites. More than 16,000 single stack operating hours, first hybrid SOFC demonstration
SOFCo (McDermott Technologies and Cummins Power Generation)	USA	0.7 kW	2000	Planar SOFC, laboratory testing, 1000 operating hours, developing 10 kW versatile SOFC unit
Sulzer Hexis	Switzerland	1 kW	1998–2002	Planar SOFC, field trials of many units
Tokyo Gas	Japan	1.7 kW	1998	Planar design, laboratory testing
TOTO/Kyushu Electric Power/Nippon Steel	Japan	2.5 kW	2000	Tubular SOFC, laboratory testing, developing 10 kW system for 2005

Figure 13.10   Siemens Westinghouse 100 kW CHP system in Westervoort, the Netherlands. (Photograph courtesy of EnergieNed.)

Table 13.3   Target and actual performance of the Siemens Westinghouse 100 kW SOFC system in Westervoort, the Netherlands [48–51]

	Nominal performance	
	Target	Actual
Average cell current (A)	150	167
Average cell voltage (V)	0.71	0.66
SOFC generator DC power (kW)	123	127
Net AC system power (kW)	103	109
Thermal power (kW)	54	64
Electrical efficiency (%)	47	46
Thermal efficiency (%)	25	27
Operating hours[a]	–	12,600

[a] Excluding 4035 hours before modification of the SOFC stack in 1999, and more than 4000 operating hours in Essen, Germany.

pads used to connect the cells together. After repair, the SOFC system was restarted in February 1999 and operated, without any measurable performance degradation, until the contractually agreed end date in November 2000. Table 13.3 shows the target and the actual performance for this system.

During the second operating period, the system underwent nine thermal cycles. The most promising of this demonstration period was the fact that the SOFC stack hardly showed any degradation at all. After the demonstration period ended in the Netherlands, the system was transferred to Essen, Germany, and restarted in August 2001, where it was run for more than 4000 operating hours, again without any significant performance degradation. This system has proven the technical feasibility of atmospheric SOFC systems for power generation; however, their high cost remains a barrier to widespread commercialisation.

13.5.1.2 220 kW Pressurised SOFC/GT Hybrid System

The main components of the 220 kW hybrid SOFC/GT system (Figure 13.11) are an SOFC stack, similar to the one in the Westervoort unit, but in a pressure vessel and a modified 75 kW Ingersoll-Rand microturbine. The micro turbine is a recuperated, double-axis system with neither inter-cooling nor reheating. The Site Acceptance Test started in May 2000. After 154 hours of operation, a power lead failure caused an automatic shutdown. Inspection disclosed that fuel, bypassing the stack, burned outside the stack and overheated the power lead. After repair, the system was restarted in January 2001 and operated for 514 hours. A low voltage in one of the stack rows required a second shutdown. The system performance during this operating period is shown in Table 13.4. There is a significant difference between the targeted and the actual performance; this is

Figure 13.11 Siemens Westinghouse 220 kW SOFC/GT hybrid system. (Photograph courtesy of Siemens Westinghouse.)

Table 13.4 Target and actual performance of the Siemens Westinghouse 220 kW SOFC/GT hybrid system at the National Fuel Cell Research Center, California [48–51]

	Nominal performance	
	Target	Actual[a]
Average cell current (A)	267	234
Average cell voltage (V)	0.61	0.64
SOFC generator DC power (kW)	187	172
Turbine AC power (kW)	47	22
Net AC system power (kW)	220	181
Electrical efficiency (%)	57	52
Operating hours	3000	667

[a] During the site acceptance test.

mainly a result of the SOFC's need for at least 1000 operating hours to reach full performance. Furthermore, the nominal power of the microturbine was too large for this system, and one of the cell rows showed abnormal voltage degradation. This system demonstrated the concept of integrating a pressurised SOFC stack with a turbine; however, much additional development is required to produce a viable, reliable SOFC/GT hybrid system.

Based on the above two systems and earlier operating experience, Siemens Westinghouse plans to develop two to three types of commercial SOFC systems in the 250 kW to 1 MW range [13], although the 1 MW hybrid system demonstrations are having difficulties finding a suitable, commercially available gas turbine. The near-future planned demonstrations include 250 kW size atmospheric SOFC systems for Kinectrics Inc. (Canada), BP (Alaska), Norske Shell (Norway) and Stadtwerke Hannover (Germany); 300 kW size SOFC/GT hybrid systems for RWE Energie (Germany) and Edison SpA (Italy); and possibly 1 MW size SOFC/GT hybrid systems for the US Environmental Protection Agency's Environmental Science Center in Maryland (USA) and for EnBW (Germany). The majority of these customers are either electric utility companies or oil companies. At the current high system cost, they represent the early market of visionaries. However, the potential for a significant cost reduction offers possibilities for entry of SOFC systems into the mainstream market.

13.5.1.3 Other Systems

Siemens Westinghouse, together with several partners, also plans to develop a 3–10 kW size CHP system for residential and other distributed power generation applications. Figure 13.12 illustrates a 5 kW size unit, about the size of a

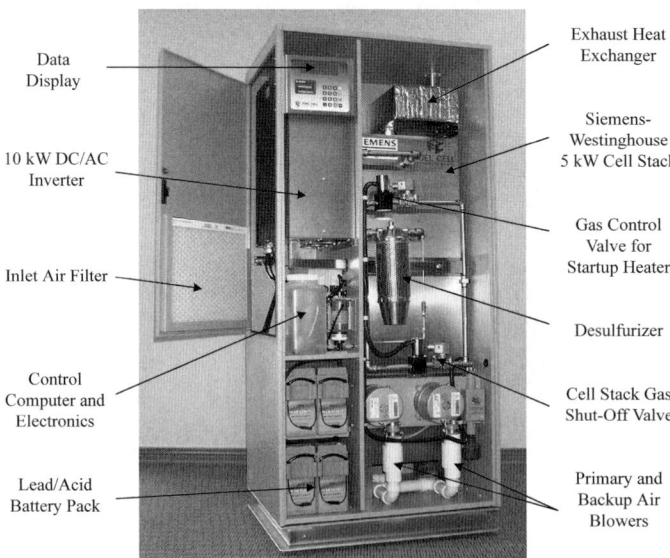

Figure 13.12   Fuel Cell Technologies' 5 kW system, built using Siemens Westinghouse tubular cell stack. (Courtesy of Fuel Cell Technologies.)

refrigerator, that Fuel Cell Technologies of Kingston, Ontario, Canada, is developing using Siemens Westinghouse cell stack. The early prototype units employ tubular cell stacks; a few of such units have been produced and delivered to customer sites; however, at present, their cost is high. The later units are expected to use flattened ribbed cells [52] and direct internal reformation of natural gas on the anode surface to drive down the system cost to about $1000 per kW.

13.5.2 Sulzer Hexis Systems

Figure 13.13 shows a 1 kW residential SOFC unit developed and marketed by Sulzer Hexis of Switzerland. Its SOFC stack consists of 70 circular planar cells separated by connecting pieces that serve as current collector, an air duct, a fuel duct and a heat exchanger. Low-pressure desulphurised natural gas is fed to the stack's inside. At the stack circumference, unreacted natural gas and air combust together to generate additional heat. A steam reformer is thermally connected to the stack. The system is equipped with a condensing boiler to generate additional heat. Other system components are a heat storage tank and

Figure 13.13   Sulzer Hexis 1 kWe residential unit based on planar SOFCs (A: thermal insulation; B: planar SOFC stack; C: heat exchanger; D: heat storage tank; E: control; F: auxiliary burner; G: DC/AC converter; H: gas desulphurisation unit; I: water treatment; J: exhaust).

heat exchanger, an inverter, an ion exchanger and a power management control system. Table 13.5 presents performance data of the Sulzer Hexis system. Laboratory stack tests have demonstrated 12,000 hours of operation with negligible degradation.

Table 13.5 Performance of Sulzer Hexis 1 kW residential CHP system [53,54]

	Performance
Average cell current (A)	27
Average cell voltage (V)	0.56
SOFC stack DC power (kW)	1.1
SOFC stack heat output (kW)	2.5
Electrical efficiency (%)	25–30[a]
Total (electric plus thermal) efficiency (%)	85
Operating hours	6400–15,700

[a] 40–50% at 0.5 kW.

Sulzer Hexis started its SOFC development in 1989 [53,54]. In the period 1989–1997, two proof-of-concept systems were demonstrated in Winterthur, Switzerland, at Sulzer Hexis facilities and in Dortmund, Germany (Dortmund Energy and Water utility company DEW). This proof-of-concept phase was followed by a field test on six customer sites in Switzerland, Japan, the Netherlands, Spain and Germany. Together, these field test units accumulated 65,600 operating hours. The field test phase was finished in 2000 and currently the market entry phase has started. Sulzer Hexis has orders for more than 400 of its HXS 1000 Premiere fuel cell units. The customers are utility companies, mainly located in Germany, Austria and Switzerland. Systems will be delivered from 2003 onward [55]. Sulzer Hexis expects to sell 10,000 units in 2005 and to increase annual sales to 260,000 units by 2010.

13.5.3 SOFC Systems of Other Companies

Acumentrics Corporation of Westwood, Massachussetts, USA, has produced several 2 kW size fully integrated rapid-start SOFC systems using microtubular cells, and delivered them to prospective customers. These systems operate on natural gas, and the intended application is as uninterruptible power supplies for broadband, computer and telecom backup.

Several other manufacturers are developing SOFC systems based on planar SOFCs. Figure 13.14 shows a photograph of a portable 50 W system, developed by GE Power Systems (formerly Honeywell) that incorporates a planar SOFC stack and other components for a self-contained power unit [56]. GE Power Systems also plans to produce larger size atmospheric and hybrid SOFC systems using planar cells.

Global Thermoelectric of Calgary, Canada, has also produced a few 2 kW size prototype residential SOFC units using planar cells for operation on natural gas or propane [57].

Systems and Applications 387

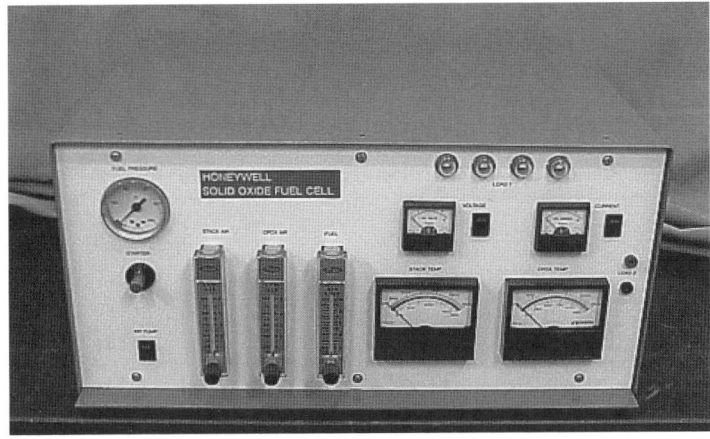

Figure 13.14   A 50 W portable unit employing planar SOFCs built by GE Power Systems [56].

Delphi/Battelle designed and built an initial 5 kW auxiliary power unit (APU) using four planar cell stacks, each 1.5 kW, from Global Thermoelectric, as is schematically illustrated in Figure 13.6. A photograph of such a unit is shown in Figure 13.15. This unit was installed in the spare tire well in the trunk of a luxury automobile and operated on gasoline. Though the testing was successful, this initial unit did not meet the specifications. An improved version (Figure 13.16) with much lower weight and volume has now been designed and built using Delphi/Battelle twin 30-cell stack. Further developments on such units are expected to decrease the start-up time and reduce cost.

Figure 13.15   A 5 kW prototype auxiliary power unit built by Delphi/Battelle using planar cells [58].

Figure 13.16 Improved version of Delphi/Battelle 5 kW APU with reduced mass and volume [59].

13.6 Summary

Solid oxide fuel cell technology is very promising because of its intrinsic simplicity and efficiency. Several markets for SOFC systems have been identified including residential, commercial and industrial CHP, distributed generation, auxiliary power units for the transportation sector, and portable power. Benefits of SOFC technology are consistent with current trends such as liberalisation of energy markets, growing environmental concerns and shift towards distributed utilities. The most promising features of SOFC systems are high efficiencies, fuel flexibility and negligible harmful emissions like particulate matter, oxides of nitrogen, oxides of sulphur, unburned CO and hydrocarbons. Competitive systems, e.g. gas engines, gas turbines and combined cycle units, are setting the base-line for economic and technical specifications. SOFC hybrids are better than existing technologies with respect to electrical efficiency, part-load efficiency and emissions; however, their control speed, control range, investment cost and lifetime have yet to be fully established. The longer start-up time of SOFC systems compared to gas engines is a clear disadvantage for certain market applications.

High SOFC system cost is the principal obstacle to successful commercialisation. A capital cost of $1000–$1500 per kW is very often quoted for SOFC systems to be competitive, but current capital costs are considerably higher although these costs are continually coming down as the development proceeds. Several manufacturers aim at development of small systems (1–10 kW) with shorter lead times and a versatile market. The US DOE SECA programme projects cost reduction to $800 per kW in 2005 and $400 per kW in 2010.

Many companies are involved in developing and testing SOFC systems. Two of them, Siemens Westinghouse Power Corporation and Sulzer Hexis, have tested many units at customer sites and appear closest to commercialisation. Other manufacturers are still testing early prototype units at their own test facilities. The majority of customers are energy companies: gas, electricity, or oil.

The major obstacle towards commercialisation is the *chasm* between the visionaries and the technology enthusiasts in energy companies, oil companies and governmental bodies on one hand and the pragmatists and conservatives on the other. Crossing this *chasm* is the main challenge for successful SOFC commercialisation which requires reducing system costs, demonstrating system reliability and lifetime, and identifying niche markets.

References

[1] D. Ghosh, M. E. Pastula, R. Boersma, D. Prediger, M. Perry, A. Horwarth and J. Devitt, in *2000 Fuel Cell Seminar Abstracts*, Courtesy Associates, Washington, DC, 2000, pp. 511–514, .
[2] G. A. Moore, *Crossing the Chasm*, HarperBusiness, 1991.
[3] *International Energy Outlook 2020*, US Department of Energy, 2000.
[4] L. A. Ruth and V. K. Der, in *Vision 21 Program Plan, Clean Energy Plants for the 21st Century*, Federal Energy Technology Center, USA, April 1999.
[5] *Ozone Transport Commission Distributed Generation Initiative*, State Service Organisation, March 2001.
[6] K. E. Yeager, *Electricity Technology Development for a Sustainable World: Bridging the Digital Divide*, World Energy Council 18th Congress, Buenos Aires, Argentina, 21–25 October 2001.
[7] L. J. M. J. Blomen and M. N. Mugerwa, *Fuel Cell Systems*, Plenum Press, 1993.
[8] R. J. F. van Gerwen and J. H. C. van der Veer, *Analysis of the Market Potential for Solid Oxide Fuel Cells in Europe*, KEMA-report no. 21065-TFO 92-3019, August 1992.
[9] A. J. Mesland and R. J. F. van Gerwen, *Global Estimate of the Market Potential for Combined Solid Oxide Fuel Cell/Gas Turbine Units in Western Europe*, KEMA-report no. 41979-KES/MLU 96-3228, November 1996.
[10] J. L. Del Monaco, *Research Needs in Distributed Generation*, National Science Foundation Workshop, Arlington, USA, 16 November 2000.
[11] L. Blum, *European Fuel Cell News*, **7**(2) (June 2000) 20–22.
[12] R. K. Dixon, *US DOE Perspectives on Distributed Energy and CHP*, Second International CHP Symposium, Amsterdam, The Netherlands, 9–10 May 2001.
[13] R. J. F. van Gerwen, *Potential for High-efficiency Hybrid Fuel Cell Systems Based on Solid Oxide Fuel Cells* (in Dutch), Netherlands Agency for Energy and Environment (Novem), report no. 50161016-KPS/SEN 01-3048, March 2002.
[14] J. Morrison, *Siemens Westinghouse Solid Oxide Fuel Cell Program*, United States Advanced Ceramics Association Annual Meeting, College Park, USA, 1 November 2001.

[15] J. A. Ciesar, *Hybrid Systems Development by the Siemens Westinghouse Power Corporation*, US Department of Energy Natural Gas/Renewable Energy Hybrid Workshop, 7 August 2001.
[16] M. Brown, *The Global Market Opportunity – and How it can be Achieved*, Second International CHP Symposium, Amsterdam, The Netherlands, 9–10 May 2001.
[17] J. Douglas, Solid futures in fuel cells, *EPRI Journal*, (March 1994) 6–13.
[18] Fuel cells look good under pressure, MPS review, *Modern Power Systems*, (June 1996) 23–26.
[19] M. Whiteley et al. *The Future of CHP in the European Market – The European Co-generation Study*, EU project no. 4.1031/P/99-169, 2001.
[20] T. E. Hoff and M. Cheney, *The Energy Journal*, **21**(3) (2000) 113–127.
[21] Westervoort SOFC: the road to commercialisation, *Modern Power Systems*, May 1998, pp. 29–32.
[22] *A High Efficiency PSOFC/ATS-Gas Turbine Power System*, Final Report DE-AC26-98FT40455, by Siemens Westinghouse Power Company for the US Department of Energy, February 2001.
[23] A. J. Appleby and F. R. Foulkes, *Fuel Cell Handbook*, Van Nostrand Reinhold, 1989.
[24] *EDB/ELSAM 100 kWe SOFC Field Test*, Executive Summary of Final Report, January 2001, EDB project no. SOFC 800, ELSAM project no. 400.3.1, Novem project no. 219.301-00.50.
[25] H. W. Venderbosch, *Survey Desulphurisation of Natural Gas for Fuel Cell Systems* (in Dutch), KEMA-report no. 34051-FPP 95-4612, 1996.
[26] *Pressurized Solid Oxide Fuel Cell/Gas Turbine Power System*, Final Report DE-AC26-98FT40355 by Siemens Westinghouse Power Company for the US Department of Energy, February 2000.
[27] J. H. Hirschenhofer et al., *Fuel Cell Handbook*, Fourth Edition, for US Department of Energy, November 1998.
[28] K. T. Chau, Y. S. Wong, C. C. Chan, *Journal of Energy Conversion and Management*, **40** (1999) 1021–1039.
[29] W. Winkler, *European Fuel Cell News*, **8**(4) (January 2002), 9–13
[30] S. Mukerjee et al., in *2000 Fuel Cell Seminar Abstracts*, Courtesy Associates, Washington, DC, 2000, pp. 530–533.
[31] E. Achenbach and E. Riensche, *Journal of Power Sources*, **52** (1994) 283–288.
[32] P. N. Dyer, R. E. Richards and S. L. Russek, in, *Solid Oxide Fuel Cells VI*, ed. S. C. Singhal and M. Dokiya, The Electrochemical Society Proceedings, Pennington, NJ, PV99-19, 1999, pp. 1173–1176.
[33] F. P. F. van Berkel, G. S. Schipper and G. M. Christie, in *Solid Oxide Fuel Cells VI*, ed. S. C. Singhal and M. Dokiya, The Electrochemical Society Proceedings, Pennington, NJ, PV99-19, 1999, pp. 1177–1184.
[34] W. K. Heidug, M. R. Haines and K. J. Li, *Economical Carbon Dioxide Recovery for Sequestration from a Solid Oxide Fuel Cell Power Plant*, Fifth International Conference on Greenhouse Gas Technologies, Australia, 2000.

[35] *Hybrid Fuel Cell Technology Overview*, National Energy Technology Laboratory, no. DOE/NETL-2001/1145, May 2001.
[36] G. J. William *et al.*, *Design Optimisation of a Hybrid Solid Oxide Fuel Cell and Gas Turbine Power Generation System*, Alstom Power Technology Centre, ETSU project no. F/01/0021/REP, 2001.
[37] J. Ouwehand *et al.*, *Applied Energy Technology* (in Dutch), Academic Service Schoonhoven.
[38] S. Campanari and E. Macchi, *Full Load and Part Load Prediction for Integrated SOFC and Microturbine Systems*, ASME turbo expo, 7–10 June 1999, Indianapolis, USA, paper 99-GT-065.
[39] C. J. Bartels, *Development of a Part Load Model for a Solid Oxide Fuel Cell System* (in Dutch), thesis, Delft University of Technology, report no. EV 2066, May 2001.
[40] J. P. Strakey, *The Solid State Energy Conversion Alliance*, Third Annual SECA Workshop, Washington, DC, 21–22 March 2002.
[41] *Assessment of Planar Solid Oxide Fuel Cell Technology*, Arthur D. Little report for DOE FETS, reference 39463-02, October 1999.
[42] J. A. F. de Ruijter, *Results of Calculations with the KEMA Greenhouse Energy Model*, January 2001.
[43] J. A. Edmonds, P. Freund and J. J. Dooley, in *Proceedings of the Fifth International Conference on Greenhouse Gas Control Technologies*, eds. D. Williams, B. Durie, P. McMullan, C. Paulson and A. Smith, CSIRO Publishing, Australia, 2001, pp. 46–51.
[44] H. Audus, in *Proceedings of the Fifth International Conference on Greenhouse Gas Control Technologies*, eds. D. Williams, B. Durie, P. McMullan, C. Paulson and A. Smith, CSIRO Publishing, Australia, 2001, pp. 91–96.
[45] B. A. Folsom and T. J. Tyson, *Combustion Modification – An Economic Alternative for Boiler NO_x Control*, GE Power Systems, report no. GER-4192, April 2001.
[46] *Fuel Cell Technology Announces Successful Prototype Operation*, press release Fuel Cell Technologies Ltd, 21 May 2002.
[47] S. E. Veyo, *Tubular SOFC Field Unit Status*, IFCC Technical Conference, Japan, 1–2 December 1999.
[48] J. Leeper, *220 kWe Solid Oxide Fuel Cell/Microturbine Generator Hybrid Proof of Concept Demonstration Report*, California Energy Commission, report no. P600-01-009, March 2001.
[49] C. A. Forbes, *Status of Major Tubular Solid Oxide Fuel Cell Demonstrations*, PowerGen International, Las Vegas, USA, 11–13 December 2001.
[50] H. Kabs *et al.*, *Experience with SOFCs for Decentralised Power Generation*, PowerGen Europe, Brussels, Belgium, 29–31 May 2001.
[51] D. Smith, Large utilities turn to solid oxide systems, *Modern Power Systems*, German Supplement, 2000, pp. 27–30.
[52] N. F. Bessette, *Siemens SOFC Technology on the Way to Economic Competitiveness*, Power Journal (Siemens Power Generation Group), January 2001 (http://www.pg.siemens.de/download).

[53] Fuel cell future, the Sulzer perspective, *Modern Power Systems*, May 2001, pp. 69–73.

[54] M. Schmidt and R. Diethelm, *Sulzer Hexis SOFC System: First Results of 3-Year Field Test*, IFCC Technical Conference, Japan, 1–2 December 1999.

[55] Fuel cell systems for Switzerland from Sulzer Hexis, *European Fuel Cell News*, **9**(3) (November 2002) 29.

[56] N. Minh, A. Anumakonda, R. Doshi, J. Guan, S. Huss, G. Lear, K. Montgomery, E. Ong and J. Yamanis, in *Solid Oxide Fuel Cellls VII*, eds. H. Yokokawa and S. C. Singhal, The Electrochemical Society Proceedings, Pennington, NJ, PV2001-16, 2001, p. 190.

[57] Global Thermoelectric delivers prototype residential solid oxide fuel cell system to Enbrigde, *European Fuel Cell News*, **8**(3) (November 2001) 27.

[58] S. Mukerjee, M. J. Grieve, K. Haltiner, M. Faville, J. Noetzel, K. Keegan, D. Schumann, D. Armstrong, D. England, J. Haller and C. DeMinco, in *Solid Oxide Fuel Cells VII*, eds. H. Yokokawa and S. C. Singhal, The Electrochemical Society Proceedings, Pennington, NJ, PV2001-16, 2001, p. 173.

[59] S. Mukerjee *et al.*, in *Solid Oxide Fuel Cells VIII*, eds. S. C. Singhal and M. Dokiya, The Electrochemical Society Proceedings, Pennington, NJ, PV2003-07, 2003, p. 88.

Editorial Index

A
ab initio electronic structure calculations 325, 326
ABB 216
activated carbon 152, 352, 371
activation energy 272, 277–282, 300, 304, 318, 325
activation polarisation 230–231, 237–250, 257, *see also* polarisation
Acumentrics Corporation 221, 223, 381, 386
Adelan 12, 15, 221, 381
afterburning 44, 370–374, 377
aging effect 90
air *see* excess air; oxidation
Allied Signal 96, 180, 277, 280
alternate geometry cells 217
alumina additives 34, 90, 101, 183, 184
alumina composite electrodes 29
alumina impurities 6
alumina porous support tubes 216
alumina test blocks 267
alumina-doped calcium titanate electrolytes 97
ammonia impurities in fuels 152
ammonium double nitrates 30
AMTEC process 78
anions 28, 35
anode-supported cells 42, 202, 208, 246, 302, 315
　anode structures 155, 163
　cathodes 142
　fabrication 96, 206
　polarisations 232–233, 235–237, 252, 257, 273, 302
　power densities 208
　testing 265, 269–270, 277, 279–282
anodes 1, 8–9, 149–169
　attributes/requirements 44, 149, 150–151, 302
　characterisation and testing 156–164, 283, 284
　composite 165–8, 169, 212, *see also* cermet anodes
　contact material interaction 188–189
　direct electrocatalytic oxidation of hydrocarbons at anode 165, 166, 335, 346, 353
　early developments 3, 37
　electrically conducting oxides 167–169, 346, 349, 353, 356
　electrolyte reactions and compatibility 58, 149, 150
　fabrication 8, 149, 153–155, 168, 212, 213, *see also* sintering and fuel/fuel impurities 16, 149, 152, 157, 250, 337, 339–341, 352–354, 356, *see also* coking and carbon formation
　interconnect reactions and compatibility 139, 149, 150, 181, 183, 186
　microtubular SOFCs 219–221, 223
　modelling 158–160, 163–164, 168, 318–325
　polarisation/overpotential 35, 135, 149, 150–151, 155–159, 161–165, 230, 232, 233, 236, 242–250, 275, 300
　stresses 315–317
　see also electrodes; nickel anodes
APM-Kanthal interconnects 186
applications 2, 18–19, 32–34, 39, 363–389, *see also specific applications*
area-specific resistance (ASR) 261, 272–282, 286
Argonne National Laboratory 42, 96, 179
Arrhenius plots 97, 99, 104, 105, 107, 158, 281
Aspen Plus 314
ASR *see* area-specific resistance (ASR)
association (binding) enthalpy 87, 88
atomic modelling 108, 325–326
austenitic steels 186
Australia 42, 381
Austria 386
automobile applications *see* vehicles
autothermal reforming 165, 337, 345–346, 350, 356
auxiliary burners in hybrid systems 75, 76
auxiliary power units (APUs) 1, 2, 15, 16, 18, 363, 364, 368, 370, 373, 387
Avogadro constant 59, 233

B

balance of plant (BOP) 14, 66–67, 165, 261–262, 370–372, 378
  modelling 314–315
ball milling 8, 221
barium doping of LSGM 104
barium oxide electrolytes 107–108, 110–112
Battelle 34, *see also* Delphi/Battelle
battery replacement 18–19, 363, *see also* auxiliary power units
bead milling 8, 221
BET technique 301
binding (association) enthalpy 87, 88
biogas 6, 15, 152, 334, 337, 354–356
bismuth oxide electrolytes 83, 100, 107–109, 244
blistering 317, *see also* delamination
boron sintering promotion 179
bottled gas 336, 343, 348, 355, *see also* propane
Boudouard reaction 30, 165, 307, 342, 347
Brown Boveri 41
brownmillerites 83, 106–108
butane 336, 343, 348, 355
Butler-Volmer equation 245, 246, 249, 254, 300, 305, 320

C

calcia doped ceria-based electrolytes 93
calcia doped lanthanum chromite 12, 139, 168, 174, 177, 179
calcia doped lanthanum manganite 123, 125, 134
calcia doped lanthanum-based electrolytes 97, 99–106
calcia doped yttrium chromite 174, 177
calcia doped zirconia 35, 84, 210
calcium in electrically conducting oxides 111, 353
calcium titanate electrolytes 96, 97
Canada 381, 384, 385
capacitors 234
carbon
  early developments 3, 23, 24, 25, 26, 27
  formation *see* coking and carbon formation
carbon dioxide
  carbon dioxide-capturing SOFC system 373, 374
  emissions 19, 337, 366, 372, 374, 379
  reactions 30, 110, 112, 165
  for reforming of hydrocarbons 342, 343, 354
carbon monoxide fuel 3, 15, 26, 152, 156, 164, 165, 168, 250, 333, 355
  oxidation within SOFC 44, 60, 306, 333
carbonate electrolyte 18
carbonyl sulphide 351
Carnot cycle 70–74, 78
cascading (series connection) 11, 32, 40, 66
casting 8, *see also* tape casting
catalysis 15–16, 44, 163, 165–166, 220, 325, 333, 352
  catalytic afterburners 44, 370–374, 377
  reforming of hydrocarbons 341–346, 347–352, *see also* electrocatalytic oxidation
  sulphur removal 351, 352
  see also specific catalysts
cathode-supported cells 142, 202, 211, 212, 218, 232, 233, 237, 252, 257, 302, 316
cathodes 1, 8, 10, 119–143
  characterisation and testing 281–286
  composite cathodes 136–138, 246, 279, 336, *see also* LSM/YSZ composite cathodes
  early developments 28, 37, 41
  electrolyte reactions and compatibility 134–136, 139–142, 187, *see also* lanthanum zirconate formation
  fabrication 133–134, 142, 246–247, 281, *see also* sintering and fuel/fuel impurities 336
  interaction with contact materials 188
  interconnect reactions and compatibility 120, 138–143, 181, 184–188
  microstructure 136, 143, 246–247, 257, *see also* electrodes, microstructure for microtubular SOFCs 219–223
  modelling 253, 318–325
  polarisation 35, 44, 133–135, 139, 156, 187, 214, 218, 230, 232–242, 257, 273, 300
  properties 119–129, 302
  reactions 10, 58, 119, 126–136
  stresses 315–317
  see also electrodes; lanthanum based cathodes
cations 28, 34
cell geometries 217–219, 252, 273, 378
  gas flow configurations 199–201, 207
  modelling 14, 297, 315, 319
  see also reference electrodes; test cells
Ceramic Fuel Cells Ltd. 13, 42, 185, 381
ceria
  additives/catalysts 10, 16, 90, 107, 112, 168, 185
  in composite electrodes 9, 37, 137, 167, 168, 170, 244, 341, 349, 353
ceria based electrolytes 83–85, 92–93, 100, 110, 112, 325, 353
  cathodes for use with 120, 129, 137–138, 187
  early developments 28, 30, 35, 44
  YSZ coating 93, 112
cermet anodes 9, 103, 149, 151, ;152, 168, 212, 219, 220, 352
  characterisation and testing 156–164
  fabrication 149, 153–155, 221, 222
  microstructure 153–154, 159, 168, *see also* electrodes, microstructure
  operation with fuels other than hydrogen 164, 165, 169, 339, 340–341

polarisation 232, 249–250
resistances and apparent activation energy 278
see also coking and carbon formation; nickel anodes
charge transfer process 237–242, 245, 246, 251, 255
anode 150, 151, 158, 160, 161, 250
charge transfer resistance 242
chemical industry applications 363, 373
chemical vapour deposition 7, 10, 205, *see also* EVD
Cheng cycle 80
chromium 12, 44
alloys 181–186, 207
chromite interconnects 173–180, 189, *see also* lanthanum chromite interconnects; yttrium chromite interconnects
chromites in gas separators 168
chromium doped electrolytes 104
poisoning/corrosion 120, 140–142, 181, 183–187, 262
see also lanthanum chromite
Chubu Electric 381
co-extrusion 11, 12, 221
coal gasification 18, 32, 37, 151, 181, 337, 363
coating methods 10, 206–207, *see also* slurry coating processes
coatings
coatings and getters for interconnects 12, 142, 182, 187–188
of YSZ for ceria based electrolytes 93, 112
cobalt
in anodes 151, 166, 168, 353
catalyst for sulphur removal 352
in cathodes 137–138, 143, *see also* lanthanum cobaltites
cobalt-doped lanthanum ferrite 137, 143
cobalt-doped LSGM 104
in lanthanum chromite interconnects 174, 179–180
cogeneration *see* combined heat and power (CHP)
coking and carbon formation 9, 16, 75, 149, 150, 336, 352–354, 377
determination method 224, 350–351
in microtubular SOFCs 223, 224
see also Boudouard reaction; reforming of hydrocarbons
combined heat and power (CHP) 334, 337, 338, 364, 369–372
industrial use 18, 337, 338, 365–366
residential use 2, 12, 78, 223, 337, 338, 363, 369–372, 380–382, 384–386
compressors 374, 377
computational fluid dynamics (CFD) 296, 297, 312
computer systems applications 223, 363

concentration polarisation 230–231, 233–237, 251, 257, *see also* polarisation
conductivity 24–27, 34, 177, *see also* electronic conductivity; ionic conductivity
conservation of momentum 295, 296
contact materials 188–189, 190, 265–267, 271
control systems 14, 376–378, 386
cooling 78, 79, *see also* excess air
copper in anodes 166, 170, 249, 349, 353
copper doping of LSGM 104
corrosion *see* chromium poisoning/corrosion
costs 363, 367, 368, 384, 388
fabrication 41, 378
interconnects 174, 181, 182, 202
maintenance 378
operational 378, 379
planar cells and stacks 224, 378
and process design modelling 314, 315
SOFCs compared with heat engines 17, 18
tubular cells and stacks 212, 217, 224
zirconia 5–6
cracking in microtubular SOFCs 219, *see also* planar cells and stacks, cracking
creep resistance in LSGM 101
Cummins Power Generation 381
current density 231, 233, 234, 249–251, 261, 273, 286
current density-voltage (I-V) curves 253, 270–272, 276, 282, 283, 286, 299–303
exchange current density 300, 304
and gas leaks 276
transfer current density 301
current measurement techniques, 239, *see also* alternating current; direct current
current thermodynamics 58, 62, 66–69, 80

D

defect chemistry 84, 85, 97, 108, 121–125, 241, 244
deflocculants 95
delamination 199, 315, 317
Delphi/Battelle 13, 15, 381, 387
Denmark (Risø National Laboratory) 277–281
design issues 12–14, 42–44, 93–96, 180, 197–225, 314
electrodes 319, *see also* geometries
hybrid systems 77–80, 314–326
interconnects 12
microtubular SOFCs 12, 13, 44, 219–225
planar SOFCs 13–14, 42, 95, 180, 197–208
process design modelling 314, 315
tubular SOFCs 13, 14, 44, 93–95, 142, 180, 197, 210–219
desulphurisation 152, 351, 371, 378, 379
diesel engines 364, 368
diesel fuel 16, 223, 336, 337, 346, 348, 355
diethyl sulphide 351, 352

diffusion, and polarisation 233–237, 243, 245, 273, 274, 277, 301
dimethyl ether fuel (DME) 336
dimethyl sulphide 351
direct current (DC) measurement methods 156, 168
direct current (DC) to alternating current (AC) inverter 372
doping 2, 5, 6, 10, 85–88, 104–106, 154, *see also specific dopants*
Dörnier 135
dry reforming of hydrocarbons 342, 343, 354
Ducrolloy 181, 182, 186
dysprosium 85, 86, 111

E
ECN 96
efficiency 13–14, 57, 103, 344
  of competing power generation systems 367–368
  in hybrid systems 17–18, 72–77, 81, 368, 372, 376, 379, 380, 388
  modelling 14, 308–315
EIS *see* impedance spectroscopy
Electric Power Development Co. 381
electrical conductivity
  ceria-based electrolytes 92, 93
  contact materials 188
  interconnects 173–177, 181, 186
  lanthanum germanium oxide 110
  perovskites 99, 123–125, 174–177
  proton-conducting oxides 110, 111
  see also electronic conductivity
electrically conducting oxides 346, 349, 353, 356
electricity production markets 365, 367
electrocatalytic oxidation of hydrocarbons at anode 165, 166, 335, 346, 353
electrochemical dissociation of water vapour 37–40
electrochemical machining 182
Electrochemical Society Proceedings 19
electrochemical thermometry 35
electrochemical vapour deposition *see* EVD
electrochemistry
  continuum-level model 294, 299–303, 308
  origins and early developments 23–26, 32
electrode-supported cell designs 232, 233, 243, 252, *see also* anode-supported cells; cathode-supported cells
electrodes
  characterisation and testing 6, 156–164, 262–267, 342
  contact layers 188–9
  early developments 23, 25–27, 37, 41
  electrode-level modelling 294, 299, 301, 311, 312, 318–325
  materials and reactions 8–11, 23, 26, 27

microstructure 135, 142, 153, 159, 234, 235, 244, 245, 301, 303, 319, 323, 324
porosity 234–235, 302
potential and overpotential 2, 30–31, 44, 230, 318, *see also* ohmic resistance/losses; polarisation
reactivity with lanthanum gallate electrolytes 103–104
see also anodes; cathodes; MIEC electrodes; reference electrodes, *and specific materials*
electrolysers 40, 134
electrolytes 1, 83–113
  anode reactions and compatibility 58, 150
  cathode reactions and compatibility 10, 58, 119, 130–138
  early developments 23, 24–40
  fabrication 3, 7–8, 93–96, 204–208, 211, 213
  and gas leakage 286
  interconnect reactions and compatibility 177, 181, 183, 221, 222
  potential and polarisation 232, 252, 263–265, *see also* electrodes; polarisations
  stresses 315–317
  see also thin-film cells; YSZ
electromotive force (Emf) 272–276, 283, 284, 286
electronic conductivity 23, 35, 84, 89, 90, 92, 93, 100, 112, 237
  electrodes 149, 152, 154, 237, 248
  electrolyte in tubular SOFCs 212
  see also electrical conductivity
electrophoretic deposition 207
electrostatic-assisted vapour deposition 207
elementary charge 59
energy balance 67, 296–299, 312, 326
energy market trends 365–367
enthalpy 53, 56–57, 59, 60, 67–69, 71, 80, 87–88, 93, 305
enthalpy effect 139
entropy effect 138, 311
entropy production 56, 60–61, 80
entropy recycling 77
environmental issues 6, 8, 16, 364–367, *see also* greenhouse gas emissions
equilibrium constant 64
equilibrium theory 304–307
Europe
  combined heat and power (CHP) 369
  electricity market 365, 366
  government funding for SOFCs 17
  Proceedings of European SOFC Forums 19
  progress on SOFCs 34, 216
EVD (electrochemical vapour deposition) 7–8, 11, 42, 90, 94–96, 119, 135, 142, 180, 211
excess air 53, 63, 65, 68
EXCO (external cooler) design 78–80
extrusion process 7, 8, 206, 211, 221, *see also* co-extrusion

F

fabrication 135, 221–222, 224
  costs 41, 378
  temperatures 133–135
  see also under specific components
failure probability 315–318
Faraday constant 59, 65, 230–231, 254, 255, 299
ferrite-based cathodes 138, 143
ferritic steel interconnects 11, 182–186, 188, 207
flat plate cells and stacks *see* planar cells and stacks
flattened ribbed cells 217–219, 225, 385
flow configurations 199–201, 207
flow and thermal models 294–299, 308, 312, 326
flue gas thermodynamics 56, 67
  in hybrid systems 69–74, 77, 78, 81
fluorides, for sintering promotion 179
fluorite oxide-perovskite interdiffusion 131, 132
fluorite structured electrolytes 83–93
  and ion conductivity 31, 34, 37, 44, 83–86, 104, 325
  see also ceria-based electrolytes; zirconia-based electrolytes
foil interconnects 185, 186
Forschungszentrum Jülich 185, 277
France 34
free (Gibbs) enthalpy 53, 57, 59, 60, 80, 93, 305
Fuel Cell Technologies 381, 385
fuel utilisation 58, 59, 68, 69, 70, 80, 308, 372, 377
  mixing effects during 53, 59, 62–66, 80
  and testing 261, 269, 272–276, 281
fuels and fuel processing 15–17, 250, 333–356
  additives and impurities 16, 17, 44, 149, 152, 157, 356, *see also* sulphur impurities in fuels
  applications using different fuels 15, 19, 223–224, 333–338, 355–356
  early developments 3, 32, 34
  effects on test procedures and modelling 271, 275, 276, 284, 286, 303
  fossil fuels 366
  fuel sources 152, 334–338
  non-SOFC fuel cells 18, 152, 333, 355
  pretreatments 152
  reforming *see* hydrocarbon fuels
  see also fuel utilisation; hydrocarbon fuels, *and other fuels*
Fuji Electric 186

G

gadolinium 92, 93, 96, 103, 104, 109, 111, 112
galvanic batteries and cells 23–24, 26, 33
gas distribution
  planar cells and stacks 180, 181, 199–201, 207
  tubular cells and stacks 213, 215, 219–221, 223
gas engines 368, 388
gas leakage 261, 271, 283–286, *see also* seals and sealants
gas manifolds *see* manifolds
gas rigs 363
gas separation 168, 173
gas turbine/SOFC hybrid systems 2, 17, 18, 214, 364, 368, 373–379
  design principles 77–80
  modelling 314
  system demonstrations 382–384
  see also hybrid systems
gas turbine/steam turbine combined cycle 368, 369
gas turbines 368
gases
  gas concentration *in situ* determination 31–33
  irreversible mixing 53, 58, 59, 62–66, 80
gasoline 16, 18, 337, 348, 355, 373, 387
General Electric Power Systems 13, 34, 37, 38, 180, 381, 386, *see also* Honeywell
geometries *see* cell geometries
germanium 109
Germany
  demonstrations 382, 384, 386
  early developments 28–31, 34
  electricity markets 366
  field tests 386
  HXS 1000 Premiere fuel cells 386
getterers/gettering 6, 12, 142, 182, 187, 188
Gibbs enthalpy *see* free (Gibbs) enthalpy
glass electrolytes 23, 27
glass sealants 207, 208, 317
Global Thermoelectric 381, 386, 387
gold
  additive in nickel catalysts 349
  contacts and seals in testing equipment 267, 271
  in electrodes 26, 149, 271, 341, 349, 351, 353
Goldschmidt tolerance factor 120, 121, 130
graphite anodes 151
greenhouse gas emissions 19, 337, 366, 372, 374, 376, 379, 388

H

h-plane plot 126
hafnia 6
halide electrolytes 28
halide impurities in biogas 355
hardware design modelling 314, 315
Hastelloy interconnects 186
Haynes interconnects 186
heat capacity 60
heat effects 310, 311

heat engines 2, 17–18, 69–77
heat exchange processes in SOFC 53
heat exchangers 12, 14, 44, 66, 78, 79–80, 202, 371, 385, 386
heat generation rate equations 303–307, 309, 311, 312
heat transfer modelling 296–299
heat treatment temperatures, perovskites 133–134
Hebb-Wagner polarisation technique 248
hexagonal structured oxides 83, 109
HEXIS concept 44, see also Sulzer Hexis
high power density solid oxide fuel cell (HPD-SOFC) 217–219
high-velocity oxygen flame (HVOF) spraying 185
hole and electron conduction
　barium indiate electrolytes 107
　lanthanum based electrolytes 101, 102
　proton-conducting oxides 110, 111
Honeywell 13, 15, see also General Electric Power Systems
HPD-SOFCs 217–219
hybrid systems 2, 74, 76–80, 376, 381, 386, 388
　costs 379
　design principles 77–80
　efficiency 71–77, 388
　modelling 314, 326
　thermodynamics 69–77
　see also gas turbine/SOFC; steam turbine/SOFC
hydrocarbon fuels 53, 152, 334–338
　additives and impurities 16, 17, 152, 355, see also sulphur impurities
　and anode 152, 164–168, 307, 335, 352–354
　early developments 34
　in hybrid systems 73–77, 334
　in microtubular SOFCs 223, 224
　pyrolysis 150, 166, 339, 342
　reforming see reforming of hydrocarbon fuels
　steam reforming 16, 152, 161, 165, 339, 342–347, 351, 354
　testing of different fuels 164, 165, 223, 224
　see also and other fuels; coking and carbon formation; fuels and fuel processing; methane fuel
hydrocarbon sensors 44
hydrodesulphurisation 351
hydrogen, pure 18, 40, 333
hydrogen dissociation 37–40, 157
hydrogen fuel 3, 15, 152, 156, 157, 208, 223, 230, 233, 237, 250, 273
　oxidation within hybrid system 71–73
　oxidation within SOFC 58, 60, 63, 168
hydrogen partial pressures 160
hydrogen sulphide 152, 351
hydroxylated surfaces 163

I
I-V curves see current density-voltage curves
ICI 8
impedance measurements 158, 324
impedance spectroscopy (EIS) 158–162, 165, 168, 239, 251–256, 282, 283
Inconel interconnects 185, 186
indium 37, 106–108, 112
industrial screening tests 271, 272
INEX (intermediate expansion) design 78–80
Ingersoll-Rand 383
interconnect-supported cells 204
interconnects 1, 11, 12, 173–190, 202, 216
　anode reactions and compatibility 150, 183
　cathode reactions and compatibility 120, 138–143, 181, 183–185, 187
　coatings and getters 12, 142, 182, 187, 188
　contact materials 188, 189
　costs 174, 181, 182, 202
　early developments 41
　electrolyte reactions and compatibility 177, 181, 183, 207, 221, 222
　fabrication 3, 4, 11, 12, 174, 179–182, 190, 213
　foil interconnects 185, 186
　metallic 11, 12, 139–142, 173, 174, 181–189, 190, 207
　for microtubular SOFCs 220, 221, 222, 224
　oxide ceramic (perovskite) 8, 11, 12, 37, 139–142, 168, 173–180, 207
　properties 173, 174, 317
　and testing 262
　transpiration experiments 187
interdiffusion 131, 132, 138, 150, 154, 189, 295
intermediate expansion (INEX) design 78–80
intermediate operating temperatures see lower/intermediate operating temperatures
inverters 372, 376, 386
ionic conductivity 23, 237
　early developments 27, 30, 31, 34–35
　electrodes 149, 152, 237, 244–248
　and fluorite structure 31, 34, 37, 44, 83–86, 91, 112
　interconnects 173, 177
　microtubular SOFCs 219
　perovskites and perovskite-related structures 83, 96–110, 126–129
　see also transport number
iron doping of electrolytes 104
iron in electrodes 3, 29, 30, 138, 143, 149, 151, 166
iron in interconnects 181, 182, see also ferritic steel interconnects
isooctane fuel 19
isotope techniques 34, 104, 126–129, 161, 238
Italy 384

J

Japan
  alternative tubular designs 216, 217
  early developments 34
  field tests 386
  government funding for SOFCs 17
  manufacturers 381
Joule (ohmic) heating 310

K

Keele University (UK) 222
kerosene 355
KIEP 96
kinetic resistance 319, 321
Kyushu Electric Power 381

L

lambda sensors *see* oxygen sensors
lamination 180
landfill gas 334, 337, 354
lanthanum
  in anodes 103, 168, 353
  in interconnects 12, 184, 185, *see also* lanthanum chromite interconnects
lanthanum based cathodes 103, 104, 119–143, *see also* lanthanum cobaltite cathodes; lanthanum ferrite cathodes; lanthanum manganite cathodes
lanthanum based electrolytes 96–106, 109, 110, 112
  doping 99–106
  early developments 30, 35, 41
  properties 100–103
  see also lanthanum gallate electrolytes
lanthanum chromite anodes 168, 353
lanthanum chromite interconnects 8, 11–12, 37, 138, 142, 173–80, 207, 213, 222
  coatings 187
lanthanum cobaltite cathodes 9, 10, 119, 126, 128, 129, 131, 137, 138, 142, 187, 188, 243, 244, 247
lanthanum cobaltite contact materials 188
lanthanum ferrite cathodes 120, 138, 143, 187, 244
lanthanum fluoride, for sintering promotion 179
lanthanum gallate electrolytes 44, 96–106, 108, 110, 112, 188, 232
lanthanum manganite anodes 353
lanthanum manganite cathodes 7, 10, 11, 119, 120–143, 184, 185, 187, 188, 220, 239, *see also* LSM
lanthanum manganite interconnect coatings 187
lanthanum manganite-based air electrodes (cathode tubes) 211
lanthanum nitrate, in interconnects 12
lanthanum strontium manganite *see* LSM
lanthanum zirconate formation 10, 119, 120, 130–136
lanthanum-doped barium indiate electrolytes 107, 112
Laplace's equation 320, 324
laser processes 180, 207
lattice structure of perovskites 120–122
Lawrence Berkeley Laboratory 277
leisure applications 363
LHV (lower heating value) 67
lighting devices 2–3, 24–26
lithium 154, 353
lower/intermediate operating temperatures 73, 87, 90–91, 97, 103, 107, 108, 202, 377
  anodes 155, 158, 163
  cathodes 120, 136–137, 257, 282
  early developments 44
  electrode-supported cells 96
  fuel issues 336, 337, 338
  hybrid systems 73, 377, 378
  and metallic interconnects 181
  and polarisation losses 302
  proton-conducting oxides 110–112
LSC *see* lanthanum cobaltite cathodes
LSCF 120, 138, 141, 143, 187
LSF *see* lanthanum ferrite cathodes
LSGM 100–104, 109, 188, 244
LSM 10, 41, 202, 221, 244, 279
  composite cathodes 120, 133–134, 136, 143, 242, 247, 254, 278, 279
  and interconnects 141–142, 182, 187
  properties 120–129
  reactions 132–136, 242, 243, 244

M

McDermott Technologies 13, 381
macrohomogeneous model 322–323
magnesia doped electrolytes 2, 97, 99–106, 112
magnesia doped lanthanum chromite interconnects 8, 11, 174, 177
magnesium ferrite electrodes 29
magnetite cathodes 3
manganese
  doping of electrolyte 104
  in electrodes 23, *see also* lanthanum based cathodes
  in interconnects 184–186
manganite interconnect coatings 187, *see also* lanthanum manganite
manifolds 199, 201, 203, 223, 224, 267, 317, 347
manufacturers of SOFCs 380–387
markets for SOFCs 363–365, 388
  energy market trends 365–367
mass balance 233, 295, 305, 306, 326
mass spectroscopy 224
mass transfer controlled steam reforming catalysts/anodes 353
mass transfer-based modelling 319, 321, 322
maximum power *see* power densities
mechanical failure probability 315–318

medical applications 363
mercaptans 152, 351, 371
mercury electrodes 23
metal-organic deposition (MOD) 246–247
methane fuel 9, 165, 181, 347, 354
    oxidation within SOFC 15–16, 60, 168, 353, 356
    reforming 74, 165, 304–307, 339, 340, 343, 351, 352, 354, 356
    see also hydrocarbon fuels; natural gas
methanol fuel 19, 336, 355
microchip industry 366, 369
microtubular SOFCs 12, 13, 44, 219–225, 381, 386
microturbines 18, 368, 383–384
MIEC (mixed ionic electronic conduction) electrodes 237, 243–248, 257, 325
migration enthalpy 88
military applications 78, 363, 378
Mitsubishi Heavy Industries (MHI) 216, 381
mixed ionic electronic conduction electrodes see MIEC electrodes
modelling 291–326
    1-D porous electrode models 322, 323
    2-D cell model 309, 110, 314, 315
    3-D models 310–312
    anode structure and thickness 163
    anode systems 158–160, 302
    atomic modelling of barium indiate 108
    cell and stack behaviour under test conditions 276
    cell and stack level-modelling 308–314, 318, 319, 326
    continuum level electrochemistry model 239, 294, 299–303, 308
    current distributions 14
    electrode-level modelling 294, 299, 301, 311, 312, 318–325
    of geometries 14, 297, 315, 319
    methane-fired combined SOFC cycle 73
    molecular-level modelling 325–326
    SOFC as power generating burner 66–69, 80
    system-level modelling 294, 314, 315
    tubular cells and stacks 14, 312–314
molar flow 59, 63–64, 80, 275
molecular sieves 352
molecular-level modelling 294, 325, 326
molybdenum 9, 185, 341, 349, 352
Monte Carlo methods 324, 325
multi-stage fuel cell generator 372

N

National Fuel Cell Research Center (USA) 380
natural gas fuel 15, 19, 164, 335
    applications 15, 19, 223, 337, 338, 342, 343, 370, 373, 386
    in hybrid systems 73, 374
    reforming 44, 216, 335, 342–345, 370, 385
    sulphur impurities 16, 152, 334

    see also coking and carbon formation; hydrocarbon fuels; methane
Navier-Stokes equations 295, 296
navigation applications 363
neodymium 108
Nernst equation 103, 156, 284, 320
Nernst lamps 24–26
Nernst mass 24, 28–29, 151
Nernst voltage (Nernst potential) 60, 63, 64, 65, 80, 160, 161, 214, 230, 277, 304
Netherlands
    electricity markets 366
    field tests 386
    tubular SOFC distributed power system 42, 370–372, 380–382
nickel anodes 9, 15, 149, 151
    coking 9, 345, 347, 349, see also coking and carbon formation
    composite see cermet anodes
    early developments 29, 37
    and sulphur fuel impurities 16
nickel catalysts 168, 342, 344, 347, 349
nickel contacts 188–189, 271
nickel doping of electrolytes 104
nickel felts for tubular stack connections 214, 380–382
nickel in interconnects 181, 184, 186
nickel wire meshes 188–189
nickel-zirconium intermetallic formation 154
nickel/YSZ substrate 96
niobium 185, 353
Nippon Steel 381
Northwestern University (USA) 277, 278, 279
Norway 384

O

odorants 152, 351, 371
ohmic resistance/losses 2, 53, 62, 96, 198, 202, 230–33, 251, 255, 273
    determination techniques 282–283
    and metallic interconnects 187–188
    and modelling 300, 310, 319, 321, 324
open circuit voltage (OCV) 103, 105, 198, 230, 231, 234, 263–267, 269, 276, 284, 286
organic solvents 8
overpotential see polarisation
oxidation 2, 58, 60, 61, 63, 71, 72
    air as oxidant 3, 4, 63–66, 137, 208, 230, 343
    electrocatalytic oxidation of hydrocarbons at anode 165, 166, 335, 346, 353
    partial oxidation of hydrocarbons 343–346, 356, 373
oxidation semiconduction see oxygen partial pressures
oxygen
    diffusion coefficients 126–129
    excess, in perovskites 121–125
    high-purity 363

in situ concentration determination
    developments 33
  isotope techniques 34, 126–129, 161, 238
  nonstoichiometry in perovskites 120–125
oxygen partial pressures 31, 35, 59, 62, 91, 92, 98, 300, 302, 317
  early investigations 31, 35
  and electrodes 24, 150, 156, 161, 236, 241, 255
  and perovskites 121–124
oxygen reduction reaction 237–242, 273, 325
oxygen sensors 4–5, 35, 40, 267, 363
oxygen-selective ceramic membrane 374

P

Pacific Northwest National Laboratory (USA) 96
paint spraying 10
partial oxidation of hydrocarbons 342–346, 356, 373, *see also* reforming
particle connectivity model 324
patents, early 26, 27, 31, 33, 34
PEFCs (polymer electrolyte fuel cells) 18, 152, 376
perovskite electrodes 37, 119–143, 167, 168, 248, 325
perovskite interconnects 37, 173–80
perovskite lattice structure 120–122
perovskite-fluorite oxide interdiffusion 131, 132
perovskite-structured electrolytes 44, 83, 96–106, 110–112
  perovskite-related structures 106–108
perturbation measurement techniques 158, 324
phosphate electrolytes 28
phosphoric acid electrolyte fuel cells 18
pipelines 19, 363
planar cells and stacks 4, 197–208, 224, 225, 381
  applications 381, 386, 387
  cell configurations 202
  corrugation 8, 197
  costs 224, 378
  cracking 8, 12, 13, 180, 199, 224, 354
  design issues 13–14, 42–44, 95, 180, 197–208, 225
  distortion 11–12
  Ducrolloy 181–182
  early developments 3–4, 33, 37
  fabrication 42, 95, 142, 180, 204–208, 378
  large planar cells 13–14
  materials 202, 203
  modelling 14, 297, 312
  performance 14, 208
  sealing 13, 14, 33, 180, 207, 208, 224, 225, 300, *see also* gas leakage; seals and sealants
  strength 90, 199
  testing 267–269, 381

and thermal conductivity of components 178
  see also anode-supported cells
Plansee AG 12, 181, 186
plasma metal organic vapour deposition 207
plasma spraying 10, 11, 40, 42, 142, 154, 180, 185, 205, 207, 212, 213, 216
plasticisers 95
platinum catalysts 152, 168, 223, 337, 342, 343, 344, 349–350
platinum contacts, use in testing 265–267, 267, 271
platinum electrodes 4, 119, 149, 188, 341
  early developments 26, 29, 34, 151
  Nernst lamp 25, 26
polarisation 133–135, 198, 218, 229–257
  determination/measurement 262–267, 273–275, 282, 283
  early developments 35, 37, 44
  modelling 299–303, 310, 318–325
  see also anodes, polarisation; cathodes, polarisation; ohmic resistance/losses
polymer electrolyte fuel cells (PEFCs) 18, 152, 376
porcelain electrolytes 24, 26–7
porous electrodes 232, 257, 273, 322, 323
porous support tubes (PST) 34, 204, 210, 216
portable applications 15, 19, 336, 355, 363, 370, 386
potassium 349
potentials and overpotentials *see* polarisation
potentiometric gas concentration *in situ* determination 31–33, 35
powder pressing 3, 4, 7, 12, 142, 179, 181, 182, 206, 221
power control systems 14, 376, 377, 386
power densities 13, 208, 219, 224, 272
  maximum 65, 103, 105, 272
  modelling 309, 315
power generating burner SOFC model 66–69, 80
power generation systems 15, 78, 95, 208, 363
  competition from other systems 368–370, 388
  costs 363, 368, 378, 379
  distributed power 365, 366, 367, 369, 370–372, 376, 384
  multi-stage fuel cell generator 372
  stationary power 2, 18, 217, 364, 378
  using microtubular SOFCs 220, 222–225
  see also and other hybrid systems; combined heat and power (CHP); gas turbine/SOFC hybrid systems
praseodymium 30, 37, 187
pressurised systems 374–376, 381, *see also* gas turbine/SOFC hybrid systems
process design modelling 314, 315
propane 15, 19, 336, 343, 348, 355, 386
proton conduction 108, 110–112
proton exchange membrane (PEM) fuel cells 333, 334, 355

PSZ (partially stabilised zirconia) 5, 133
pyrochlores 96
pyrolusite 23

Q
quantum mechanical simulation techniques 325, 326
quartz electrolytes 26

R
radioactive impurities 6
radionuclides 35
rare earths 34, 110–112, 129, 167, 173, 174, *see also specific elements*
rate equations 303–307, 309, 311, 312
reaction enthalpy 56, 60, 67, 71, 157
reaction entropy 56, 60, 60–61, 81
reactions modelling 253, 294, 303–314, 322–325
reactive power 372, 376
recuperators 371, 375, 376
reduced operating temperatures *see* lower/intermediate operating temperatures
reduction semiconduction 35
reference electrodes 156, 160, 252, 253, 261, 262–267, 324
reforming of hydrocarbons 16, 73–77, 81, 152, 163, 165, 169, 216, 333, 336, 338–351
   carbon dioxide method (dry reforming) 342, 343, 354
   direct internal reforming 44, 169, 333, 338, 341, 352, 353, 356, 385
   indirect internal 333, 338, 339, 341, ;342
   partial oxidation method 342–346, 356, 373
   pre-reforming 169, 216, 223, 370, 371
   see also electrocatalytic oxidation of hydrocarbons at anode; methane fuel, reforming; natural gas, reforming; steam reforming
renewable energy production 366, 367
renewable fuel sources 152, 334, 337, 354, 355, 356
resistances 101, 242, 278, 319, 321, *see also* ohmic resistance/losses
resistors and capacitors (Warburg element) 234, 251
reversible efficiency 57
reversible heat 56, 69–70
reversible heat engines 69–71, 74, 76
reversible power 59
reversible SOFC thermodynamics 56–62, 69–77
reversible voltage 59, 61, 65, 80
reversible work 53, 56, 62, 70–72, 80
rhodium 341, 342, 343, 344, 349
ribbed and flattened cell design 217–219, 225, 385
Risø National Laboratory (Denmark) 277–281
Rolls-Royce 216, 381
ruthenium 163, 168, 349, 350, 353

S
samarium 86, 92, 93, 103, 111, 141
Sanyo Electric 185, 186
scandia doping 6, 85, 86, 89, 97
scandia-doped zirconia (SSZ) 89, 90, 96, 112
Schrödinger equation 326
screen printing 4–5, 8, 10, 42, 96, 243
screening tests 271, 272
SDC (samaria doped ceria) 141
seals and sealants 207, 208, 317, *see also* gas leakage; planar cells and stacks, sealing
SECA (Solid State Energy Conversion Alliance) 378, 388
sedimentation fabrication method 207
Seebeck coefficient 125
segmented-in-series cell design 217
semiconductor history 27
sensors 4–5, 44, *see also* oxygen sensors
series connection (cascading) 11, 32, 40, 66
Shell mileage marathon (1996) 223
Siemens and Halske 27
Siemens-Westinghouse 15, 96, 181, 185, 190, 210, 211, 370, 378–385, 388, *see also* Westinghouse
silica in interconnects 183, 184
silicate production from dust in fuel 152
silver in electrodes 26, 166
silver pins in interconnects 186
silver sulphide conductivity 23
silver wire interconnects 220
SIMS (secondary ion mass spectroscopy) 104, 126, 128, 238
sintering 119
   anode fabrication 9, 153, 154, 166, 180, 213, 246
   at anode after direct reforming 341, 352
   cathode-supported tubular cells 142, 211, 216
   cathodes 9, 37, 119, 180, 281
   contact material 188
   electrolytes 30, 31, 37, 40, 95, 119, 246
   interconnects 12, 179–182, 207, 213
   promotion aids 153, 179–180
slip casting 207
slurry coating processes 95, 96, 119, 142, 154, 180, 213, 216
sodium ion conductivity 34
SOFCo 13, 381
solid polymer electrolyte fuel cells *see* PEFCs
Solid State Energy Conversion Alliance (SECA) 378, 388
solvent extraction techniques 8
Southern California Edison 380
spacecraft power generation 37
Spain 386
spinel formation 184, 186
spray pyrolysis 205, 206

sputtering 206
SSZ (scandia-doped zirconia) 6, 89, 97
STAR-CD 297
start-up issues 19, 150, 368, 369, 388
   hybrid systems 377–8
   planar designs 13, 199
   tubular SOFCs 219, 222–224, 386
steam reforming 16, 152, 165, 304–307, 339, 342–347, 351, 353, 385
steam turbine/SOFC hybrid systems 78
steam turbines 368, 369
steel alloy interconnects 11, 181–186, 190, 207, 262
steel foil interconnects 185, 186
Stirling heat engine/SOFC hybrid system 2, 78
stochastic electrode structure model 324
stoichiometric air demand 68
stoichiometric oxygen demand 68
strain, nonstoichometry-induced 317, 318
strontium, substituent in anodes 168, 353
strontium based proton conducting oxides 110–112
strontium doped lanthanum chromite interconnects 11–12, 37, 174, 177
strontium doped lanthanum ferrite cathodes 120, 138, 143, see also LSCF
strontium doped lanthanum gallates 99–104, 112, see also LSGM
strontium doped lanthanum manganite cathodes see LSM
strontium fluoride for sintering promotion 179
strontium nitrate in interconnects 12
sulphate electrolytes 28
sulphur impurities in fuels 16, 150, 152, 157, 335–337, 342, 351, 352, 355, 371
Sulzer Hexis 12, 13, 14, 15, 44, 185, 337, 338, 385, 386, 388
   HXS 1000 Premiere fuel cells 386
Switzerland
   demonstrations 386
   early developments 34
   field tests 386
   HXS 1000 Premiere fuel cells 386
   manufacturers 381
synthesis gas (syngas) 333, 336, 346, 356, 363, 373
systems 363–369
   control and dynamics 14, 376–378, 386
   demonstrations 380–387
   modelling 294, 314, 315
   see also applications; design issues; hybrid systems; power generation systems; testing

T
Tafel equation 157, 160, 241, 249
tape calendering 8, 11, 12, 180, 205, 206
tape casting 3, 4, 7, 8, 42, 90, 95, 96, 142, 180, 205–207
telecommunications applications 363

temperature
   and binding enthalpy 87, 88
   and conductivity 24, 89, 102
   during cathode fabrication 133–135
   effect on anodes 160, 162, 163
   electrolyte dependence 89, 103
   gradients 14, 198, 224, 311, 340, 353, 356
   heat treatment temperatures for perovskites 133, 134
   in hybrid systems 72, 73, 74, 377, 378
   measurement and testing 14, 35, 261, 272, 279–281, 284
   and reversible cell voltage 60, 61, 65
   see also lower/intermediate operating temperatures
temperature-programmed oxidation (TPO) technique 224, 350, 351
terbium 353
testing 3, 213, 214, 261–86
   field testing of systems 380–385
   industrial screening tests 271, 272
   laboratory testing of systems 381
   test cells 265–267, 324
thermal expansion and expansion coefficients 315–318
   anodes 9, 149
   cathodes 10
   contact materials 188
   electrolytes 13, 90, 104, 177, 181
   interconnects 11, 12, 173, 177, 181, 183, 185, 186
   perovskites 125–6, 137
thermal and flow models 294–299, 308, 312, 326
thermal insulation 14, 315
thermal shock 8, 13, 14, 19, 44, 199, 219, 220, 221, 371, 375, 377
thermal spraying 213
thermal stresses 294, 297–299
   modelling 315–318
thermodynamics 53–80
   early developments 23–26
   first and second laws 56
   of ideal reversible SOFC 56–62
   terms and definitions 54, 55
thermoelectric conversion 78
thermoelectric generators 363
thermoelectricity, early developments 23, 24
thermomechanical model 315–318
thermoneutral voltage 309
thin film cathodes 247
thin film cells
   early developments 34, 41
   fabrication 205, 206
   polarisation and polarisation measurement 252, 302
   and reference electrodes 262
   test results 279
   see also electrolytes; microtubular SOFCs

thiophene 152, 351, 371
thorium 30, 31
three-electrode configuration 156, 262–5
three-phase boundaries 237–242, 244–245, 301, 320, 325, 326
  anodes 151, 163, 165, 250
  cathodes 10, 37, 128, 134, 137, 139, 237–243, 253, 257
  time dependence, and polarisation 232, 234, 251, 255, 256
titanate anodes 167, 353
titanate electrolytes 96, 97
titania doping of anodes 154, 353
titanium in interconnects 184, 185
Toho Gas 96
Tokyo Gas 381
tolerance factor 120, 121, 130
tortuosity 234, 235, 237, 302
Toto Ltd 95, 96, 216, 381
toxic hazards 6, see also environmental issues
transfer coefficient 304
transfer printing 207
transmission electron microscopy (TEM) 133
transport numbers 27, 30, 35, 97, 101, 109
transportation applications see vehicles
tubular cells and stacks 14, 42, 197, 210–219, 224, 381, 385
  alternative designs 216–219, 224
  chromite interconnects 190
  costs 212, 216, 224
  design issues 13, 14, 44, 93–95, 142, 180, 197, 210–219
  early developments 32, 33, 37, 40, 42, 210
  fabrication processes 7–8, 11, 12, 42, 93–95, 119, 142, 180, 210–213, 224
  modelling 14, 312–314
  operation and performance 213–214
  running on biogas 354, 355
  stack construction 214–216
  testing 213, 214, 276, 381
  see also microtubular SOFCs
tungsten 25, 28, 185
turbines see gas turbines; microturbines; steam turbines

U

United Kingdom 34, 222, 381
urania 37
USA
  combined heat and power (CHP) 369, 370
  demonstrations 384
  early progress on SOFCs 33
  electricity markets 365
  government support for SOFCs 17, 378, 384, 388
  manufacturers 381
  National Fuel Cell Research Center 380
  research into cathode/electrolyte reactions 133
USSR 34

V

vacuum evaporation 207
vanadium 131, 168
vapour deposition methods 7, 10, 205, 207, see also EVD
vegetable matter as fuel see biogas
vehicles 336, 355, 370, 378
  air conditioning 15
  auxiliary power supplies 2, 15, 16, 18, 363, 364, 368, 370, 373, 387
  microtubular cell stack-powered 223
  oxygen sensors 4–5, 35, 40, 267, 363
  zero emission electric vehicles 18
Viking Chemicals 8
voltage
  early investigations 26, 30–31
  and modelling 308, 309
  see also current density-voltage curves; Nernst voltage; open circuit voltage (OCV); reversible voltage; voltage reductions/losses
voltage reductions/losses 53, 59, 62–66, 257, 380–382, 384
  and fuel utilisation 62, 62–66, 371
  see also ohmic resistance/losses; polarisation

W

Warburg element 234, 251
water vapour effects 141, 142, 161–163, 187, 273, 277, 283
water-based casting 8
Weibull function 316, 317
Westinghouse 7, 11, 13, 14, 32, 42, 94, 95, 135, see also Siemens-Westinghouse
wet spraying 42, 142, 185
wetting agents 95
work 70–72, 231

X

X-ray diffraction analysis 100

Y

YSZ 1, 5–6, 90–91, 202, 209, 222, 223
  for coating of ceria based electrolytes 93, 112
  compared with other electrolytes 100, 103, 104, 107, 108
  compatibility with interconnects 177, 181
  early developments 2, 3, 24, 25, 83, 219
  fabrication 7–8, 12, 90–91, 94–96, 142, 155, 221
  in oxygen sensors 4–5, 363
  polarisation 232, 242, 244, 246, 257
  reactions with cathodes 119, 129, 130–136, 242
  YSZ/LSM composite cathodes see LSM/YSZ composite cathodes
  YSZ/nickel composite anodes see cermet anodes
  see also electrolytes

ytterbia 6, 34, 90, 110–112
ytterbia doped zirconia (YbSZ) 90
yttria 6, 30, 84–85, 353
yttria doped ceria electrolytes 93
yttria doped zirconia *see* YSZ
yttrium
   in interconnects 173–180, 184, 190
   and proton-conducting oxides 110–112
yttrium manganite interconnect coating 187
yttrium zirconate electrolyte 96

Z

zinc amalgam electrodes 23
zinc oxide desulphurisation 152, 371
zircon resources 6
zirconate(s) *see* lanthanum zirconate formation;
   yttrium zirconate electrolyte
zirconia 2–6, 325
   partially stabilised (PSZ) 5, 133
zirconia based electrolytes 90–91, 94, 112
   doping 5, 6, 7–8, 35, 90, 90–91
   early developments 24, 28, 30, 31, 32–33, 44, 210
   properties 90, 112, 325
   reactions with cathodes 119, 129, 130–138
   see also calcia-stabilised zirconia; YSZ (yttria-stabilised zirconia)
zirconia lighting filaments 2–3, 25, 26
zirconia sensors 4–5
zirconium in interconnects 184, 185